AF389416

Biomarkers of
Environmental Toxicants

Biomarkers of Environmental Toxicants

Editors

Kun LU
Robert J. Turesky

MDPI • Basel • Beijing • Wuhan • Barcelona • Belgrade • Manchester • Tokyo • Cluj • Tianjin

Editors
Kun LU
University of North Carolina
USA

Robert J. Turesky
University of Minnesota
USA

Editorial Office
MDPI
St. Alban-Anlage 66
4052 Basel, Switzerland

This is a reprint of articles from the Special Issue published online in the open access journal *Toxics* (ISSN 2305-6304) (available at: https://www.mdpi.com/journal/toxics/special_issues/Toxicants-Biomarker).

For citation purposes, cite each article independently as indicated on the article page online and as indicated below:

LastName, A.A.; LastName, B.B.; LastName, C.C. Article Title. *Journal Name* **Year**, *Article Number*, Page Range.

ISBN 978-3-03936-736-8 (Hbk)
ISBN 978-3-03936-737-5 (PDF)

Contents

About the Editors

Kun LU is an associate professor at the Gillings School of Global Public Health, University of North Carolina at Chapel Hill. He has extensive experience in gut microbiome, metabolomics profiling, DNA adducts, and biomarker development. His research currently has an emphasis on microbiome research, biomarker discovery and exposome mapping. Dr. Lu's lab aims to answer how the gut microbiome interacts with environmental exposure and affects disease susceptibility, and how host factors crosstalk with the microbiome to influence its response. His lab first discovered that exposure to environmental toxins, such as organophosphate pesticides and arsenic, perturbed the gut microbiome community and its associated functions. His lab demonstrated that gut microbiome phenotypes significantly influence toxicological responses to environmental exposure. Likewise, Dr. Lu's lab maps the exposome for human disease, with the goals of characterizing all exposures over the lifespan via high-resolution mass spectrometry, understanding the health impact of the exposome, and designing strategies to reduce exposure-associated adverse effects. Dr. Lu's lab has also been developing protein/DNA adducts as novel and sensitive biomarkers to evaluate the health effects of different environmental chemicals. He has published over 70 peer-reviewed articles and served on a number of external scientific review panels/committees.

Robert J. Turesky is the Masonic Chair in Cancer Causation and a professor at the Department of Medicinal Chemistry. His research is devoted to cancer etiology programs at the University of Minnesota. Dr. Turesky received his PhD in nutrition and food science from M.I.T. Prior to his current position, Dr. Turesky served as: the group leader of the Biomarkers Unit, Nestlé Research Center, Lausanne, Switzerland (1986–2000); Division Director of Chemistry, National Center for Toxicological Research, U.S. Food and Drug Administration, Jefferson, AR, (2000–2004); and Principal Investigator, Wadsworth Center, New York State Department of Health (2004–2013). He investigates the biochemical toxicology of dietary and environmental genotoxicants, and applies mass spectrometry methods to identify and measure biomarkers of these chemicals in molecular epidemiology studies that seek to understand the role of chemical exposures in the etiology of cancer.

Editorial

Biomarkers of Environmental Toxicants: Exposure and Biological Effects

Robert J. Turesky [1,*] and Kun Lu [2,*]

[1] Masonic Cancer Center and Department of Medicinal Chemistry, University of Minnesota, 2231 6th St. SE, Minneapolis, MN 55455, USA
[2] Department of Environmental Sciences and Engineering, University of North Carolina at Chapel Hill, Chapel Hill, NC 27599, USA
* Correspondence: rturesky@umn.edu (R.J.T.); kunlu@unc.edu (K.L.)

Received: 8 May 2020; Accepted: 22 May 2020; Published: 22 May 2020

Biomarkers of environmental toxicants are measures of exposures and effects, some of which can serve to assess disease risk and interindividual susceptibilities. Metabolites, protein, and DNA adducts also serve to elucidate mechanisms of bioactivation and detoxication of reactive toxicant intermediates. Some environmental chemicals act as modulators of gene and protein activity and induce dysbiosis of the microbiome, which impacts the metabolome and overall health. In this Special Issue on "Biomarkers of Environmental Toxicants", original research and review articles are reported on the latest biochemical, bioanalytical, and mass spectrometry-based technologies to monitor exposures through targeted and nontargeted methods, and on mechanistic studies that examine the biological effects of environmental toxicants in cells and humans.

The exposome, or the totality of environmental exposures, represents both internal exposures originating from host physiology and external exposures deriving from diverse toxicants and chemicals, infectious agents, as well as diet and drugs. Exposome has been proposed to be a critical entity of human disease. Xue et al. reviewed the current chemical exposome measurement approaches with a focus on those based on the mass spectrometry [1]. They further explored the strategies in implementing the concept of chemical exposome and discussed the available chemical exposome studies. Early progress in chemical exposome research was outlined, and major challenges were highlighted. Apparently, efforts towards chemical exposome have only uncovered the tip of the iceberg, and further advancement in measurement techniques, computational tools, high-throughput data analysis, and standardization would allow for more exciting discoveries to be made concerning the role of exposome in human health and disease.

The gut microbiome has emerged as a new mediator in human health. On the other hand, the human gut microbiome can be easily disturbed upon exposure to a range of toxic environmental agents. Environmentally induced perturbation in the gut microbiome is strongly associated with human disease risk. Functional gut microbiome alterations that may adversely influence human health is an increasingly appreciated mechanism by which environmental chemicals exert their toxic effects. Tu et al. defined the functional damage driven by environmental exposure in the gut microbiome as gut microbiome toxicity [2]. The establishment of gut microbiome toxicity links the toxic effects of various environmental agents and microbiota-associated diseases, calling for a more comprehensive toxicity evaluation with an extended consideration of gut microbiome toxicity.

Living organisms respond to environmental changes and xenobiotic exposures via diverse mechanisms. tRNA-mediated mechanisms are only recently emerging as important modulators of cellular stress responses. Huber et al. discussed many ways that nucleoside modifications confer high functional diversity to tRNAs, with a focus on tRNA modification-mediated regulation of the eukaryotic response to environmental stress and toxicant exposures [3]. Additionally, the potential applications of tRNA modification biology in the development of early biomarkers of pathology are also highlighted. Their review highlights a high functional diversity, ranging from the control of tRNA maturation and translation initiation, to translational enhancement through modification-mediated codon-biased translation of mRNAs encoding stress response proteins, and translational repression by stress-induced tRNA fragments. Future work in this area would provide more exciting discoveries to better understand the role of tRNA modification in exposure-induced human disease.

Exposure to heavy metals is ubiquitous and has been associated with a number of human diseases. Biomarkers of heavy metal exposure in children are particularly important for monitoring exposure and health risk assessment. The study by Jursa et al. measured hair Mn, Pb, Cd, and As levels in children from the Mid-Ohio Valley to determine within and between-subject predictors of hair metal levels. Specifically, occipital scalp hair was collected in 2009–2010 from 222 children aged 6–12 years (169 females, 53 males) participating in a study of chemical exposure and neurodevelopment in an industrial region of the Mid-Ohio Valley [4]. They found that hair Mn and Pb levels were comparable (median 0.11 and 0.15 µg/g, respectively) and were ~10-fold higher than hair Cd and As levels (0.007 and 0.018 µg/g, respectively). In addition, metal levels were different between male and female subjects and showed different profiles along hair segments.

Benzene is a known carcinogen and causes hematotoxicity. Benzene exposure occurs through factory occupations, and from emissions of burning coal and oil in the air. Cigarette smoking is another important source of exposure to benzene. In this research article, Tranfo and colleagues employed a targeted LC-MS/MS to biomonitor benzene through the biomarker S-phenyl-mercapturic acid (SPMA), which is formed from the benzene oxide metabolite conjugated with glutathione (GSH), and then excreted in urine as SPMA [5]. The findings reveal that the main source of benzene exposure in a cohort not occupationally exposed to benzene in central Italy was through active smoking; however, nonsmokers were also exposed to airborne concentrations of this carcinogen.

Phthalates are used as plasticizers and additives in many consumer products. Human exposure to phthalates is prevalent and occurs mainly through dietary sources, dermal absorption, and inhalation. Laboratory animal studies reveal endocrine-disrupting and reproductive effects of phthalates. Thus, human exposure to phthalates is a public health concern. Wang and colleagues have compiled a review on the biomonitoring studies of phthalates in populations across the globe and associated adverse health effects. Urine, serum, amniotic fluid, breast milk, semen, and saliva serve as biospecimens to screen phthalate metabolites. Epidemiological studies have linked high exposure to phthalates with sex anomalies, endometriosis, altered reproductive development, early puberty and fertility, breast and skin cancer, allergy and asthma, overweight and obesity, insulin resistance, and type II diabetes [6].

Aldehydes are ubiquitous in the environment, originating from man-made sources, tobacco smoke, and natural processes. Some aldehydes are implicated in diseases, including diabetes, cardiovascular diseases, neurodegenerative disorders (i.e., Alzheimer's and Parkinson's Diseases), and cancer. Aldehydes are strong electrophiles that react with nucleophilic sites in DNA and proteins to form reversible and irreversible modifications. These modifications, if not eliminated or repaired, can lead to alteration in cellular homeostasis, cell death, and contribute to disease pathogenesis. In this review, Dator and colleagues describe the metabolism of aldehydes in vivo, and the bioanalytical, and mass spectrometry-based approaches to characterize aldehydes in cells and biomonitoring in humans [7].

Hemoglobin (Hb) and albumin (Alb) are the most abundant proteins in the blood and form covalent adducts with toxicants and endogenous electrophiles. In this review, Preston and Phillips describe targeted and nontargeted mass spectrometry-based strategies to measure exposures to a wide range of toxicants that alkylate the N-terminal valine residues of the α-chain of Hb, and the cysteine residue (Cys-β93) of the β-chain of Hb, which reacts with many electrophiles, including carcinogenic aromatic amines [8]. The histidine residues of Alb react with epoxides of polycyclic aromatic hydrocarbons, and lysine residues form adducts with aflatoxin B_1 dialdehyde. The highly nucleophilic Cys-34 residue of Alb is the only site for which untargeted adductomic methods have served to screen for electrophiles in humans. Protein adductomics techniques can screen for harmful exposures to causative agents of chronic disease and identifying individuals at risk.

Aasa and colleagues employed the *N*-(2,3-dihydroxypropyl)valine hemoglobin adduct formed with glycidol, a carcinogen present in refined edible oils, to assess internal doses of this genotoxicant in a cohort of children [9]. In this research article, the investigators report the adduct showed a fivefold variation between the children. The estimated mean intake of glycidol (1.4 µg/kg/day) was about two times higher than the estimated intake for children by the European Food Safety Authority. The estimated lifetime cancer risk ($200/10^5$) was calculated by a multiplicative risk model from the lifetime in vivo doses of glycidol in the children and exceeded the acceptable cancer risk estimate. The protein adduct biomarker data, calculated intakes, and corresponding estimated cancer risks emphasize the importance of identifying and mitigating the sources of background exposure to glycidol from foods and other possible sources.

Exposure to environmental chemicals often leads to diverse DNA damage, and the formation of DNA adducts is one of the key events in chemical-induced carcinogenesis. It is critical to determine whether DNA adducts cause mutagenesis. Chemical incorporation of a modification at a specific site within a vector (site-specific mutagenesis) has been a useful tool to deconvolute what types of damage quantified in biologically relevant systems may lead to toxicity and/or mutagenicity, thereby allowing researchers to focus on the most relevant biomarkers that may impact human health. Here, Bian et al. introduced shuttle vector-based methods and reviewed a sampling of the DNA modifications that have been studied by shuttle vector techniques [10].

A significant limitation in biomonitoring cancer-causing agents is the paucity of fresh frozen tissues available for DNA adduct biomarker research. In this review, Yun and colleagues report on the methods commonly used to biomonitor DNA adducts, and the use of formalin-fixed paraffin-embedded (FFPE) tissues for the measurements of DNA adducts of genotoxicants found in the diet and tobacco smoke [11]. The authors developed a technique to retrieve the DNA adducts from FFPE tissues under mild conditions that completely reverses the DNA crosslinks while preserving the structures of the DNA lesions. FFPE tissues for which there is a clinical diagnosis of disease present a previously untapped source of biospecimens for molecular epidemiology studies that seeks to assess the causal role of environmental chemicals in cancer etiology.

DNA adducts are believed to play a central role in the induction of cancer in cigarette smokers. Ma and colleagues have summarized the research on DNA adducts formed with carcinogens in tobacco smoke and from oxidative DNA damage [12]. The analytical approaches most commonly used are mass spectrometry (MS), ^{32}P-postlabeling, and immunohistochemistry. Because of the high selectivity and sensitivity, MS methods are the preferred technique and have largely supplanted immunochemical and postlabeling techniques over the past decade. DNA adducts of different classes of tobacco carcinogens have been identified in human biospecimens. Issues pertaining to the validation of DNA adducts such as biomarkers, mitigation of artifacts, and caveats in the designs of human studies are highlighted.

Aristolochic acids (AAs) are found in *Aristolochia* plants, some of which have been used in the preparation of traditional herbal medicines worldwide. AAs are highly nephrotoxic and carcinogenic to humans and implicated as causative agents of the Balkan endemic nephropathy (BEN) and "Chinese herbs nephropathy" in Asia. In this article, Chan and colleagues provide an overview of the exposure of AAs in the Balkan Peninsula, where the comingling of *Aristolochia* plants with grains and the release of AAs from decayed seeds of *Aristolochia* plants contaminate the agricultural soil, the food crops, and the water supply [13]. The links between exposure to AAs and their biomarkers of DNA damage, mutations in cancer driver genes, and mechanisms of kidney fibrosis in Asian cohorts are reported.

Accelerator mass spectrometry (AMS) is an exquisitely sensitive technique to measure long-lived radionuclides that occur naturally in the environment. AMS has been used for many years in the earth sciences, such as for radiocarbon dating in archaeology. In this review, Malfatti and colleagues describe the approaches and advances employing AMS in human health and risk assessment. The applications of radiocarbon tracer technology in cancer-related studies assessing low-dose toxicology studies of naphthalene-DNA adduct formation, benzo[a]pyrene pharmacokinetics in humans, and the antibacterial triclocarban exposure and impact on the endocrine system are reported [14]. AMS applications in precision medicine include the use of radiocarbon-labeled cells for better defining mechanisms of metastasis and the use of drug-DNA adducts in the in vivo and ex vivo microdosing strategy of chemotherapeutics as predictive biomarkers of interindividual response to chemotherapy.

In summary, this collection of original research and review articles provides a valuable update of the most recent biochemical and analytical tools that employ biomarkers in toxicology research, biomarker discovery, and exposure and risk assessment in population-based studies.

Conflicts of Interest: The authors declare no conflict of interest.

References

1. Xue, J.; Lai, Y.; Liu, C.; Ru, H. Towards Mass Spectrometry-Based Chemical Exposome: Current Approaches, Challenges, and Future Directions. *Toxics* **2019**, *7*, 41. [CrossRef] [PubMed]
2. Tu, P.; Chi, L.; Bodnar, W.; Zhang, Z.; Gao, B.; Bian, X.; Stewart, J.; Fry, R.; Lu, K. Gut Microbiome Toxicity: Connecting the Environment and Gut Microbiome-Associated Diseases. *Toxics* **2020**, *8*, 19. [CrossRef] [PubMed]
3. Huber, S.; Leonardi, A.; Dedon, P.; Begley, T. The Versatile Roles of the tRNA Epitranscriptome during Cellular Responses to Toxic Exposures and Environmental Stress. *Toxics* **2019**, *7*, 17. [CrossRef] [PubMed]
4. Jursa, T.; Stein, C.; Smith, D. Determinants of Hair Manganese, Lead, Cadmium and Arsenic Levels in Environmentally Exposed Children. *Toxics* **2018**, *6*, 19. [CrossRef] [PubMed]
5. Tranfo, G.; Pigini, D.; Paci, E.; Bauleo, L.; Forastiere, F.; Ancona, C. Biomonitoring of Urinary Benzene Metabolite SPMA in the General Population in Central Italy. *Toxics* **2018**, *6*, 37. [CrossRef] [PubMed]
6. Wang, Y.; Zhu, H.; Kannan, K. A Review of Biomonitoring of Phthalate Exposures. *Toxics* **2019**, *7*, 21. [CrossRef] [PubMed]
7. Dator, R.P.; Solivio, M.J.; Villalta, P.W.; Balbo, S. Bioanalytical and Mass Spectrometric Methods for Aldehyde Profiling in Biological Fluids. *Toxics* **2019**, *7*, 32. [CrossRef] [PubMed]
8. Preston, G.W.; Phillips, D.H. Protein Adductomics: Analytical Developments and Applications in Human Biomonitoring. *Toxics* **2019**, *7*, 29. [CrossRef] [PubMed]
9. Aasa, J.; Vryonidis, E.; Abramsson-Zetterberg, L.; Törnqvist, M. Internal Doses of Glycidol in Children and Estimation of Associated Cancer Risk. *Toxics* **2019**, *7*, 7. [CrossRef] [PubMed]
10. Bian, K.; Delaney, J.; Zhou, X.; Li, D. Biological Evaluation of DNA Biomarkers in a Chemically Defined and Site-Specific Manner. *Toxics* **2019**, *7*, 36. [CrossRef] [PubMed]
11. Yun, B.H.; Guo, J.; Turesky, R.J. Formalin-Fixed Paraffin-Embedded Tissues—An Untapped Biospecimen for Biomonitoring DNA Adducts by Mass Spectrometry. *Toxics* **2018**, *6*, 30. [CrossRef] [PubMed]
12. Ma, B.; Stepanov, I.; Hecht, S.S. Recent Studies on DNA Adducts Resulting from Human Exposure to Tobacco Smoke. *Toxics* **2019**, *7*, 16. [CrossRef] [PubMed]
13. Chan, C.-K.; Liu, Y.; Pavlović, N.M.; Chan, W. Aristolochic Acids: Newly Identified Exposure Pathways of this Class of Environmental and Food-Borne Contaminants and its Potential Link to Chronic Kidney Diseases. *Toxics* **2019**, *7*, 14. [CrossRef]
14. Malfatti, M.A.; Buchholz, B.A.; Enright, H.A.; Stewart, B.J.; Ognibene, T.J.; McCartt, A.D.; Loots, G.G.; Zimmermann, M.; Scharadin, T.M.; Cimino, G.D.; et al. Radiocarbon Tracers in Toxicology and Medicine: Recent Advances in Technology and Science. *Toxics* **2019**, *7*, 27. [CrossRef]

Review

Towards Mass Spectrometry-Based Chemical Exposome: Current Approaches, Challenges, and Future Directions

Jingchuan Xue [1], Yunjia Lai [1], Chih-Wei Liu [1] and Hongyu Ru [2,*]

[1]   Center for Environmental Health and Susceptibility, University of North Carolina at Chapel Hill, Chapel Hill, NC 27599, USA
[2]   Department of Population Health and Pathobiology, North Carolina State University, Raleigh, NC 27607, USA
*   Correspondence: hru@ncsu.edu

Received: 6 June 2019; Accepted: 14 August 2019; Published: 18 August 2019

Abstract: The proposal of the "exposome" concept represents a shift of the research paradigm in studying exposure-disease relationships from an isolated and partial way to a systematic and agnostic approach. Nevertheless, exposome implementation is facing a variety of challenges including measurement techniques and data analysis. Here we focus on the chemical exposome, which refers to the mixtures of chemical pollutants people are exposed to from embryo onwards. We review the current chemical exposome measurement approaches with a focus on those based on the mass spectrometry. We further explore the strategies in implementing the concept of chemical exposome and discuss the available chemical exposome studies. Early progresses in the chemical exposome research are outlined, and major challenges are highlighted. In conclusion, efforts towards chemical exposome have only uncovered the tip of the iceberg, and further advancement in measurement techniques, computational tools, high-throughput data analysis, and standardization may allow more exciting discoveries concerning the role of exposome in human health and disease.

Keywords: chemical exposome; biomonitoring; environmental monitoring; mass spectrometry; disease; bioinformatics

---

1. Introduction

Studies have demonstrated that environmental factors play an equal or even more significant role in the pathogenesis of human chronic diseases compared with other risk factors such as genetic variants [1–4]. Environmental factors can induce changes in the human genome, transcriptome, epigenome, proteome, and metabolome. In 2005, the concept of exposome was first proposed by Christopher Wild to account for the unexplained risk factors underlying human diseases [5]. The "exposome" concept has shaped the thinking of scientists when studying environment-disease associations by switching the research paradigm from a single exposure-disease model to an agonistic analysis of environmental influences on human health [6–8]. Many research areas are benefiting from this mindset shift, including environmental epidemiology, health risk assessment, biomonitoring and environmental monitoring, and mechanistic biology. A considerable amount of published commentaries and reviews have highlighted the potential benefits of the characterization and integration of exposome in future studies [9–17].

The definition of exposome has been evolving ever since its birth. Wild originally defined it as "the totality of environmental exposures from birth onwards" [5]. Later, he redefined the scope of exposome to include three broad categories of non-genetic exposure: internal (e.g., metabolism, gut microbiome, inflammation), specific external (e.g., environmental pollutants, diet, occupation), and general external (e.g., socioeconomic status, education, and climate) [18]. In 2014, Miller and Jones expanded the Wild definition to include the measures of biological responses to these exposures [19].

Exposome studies aim to accomplish two critical goals: (1) to measure the cumulative exposures throughout the entire life of humans and (2) to evaluate the associations or causal relationships between these exposures and any biological changes. Depending on the specific types of exposure, exposome can be measured through a wide array of techniques including remote sensors, questionnaires, geography information systems, biomonitoring and environmental monitoring, and metabolomics [13,20,21]. As a comprehensive review of exposome is extremely large in scope, in this paper, we focus on the entire chemical exposures from embryo onwards, "chemical exposome".

There is still a long way to go to completely characterize the human chemical exposome. However, efforts are being made to capture a critical portion of it [22–24]. The aim of this paper is to undertake a systematic review of published evidences regarding the measurement techniques for the chemical exposome and available studies linking chemical exposome to human health. We further discuss the challenges confronted in implementing the exposome concept. Furthermore, we propose potential solutions to address the scientific and technological barriers to advancing exposome research.

2. Exposome Measurement Approaches

Measuring the exposome has been a challenging task because of its complex and dynamic nature. To capture the full spectrum of exposures of interest in an individual's life, many approaches have either been newly developed or transferred from other fields. For ease of description, we group the available approaches into three categories: chemical approaches (directly measure the exposures or early biomarkers using chemistry techniques), biological approaches (measure the biological changes induced by exposures using molecular biology techniques), and other approaches (those techniques which do not belong to either chemical or biological approaches, such as a personal wearable device). This paper focuses on the available mass spectrometry-based measurement techniques, but briefly discusses other approaches.

2.1. Chemical Approaches

Mass spectrometry based analytical techniques are widely used in biomonitoring and environmental monitoring studies and have become the predominant chemical approach in characterizing chemical exposome [9,25]. Mass spectrometry, coupled with a separation technique such as liquid or gas chromatography, has been the most popular method used in the direct measurement of chemicals (xenobiotics, their metabolites, and the early biomarkers) in biological or environmental samples due to the superiority in sensitivity, specificity, and dynamic range [9,25]. Low-resolution mass spectrometry has been traditionally used in targeted analytical methods to measure one or several classes of known chemicals in the sample [26–28]. High-resolution mass spectrometry-based biomonitoring techniques are considered very promising tools in achieving a more complete understanding of the biological significance of exposome [8,9,25].

2.1.1. Low-Resolution Mass Spectrometry

Low-resolution mass spectrometers (LRMS), such as the triple quadrupole mass analyzer (QqQ), is only able to achieve m/z accuracy at unit level (~1 amu mass window) and is not capable of distinguishing compounds with very similar molecular mass. Also, LRMS has shown a low sensitivity in full scan mode. Both restrictions limit the ability of LRMS to detect unknowns. However, the selected reaction monitoring (SRM) mode of LRMS allows the quantitative analyses of a list of target precursor-product ion transitions with high sensitivity and broad linear dynamic range. To ensure the identification of a compound with LRMS, the retention time, at least two transitions (two product ions of the precursor ion), and their ratio of intensity are normally required [29]. False positive identifications are possible when only one transition is used [30].

Traditional biological measurement (targeted analysis) of exposures heavily relies on LRMS, which measures target xenobiotic compounds or their metabolites in the biological samples. This analytical platform typically provides validated and reliable quantification for analytes even at trace levels. This is of critical value for exposome studies because concentrations of xenobiotics are normally low in biospecimens [22]. Biomonitoring data in current databases or health surveys are primarily obtained via this approach because of its availability and maturity. For instance, the National Biomonitoring Program (NBP), launched by the Centers for Disease Control and Prevention (CDC) in the U.S., is routinely measuring approximately 300 chemicals which are known to be toxic to human beings [31]. New chemicals are added to the list when sufficient evidence supports the toxicological relevance and occurrence in the human body.

However, LRMS based targeted analysis has several limitations when it comes to exposome: (1) the inability to cover a wide range of chemicals in a single run; (2) the high possibility of missing the compounds which are at high levels in the biospecimens but not in the target analyte list; (3) the limited availability of commercial standards to quantify the target analytes in the samples; (4) the potential of false positives due to certain precursor ions only generating one fragment ion; and (5) the limiting of detection of certain chemicals in analysis due to matrix interferences.

Efforts have been taken to capture the exposome with this targeted approach. To increase the coverage, scientists seek to measure as many chemicals as possible in one sample through multiple runs. In a recent study, 128 persistent and non-persistent endocrine disruptors belonging to 13 chemical classes were measured with both liquid chromatography-tandem mass spectrometry (LC-MS/MS) and gas chromatography mass spectrometry (GC-MS) [32]. Although this approach is costly, laborious, and needs a large volume of biological samples, it can provide validated information about the quantities of a relatively large number of chemicals in the human body. Further, with the technical advances in mass analyzers, the latest generation of QqQ instruments allows a notable increase in the number of transitions acquired within the same run. For instance, based on the triggered multiple reaction monitoring (tMRM) function in Agilent 6400 series QqQ instruments, a multiresidue LC-MS/MS method was established with a coverage of about 450 globally important pesticides within 10 min [33]. Several multi-target screening LC-MS/MS methods have also been established to simultaneously measure up to 700 drugs with the hybrid triple quadrupole linear ion trap technology (QTrap) [34,35].

2.1.2. High-Resolution Mass Spectrometry

High-resolution mass spectrometers (HRMS) overcome the drawbacks of LRMS by providing high-quality mass resolution, exact molecular mass, and high sensitivity in full scan mode [36,37], which allows less strict requirements for the chromatography separation and improves the capability to detect low abundance chemicals in complex samples. Common HRMS typically possess mass-resolving power > 10, 000 (R, defined at full width at half maximum, FWHM), including time of flight (TOF), Fourier Transform Orbitrap (FT-Orbitrap), and Fourier transform-ion cyclotron resonance-mass spectrometer (FT-ICR-MS). FT-ICR-MS is costly and has limited availability and user feasibility, followed by FT-Orbitrap and TOF instruments. Currently, a vast majority of biomonitoring and environmental monitoring studies are based on TOF instruments, followed by FT-Orbitrap, with FT-ICR-MS platform generally not used in this field [25]. The pros and cons of TOF and FT-Orbitrap have been reviewed in multiple review articles [36–38]. Firstly, the FT-Orbitrap typically possesses higher mass-resolving power and mass accuracy than TOF instruments [38]. Mass-resolving power is highly dependent on the molecular composition, molecular mass, and scan speed [38]. Q Exactive Hybrid Quadrupole-Orbitrap is one of the commonly used FT-Orbitrap instruments and its resolution can be up to 1000 K at m/z of 200. Secondly, FT-Orbitrap instruments under full scan mode can achieve similar sensitivity down to a femtogram as QqQ instruments. As one of the newest versions of TOF instruments, Agilent 6550 iFunnel Q-TOF LC/MS can achieve a mass resolving power greater than 25K at 322 m/z. However, the sensitivity of TOF instruments under full scan mode is compromised with 1–2 orders of magnitude lower than that of QqQ operating in SRM mode [38]. This limits the capability of TOF instruments to measure chemicals at trace levels in samples. Thirdly, FT-Orbitrap instruments show an inverse relationship between scanning speed and mass resolution, while TOF instruments can maintain a rapid scanning rate regardless of the resolving power. Thus, the FT-Orbitrap based instrumental method needs to be well optimized to achieve a compromise between required mass resolution and adequate chromatography to adequately serve the research purpose [39]. Further, it needs to be noted that the application of an automatic gain control (AGC) in Orbitrap instruments is employed to prevent the overfilling of C-trap, but it also can affect the detection capability of trace compounds in complex matrices by reducing the absolute number of analyte ions that enter the ion trap [40]. The multiplexing feature of Q-Orbitrap instruments can help overcome this problem and achieve a higher sensitivity and an extended intrascan dynamic range for low-abundance chemicals in the complex matrices [40].

HRMS is typically combined with other mass analyzers to form hybrid instruments in real applications, including quadrupole (Q)-TOF, ion trap (IT)-TOF, Q Exactive, LTQ-Orbitrap, and Orbitrap Fusion Lumos Tribrid MS [36]. These hybrid mass analyzers can offer additional advantages such as increased sensitivity and fragments information for structural elucidation. In general, three analytical approaches have been developed with these hybrid mass spectrometers for exposome studies.

Targeted analysis. In HRMS based targeted analysis, standards and information of target analytes are available. There are several advantages of HRMS based targeted analysis over LRMS based analysis. Firstly, the confidence in compound identification increases significantly by using exact mass, especially for those chemicals that have only one transition and non-specific transitions (e.g., neutral loss of H_2O or CO_2, which are also common for matrix interferences). It has been reported that an excellent selectivity regardless of the matrices could be achieved for most compounds with an R = 30,000, by which the target analytes could be distinguished from interferences of the same nominal *m/z* [38]. The separation of isobars generally requires higher mass resolving power, for instance, the isobars glutamine and lysine are separated with a multiple-reflection TOF MS in excess of 70 K after a flight time of 0.2 ms [41]. Except for the transitions and retention times, isotopic pattern and monoisotopic mass of analytes can also be used for the identification. Secondly, a higher coverage of analytes can be achieved in a single run. Many hybrid mass analyzers, like Q-TOF and Q Exactive, offer data-dependent MS/MS acquisitions. For example, if a compound in a target ion list is detected in the full scan mode, an MS/MS analysis will be triggered accordingly. This allows for full scan product ion spectra recording within the same run for a significant number of compounds at superior sensitivity than the limit of most LRMS. The limitation of TOF instruments based targeted analysis lies in the lower sensitivity and linear dynamic range compared with the QqQ [42,43]. However, many studies have achieved better sensitivity employing full scan or full MS-data dependent MS^2 mode with FT-Orbitrap instruments compared with QqQ in targeted analysis [44–46].

Suspect screening (biased non-targeted analysis). This approach is used when two conditions are met: compound-specific information (e.g., molecular formula, chemical structure, and physicochemical properties) of the suspects are known and the reference standards for the suspects are not available. M+1 and M+2 isotopes of precursor ion are critical in identifying the chemical formula and MS/MS spectral information is critical in elucidating the chemical structure. Using either Q-TOFMS or Orbitrap instruments, many studies have employed this analytical approach in successfully identifying suspect transformation products/metabolites of parent compounds in environmental media, such as natural waters and wastewater [47,48]. Recently this suspect screening technique was used for human samples. One study used GC-TOF/MS platform and found the presence of pentachlorobiphenyl in child brain tissue [49]. Another study used LC-QTOF/MS platform for the identification of new environmental organic acids (EOAs) in maternal plasma and discovered 65 suspect EOAs including benzophenone-1 and bisphenol S [50]. The available information of the compounds can help facilitate the acquisition and identification of suspects. Data-dependent acquisition mode is normally used in suspect screening, in which the sample analysis starts with full scan MS acquisition and switches to MS/MS mode when an analyte of interest appears in the run and is recognized by the data software based on pre-determined criteria [25]. Triggers normally used in the data-dependent acquisition mode have been reviewed in details, including ion intensity, accurate mass inclusion, isotope pattern, pseudo-neutral loss, and mass defect criteria [25]. When a suspect ion is found in the samples, chemical structure and physicochemical property-derived information can be used in the identification and confirmation of the compounds [47].

Unknown screening (unbiased non-targeted analysis). When starting without any prior information about the compounds to be detected, an unknown screening approach is employed. Unknown screening theoretically enables the measurement of an unlimited number of compounds in the sample, making it a promising technique for the full characterization of the exposome. Workflows in analyzing HRMS data in non-targeted metabolomics studies have been reviewed elsewhere [25,51–53]. In exposome studies, the workflow can be borrowed from metabolomics with minor modifications, as detailed below.

(i)   grouping all the features (ions or peaks) from the same compound based on the full scan MS chromatogram and determining monoisotopic or neutral molecular mass. An enormous number of ions are present in the chromatogram, but not every ion represents an individual compound. One compound may form different adducts (e.g., protonated and deprotonated ions, $[M + Na]^+$, and $[M + NH_4]^+$), neutral losses (e.g., $[M + H-H_2O]^+$), isotopes (e.g., M+1 and M+2 isotope of precursor ion), and even in-source fragments. A variety of strategies have been proposed to group peaks including comparing expected theoretical distances between known ion adduct masses with experimental distances. One recent study suggested to extract MS pseudospectra based on peak shape and peak abundance based on the assumption that all the peaks with different m/z ratios from the same compounds ideally have a similar peak shape and strong linear relation in relative abundance across samples [51].

(ii) acquiring a list of potential chemical candidates by searching the monoisotopic mass or the molecular formula assigned against the databases. It has been reported that monoisotopic mass-based searching resulted in a higher percentage of chemicals in the number one rank position than chemical formula-based searching [54]. Available chemical substance databases such as PubChem and ChemSpider have been reviewed [52]. Recently, a few more databases have been developed for search including CompTox Chemistry Dashboard, Exposome-Explorer, and Toxic Exposome Database (T3DB) [55–57].

(iii) ranking the candidate list based on other information of the unknown chemical, including MS/MS spectral information, retention time, biochemical pathway and environmental chemistry knowledge. Fragments information can differentiate molecules with the same neutral mass in most cases. There are databases with experimental or in silico MS/MS spectral information available for reference, such as the Human Metabolome Database (HMDB) and METLIN. Currently available mass spectral databases have been reviewed elsewhere [52,58]. Retention time information can be obtained through quantitative structure-retention relationships (QSRR) models when reference standards are not available [59,60]. Biochemical pathway and environmental chemistry knowledge can also be used to narrow down putative identified compounds. In metabolomics, many bioinformatics tools are using biochemical pathways to filter and rank lists of candidates such as XCMS, xMSannotator, and mummichog [61–63]. It is expected that environmental chemistry knowledge will be incorporated into bioinformatic tools to facilitate the identification of xenobiotics.

The workflow mentioned above works well for identifying chemicals in one single sample. When the comparison across different samples is involved, other analyses are needed, such as peak aligning across samples, data pretreatment, and statistical analyses [25]. There are a wide range of chemometric and bioinformatic tools available at each stage in HRMS data analyses: preprocessing, annotation, and statistical analysis [52,64,65]. A variety of workflows, which encompass all stages of HRMS data analyses, are also available, including workflow4metabolomics, galaxy-M, XCMS online, MetaboAnalyst, MAVEN, MAIT, and MZmine 2 [64]. To standardize the confidence level in unknown identification, a variety of systems have been established to communicate the confidence [66,67]. In general, five levels are present and listed as follows with the increase of confidence, as shown in Figure 1: exact mass only, unequivocal molecular formula, multiple putative chemical structures available, one putative chemical structure, and validated structure with reference standard match in both retention time and mass spectra [68].

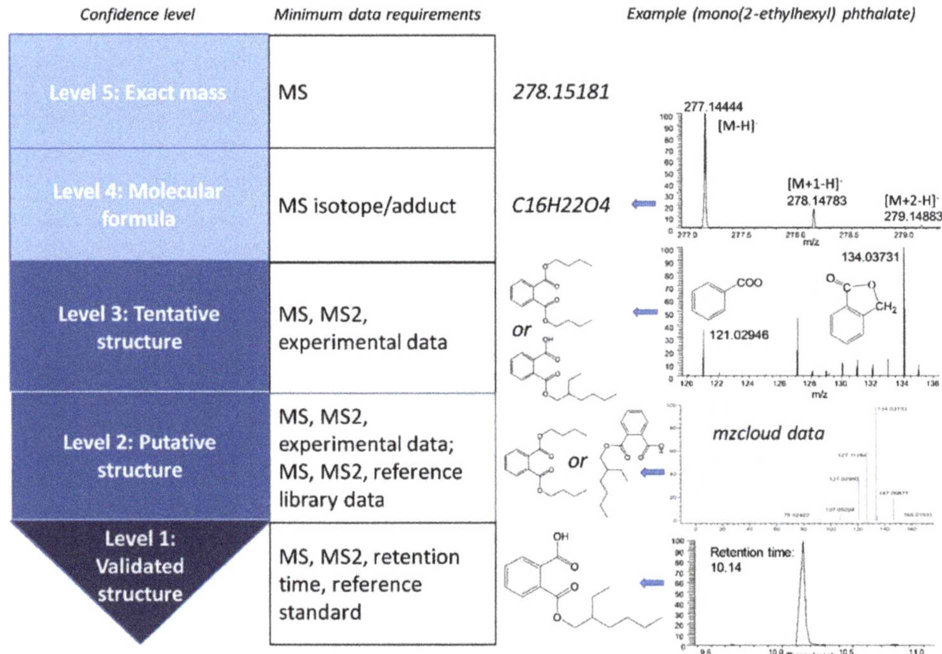

Figure 1. Proposed confidence levels in unknown xenobiotic identification with high resolution mass spectrometric analysis and exemplified with mono(2-ethylhexyl) phthalate.

Until now, targeted analysis, suspect screening, and unknown screening based on TOF instruments (mainly QTOF) have been used on a widespread basis to measure human exposures to a broad range of chemicals, including both persistent (e.g., organochlorine pesticides, polychlorinated biphenyls, polybrominated diphenyl ethers) and non-persistent (e.g., drugs, pesticides, surfactants, personal care products) chemicals [25,69]. One study developed a LC-QTOF/MS based suspect screening technique with a library of collision-induced dissociation accurate mass spectra of more than 2500 toxic compounds, including illegal and therapeutic drugs, pesticides, and alkaloids [70]. A large number of studies also reported the application of FT-Orbitrap instruments in xenobiotic screening including drugs and pesticides in both human and environmental samples [71–76]. Most studies are using the three analytical approaches together to capture a wide range of chemicals [39,76].

One research area in which HRMS has gained widespread application is metabolomics. Metabolome refers to the sum of all low molecular weight metabolites present in a living system [77]. Non-targeted metabolomics based on HRMS can simultaneously detect the endo- and exogenous chemicals, directly linking exposure to internal dose, biological effects, and disease pathobiology, which is a critical component of the exposome [78]. Thus, metabolomics has become a critical platform in exposome research.

2.2. Biological Approaches

Human exposure to xenobiotics can induce changes of the biological functions in many respects, such as gene transcription and protein synthesis. Instead of directly measuring the exposure itself, biological approaches measure these biological changes to understand the influences of exposure on human health. This approach is particularly helpful for those exposures which cannot be measured directly, such as reactive agents. For instance, reactive electrophiles, including reactive oxygen and nitrogen species, aldehydes, oxiranes, and quinones, can rapidly react with DNA and protein once absorbed into the organism [79]. Measuring adducts of these electrophiles with blood electrophiles such as hemoglobin (Hb) and human serum albumin (HAS) can help assess the human exposure to reactive electrophiles [79].

Analytical techniques used in biological approaches can be either instrument-based (e.g., mass spectrometry), or effect-based (e.g., bioassays), or the combination of both [80–82]. As the instrument-based techniques have been discussed earlier, here we focus on the effect-based and combined approaches. Advantages of effect-based approaches are listed as follows: (1) covers a wide spectrum of modes of action (MOA); (2) provides the opportunity to investigate the actual molecular targets; (3) allows prioritization of individual organisms for further investigation as well as chemical groups for identifying relevant mixture components; and (4) is time/cost effective and can be applied in high through-put screening scenarios [80,83,84]. One recent study screened and evaluated internal exposures of turtles to chemical mixtures based on an in vitro effect-based approach including a battery of sensitive bioassays with different modes of action, including aryl hydrocarbon receptor (AhR)-mediated xenobiotics (AhR-CAFLUX), NrF2-mediated oxidative stress (AREc32), NFκB mediated response to inflammation (NFκB-bla), estrogen binding (VM7Luc4E2), and baseline toxicity (Microtox) [80].

In most cases, routinely measured target compounds only partly account for the biological effects of concern. Thus, a hybrid approach, or effect-directed analysis (EDA), a combination of biotesting, fractionation procedures, and chemical analytical methods, is proposed to identify non-target compounds that cause biological toxicity in the sample [85]. General procedures of EDA are as follows: (1) extract the compounds of interest in an effective and non-selective way; (2) clean-up the extracted samples fraction; (3) conduct bioassay to select the interesting sample fraction; (4) fractionate the selected sample fraction to reduce the complexity; and (5) identify the suspected compounds with instrumental analysis methods. EDA has become a successful strategy in identifying biologically active compounds in environmental samples [82,86]. Although this approach has scarcely been applied in biological samples because of the difficulty in sample preparation, it is a promising approach in identifying the unknown chemical exposure leading to certain specific toxicity endpoint [87]. For instance, by using EDA approach, one study revealed the presence of new environmental contaminants, including di- and two monohydroxylated octachlorinated biphenyls (octaCBs) and linear and branched nonylphenol (NP), in polar bear serum with transthyretin-binding potency bioassay [23].

One well known example of the biological approaches in exposome research is omics profiling, which has found applications in large scale studies at population level [78]. High-dimensional analytical platforms were usually employed in omics profiling [78]. Besides metabolomics, omics profiling approaches also include genomics, epigenomics, transcriptomics, and proteomics [78]. Within the exposome framework, these approaches can provide a deep understanding at system biology-level of how chemical exposure influences the human health [78]. An unprecedented source of information has been produced with respect to the effective biological effects of exposures at omic-level. This omics data can be used in the generation of novel hypothesis to discover the disease etiologies of chemical exposure.

2.3. Other Approaches

Human exposome encompasses all types of exposures throughout the life course, including those from exo- and endogenous processes at individual level (e.g., environmental contaminants and infection) and general exposures at global level (e.g., climate and social economic status) [20,78]. Biological approaches can identify the influences of these exposures on human health, but are not able to characterize the exposure source, identify the route of exposure, and provide a picture of spatial and temporal variability of the exposure, which are critical in establishing links between exposome and biological significance [20,78]. While a few external exposures (e.g., environmental contaminants exposure) can be assessed through chemical approaches most external exposures (e.g., air quality and social economic status) need to be measured through other approaches such as questionnaires and static monitors [20,78]. Recently, a variety of novel assessment methods have been employed to quantify the external exposures, including those methods based on geographic information systems, environmental sensors, and personal sensing technology [20,78]. Although this review focuses on the analytical approaches employed in characterizing the chemical exposome, it is highly stressed that these new external exposure measurement methods are an integral part of the entire exposome measurement. With the rapid advancement of technologies, increasing popular applications of these novel methods are expected in exposome studies at both individual and population levels.

3. Measurement-Based Exposome Studies

Commonly-used strategies for chemical exposome study include top-down and bottom-up approaches [1]. A top-down exposome strategy starts with measuring all chemicals in a subject's (cases and controls) biospecimen, such as blood, at each life stage either through direct measurement or by investigating the physiological effects of exposures. After identifying the variant with significance by a series of data analysis, biological annotation is followed to determine the biological significance of the critical variable. This approach covers both exogenous and endogenous chemicals in the internal chemical environment, offers an efficient means for profiling individual exposures, and is the predominant strategy used in chemical exposome studies to date. A bottom-up strategy starts with a complete measurement of all the chemicals present in each external source (e.g., air, food, water, etc.) of a subject's exposome at each time point. After determining the analyte possessing significant association with health outcomes, uptake and metabolism of the analyte in the human body is evaluated. This approach requires enormous efforts in identifying the important chemicals in various environmental media and would miss important endogenous components in the body generated due to non-chemical factors such as physical activity, noise, inflammation, and social stress. One significant advantage of this approach is that it can provide valuable information regarding the critical chemicals present in each external source, making it complementary to top-down approach. Details of the two approaches as well as the available exposome studies based on these two approaches are discussed below.

3.1. Top-Down Exposome Approach

The research process of a chemical exposome study employing a top-down approach can be divided into five steps theoretically. (1) sample preparation, including both the collection of samples from groups of population (e.g., health and diseased) and the pretreatment of samples for instrumental analysis; (2) sample analysis, including the selection of instrumental types and methods; (3) data analysis, including chemometric analysis of the raw data and statistical analysis of the processed data; (4) biological annotation, determining the biological significance of the critical exposure identified; and (5) source identification, characterizing the source of the critical exposures identified.

Biomonitoring of biospecimens such as blood has several advantages for the exposome research. Firstly, it allows for the simultaneous measurement of exposures and the metabolic phenotypes. Secondly, it allows for the development and measurement of biomarkers for historic and current exposures. Any exposure including both chemical and non-chemical factors can leave unique signatures in the human body, even in the case of an exposure that has passed a long time ago [88,89]. If the signature is persistent and irreversible, we can assess the health outcomes of historic exposure. Thirdly, it allows for the development and measurement of biomarkers for disease at any stages. There is a long way to go from exposure to observed effects: external dose; internal dose; target organ dose; target organ metabolism; target organ responses; cellular/subcellular dose and interaction; toxic response; observed effects [90]. The biomarker development of diseases can help explain the roles of environmental exposures in the pathogenesis of diseases as well as facilitate the diagnostic and prognostic of diseases. Lastly, it allows for the retrospective analysis of samples. Biospecimen samples can be stored for a long period of time in appropriate conditions.

Many large cohort studies are using targeted approaches or the combination of both targeted and non-targeted approaches to collect a wide range of exposures, followed by statistical analyses to investigate the relationships between important groups of environmental exposure as well as the associations between exposure and disease. For instance, INfancia y Medio Ambiente (The INMA), a birth cohort study in Spain, investigated the levels of exposure to a wide array of pollutants during pregnancy such as brominated flame retardants, perfluoroalkyl substances, and metals, aiming to examine the role of environmental pollutants during pregnancy and early childhood in relation to child growth and development [91]. The existing exposome initiative in the U.S., the Children's Health Exposure Analysis Resource (CHEAR), is using both targeted and non-targeted analyses through a network of laboratories to provide a comprehensive measurement of myriads of environmental xenobiotics and biological response indicators in various biological samples to better understand the roles of environmental factors in children's health [7].

In addition to the national efforts towards exposome, many individual laboratories in the academic field are also dedicated to the exposome research. Dr. Dean Jones' group established an automated workflow for non-targeted exposome analysis with a dual chromatography (DC)-FTMS, combining a reverse phase C18 chromatography and anion exchange (AE) chromatography [24]. An adaptive processing software package, apLCMS, was exclusively designed for LC-FTMS data analyses [24]. The addition of C18 column increased the m/z feature detection by 23–36%, yielding a total number of features up to 7000 for individual samples [24]. This exposome workflow was capable of detecting environmental chemicals in the nanomolar and sub-nanomolar concentration ranges [92]. It has been extensively used in multiple studies to simultaneously detect endogenous metabolites with plasticizers, insecticides, fungicides, herbicides, drugs, bacterial products, and correlate environmental chemical exposure with a variety of health outcomes, including tuberculosis disease and neurological development [93–99].

By combining redesigned METLIN Exposome database with XCMS platform and cognitive computing, a newly established nontargeted workflow allows the detection of endocrine disrupting chemicals at low-nanomolar concentrations in human serum and urine and also allows the readout of the biological effect of a chemical [100]. An innovation of this workflow is using artificial intelligence as a potential tool to prioritize findings in exposome studies [100]. It is expected that this workflow will be more extensively used in future exposome research.

3.2. Bottom-Up Exposome Approach

The complete research process of bottom-up approach-based exposome studies include the following procedures: (1) sample preparation, including sample collection from all the potential external exposure sources and sample pretreatment prior to instrumental analysis; (2) sample analysis, identifying the chemicals present in the samples; (3) data analysis to identify the compounds associated with the disease of interest; and (4) biological validation, studying the metabolism and toxicity of the selected analytes in animal or cell-based models to confirm the relationship hypothesized.

This approach helps identify the critical exposure sources, therefore facilitating the authority or person to take actions to mitigate exposures. It is beneficial for studying those health outcomes which can be mainly ascribed to external exposure, such as allergies [101]. Although genetic factors also contribute to the incidences of allergies, it has been recognized that the increase in allergies observed in the past decades can be explained exclusively by environmental changes occurring in the same time period [101]. A number of studies have suggested that a variety of air pollutants, including volatile organic compounds, formaldehyde, toluene, and polycyclic aromatic hydrocarbons, not only exacerbate but also cause many types of allergies such as atopic dermatitis [102–104]. However, known environmental factors found with a traditional research paradigm cannot explain the increase in the prevalence of allergic diseases worldwide. Therefore, a recent study called for the integration of the external exposome in the etiopathogenesis of these diseases since a wide range of environmental factors were involved [101].

This approach can also be used to understand the metabolism of xenobiotics in biological organisms. By comparing the mass spectra arising from the environmental media and the organisms living in it, for instance, water and fish, it is possible to differentiate the xenobiotics, metabolized xenobiotics, and endogenous metabolites in the organism [105].

The major limitation of this approach is the laborious workload needed in acquiring the complete exposome from endless external sources. Therefore, careful selection of exposure sources is crucial in the successful identification of the critical exposure related with the target disease. Knowledge of potential mechanisms of disease and the differences between cases and controls are helpful in the selection of appropriate exposure sources.

4. Publicly Accessible Data-Based Exposome Studies

The ultimate goal of exposome study is to investigate the role of exposome in the pathogeneses of human diseases. Because of the huge challenge in the measurement of exposome, a few studies focused on those human exposures with publicly-available data [14]. Although this data is neither comprehensive across all exposure domains nor longitudinal, it provides an important platform for generating and testing hypothesis between exposome and human health. Such platforms allow the simultaneous analysis of multiple types of exposure (e.g., chemical exposure, social economic status, etc.). One successful example is the National Health and Nutrition Examination Survey (NHANES), a biannual health survey conducted by the U.S. Centers for Disease Control and Prevention, which provides information about the range of representative exposures across the general population. The NHANES includes environmental exposures such as chemicals, nutrients, and infectious agents and the measurement tools include LC/GC-MS, immunological assays, and questionnaires.

Using cross-sectional data from NHANES, Patel et al. employed an Environmental-Wide Association Study (EWAS) approach to investigate the relationships between 266 unique environmental factors and the clinical status for type 2 diabetes [106]. EWAS relies on linear regression models fitted independently for each covariate to separately examine the association between single exposure factor and the health outcome [106]. Then the factors with significant associations were validated across all models [106]. EWAS approach has been used on a widespread basis to assess the comprehensive relationships between a broad range of environmental/behavioral/clinical factors and various types of diseases, including blood pressure [107], type 2 diabetes in the Marshfield Personalized Medicine Research Project Biobank [108], all-cause mortality [109], and telomere length [110]. Besides EWAS, other approaches were also proposed to assess the effects of multiple chemical and non-chemical environmental stressors on health outcomes. For instance, by combining "big data", computational tools, and traditional biostatistics, one research evaluated putative relationships between exposures from natural, built, and social environment domains and lung cancer mortality and mortality disparities across four race and gender groups [111]. A total of 2162 chemical and nonchemical environmental stressor was involved in the study [111].

To move from a single exposure-disease analysis paradigm to cumulative exposure-disease models, advanced biostatistics are required to tackle large, multiple, heterogeneous, and secondary datasets. One recent study compared the performance of commonly-used multiple linear regression statistical methods within exposome context (237 exposure covariates), including EWAS, EWAS-multiple linear regression (MLR), Elastic net, sparse partial least squares regression, Graphical Unit Evolutionary Stochastic Search, and Deletion-Substitution-Addition algorithm (DSA) [112]. Authors found that none of the statistical methods outperformed others across all scenarios and properties examined. However, overall, multivariate methods outperformed univariate approaches in investigating the exposome [112]. Barrera-Gomez et al. extended this work by considering scenarios with statistical interactions and by providing a systematic comparison of methods that have been recommended to search for interactions [113]. In this study, Group-Lasso INTERaction-NET (GLINTERNET) and DSA had better overall performance than the other methods in detecting two-way interactions, but the sensitivity and false discovery rate was compromised [113]. Therefore, none of the statistical methods has outperformed others in analyzing such a large number of exposures so far. Many factors can affect the selection of the right exposure variant, such as highly correlated exposures and multiplicity. Patel and Ioannidis argued that effects that survive multiplicity considerations and that are large may be prioritized for future scrutiny in the exposome studies [114]. Future efforts may focus on other statistical methodologies such as profile regression, cluster analysis, and even machine learning methods in tackling large exposome datasets [112].

Several metabolome databases also provide essential information about the exposures, such as levels of xenobiotics and their metabolites in the biospecimens. HMDB, currently one of the world's most comprehensive metabolome database, is a great example [115]. Its latest version contains 114,100 metabolites and 21,834 xenobiotics and their metabolites [115]. Many studies have been conducted to interrelate the xenobiotics and their metabolites with endogenous metabolites, metabolic pathway, and health outcomes based on HMDB and other publicly available information. Rappaport et al. obtained human blood concentrations of 1561 small molecules and metals derived from foods, drugs, pollutants, and endogenous processes from the literature [compiled by the Human Metabolome Database (HMDB) and NHANES], and mapped chemical similarities after weighting by blood concentrations, disease-risk citations, and numbers of human metabolic pathways [22]. The results showed that endogenous chemicals, drugs, and food chemicals have similar concentration ranges in human blood, whereas those of pollutants were 1000 times lower [22]. While chemicals in the four classes were equally studied in terms of disease risks, studies of metabolic pathways were dominated by endogenous molecules and essential nutrients [22]. Bessonneau et al. obtained concentrations of 1233 chemicals that had been detected in saliva from the literature integrated into the HMDB, then connected salivary metabolites with human metabolic pathways and PubMed Medical Subject Headings (MeSH) terms, followed by pathway enrichment and pathway topology analyses [116]. The study found that 196 salivary metabolites with KEGG id were mapped into 49 metabolic pathways and associated with human metabolic diseases, central nervous system diseases, and neoplasms [116]. Saliva exposome represents at least 14 metabolic pathways, including amino acid metabolism, TCA cycle, gluconeogenesis, glutathione metabolism, pantothenate and CoA biosynthesis, and butanoate metabolism [116]. These studies offer insights into the roles of environmental factors in the etiology of diseases from a systematic perspective.

5. Challenges in Exposome Research

The introduction of exposome is expected to have breakthrough changes in uncovering the secrets of human diseases. However, a variety of challenges are present at each step of exposome research, measuring the exposome and linking it with health outcomes [17,117]. These challenges as well as the potential solutions are briefly discussed in this section with a focus on the chemical exposome research.

5.1. Challenges in Measuring the Exposome

The complexity and heterogeneity nature of exposome and its dynamic variation in both time and space presents a huge challenge in measuring the exposome [18]. The current most comprehensive approach is constructing epidemiological cohort studies with large sample size and long-term follow-ups. Human samples such as blood are collected at each critical stage, including fetal, early postnatal, childhood, teenage, and adult, from the same population. EXPOsOMICS and HELIX projects in the European Union and CHEAR project in the U.S. are exemplary studies regarding exposome research [7,118,119]. However, such studies are usually costly and laborious, and unaffordable for individual laboratories. There is an urgent need for alternative approaches to lower the cost of exposome research. Tooth and hair matrix biomarkers can incorporate the intensity and timing of exposure and has been referred to as "retrospective temporal exposome" [120–122]. This provides an effective approach to study the historical exposures in human beings for individual principal investigators.

Another challenge is that no single analytical technique can exhaust the chemical exposome in one sample because of the remarkable differences of chemicals in a wide range of physicochemical properties including mass, polarity, abundance, lipophilicity and pKa [89,123]. Even with the same technique, different sample processing methods and parameter settings can influence the results significantly. To measure as many chemicals as possible, samples should be carefully portioned and appropriately processed to fit different analytical techniques, which is costly and laborious.

Xenobiotics and their metabolites in the biological samples are usually at trace levels with several orders of magnitude lower than that of endogenous metabolites [22]. This demands high sensitivity for analytical instruments, which is also one of the reasons that mass spectrometry-based analytical platforms are gaining popularity in chemical exposome measurement. Low abundance mass spectra are significantly affected by the instrumental noise. Further, isotope pattern observed in high abundance mass spectra is usually not available for xenobiotics, which increases the difficulty in the identification of these compounds. Recently, efforts are being made to increase the intensity of these low abundance signals. For instance, one research group from Singapore isotopically labelled those xenobiotic biomarkers with common functional groups including phenolic hydroxy, carboxyl, and primary amine [124]. This method has improved sensitivity of 2–1184 fold for xenobiotics compared with other mass spectrometry based methods [124]. It has also been reported that increasing the number of replicate injections can help improve the reliability in low abundance chemicals measurement in high resolution metabolomics [125]. One recent study recommends integrating ion mobility spectrometry into mass spectrometry-based exposome measurements, which can provide increased overall measurement dynamic range and thus result in frequent detections of lower abundance molecules that are previously undetected [68].

HRMS data analysis such as unknown identification is also a huge challenge in HRMS based chemical exposome studies. None of the chemometric and bioinformatic tools available can successfully group and align all the features correctly. Every algorithm has its own pros and cons. In addition, although the compound databases are increasing the coverage annually, they are still far behind the number of chemicals available. More than 60 million chemicals are present in PubChem, however, only around 220,000 MS/MS spectra from 20,000 molecules or so are accessible in the databases [126].

5.2. Challenges in Associating Exposome with Diseases

In exposome, we are dealing with thousands of environmental risk factors that vary by source, place, and time. These factors affect human health differently depending on the exposure route, exposure timing window, dose, and specific target organ. In addition, these factors can also interact with each other to have synergistic, additive, or antagonistic effects when exerting effects on certain health outcome. To fully understand the effects of exposome on human health, it is necessary to integrate all the exposure factors and evaluate their effects systematically, which poses a great challenge.

In addition, exposome mapping could discover hundreds or even thousands of altered molecular features associated with disease endpoints. However, it is difficult to identify key exposome features that may drive disease or contribute to disease etiology. To address this, advanced statistical methodologies, such as machine learning and artificial intelligence, hold the promise of pinning down molecular features that play a key role in the pathogenesis of human disease.

6. Future Directions

Many factors can contribute to the further development of exposome research, including advancement in analytical platforms, high-throughput statistical analysis, HRMS data mining algorithms, large database of chemical substances and mass spectra, and biochemical pathways. It is the rapid advancement in the high-resolution metabolomics techniques that provides a high-throughput and affordable platform for the monitoring of environmental exposures in human beings. Instruments with high sensitivity, broad dynamic range, high resolution, high mass accuracy and a low cost are desirable in measuring the chemical exposome. Exciting achievements are also expected from the development of effective bioinformatic computation. Excellent algorithms are needed to help remove the noise in the mass spectra, group peaks, generate MS/MS spectra and retention time information, etc. It is time for exposome to embrace techniques such as machine learning.

With the establishment of exposome ontology through national studies such as CHEAR, more exposome researches are expected from the academic field. Thus, it is necessary to establish standards to allow cross comparison and validation. If every study follows the established standards, the data can be used for the integrative and systematic study in the future contributing to the complete picture of the exposome research. To establish the relationship between exposome and health outcomes, traditional statistical analysis needs to embrace big data analysis techniques to pinpoint the critical exposures. To better correlate the exposome with biological effects, high-throughput analytical techniques are needed to integrate data in exposome with other omics analyses including genome, transcriptome, proteome, and epigenome. This will help the principal investigators to study the mechanistic basis of the exposome in an affordable way.

7. Conclusions

Exposome research paradigm provides a great opportunity to identify critical non-genetic factors that contribute to the onset and progress of various diseases. This paper discussed the commonly used measurement techniques in chemical exposome research and reviewed the available chemical exposome studies. As the technologies moving forward along with the establishment of exposome ontology, more exciting discoveries are waiting in the journey to uncover the roles of non-genetic factors in the pathogeneses of human diseases.

Author Contributions: Conceptualization, H.R.; original draft preparation, J.X.; review and editing, all.

Funding: This research was funded in part by National Institute of Environmental Health Sciences under grant number P30ES025,128, R01ES024950, P30ES010126 and the North Carolina State University.

Conflicts of Interest: The authors declare no conflict of interest.

References

1. Rappaport, S.M.; Smith, M.T. Epidemiology. Environment and disease risks. *Science* **2010**, *330*, 460–461. [CrossRef] [PubMed]
2. Willett, W.C. Balancing life-style and genomics research for disease prevention. *Science* **2002**, *296*, 695–698. [CrossRef] [PubMed]
3. Polderman, T.J.; Benyamin, B.; de Leeuw, C.A.; Sullivan, P.F.; van Bochoven, A.; Visscher, P.M.; Posthuma, D. Meta-analysis of the heritability of human traits based on fifty years of twin studies. *Nat. Genet.* **2015**, *47*, 702–709. [CrossRef] [PubMed]
4. Rappaport, S.M. Genetic Factors Are Not the Major Causes of Chronic Diseases. *PLoS ONE* **2016**, *11*, e0154387. [CrossRef] [PubMed]
5. Wild, C.P. Complementing the genome with an "exposome": The outstanding challenge of environmental exposure measurement in molecular epidemiology. *Cancer Epidemiol. Biomark. Prev.* **2005**, *14*, 1847–1850. [CrossRef] [PubMed]
6. Niedzwiecki, M.M.; Miller, G.W. The Exposome Paradigm in Human Health: Lessons from the Emory Exposome Summer Course. *Environ. Health Perspect.* **2017**, *125*, 064502. [CrossRef] [PubMed]
7. Cui, Y.; Balshaw, D.M.; Kwok, R.K.; Thompson, C.L.; Collman, G.W.; Birnbaum, L.S. The Exposome: Embracing the Complexity for Discovery in Environmental Health. *Environ. Health Perspect.* **2016**, *124*, A137–A140. [CrossRef] [PubMed]
8. Johnson, C.H.; Athersuch, T.J.; Collman, G.W.; Dhungana, S.; Grant, D.F.; Jones, D.P.; Patel, C.J.; Vasiliou, V. Yale school of public health symposium on lifetime exposures and human health: The exposome; summary and future reflections. *Hum. Genom.* **2017**, *11*, 32. [CrossRef] [PubMed]
9. Dennis, K.K.; Marder, E.; Balshaw, D.M.; Cui, Y.; Lynes, M.A.; Patti, G.J.; Rappaport, S.M.; Shaughnessy, D.T.; Vrijheid, M.; Barr, D.B. Biomonitoring in the Era of the Exposome. *Environ. Health Perspect.* **2017**, *125*, 502–510. [CrossRef]

10. Al-Chalabi, A.; Pearce, N. Commentary: Mapping the Human Exposome: Without It, How Can We Find Environmental Risk Factors for ALS? *Epidemiology* **2015**, *26*, 821–823. [CrossRef]

11. Buck Louis, G.M.; Smarr, M.M.; Patel, C.J. The Exposome Research Paradigm: An Opportunity to Understand the Environmental Basis for Human Health and Disease. *Curr. Environ. Health Rep.* **2017**, *4*, 89–98. [CrossRef]

12. Lioy, P.J.; Rappaport, S.M. Exposure science and the exposome: An opportunity for coherence in the environmental health sciences. *Environ. Health Perspect.* **2011**, *119*, A466–A467. [CrossRef]

13. Rappaport, S.M. Implications of the exposome for exposure science. *J. Expo. Sci. Environ. Epidemiol.* **2011**, *21*, 5–9. [CrossRef]

14. Stingone, J.A.; Buck Louis, G.M.; Nakayama, S.F.; Vermeulen, R.C.; Kwok, R.K.; Cui, Y.; Balshaw, D.M.; Teitelbaum, S.L. Toward Greater Implementation of the Exposome Research Paradigm within Environmental Epidemiology. *Annu. Rev. Public Health* **2017**, *38*, 315–327. [CrossRef]

15. Buck Louis, G.M.; Yeung, E.; Sundaram, R.; Laughon, S.K.; Zhang, C. The exposome—Exciting opportunities for discoveries in reproductive and perinatal epidemiology. *Paediatr. Perinat. Epidemiol.* **2013**, *27*, 229–236. [CrossRef]

16. Escher, B.I.; Hackermuller, J.; Polte, T.; Scholz, S.; Aigner, A.; Altenburger, R.; Bohme, A.; Bopp, S.K.; Brack, W.; Busch, W.; et al. From the exposome to mechanistic understanding of chemical-induced adverse effects. *Environ. Int.* **2017**, *99*, 97–106. [CrossRef]

17. Siroux, V.; Agier, L.; Slama, R. The exposome concept: A challenge and a potential driver for environmental health research. *Eur. Respir. Rev.* **2016**, *25*, 124–129. [CrossRef]

18. Wild, C.P. The exposome: From concept to utility. *Int. J. Epidemiol.* **2012**, *41*, 24–32. [CrossRef]

19. Miller, G.W.; Jones, D.P. The nature of nurture: Refining the definition of the exposome. *Toxicol. Sci.* **2014**, *137*, 1–2. [CrossRef]

20. Turner, M.C.; Nieuwenhuijsen, M.; Anderson, K.; Balshaw, D.; Cui, Y.; Dunton, G.; Hoppin, J.A.; Koutrakis, P.; Jerrett, M. Assessing the Exposome with External Measures: Commentary on the State of the Science and Research Recommendations. *Annu. Rev. Public Health* **2017**, *38*, 215–239. [CrossRef]

21. van Tongeren, M.; Cherrie, J.W. An integrated approach to the exposome. *Environ. Health Perspect.* **2012**, *120*, A103–A104, author reply A104. [CrossRef]

22. Rappaport, S.M.; Barupal, D.K.; Wishart, D.; Vineis, P.; Scalbert, A. The blood exposome and its role in discovering causes of disease. *Environ. Health Perspect.* **2014**, *122*, 769–774. [CrossRef]

23. Simon, E.; van Velzen, M.; Brandsma, S.H.; Lie, E.; Loken, K.; de Boer, J.; Bytingsvik, J.; Jenssen, B.M.; Aars, J.; Hamers, T.; et al. Effect-directed analysis to explore the polar bear exposome: Identification of thyroid hormone disrupting compounds in plasma. *Environ. Sci. Technol.* **2013**, *47*, 8902–8912. [CrossRef]

24. Soltow, Q.A.; Strobel, F.H.; Mansfield, K.G.; Wachtman, L.; Park, Y.; Jones, D.P. High-performance metabolic profiling with dual chromatography-Fourier-transform mass spectrometry (DC-FTMS) for study of the exposome. *Metabolomics* **2013**, *9*, S132–S143. [CrossRef]

25. Andra, S.S.; Austin, C.; Patel, D.; Dolios, G.; Awawda, M.; Arora, M. Trends in the application of high-resolution mass spectrometry for human biomonitoring: An analytical primer to studying the environmental chemical space of the human exposome. *Environ. Int.* **2017**, *100*, 32–61. [CrossRef]

26. Asimakopoulos, A.G.; Xue, J.; De Carvalho, B.P.; Iyer, A.; Abualnaja, K.O.; Yaghmoor, S.S.; Kumosani, T.A.; Kannan, K. Urinary biomarkers of exposure to 57 xenobiotics and its association with oxidative stress in a population in Jeddah, Saudi Arabia. *Environ. Res.* **2016**, *150*, 573–581. [CrossRef]

27. Xue, J.; Wu, Q.; Sakthivel, S.; Pavithran, P.V.; Vasukutty, J.R.; Kannan, K. Urinary levels of endocrine-disrupting chemicals, including bisphenols, bisphenol A diglycidyl ethers, benzophenones, parabens, and triclosan in obese and non-obese Indian children. *Environ. Res.* **2015**, *137*, 120–128. [CrossRef]

28. Lenters, V.; Portengen, L.; Smit, L.A.; Jonsson, B.A.; Giwercman, A.; Rylander, L.; Lindh, C.H.; Spano, M.; Pedersen, H.S.; Ludwicki, J.K.; et al. Phthalates, perfluoroalkyl acids, metals and organochlorines and reproductive function: A multipollutant assessment in Greenlandic, Polish and Ukrainian men. *Occup. Environ. Med.* **2015**, *72*, 385–393. [CrossRef]

29. Shoemaker, J.; Dietrich, W. *Single Laboratory Validated Method for Determination of Cylindrospermopsin and Anatoxin-a in Ambient Water by Liquid Chromatography/Tandem Mass Spectrometry (LC/MS/MS)*; US EPA Office of Research and Development: Washington, DC, USA, 2017.

30. Schlittenbauer, L.; Seiwert, B.; Reemtsma, T. A false positive finding in liquid chromatography/triple quadrupole mass spectrometry analysis by a non-isobaric matrix component: The case of benzotriazole in urine for human biomonitoring. *Rapid. Commun. Mass Spectrom.* **2016**, *30*, 1560–1566. [CrossRef]

31. CDC, Center for Disease Control and Prevention. National Biomonitoring Program. Available online: https://www.cdc.gov/biomonitoring/about.html (accessed on 1 August 2017).

32. Chung, M.K.; Kannan, K.; Louis, G.M.; Patel, C.J. Toward Capturing the Exposome: Exposure Biomarker Variability and Co-Exposure Patterns in the Shared Environment. *Environ. Sci. Technol.* **2018**, *52*, 8801–8810. [CrossRef]

33. Improved LC/MS/MS Pesticide Multiresidue Analysis Using Triggered MRM and Online Dilution. Available online: https://www.agilent.com/cs/library/applications/5991-7193EN.pdf (accessed on 18 August 2019).

34. Dresen, S.; Ferreiros, N.; Gnann, H.; Zimmermann, R.; Weinmann, W. Detection and identification of 700 drugs by multi-target screening with a 3200 Q TRAP LC-MS/MS system and library searching. *Anal. Bioanal. Chem.* **2010**, *396*, 2425–2434. [CrossRef]

35. Mueller, C.A.; Weinmann, W.; Dresen, S.; Schreiber, A.; Gergov, M. Development of a multi-target screening analysis for 301 drugs using a QTrap liquid chromatography/tandem mass spectrometry system and automated library searching. *Rapid Commun. Mass Spectrom.* **2005**, *19*, 1332–1338. [CrossRef]

36. Lin, L.; Lin, H.; Zhang, M.; Dong, X.; Yin, X.; Qu, C.; Ni, J. Types, principle, and characteristics of tandem high-resolution mass spectrometry and its applications. *RSC Adv.* **2015**, *5*, 107623–107636. [CrossRef]

37. Marshall, A.G.; Hendrickson, C.L. High-resolution mass spectrometers. *Annu. Rev. Anal. Chem.* **2008**, *1*, 579–599. [CrossRef]

38. Krauss, M.; Singer, H.; Hollender, J. LC-high resolution MS in environmental analysis: From target screening to the identification of unknowns. *Anal. Bioanal. Chem.* **2010**, *397*, 943–951. [CrossRef]

39. Hernández, F.; Ibáñez, M.; Bade, R.; Bijlsma, L.; Sancho, J.V. Investigation of pharmaceuticals and illicit drugs in waters by liquid chromatography-high-resolution mass spectrometry. *TrAC Trends Anal. Chem.* **2014**, *63*, 140–157.

40. Romero-González, R.; Frenich, A.G. *Applications in High Resolution Mass Spectrometry: Food Safety and Pesticide*; Elsevier: Amsterdam, The Netherlands, 2017.

41. Dickel, T.; Plass, W.R.; Lippert, W.; Lang, J.; Yavor, M.I.; Geissel, H.; Scheidenberger, C. Isobar Separation in a Multiple-Reflection Time-of-Flight Mass Spectrometer by Mass-Selective Re-Trapping. *J. Am. Soc. Mass Spectrom.* **2017**, *28*, 1079–1090. [CrossRef]

42. Lacorte, S.; Fernandez-Alba, A.R. Time of flight mass spectrometry applied to the liquid chromatographic analysis of pesticides in water and food. *Mass Spectrom. Rev.* **2006**, *25*, 866–880. [CrossRef]

43. Sancho, J.V.; Pozo, O.J.; Ibanez, M.; Hernandez, F. Potential of liquid chromatography/time-of-flight mass spectrometry for the determination of pesticides and transformation products in water. *Anal. Bioanal. Chem.* **2006**, *386*, 987–997. [CrossRef]

44. Zhang, N.R.; Yu, S.; Tiller, P.; Yeh, S.; Mahan, E.; Emary, W.B. Quantitation of small molecules using high-resolution accurate mass spectrometers—A different approach for analysis of biological samples. *Rapid Commun. Mass Spectrom.* **2009**, *23*, 1085–1094. [CrossRef]

45. Henry, H.; Sobhi, H.R.; Scheibner, O.; Bromirski, M.; Nimkar, S.B.; Rochat, B. Comparison between a high-resolution single-stage Orbitrap and a triple quadrupole mass spectrometer for quantitative analyses of drugs. *Rapid Commun. Mass Spectrom.* **2012**, *26*, 499–509. [CrossRef]

46. Reinholds, I.; Pugajeva, I.; Bartkevičs, V. Comparison of Tandem Quadrupole Mass Spectrometry and Orbitrap High Resolution Mass Spectrometry for Analysis of Pharmaceutical Residues in Biota Samples. *Mat. Sci. Appl. Chem.* **2016**, *33*, 5–10. [CrossRef]

47. Kern, S.; Fenner, K.; Singer, H.P.; Schwarzenbach, R.P.; Hollender, J. Identification of transformation products of organic contaminants in natural waters by computer-aided prediction and high-resolution mass spectrometry. *Environ. Sci. Technol.* **2009**, *43*, 7039–7046. [CrossRef]

48. Hernandez, F.; Ibanez, M.; Gracia-Lor, E.; Sancho, J.V. Retrospective LC-QTOF-MS analysis searching for pharmaceutical metabolites in urban wastewater. *J. Sep. Sci* **2011**, *34*, 3517–3526. [CrossRef]

49. Cappiello, A.; Famiglini, G.; Palma, P.; Termopoli, V.; Lavezzi, A.M.; Matturri, L. Determination of selected endocrine disrupting compounds in human fetal and newborn tissues by GC-MS. *Anal. Bioanal. Chem.* **2014**, *406*, 2779–2788. [CrossRef]

50. Gerona, R.R.; Schwartz, J.M.; Pan, J.; Friesen, M.M.; Lin, T.; Woodruff, T.J. Suspect screening of maternal serum to identify new environmental chemical biomonitoring targets using liquid chromatography-quadrupole time-of-flight mass spectrometry. *J. Expo. Sci. Environ. Epidemiol.* **2018**, *28*, 101–108. [CrossRef]

51. Domingo-Almenara, X.; Montenegro-Burke, J.R.; Benton, H.P.; Siuzdak, G. Annotation: A Computational Solution for Streamlining Metabolomics Analysis. *Anal. Chem* **2018**, *90*, 480–489. [CrossRef]

52. Yi, L.; Dong, N.; Yun, Y.; Deng, B.; Ren, D.; Liu, S.; Liang, Y. Chemometric methods in data processing of mass spectrometry-based metabolomics: A review. *Anal. Chim Acta* **2016**, *914*, 17–34. [CrossRef]

53. Gorrochategui, E.; Jaumot, J.; Lacorte, S.; Tauler, R. Data analysis strategies for targeted and untargeted LC-MS metabolomic studies: Overview and workflow. *TrAC Trends Anal. Chem.* **2016**, *82*, 425–442. [CrossRef]

54. McEachran, A.D.; Sobus, J.R.; Williams, A.J. Identifying known unknowns using the US EPA's CompTox Chemistry Dashboard. *Anal. Bioanal. Chem.* **2017**, *409*, 1729–1735. [CrossRef]

55. Williams, A.J.; Grulke, C.M.; Edwards, J.; McEachran, A.D.; Mansouri, K.; Baker, N.C.; Patlewicz, G.; Shah, I.; Wambaugh, J.F.; Judson, R.S.; et al. The CompTox Chemistry Dashboard: A community data resource for environmental chemistry. *J. Cheminformatics* **2017**, *9*, 61. [CrossRef]

56. Neveu, V.; Moussy, A.; Rouaix, H.; Wedekind, R.; Pon, A.; Knox, C.; Wishart, D.S.; Scalbert, A. Exposome-Explorer: A manually-curated database on biomarkers of exposure to dietary and environmental factors. *Nucleic Acids Res.* **2017**, *45*, D979–D984. [CrossRef]

57. Wishart, D.; Arndt, D.; Pon, A.; Sajed, T.; Guo, A.C.; Djoumbou, Y.; Knox, C.; Wilson, M.; Liang, Y.; Grant, J.; et al. T3DB: The toxic exposome database. *Nucleic Acids Res.* **2015**, *43*, D928–D934. [CrossRef]

58. Kind, T.; Tsugawa, H.; Cajka, T.; Ma, Y.; Lai, Z.; Mehta, S.S.; Wohlgemuth, G.; Barupal, D.K.; Showalter, M.R.; Arita, M.; et al. Identification of small molecules using accurate mass MS/MS search. *Mass Spectrom. Rev.* **2016**, *9999*, 1–20. [CrossRef]

59. Randazzo, G.M.; Tonoli, D.; Strajhar, P.; Xenarios, I.; Odermatt, A.; Boccard, J.; Rudaz, S. Enhanced metabolite annotation via dynamic retention time prediction: Steroidogenesis alterations as a case study. *J. Chromatogr. B Anal. Technol. Biomed. Life Sci.* **2017**, *1071*, 11–18. [CrossRef]

60. Randazzo, G.M.; Tonoli, D.; Hambye, S.; Guillarme, D.; Jeanneret, F.; Nurisso, A.; Goracci, L.; Boccard, J.; Rudaz, S. Prediction of retention time in reversed-phase liquid chromatography as a tool for steroid identification. *Anal. Chim. Acta* **2016**, *916*, 8–16. [CrossRef]

61. Li, S.; Park, Y.; Duraisingham, S.; Strobel, F.H.; Khan, N.; Soltow, Q.A.; Jones, D.P.; Pulendran, B. Predicting network activity from high throughput metabolomics. *PLoS Comput. Biol.* **2013**, *9*, e1003123. [CrossRef]

62. Uppal, K.; Walker, D.I.; Jones, D.P. xMSannotator: An R Package for Network-Based Annotation of High-Resolution Metabolomics Data. *Anal. Chem.* **2017**, *89*, 1063–1067. [CrossRef]

63. Forsberg, E.M.; Huan, T.; Rinehart, D.; Benton, H.P.; Warth, B.; Hilmers, B.; Siuzdak, G. Data processing, multi-omic pathway mapping, and metabolite activity analysis using XCMS Online. *Nat. Protoc.* **2018**, *13*, 633–651. [CrossRef]

64. Spicer, R.; Salek, R.M.; Moreno, P.; Canueto, D.; Steinbeck, C. Navigating freely-available software tools for metabolomics analysis. *Metabolomics* **2017**, *13*, 106. [CrossRef]

65. Misra, B.B.; Fahrmann, J.F.; Grapov, D. Review of emerging metabolomic tools and resources: 2015-2016. *Electrophoresis* **2017**, *38*, 2257–2274. [CrossRef]

66. Sumner, L.W.; Amberg, A.; Barrett, D.; Beale, M.H.; Beger, R.; Daykin, C.A.; Fan, T.W.; Fiehn, O.; Goodacre, R.; Griffin, J.L.; et al. Proposed minimum reporting standards for chemical analysis Chemical Analysis Working Group (CAWG) Metabolomics Standards Initiative (MSI). *Metabolomics* **2007**, *3*, 211–221. [CrossRef]

67. Schymanski, E.L.; Jeon, J.; Gulde, R.; Fenner, K.; Ruff, M.; Singer, H.P.; Hollender, J. Identifying small molecules via high resolution mass spectrometry: Communicating confidence. *Environ. Sci. Technol.* **2014**, *48*, 2097–2098. [CrossRef]

68. Metz, T.O.; Baker, E.S.; Schymanski, E.L.; Renslow, R.S.; Thomas, D.G.; Causon, T.J.; Webb, I.K.; Hann, S.; Smith, R.D.; Teeguarden, J.G. Integrating ion mobility spectrometry into mass spectrometry-based exposome measurements: What can it add and how far can it go? *Bioanalysis* **2017**, *9*, 81–98. [CrossRef]

69. Hernandez, F.; Portoles, T.; Pitarch, E.; Lopez, F.J. Searching for anthropogenic contaminants in human breast adipose tissues using gas chromatography-time-of-flight mass spectrometry. *J. Mass Spectrom.* **2009**, *44*, 1–11. [CrossRef]

70. Broecker, S.; Herre, S.; Wust, B.; Zweigenbaum, J.; Pragst, F. Development and practical application of a library of CID accurate mass spectra of more than 2,500 toxic compounds for systematic toxicological analysis by LC-QTOF-MS with data-dependent acquisition. *Anal. Bioanal. Chem.* **2011**, *400*, 101–117. [CrossRef]

71. Roca, M.; Leon, N.; Pastor, A.; Yusa, V. Comprehensive analytical strategy for biomonitoring of pesticides in urine by liquid chromatography-orbitrap high resolution masss pectrometry. *J. Chromatogr. A* **2014**, *1374*, 66–76. [CrossRef]

72. Li, X.; Shen, B.; Jiang, Z.; Huang, Y.; Zhuo, X. Rapid screening of drugs of abuse in human urine by high-performance liquid chromatography coupled with high resolution and high mass accuracy hybrid linear ion trap-Orbitrap mass spectrometry. *J. Chromatogr. A* **2013**, *1302*, 95–104. [CrossRef]

73. Helfer, A.G.; Michely, J.A.; Weber, A.A.; Meyer, M.R.; Maurer, H.H. Orbitrap technology for comprehensive metabolite-based liquid chromatographic-high resolution-tandem mass spectrometric urine drug screening—Exemplified for cardiovascular drugs. *Anal. Chim. Acta* **2015**, *891*, 221–233. [CrossRef]

74. Plassmann, M.M.; Brack, W.; Krauss, M. Extending analysis of environmental pollutants in human urine towards screening for suspected compounds. *J. Chromatogr. A* **2015**, *1394*, 18–25. [CrossRef]

75. Senyuva, H.Z.; Gokmen, V.; Sarikaya, E.A. Future perspectives in Orbitrap-high-resolution mass spectrometry in food analysis: A review. *Food Addit. Contam. Part. A Chem. Anal. Control. Expo. Risk Assess.* **2015**, *32*, 1568–1606. [CrossRef]

76. Schymanski, E.L.; Singer, H.P.; Longree, P.; Loos, M.; Ruff, M.; Stravs, M.A.; Ripolles Vidal, C.; Hollender, J. Strategies to characterize polar organic contamination in wastewater: Exploring the capability of high resolution mass spectrometry. *Environ. Sci. Technol.* **2014**, *48*, 1811–1818. [CrossRef]

77. Wild, C.P.; Scalbert, A.; Herceg, Z. Measuring the exposome: A powerful basis for evaluating environmental exposures and cancer risk. *Environ. Mol. Mutagen.* **2013**, *54*, 480–499. [CrossRef]

78. Niedzwiecki, M.M.; Walker, D.I.; Vermeulen, R.; Chadeau-Hyam, M.; Jones, D.P.; Miller, G.W. The Exposome: Molecules to Populations. *Annu. Rev. Pharmacol. Toxicol.* **2019**, *59*, 107–127. [CrossRef]

79. Rappaport, S.M.; Li, H.; Grigoryan, H.; Funk, W.E.; Williams, E.R. Adductomics: Characterizing exposures to reactive electrophiles. *Toxicol. Lett.* **2012**, *213*, 83–90. [CrossRef]

80. Dogruer, G.; Weijs, L.; Tang, J.Y.; Hollert, H.; Kock, M.; Bell, I.; Madden Hof, C.A.; Gaus, C. Effect-based approach for screening of chemical mixtures in whole blood of green turtles from the Great Barrier Reef. *Sci. Total Environ.* **2018**, *612*, 321–329. [CrossRef]

81. Tang, J.Y.; McCarty, S.; Glenn, E.; Neale, P.A.; Warne, M.S.; Escher, B.I. Mixture effects of organic micropollutants present in water: Towards the development of effect-based water quality trigger values for baseline toxicity. *Water Res.* **2013**, *47*, 3300–3314. [CrossRef]

82. Brack, W.; Ait-Aissa, S.; Burgess, R.M.; Busch, W.; Creusot, N.; Di Paolo, C.; Escher, B.I.; Mark Hewitt, L.; Hilscherova, K.; Hollender, J.; et al. Effect-directed analysis supporting monitoring of aquatic environments—An in-depth overview. *Sci. Total Environ.* **2016**, *544*, 1073–1118. [CrossRef]

83. Escher, B.I.; Dutt, M.; Maylin, E.; Tang, J.Y.M.; Toze, S.; Wolf, C.R.; Lang, M. Water quality assessment using the AREc32 reporter gene assay indicative of the oxidative stress response pathway. *J. Environ. Monitor.* **2012**, *14*, 2877–2885. [CrossRef]

84. Tang, J.Y.; Aryal, R.; Deletic, A.; Gernjak, W.; Glenn, E.; McCarthy, D.; Escher, B.I. Toxicity characterization of urban stormwater with bioanalytical tools. *Water Res.* **2013**, *47*, 5594–5606. [CrossRef]

85. Brack, W. Effect-directed analysis: A promising tool for the identification of organic toxicants in complex mixtures? *Anal. Bioanal. Chem.* **2003**, *377*, 397–407. [CrossRef]

86. Tian, Z.; Gold, A.; Nakamura, J.; Zhang, Z.; Vila, J.; Singleton, D.R.; Collins, L.B.; Aitken, M.D. Nontarget Analysis Reveals a Bacterial Metabolite of Pyrene Implicated in the Genotoxicity of Contaminated Soil after Bioremediation. *Environ. Sci. Technol.* **2017**, *51*, 7091–7100. [CrossRef]

87. Simon, E.; Lamoree, M.H.; Hamers, T.; de Boer, J. Challenges in effect-directed analysis with a focus on biological samples. *Trends Anal. Chem.* **2015**, *67*, 179–191. [CrossRef]

88. Van Breda, S.G.; Wilms, L.C.; Gaj, S.; Jennen, D.G.; Briede, J.J.; Kleinjans, J.C.; de Kok, T.M. The exposome concept in a human nutrigenomics study: Evaluating the impact of exposure to a complex mixture of phytochemicals using transcriptomics signatures. *Mutagenesis* **2015**, *30*, 723–731. [CrossRef]

89. Pleil, J.D.; Stiegel, M.A. Evolution of environmental exposure science: Using breath-borne biomarkers for "discovery" of the human exposome. *Anal. Chem.* **2013**, *85*, 9984–9990. [CrossRef]

90. Asante-Duah, K. *Public Health Risk Assessment for Human Exposure to Chemicals*; Springer: Washington, DC, USA, 2017.

91. Robinson, O.; Basagana, X.; Agier, L.; de Castro, M.; Hernandez-Ferrer, C.; Gonzalez, J.R.; Grimalt, J.O.; Nieuwenhuijsen, M.; Sunyer, J.; Slama, R.; et al. The Pregnancy Exposome: Multiple Environmental Exposures in the INMA-Sabadell Birth Cohort. *Environ. Sci. Technol.* **2015**, *49*, 10632–10641. [CrossRef]

92. Go, Y.M.; Walker, D.I.; Liang, Y.; Uppal, K.; Soltow, Q.A.; Tran, V.; Strobel, F.; Quyyumi, A.A.; Ziegler, T.R.; Pennell, K.D.; et al. Reference Standardization for Mass Spectrometry and High-resolution Metabolomics Applications to Exposome Research. *Toxicol. Sci.* **2015**, *148*, 531–543. [CrossRef]
93. Neujahr, D.C.; Uppal, K.; Force, S.D.; Fernandez, F.; Lawrence, C.; Pickens, A.; Bag, R.; Lockard, C.; Kirk, A.D.; Tran, V.; et al. Bile acid aspiration associated with lung chemical profile linked to other biomarkers of injury after lung transplantation. *Am. J. Transplant.* **2014**, *14*, 841–848. [CrossRef]
94. Park, Y.H.; Lee, K.; Soltow, Q.A.; Strobel, F.H.; Brigham, K.L.; Parker, R.E.; Wilson, M.E.; Sutliff, R.L.; Mansfield, K.G.; Wachtman, L.M.; et al. High-performance metabolic profiling of plasma from seven mammalian species for simultaneous environmental chemical surveillance and bioeffect monitoring. *Toxicology* **2012**, *295*, 47–55. [CrossRef]
95. Osborn, M.P.; Park, Y.; Parks, M.B.; Burgess, L.G.; Uppal, K.; Lee, K.; Jones, D.P.; Brantley, M.A., Jr. Metabolome-wide association study of neovascular age-related macular degeneration. *PLoS ONE* **2013**, *8*, e72737. [CrossRef]
96. Cribbs, S.K.; Park, Y.; Guidot, D.M.; Martin, G.S.; Brown, L.A.; Lennox, J.; Jones, D.P. Metabolomics of bronchoalveolar lavage differentiate healthy HIV-1-infected subjects from controls. *AIDS Res. Hum. Retroviruses* **2014**, *30*, 579–585. [CrossRef]
97. Roede, J.R.; Uppal, K.; Park, Y.; Lee, K.; Tran, V.; Walker, D.; Strobel, F.H.; Rhodes, S.L.; Ritz, B.; Jones, D.P. Serum metabolomics of slow vs. rapid motor progression Parkinson's disease: A pilot study. *PLoS ONE* **2013**, *8*, e77629. [CrossRef]
98. Go, Y.M.; Walker, D.I.; Soltow, Q.A.; Uppal, K.; Wachtman, L.M.; Strobel, F.H.; Pennell, K.; Promislow, D.E.; Jones, D.P. Metabolome-wide association study of phenylalanine in plasma of common marmosets. *Amino Acids* **2015**, *47*, 589–601. [CrossRef]
99. Frediani, J.K.; Jones, D.P.; Tukvadze, N.; Uppal, K.; Sanikidze, E.; Kipiani, M.; Tran, V.T.; Hebbar, G.; Walker, D.I.; Kempker, R.R.; et al. Plasma metabolomics in human pulmonary tuberculosis disease: A pilot study. *PLoS ONE* **2014**, *9*, e108854. [CrossRef]
100. Warth, B.; Spangler, S.; Fang, M.; Johnson, C.H.; Forsberg, E.M.; Granados, A.; Martin, R.L.; Domingo-Almenara, X.; Huan, T.; Rinehart, D.; et al. Exposome-Scale Investigations Guided by Global Metabolomics, Pathway Analysis, and Cognitive Computing. *Anal. Chem* **2017**, *89*, 11505–11513. [CrossRef]
101. Cecchi, L.; D'Amato, G.; Annesi-Maesano, I. External exposome and allergic respiratory and skin diseases. *J. Allergy Clin. Immunol.* **2018**, *141*, 846–857. [CrossRef]
102. Gehring, U.; Wijga, A.H.; Brauer, M.; Fischer, P.; de Jongste, J.C.; Kerkhof, M.; Oldenwening, M.; Smit, H.A.; Brunekreef, B. Traffic-related air pollution and the development of asthma and allergies during the first 8 years of life. *Am. J. Respir. Crit. Care Med.* **2010**, *181*, 596–603. [CrossRef]
103. Kramer, U.; Sugiri, D.; Ranft, U.; Krutmann, J.; von Berg, A.; Berdel, D.; Behrendt, H.; Kuhlbusch, T.; Hochadel, M.; Wichmann, H.E.; et al. Eczema, respiratory allergies, and traffic-related air pollution in birth cohorts from small-town areas. *J. Dermatol. Sci.* **2009**, *56*, 99–105. [CrossRef]
104. Huang, C.C.; Wen, H.J.; Chen, P.C.; Chiang, T.L.; Lin, S.J.; Guo, Y.L. Prenatal air pollutant exposure and occurrence of atopic dermatitis. *Br. J. Dermatol.* **2015**, *173*, 981–988. [CrossRef]
105. Southam, A.D.; Lange, A.; Al-Salhi, R.; Hill, E.M.; Tyler, C.R.; Viant, M.R. Distinguishing between the metabolome and xenobiotic exposome in environmental field samples analysed by direct-infusion mass spectrometry based metabolomics and lipidomics. *Metabolomics* **2014**, *10*, 1050–1058. [CrossRef]
106. Patel, C.J.; Bhattacharya, J.; Butte, A.J. An Environment-Wide Association Study (EWAS) on type 2 diabetes mellitus. *PLoS ONE* **2010**, *5*, e10746. [CrossRef]
107. Tzoulaki, I.; Patel, C.J.; Okamura, T.; Chan, Q.; Brown, I.J.; Miura, K.; Ueshima, H.; Zhao, L.; Van Horn, L.; Daviglus, M.L.; et al. A nutrient-wide association study on blood pressure. *Circulation* **2012**, *126*, 2456–2464. [CrossRef]
108. Hall, M.A.; Dudek, S.M.; Goodloe, R.; Crawford, D.C.; Pendergrass, S.A.; Peissig, P.; Brilliant, M.; McCarty, C.A.; Ritchie, M.D. Environment-wide association study (EWAS) for type 2 diabetes in the Marshfield Personalized Medicine Research Project Biobank. *Pac. Symp. Biocomput.* **2014**, 200–211.
109. Patel, C.J.; Rehkopf, D.H.; Leppert, J.T.; Bortz, W.M.; Cullen, M.R.; Chertow, G.M.; Ioannidis, J.P. Systematic evaluation of environmental and behavioural factors associated with all-cause mortality in the United States national health and nutrition examination survey. *Int. J. Epidemiol.* **2013**, *42*, 1795–1810. [CrossRef]

110. Patel, C.J.; Manrai, A.K.; Corona, E.; Kohane, I.S. Systematic correlation of environmental exposure and physiological and self-reported behaviour factors with leukocyte telomere length. *Int. J. Epidemiol.* **2017**, *46*, 44–56. [CrossRef]

111. Juarez, P.D.; Hood, D.B.; Rogers, G.L.; Baktash, S.H.; Saxton, A.M.; Matthews-Juarez, P.; Im, W.; Cifuentes, M.P.; Phillips, C.A.; Lichtveld, M.Y.; et al. A novel approach to analyzing lung cancer mortality disparities: Using the exposome and a graph-theoretical toolchain. *Environ. Dis.* **2017**, *2*, 33–44.

112. Agier, L.; Portengen, L.; Chadeau-Hyam, M.; Basagana, X.; Giorgis-Allemand, L.; Siroux, V.; Robinson, O.; Vlaanderen, J.; Gonzalez, J.R.; Nieuwenhuijsen, M.J.; et al. A Systematic Comparison of Linear Regression-Based Statistical Methods to Assess Exposome-Health Associations. *Environ. Health Perspect.* **2016**, *124*, 1848–1856. [CrossRef]

113. Barrera-Gomez, J.; Agier, L.; Portengen, L.; Chadeau-Hyam, M.; Giorgis-Allemand, L.; Siroux, V.; Robinson, O.; Vlaanderen, J.; Gonzalez, J.R.; Nieuwenhuijsen, M.; et al. A systematic comparison of statistical methods to detect interactions in exposome-health associations. *Environ. Health* **2017**, *16*, 74. [CrossRef]

114. Patel, C.J.; Ioannidis, J.P. Placing epidemiological results in the context of multiplicity and typical correlations of exposures. *J. Epidemiol. Community Health* **2014**, *68*, 1096–1100. [CrossRef]

115. Wishart, D.S.; Feunang, Y.D.; Marcu, A.; Guo, A.C.; Liang, K.; Vazquez-Fresno, R.; Sajed, T.; Johnson, D.; Li, C.; Karu, N.; et al. HMDB 4.0: The human metabolome database for 2018. *Nucleic Acids Res.* **2018**, *46*, D608–D617. [CrossRef]

116. Bessonneau, V.; Pawliszyn, J.; Rappaport, S.M. The Saliva Exposome for Monitoring of Individuals' Health Trajectories. *Environ. Health Perspect.* **2017**, *125*, 077014. [CrossRef]

117. Slama, R.; Vrijheid, M. Some challenges of studies aiming to relate the Exposome to human health. *Occup. Environ. Med.* **2015**, *72*, 383–384. [CrossRef]

118. Vineis, P.; Chadeau-Hyam, M.; Gmuender, H.; Gulliver, J.; Herceg, Z.; Kleinjans, J.; Kogevinas, M.; Kyrtopoulos, S.; Nieuwenhuijsen, M.; Phillips, D.H.; et al. The exposome in practice: Design of the EXPOsOMICS project. *Int. J. Hyg. Environ. Health* **2017**, *220*, 142–151. [CrossRef]

119. Vrijheid, M.; Slama, R.; Robinson, O.; Chatzi, L.; Coen, M.; van den Hazel, P.; Thomsen, C.; Wright, J.; Athersuch, T.J.; Avellana, N.; et al. The human early-life exposome (HELIX): Project rationale and design. *Environ. Health Perspect.* **2014**, *122*, 535–544. [CrossRef]

120. Pragst, F.; Broecker, S.; Hastedt, M.; Herre, S.; Andresen-Streichert, H.; Sachs, H.; Tsokos, M. Methadone and illegal drugs in hair from children with parents in maintenance treatment or suspected for drug abuse in a German community. *Ther. Drug Monit.* **2013**, *35*, 737–752. [CrossRef]

121. Andra, S.S.; Austin, C.; Wright, R.O.; Arora, M. Reconstructing pre-natal and early childhood exposure to multi-class organic chemicals using teeth: Towards a retrospective temporal exposome. *Environ. Int.* **2015**, *83*, 137–145. [CrossRef]

122. Andra, S.S.; Austin, C.; Arora, M. The tooth exposome in children's health research. *Curr. Opin. Pediatr.* **2016**, *28*, 221–227. [CrossRef]

123. Liu, K.H.; Walker, D.I.; Uppal, K.; Tran, V.; Rohrbeck, P.; Mallon, T.M.; Jones, D.P. High-Resolution Metabolomics Assessment of Military Personnel: Evaluating Analytical Strategies for Chemical Detection. *J. Occup. Environ. Med.* **2016**, *58*, S53–S61. [CrossRef]

124. Jia, S.; Xu, T.; Huan, T.; Chong, M.; Liu, M.; Fang, W.; Fang, M. Chemical Isotope Labeling Exposome (CIL-EXPOSOME): One High-Throughput Platform for Human Urinary Global Exposome Characterization. *Environ. Sci. Technol.* **2019**, *53*, 5445–5453. [CrossRef]

125. Walker, D.I.; Mallon, C.T.; Hopke, P.K.; Uppal, K.; Go, Y.M.; Rohrbeck, P.; Pennell, K.D.; Jones, D.P. Deployment-Associated Exposure Surveillance With High-Resolution Metabolomics. *J. Occup. Environ. Med.* **2016**, *58*, S12–S21. [CrossRef]

126. Johnson, S.R.; Lange, B.M. Open-access metabolomics databases for natural product research: Present capabilities and future potential. *Front. Bioeng. Biotechnol.* **2015**, *3*, 22. [CrossRef]

 *toxics*

Review

Gut Microbiome Toxicity: Connecting the Environment and Gut Microbiome-Associated Diseases

Pengcheng Tu, Liang Chi, Wanda Bodnar, Zhenfa Zhang, Bei Gao, Xiaoming Bian, Jill Stewart, Rebecca Fry and Kun Lu *

Department of Environmental Sciences and Engineering, University of North Carolina at Chapel Hill, Chapel Hill, NC 27599, USA; ptu@live.unc.edu (P.T.); liang16@live.unc.edu (L.C.); wanda_bodnar@unc.edu (W.B.); zhenfaz@email.unc.edu (Z.Z.); wintergb2012@gmail.com (B.G.); bxmroly@uga.edu (X.B.); jill.stewart@unc.edu (J.S.); rfry@unc.edu (R.F.)
* Correspondence: kunlu@unc.edu; Tel.: +1-919-966-7337

Received: 4 September 2019; Accepted: 6 March 2020; Published: 12 March 2020

Abstract: The human gut microbiome can be easily disturbed upon exposure to a range of toxic environmental agents. Environmentally induced perturbation in the gut microbiome is strongly associated with human disease risk. Functional gut microbiome alterations that may adversely influence human health is an increasingly appreciated mechanism by which environmental chemicals exert their toxic effects. In this review, we define the functional damage driven by environmental exposure in the gut microbiome as gut microbiome toxicity. The establishment of gut microbiome toxicity links the toxic effects of various environmental agents and microbiota-associated diseases, calling for more comprehensive toxicity evaluation with extended consideration of gut microbiome toxicity.

Keywords: gut microbiome; environment; chemical toxicity

---

1. Introduction

The human gut microbiome including the microorganisms, their genomes, and the surrounding environment in the gut, has received unprecedented attention over the past decade [1]. Mounting evidence suggests that the metabolic activities in the gut microbiome are profoundly intertwined with human health and disease [2]. A number of important functions performed by the gut microbiome are well recognized including the digestion of polysaccharides, biosynthesis of vitamins and nutrients, colonization resistance, and immune system modulation [3–5]. Moreover, the effects of the gut microbiome on host metabolism and physiology extend beyond the gut to distant organs such as the liver, muscle, and brain [2,6]. Due to its crucial role in human fitness, the gut microbiome is now considered as a new organ in the human body [7–9]. It is unequivocal that the gut microbiome functions properly on the premise that a normal gut microbial homeostasis is maintained [2]. However, the constitution and functionality of the gut microbiome can be readily influenced by diverse intrinsic and extrinsic factors [10]. For example, exposure to various xenobiotics leads to functional perturbation in the gut microbiome [11–15]. Mounting studies suggest that these environmentally induced perturbations are potentially linked to elevated disease risks [2,16]. Adverse health outcomes including inflammatory bowel disease (IBD), obesity, diabetes, cardiovascular disease, liver disease, colorectal cancer, and neurological disorders can be at least in part attributed to undesirable functional alterations in the gut microbiome [17–23].

Certain environmental toxins can induce damage and dysfunction in the liver, which is termed as liver toxicity. Exposure to these toxins changes the morphology and functionality of the liver, hence leading to liver diseases. Similarly, exposure to some environmental chemicals causes structural differences and functional alterations in the gut microbiome, which probably results in a series of adverse health outcomes. For instance, arsenic exposure can perturb the composition and metabolites of the mouse gut microbiome, which potentially contributes to its toxicity [12]. Given the increasingly recognized role of the gut microbiome in human health coupled with its susceptibility to environmental insults, it is of significance to define gut microbiome toxicity. As the gut microbiome is viewed as a new human organ by its importance, we accordingly define the environmentally driven functional damage in the gut microbiome as gut microbiome toxicity. The development of gut microbiota-related diseases can be generalized as environmental factors leading to deleterious alterations in the gut microbiome, which adversely affect human health via host–gut microbiota interactions. In other words, gut microbiome toxicity triggered by environmental exposures contributes to gut microbiota-related adverse outcomes. By including gut microbiome toxicity into the organ toxicity family, we can now discuss the relationship between the environment and gut microbiota-related diseases in the context of toxicology (Figure 1).

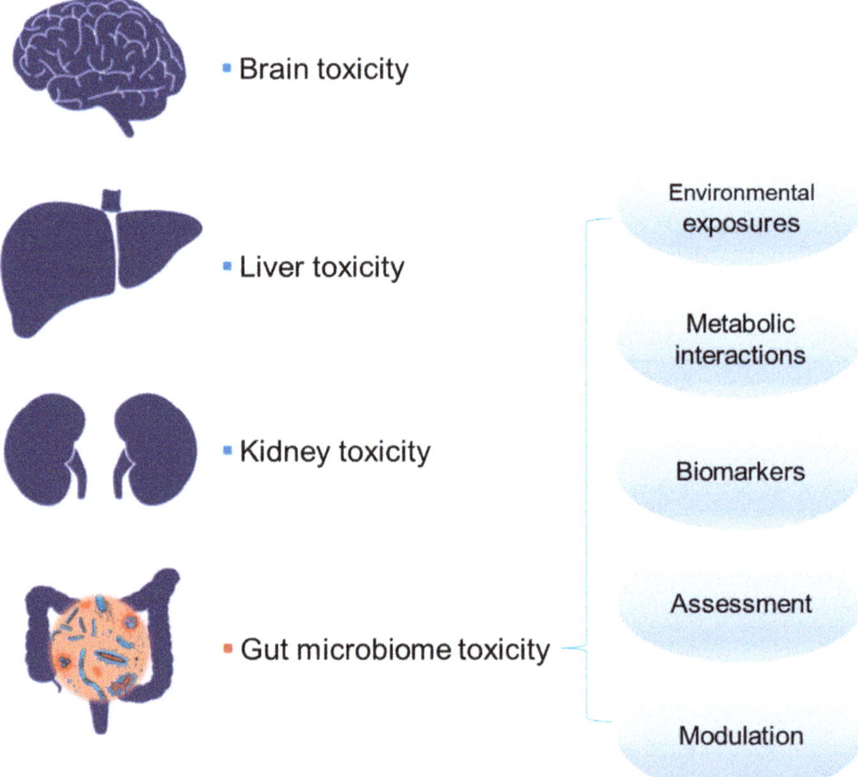

Figure 1. A potential new member of the organ toxicity family: gut microbiome toxicity. Toxicity of organs including brain, liver, and kidney is well defined and acknowledged. Similarly, the discussion of gut microbiome toxicity encompasses environmental exposures (causes), interactions between the gut microbiome toxicity and human diseases (mechanisms of gut microbiota-related diseases), biomarker and assessment (diagnosis), and modulation (treatment).

Environmental exposure is a significant risk factor for a series of human diseases, overlapping those diseases that are associated with the gut microbiome [24–26]. Thus, gut microbiome toxicity may be the missing link between environmental exposure and microbiome-related human diseases.

Moreover, the current toxicity testing system does not include toxic endpoints regarding the effects of environmental chemicals on the gut microbiome [27,28]. Considering the potential involvement of the gut microbiome in human disease, it is imperative to integrate gut microbiome toxicity into the toxicity assessment of environmental exposure. Thus, the establishment of gut microbiome toxicity may offer insights regarding the mechanistic basis underlying the toxicity of environmental chemicals, calling for more comprehensive risk assessment with the integration of gut microbiome toxicity. Additionally, environmentally driven alterations in the gut microbiome are not necessarily always adverse. By functional damage, we refer in particular to those that potentially contribute to adverse health outcomes, for instance, the production of pro-inflammatory metabolites. With the introduction of 'gut microbiome toxicity', we highlight the underappreciated mechanisms by which environmental factors lead to or exacerbate diseases through perturbing the gut microbiome functions.

In this review, we carefully define gut microbiome toxicity as environmentally driven functional damage in the gut microbiome. Functional damage may include changes in bacterial metabolites, loss of bacterial diversity, or effects on energy metabolism and balance. We focus on recent studies in support of the establishment of gut microbiome toxicity, and we accordingly discuss the environmental exposures, metabolic interactions in human disease, biomarkers and assessment, and modulation (Figure 1). Specifically, we review recent studies demonstrating the functional perturbation in the gut microbiome driven by various xenobiotics such as antibiotics, heavy metals, pesticides, and artificial sweeteners. These functional changes include, but are not limited to, alterations in the bacterial production of metabolites, diversity loss in the bacterial community, and interference in energy metabolism, which are further linked to the development of gut microbiota-related diseases. Moreover, microbiome changes including compositional and functional changes can serve as biomarkers for gut microbiome toxicity. Additionally, we briefly summarize current approaches for the assessment of gut microbiome toxicity as well as effective gut microbiome modulation.

2. Environmental Exposures

The fact that a number of xenobiotics can trigger gut microbiome toxicity suggests the underestimation of the toxic effects of specific chemicals. On one hand, the induction of gut microbiome toxicity may be considered a potential new mechanism by which known toxic chemicals (e.g., heavy metals, pesticides) lead to or exacerbate human diseases. On the other hand, it is of necessity to reconsider the health effects and acceptable daily intake (ADI) of widely-used chemicals such as food additives in the context of their contribution to gut microbiome toxicity. The impact of xenobiotics on the human gut microbiome can be direct or indirect. The human gut microbiome encodes more diverse metabolic enzymes, which greatly expands the repertoire of biochemical reactions within the human body [29]. Some environmental chemicals can directly affect the gut bacteria by interrupting specific metabolic pathway or gene expression, leading to distinct selection pressures, hence shaping the gut microbial community due to the uniqueness of the set of metabolic pathways and genome possessed by different bacterial species [29]. Therefore, the selection of resistant bacteria upon certain exposure could lead to an unbalanced gut eco-system. Additionally, some environmental chemicals can indirectly impact the gut microbiome through the influence on host physiology (e.g., gut mucosa [30]) and cell-to-cell communications of bacteria (e.g., quorum sensing [31]). That being said, the mechanistic basis underlying the microbial perturbation induced by specific chemical exposure remains elusive. Here, we highlight representative xenobiotics such as antibiotics, heavy metals, pesticides, and artificial sweeteners that cause gut microbiome toxicity with significant functional alterations.

2.1. Antibiotics/Drugs

It is commonly accepted that antibiotic administration, especially broad-spectrum antibiotics, severely impacts commensal bacteria. Both short-term and long-term antibiotic treatments lead to gut microbiome toxicity, although partial recovery may occur [32,33]. In many cases, effects of antibiotics on bacterial communities result in diversity loss and compositional imbalance [34]. Moreover, antibiotics not only disturb the gut microbiome at the compositional level, but also substantially change its functional profiles. For example, a recent study used a multi-omics approach to resolve the changes induced by beta-lactam in human gut microbiome [35]. The results showed that beta-lactam treatment caused both taxonomic and functional alterations in gut microbiome supported by alterations at the metagenomic, metatranscriptomic, metametabolomic, and metaproteomic levels. Antibiotic exposure in mice has been linked to diseases such as obesity and diabetes [36,37]. Aside from antibiotics, non-antibiotic drugs also affect the gut microbiome [38]. For instance, metformin [39], non-steroidal anti-inflammatory drugs [40], proton pump inhibitors [41], and atypical antipsychotics [42] are reported to have effects on the gut microbiome, although the health consequences remain underexplored. A most recent study tested 1200 marketed drugs by in vitro screening to investigate their effects on the gut microbiome [43]. A quarter of human-targeted drugs were discovered to have effects on the gut bacteria to some degree, indicating the potential of medication to induce gut microbiome toxicity.

2.2. Heavy Metals

Heavy metals continue to be a class of intensely-studied environmental contaminants. However, the role of heavy metals in gut microbiome toxicity still remains underappreciated. In fact, the gut bacteria play an important role in the biotransformation of heavy metals, which may promote or attenuate their toxicity. For example, human gut bacteria are able to transform inorganic arsenic into less toxic organic arsenic species [44], and demethylation of methyl-mercury by gut bacteria can generate more toxic inorganic mercury [45]. Rats exposed to heavy metals including arsenic, cadmium, cobalt, chromium, and nickel exhibited significant changes in their gut microbial compositions [46]. Moreover, the functional profiles in the gut microbiome can be perturbed by heavy metals. Four weeks of arsenic exposure in drinking water (10 ppm) caused significantly different metabolite profiles in the mouse gut microbiome [12]. Likewise, 13 weeks of arsenic exposure with an environment-relevant dose (100 ppb) also perturbed diverse bacterial metabolic pathways [47]. Arsenic-induced gut microbiome toxicity provided a new angle to look at the mechanism of arsenic toxicity. Follow-up studies evaluating arsenic metabolism further demonstrated that arsenic-induced gut microbiome toxicity can be affected by factors including host genetics [48], gender [49], and bacterial infection [50]. Moreover, different arsenic doses (10 ppm and 100 ppb) induced different levels of perturbation in the mouse gut microbiome, indicating the dose-dependent effects of arsenic exposure, which together with toxicity response thresholds of arsenic-induced gut microbiome toxicity need to be further defined. In addition, exposure to manganese and lead disturbs the gut microbial functions of mice with perturbed pathways and metabolites [51,52].

2.3. Pesticides

Excessive use of pesticides in agriculture has raised concern about their health effects. The argument that certain pesticides are safe to humans because their targeted pathways do not even exist in the human body fails to consider the microbes in the gut [53]. For example, herbicides like 2,4-dichlorophenoxyacetic acid (2,4-D), which impact plant hormones, may affect gut bacteria because not only plants but also bacteria can synthesize plant hormones [54]. Likewise, the shikimate pathway, the target of herbicide glyphosate, is commonly present in human gut bacteria [55,56]. In bacteria, this pathway has an important function linking the metabolism of carbohydrates to the biosynthesis of folates and aromatic amino acids. Several studies have demonstrated the association between gut microbiome toxicity and pesticide exposure. For example, the fungicide imazalil changed the composition of gut microbiome in zebrafish and mice [57,58]. Of interest, a recent study showed that the mouse gut microbiome was perturbed by 13 weeks of diazinon exposure (4 ppm) [13]. Bacterial genes and metabolites involved in neurotransmitter synthesis were significantly perturbed, suggesting that diazinon-induced gut microbiome toxicity with altered bacterial biosynthesis of the neurotransmitter may be partially responsible for the neurotoxicity of diazinon [59,60]. In addition, exposure to diazinon and malathion impacts the quorum sensing of gut bacteria, providing evidence that affecting bacterial communications may be one of the underlying mechanisms of gut microbial perturbations [61,62].

2.4. Artificial Sweeteners

Food additives have facilitated the development of the modern food industry. Normally, food additives (e.g., artificial sweeteners, emulsifiers, preservatives) are added in food products with an approved safe amount. Nevertheless, gut microbiome toxicity was not taken into consideration when the related standards were determined. Many artificial sweeteners are considered safe because they are poorly metabolized by the human body [29]. However, the gut bacteria are actively involved in the biotransformation. For example, cyclamate, which is currently banned in the USA, can be metabolized by gut bacteria into cyclohexylamine, which is carcinogenic [63]. Artificial sweeteners stevioside and xylitol can also be metabolized by the gut bacteria [64,65]. Several studies have demonstrated that some artificial sweeteners and emulsifiers were able to induce gut microbiome toxicity with potential gut microbiota-related health consequences. For instance, in an elegantly conducted study by Suez and colleagues, consumption of saccharin induced both compositional and functional changes in mouse gut microbiome that might be involved in the development of glucose intolerance [66]. Another study reported similar results with increased inflammatory levels in addition to gut microbial perturbation induced by saccharin in mice [15]. Additionally, artificial sweeteners acesulfame potassium [61], sucralose [67], aspartame [68], and neotame [69] can also perturb bacterial metabolites in concert with health implications including obesity and inflammation. Moreover, another study found that two commonly used emulsifiers altered mouse gut microbial composition together with elevated inflammatory levels [70].

2.5. Others

The above discussion is not intended to be exhaustive. More information could be referred to in recent reviews regarding the relationship between xenobiotics and the gut microbiome [11,29,71]. We emphasize functional changes in the gut microbiome induced by environmental exposure in the current review, therefore studies were included documenting not only the compositional shifts after exposure, but also functional alterations manifested by functional metagenomics and metabolomics. A rapidly-increasing list of xenobiotics is linked to gut microbiome toxicity. Some are commonly present in our daily life; a typical example is the antibacterial and antifungal agent triclosan. It has been repeatedly reported that triclosan induced changes in the gut microbiome using multiple animal models [14,72–74]. However, the effects of triclosan on human gut microbiome remain controversial [75]. Furthermore, exposure to nicotine (a major toxic component of tobacco smoke) also perturbed the gut microbiome, affecting bacterial production of neurotransmitters in mice [76]. Such a large range of chemicals that may induce gut microbiome toxicity supports the necessity of considering gut microbiome toxicity regarding the toxicity evaluation of environmental agents.

3. Relationship between Gut Microbiome Toxicity and Human Diseases

The mutually beneficial relationship between the gut microbiome and the host is built on the premise that a well-balanced gut microbiota is maintained [2]. However, when afflicted with gut microbiome toxicity, functional alterations occur in the gut microbiome. Although it is difficult to disentangle these alterations, changes in microbial metabolites, diversity loss, and interference in energy metabolism are three major types of microbial disturbances that may adversely impact the host health via multiple host–microbiota axes, potentially leading to increased disease risks. Therefore, gut microbiome toxicity is a new link between the environment and human diseases. It should be noted that not all changes in the gut microbiome associated with environmental exposure are necessarily adverse. Nevertheless, it is of significance to establish the role of the gut microbiome in the toxicity of a number of environmental toxic agents, which has been largely underappreciated in the chemical research of toxicity. In this part, we discuss the connection between gut microbiome toxicity and human diseases, providing some mechanistic insights regarding environmentally driven gut microbiome-associated diseases.

3.1. Changes in Microbial Metabolites

Production of functional metabolites by bacteria plays a key role in human health and disease [4]. Gut microbiome toxicity has an altered bacterial metabolite profile, which influences host metabolism and physiology in a significant way. First, numerous bacterial metabolites act as signaling molecules through binding to receptors and activating diverse signaling cascades. Pathogen-associated molecular patterns (PAMPs) including lipopolysaccharide (LPS) and peptidoglycan can bind to Toll-like receptor 4 and nucleotide-binding oligomerization domain, respectively; both of which lead to pro-inflammatory effects [77–79]. Classic metabolites of gut bacteria, short-chain fatty acids (SCFAs), and bile acids can also function as signaling molecules and bind to cellular receptors. Specifically, SCFAs can bind to G-protein-coupled receptors (GPCRs), and bile acids can bind to GPCR TGR5 and nuclear receptor farnesoid X receptor (FXR) [80]. Activation of signaling pathways is implicated in important biological functions; the gut microbiome may therefore contribute to human health and disease by regulating metabolic activities involved in the production of SCFAs and bile acids. For instance, SCFAs and bile acids can modulate the secretion of glucagon-like peptide-1 (GLP-1) by binding to GPR43 [81] and TGR5 [82], respectively, which affects insulin secretion and glucose homeostasis. Perturbation in those bacterial activities may affect the risk of diabetes. In addition, tryptophan metabolites produced by bacteria such as indole 3-propionic acid and indole-3-acetic acid regulate intestinal immune cells and barrier functions through the activation of aryl hydrocarbon receptor (AHR) and the pregnane X receptor (PXR) [83–85]. AHR activation is involved in inflammatory bowel disease (IBD) among other diseases, and it is suggested that a reduction in bacterial tryptophan metabolism may contribute to IBD [85]. Second, some bacterial metabolites are strongly associated with specific diseases and phenotypes. A compelling example is the association of trimethylamine N-oxide (TMAO) and cardiovascular disease [19]. The gut bacteria can convert dietary components choline and L-carnitine to trimethylamine (TMA), which is further metabolized into TMAO in the liver. Gut microbiome-derived TMAO is highly correlated with cardiovascular disease risks. Likewise, products of protein fermentation (e.g., N-nitroso compounds, polyamines) derived by gut bacteria exert carcinogenetic effects and promote colorectal cancer [21]. Third, microbiome-derived metabolites play a role in brain functions through the gut–brain axis, many of which are neurotransmitters or their precursors (e.g., serotonin, gamma-aminobutyric acid) [4]. As mentioned previously, bacterial metabolites that are neurotransmitters were perturbed by environmental chemicals such as organophosphate diazinon and nicotine, which may partially explain their neurotoxicity. Additionally, the gut microbiome is an important source of beneficial vitamins and nutrients, therefore reduction in the bacterial production of those beneficial metabolites could be detrimental to human health [86]. Taken together, these examples support that gut microbiome toxicity can lead to diseases via altered metabolite profiles.

3.2. Diversity Loss

Diversity loss has been associated with many microbiota-related diseases such as IBD [87,88], irritable bowel syndrome (IBS) [89], acute diarrhea [90], and *Clostridium difficile*-associated disease (CDAD) [91]. Trillions of microorganisms residing in the human gut form a complex microbial ecosystem, which is deeply intertwined with human biology [34]. Therefore, it is important to view the gut microbiome from an ecological perspective, although it is formidable due to fluctuations over time and variations between individuals [34]. Resilience is the extent of perturbation that an ecosystem can tolerate before it equilibrates toward a different state [92]. Resilience of the gut microbiome is crucial to colonization resistance to pathogens [34,93]. Species richness and evenness is key to the resilience of the gut microbial community. Gut microbiome with species-rich communities is less susceptible to perturbation and stress because different species are specialized to each potentially-limiting resources [94]. Moreover, high species richness enables alternative species with similar functions to fill a niche and maintain the diversity when the original species is compromised [95]. The diversity of the gut microbial ecosystem can be compromised by environmental factors (e.g., antibiotics), which makes it less resilient and more susceptible to pathogen invasion. For instance, antibiotics can induce changes in the gut microbiome and metabolic features that increased its susceptibility to *Clostridium difficile* infection [96]. In addition, a core set of gut microbial species across individuals does not exist. However, a functional core microbiome is shared with similar functional gene profiles [97]. Maintaining the functional core of the gut microbiome is indispensable because normal functioning of the human biology relies in part on the essential functions performed by the gut microbiome. However, exposure to toxic environmental chemicals possibly reduces species richness and diversity of the gut microbiome, leading to potential dysfunction.

3.3. Interference in Energy Metabolism

Accumulating evidence suggests that the gut microbiome plays a crucial role in energy metabolism. Humans cannot degrade most plant polysaccharides, which instead, can be utilized by the gut bacteria, producing SCFAs that are important energy substrates [80]. Direct evidence supporting the role of the gut microbiome in energy balance is that germ-free rats have reduced intestinal levels of SFCAs and doubled excretion of calories through urine and feces [98,99]. It is suggested that the capacity for the energy harvest of the gut microbiome is correlated with its microbial composition [100], specifically, the ratio of two major phyla *Firmicutes* and *Bacteroidetes*. Moreover, enriched genes encoding enzymes that are important for the initial steps of complex carbohydrate metabolism were found in the gut microbiome of obese mice [101]. Studies showed that the energy balance and body weight of the host is associated with the gut microbiome types. For instance, germ-free mice with fecal microbiota transplantation from obese mice gained more weight than that from lean mice [100]. Likewise, mice with the microbiota from people afflicted with Kwashiorkor, a form of malnutrition, suffered severe weight loss [102]. Thus, it is possible that gut microbiome toxicity interferes with the energy extraction and harvest, leading to diseases such as obesity or malnutrition.

4. Biomarkers and Assessment of Gut Microbiome Toxicity

Routine toxicity screening and evaluation of environmental chemicals fail to consider gut microbiome toxicity. There is no toxic endpoint currently established to report the relative toxic effects of certain chemicals on the gut microbiome. Thus, it is imperative to assess the functional alterations induced by various environmental chemicals, or at least the chemicals of frequent and long-term exposure (e.g., artificial sweeteners). Current approaches for the assessment of gut microbiome toxicity mainly comprise an integration of animal models (e.g., mouse, rat, and germ-free animals) and the meta-omics toolkit [10]. The use of animal models enables us to mimic the progress of gut microbiome toxicity under environmental exposures; the meta-omics toolkit comprises sequencing-based gene profiling and mass spectrometry-based metabolite profiling. Meta-omics comprise approaches that reveal both compositional levels and functional levels. Compositional profiling, that is, taxonomic profiling, provides details of the microbial constitution and diversity. However, knowing the taxonomic information alone does not necessarily lead to an accurate understanding of microbiome functions due to the existence of functional redundancy in the microbiota [34]. In the context of gut microbiome toxicity, the functional changes including the genes, mRNAs, proteins, and metabolites are what we should emphasize. Furthermore, humanized gnotobiotic mice with gut microbiota more similar to that of humans allow for better elucidation of the interactions between human gut microbiome and the environment [103]. The use of germ-free mice and in vitro techniques extends the observational studies to causality [10]. The accurate assessment of gut microbiome toxicity provides knowledge of how gut microbes react to environmental exposures, offering insights into the mechanistic basis of chemical-induced microbial perturbations and diagnostic markers for microbiota-associated diseases.

In order to diagnose gut microbiome toxicity, specific and effective biomarkers are needed. The gut microbiome and its functions will change under various environmental pressure at almost all times, however, not all changes are necessarily adverse and lead to adverse outcomes. Therefore, it is imperative to develop strategies identifying alterations that adversely influence human health. Currently the techniques and approaches used for gut microbiome assessment are usually at the meta-level; thus, the pinpoint of specific bacterial genes or metabolites that can be used to sensitively indicate environmentally induced dysfunction in the gut microbiome is warranted. Additionally, it should be noted that biomarker development may be on a case-by-case basis. Different xenobiotics would induce distinct gut microbiome changes. The elucidation of the role of the gut microbiome in the toxicity of certain exposure is the premise of biomarker development of gut microbiome toxicity.

Biomarkers are commonly used as primary end points in basic and clinical research, connecting environmental exposures to health outcomes [104]. Incorporating gut microbiome toxicity, our understanding of biomarkers should include functional changes in the gut microbiome as critical indicators in progressions from exposure to microbiome-associated diseases [105]. The functional role of the gut microbiome in host metabolism and physiology is largely determined by microbiome metabolic profiles, especially metabolic pathways and products of gut bacteria. Exposure to a range of xenobiotics would lead to perturbation in microbiome profiles, thereby resulting in functional alterations and gut microbiome toxicity. An in-depth look at microbiome changes upon various environmental exposures will provide insights regarding biomarkers of gut microbiome toxicity induced by specific environmental chemicals. While keeping the host in the picture, development and characterization of sensitive and robust biomarkers of gut microbiome toxicity could spur new advances in environment–microbiome interactions and microbiome-related diseases.

Biomarkers of gut microbiome toxicity could be bacterial species, genes, or metabolites, even a combination of several these markers. Signature changes in the gut microbiome upon exposure to certain chemical could be used to indicate an exposure to or the effect of specific xenobiotics, which provides a novel and potentially less invasive method for environmental health monitoring. More importantly, if the underlying mechanisms of chemical toxicity involves perturbation of the gut microbiome with specific functional alterations, then these alterations can also be used as potential biomarkers of environmentally driven health conditions.

Recent studies have documented the functional changes in the gut microbiome upon exposure to diverse xenobiotics (Table 1) [12–15,47,51,52,61,62,66–70,76,106–109]. Consistent changes in gut microbial profiles could be potential biomarkers of gut microbiome toxicity associated with specific chemical exposures. Outlining changing patterns and trajectories of microbial composition offers a sketch of biomarkers for gut microbiome toxicity. The *Firmicutes/Bacteroidetes* ratio is suggested to be indicative of energy harvesting capacity in the gut microbiome that is associated with host adiposity [100]. Likewise, *Enterobacteriaceae* is associated with gut inflammation [110]. The ratio of *Firmicutes* and *Bacteroidetes* as well as the abundance of *Enterobacteriaceae* in the gut can be readily changed by chemicals such as carbendazim [108] and aspartame [68]. Thus, such taxonomic characteristics can serve as biomarkers of gut microbiome toxicity associated with health outcomes such as inflammation and obesity. Moreover, distinctive changes in functional profiles such as key metabolites and metabolic pathways could serve as more relevant biomarkers because alterations in functional profiles directly influence the host. For example, arsenic exposure perturbed the gut microbial metabolite profiles, especially indole-containing metabolites, isoflavone metabolites, and bile acids [12]. Alterations in these functional metabolites could be a potential new mechanism of arsenic toxicity, and particularly, changes of these metabolites (e.g., bile acids and indole-containing compounds) can be used as biomarkers of arsenic-induced gut microbiome toxicity. Likewise, consumption of artificial sweeteners is associated with increased levels of pro-inflammatory metabolites and genes in the gut microbiome. This may be used as bioindicators of artificial sweeteners-induced gut microbiome toxicity that consequently leads to inflammation [15,67]. In addition, diazinon changed the bacterial pathways and metabolites involved in neurotransmitters in a gender-dependent manner, indicating that those bacteria-derived neurotransmitters can be biomarkers to probe gut microbiome toxicity arising from chemicals that have neurological toxicity [13]. The gender-dependent effect also indicates individual variation in biomarkers of gut microbiome toxicity resulting from gender differences in the gut microbiome.

Table 1. Microbiome changes associated with specific chemical exposure that can serve as potential biomarkers of gut microbiome toxicity.

Class	Chemical	Potential Biomarker of Gut Microbiome Toxicity			Adverse Outcome	Reference
		Gut Bacteria	Gene and Pathway	Metabolite		
Heavy metal	Arsenic	*Catabacteriaceae* ↓ *Clostridiaceae* ↓ *Clostridiales Family XIII Incertae Sedis* ↑ *Erysipelotrichaceae* ↓	LPS, DNA repair, multi-drug resistance ↑	Indolelatic acid, Indole-3-carbinol, daidzein, glycocholic acid↓		[12,47]
	Cadmium Lead	*γ-Proteobacteria* ↓ *Firmicutes/Bacteroidetes ratio* ↓ *Ruminococcus* ↓ *Oscillospira* ↓	Oxidative stress ↑	LPS ↑ Vitamin E and bile acids ↓	Inflammation	[106] [52]
	Manganese	*Firmicutes*↑ *Bacterodetes* ↓ in males *Firmicutes* ↓ in females	Phenylalanine synthesis ↑ in females, phenylalanine synthesis ↓ in males	Vitamin E↓, phenylalanine ↑ in females		[51]
Pesticide	Carbamate aldicarb	*Clostridium* ↑ *Anaerostipes* ↓	Virulence, adhesion and bacteriocins ↑	1-Methylnicotinamide ↑		[109]
	Carbendazim	*Firmicutes/Bacteroidetes* ratio ↑			Inflammation	[108]
	Diazinon	*Lachnospiraceae Johnsonella* ↑ in females, *Lachnospiraceae Johnsonella* ↓ in males	Tryptophase ↑ in males	Taurine and glycine ↓ in males		[13]
	Imazalil	*Clostridiales* ↑ *Lachnospiraceae* ↑			Inflammation	[107]
	Malathion	*Mogibacteriaceae* ↑	Quorum sensing, virulence and pathogenicity ↑			[62]
Artificial sweetener	Ace-K	*Bacteroides, Anaerostipes* and *Sutterella* ↑ in males	Carbohydrate metabolism ↑ in males, LPS ↑	Pyruvic acid ↑ in males	Obesity in males	[61]
	Aspartame	*Enterobacteriaceae* ↑ *Clostridium leptum* ↑ *Firmicutes/Bacteroidetes* ratio ↓		Propionate ↑	Diabetes	[68]
	Neotame	*Firmicutes/Bacteroidetes* ratio ↓	Butyrate biosynthesis ↓	Malic acid and glyceric acid ↓ cholesterol ↑		[69]
	Saccharin	*Bacteroides* ↑ *Clostridiales* ↑	Glycan degradation ↑	Acetate and propionate ↑	Diabetes	[66]
	Saccharin	*Corynebacterium* ↑ *Roseburia* ↑ *Ruminococcus* ↓ *Turicibacter* ↑	LPS biosynthesis, fimbrial proteins and bacterial toxins ↑	Equol ↓, quinolinic acid ↑	Inflammation	[15]
	Sucralose	*Christensenellaceae* ↑ *Clostridiaceae* ↑ *Erysipelotrichaceae* ↓	LPS biosynthesis, fimbrial proteins and bacterial toxins ↑	Quorum sensing molecules ↓, bile acids ↓	Inflammation	[67]
Other	Emulsifiers (P80 and CMC)	*Akkermansia, Proteobacteria* ↑		LPS and flagellin ↑	Colitis	[70]
	Nicotine	*F16*↓ in females, *F16* ↑ in males	Oxidative stress and DNA repair ↑ in male	Serine and glycine ↑ in females, serine and glycine ↓ in males		[76]
	Triclosan	*Turicibacteraceae* ↓ *Clostridiaceae* ↓ *Lactobacillaceae* ↑ *Streptococcaceae* ↑ *Christensenellaceae*↓	Antibiotic resistance and metal resistance ↑ Stress response ↑			[14]

More efforts should be put into the search and validation of biomarkers of gut microbiome toxicity, which would further elucidate the link between environmental chemicals and microbiome-related disease. Delineating these microbial changes and elucidating their biological effects is undoubtedly challenging due to the complexities within the gut microbiome as well as the intertwinement between the gut microbiome and other systems including immune, endocrine, and nervous systems. However, recent advances and emerging approaches are enabling progress toward a better understanding of gut microbiome toxicity biomarkers, which will inform toxicology risk assessment and development of therapeutic interventions via modulation of the gut microbiome.

5. Gut Microbiome Modulation

It is increasingly acknowledged that one of the crucial mechanisms underlying chemical toxicity is perturbation of the gut microbiome functions. The inclusion of the modulation section corresponds to 'treatment' in traditional organ toxicity and related diseases. Therefore, it is reasonable to include the treatment of gut microbiome-associated diseases—gut microbiome modulation.

The gut microbiome is becoming an attractive therapeutic target, especially now with its role well recognized in human health and disease. Current approaches for gut microbiome modulation including fecal microbiota transplantation (FMT), probiotics, and prebiotics are mainly untargeted without predictable outcomes [10]. To move from untargeted toward targeted modulation, a healthy gut microbiome needs to be defined. A consensus on the healthy endpoints of gut microbiome modulation remains elusive, which is a major challenge [111]. Nevertheless, the potential of targeted, hypothesis-driven gut microbiome modulation has been demonstrated in some recent studies. Use of whole foods or food components as dietary intervention to modulate the gut microbiome has received increasing attention due to their low toxicity profiles and high patient compliance [112]. Even with well acknowledged health benefits and capacity of gut microbiome modulation, it should be noted that there is evidence that dietary fiber could also possibly exacerbate gut conditions [113,114]. Here, we use *Akkermansia muciniphila* (*A. muciniphila*) as an example to review recent progress on attempts at targeted microbiome modulation.

A. muciniphila, a mucin-degrading bacterium commonly present in human and mouse gut microbiome, has many probiotic effects in gut barrier function, glucose homeostasis, and inflammation in humans and diverse animal models [115–119]. Several studies reported targeted gut microbiome modulation with increased *A. muciniphila* population via consumption of whole foods or food components. For example, consumption of several berry fruits including cranberries and raspberries promoted increased content and enhanced function of *A. muciniphila* in the gut microbiome in rodent studies. Specifically, cranberry extract improved insulin sensitivity and reduced weight gain in concert with a significant increase of *A. muciniphila* in diet-induced obese mice [120]. Likewise, black raspberries boosted *A. muciniphila* population in the gut microbiome together with profound changes in microbial functions and metabolites [121–123]. The polyphenols abundant in berry fruits could be a reason that *A. muciniphila* thrives. Feeding polyphenols from grapes to mice showed similar results with a drastic increase of *A. muciniphila* [124]. The gut microbiome offers a link between polyphenols and their diverse beneficial effects because polyphenols are poorly absorbed and metabolized by the human body [124]. Meanwhile, *A. muciniphila* uses mucin as carbon, nitrogen, and energy sources [125]. Goblet cells are the major producer of mucin in the intestinal epithelium [126]. It is reported that the number of goblet cells and the thickness of intestinal mucosa were increased in rats fed oligofructose [127]. Therefore, oligofructose may be an alternative factor for the increase of *A. muciniphila* in mice fed berries, which is supported by the study that administration of oligofructose did increase the *A. muciniphila* population in the gut microbiome of mice [128]. Of particular interest, metformin, medication to treat type 2 diabetes, also promotes *A. muciniphila* population in the gut microbiome, which is believed to contribute to its therapeutic effects [129,130]. Perturbation by environmental toxic chemicals and modulation by dietary components regarding the gut microbiome are fundamentally similar, except with different expected outcomes. Knowledge of how gut microbes react to xenobiotics and dietary components will address gaps in our understanding of both perturbation and modulation of the gut microbiome.

6. Conclusions

To summarize, exposure to xenobiotics such as antibiotics, heavy metals, and artificial sweeteners induces gut microbiome toxicity. Compositional alterations and functional changes occur along with this process in the gut microbiome, which can serve as potential biomarkers of gut microbiome toxicity. These chemical-induced perturbations lead to human diseases via several mechanisms including changes in the metabolite profiles, diversity loss, and altered energy metabolism (Figure 2).

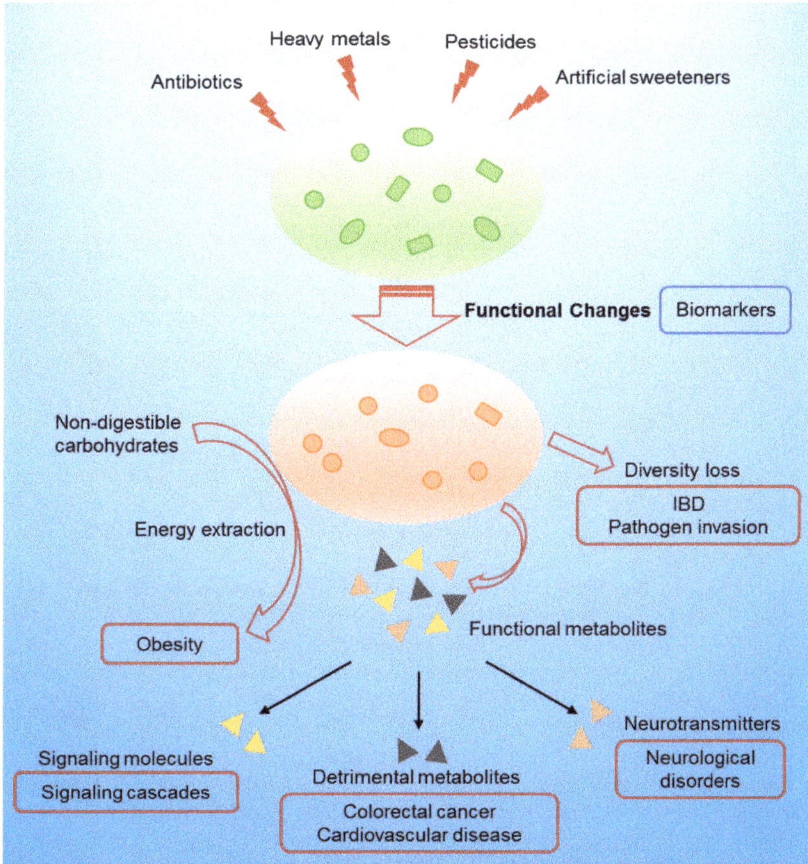

Figure 2. Schematic representation of how gut microbiome toxicity connects the environment and microbiota-associated diseases. Triangles of different colors at the bottom represent functional metabolites produced by a perturbed gut microbiome such as signaling molecules, detrimental metabolites, and neurotransmitters, which could potentially contribute to adverse health outcomes.

Given the continued enthusiasm in gut microbiome research, it is now an opportune time to examine environmentally induced gut microbiome alterations through the lens of toxicology. Although strong connection has already been established between gut microbiome disturbances and environmental exposure, the mechanisms of these disturbances and health implications await future studies. The goal of this paper was to establish and emphasize gut microbiome toxicity with the definition of chemical-driven functional damage in the gut microbiome and to review the current state of knowledge regarding biomarker, assessment, and modulation of gut microbiome toxicity. Toxic effects of various environmental agents on the gut microbiome must not be underappreciated. The integration of gut microbiome toxicity endpoints into the evaluation of chemical toxicity will provide a better understanding of the associations between the environment and human health and disease, and will facilitate the development of diagnostic markers and therapeutic interventions.

Funding: This research was funded by the University of Georgia, the University of North Carolina at Chapel Hill, and the NIH/NIEHS (R01ES024950).

Conflicts of Interest: The authors declare no conflicts of interest.

References

1. Marchesi, J.R.; Ravel, J. The vocabulary of microbiome research: A proposal. *Microbiome* **2015**, *3*, 31. [CrossRef]
2. Nicholson, J.; Holmes, E.; Kinross, J.; Burcelin, R.; Gibson, G.; Jia, W.; Pettersson, S. Host-Gut Microbiota Metabolic Interactions. *Science* **2012**, *336*, 1262–1267. [CrossRef] [PubMed]
3. Buffie, C.G.; Bucci, V.; Stein, R.R.; McKenney, P.T.; Ling, L.; Gobourne, A.; No, D.; Liu, H.; Kinnebrew, M.; Viale, A.; et al. Precision microbiome reconstitution restores bile acid mediated resistance to *Clostridium difficile*. *Nature* **2015**, *517*, 205–208. [CrossRef] [PubMed]
4. Sharon, G.; Garg, N.; Debelius, J.; Knight, R.; Dorrestein, P.C.; Mazmanian, S.K. Specialized metabolites from the microbiome in health and disease. *Cell Metab.* **2014**, *20*, 719–730. [CrossRef] [PubMed]
5. Koh, A.; De Vadder, F.; Kovatcheva-Datchary, P.; Bäckhed, F. From Dietary Fiber to Host Physiology: Short-Chain Fatty Acids as Key Bacterial Metabolites. *Cell* **2016**, *165*, 1332–1345. [CrossRef] [PubMed]
6. Schroeder, B.O.; Bäckhed, F. Signals from the gut microbiota to distant organs in physiology and disease. *Nat. Med.* **2016**, *22*, 1079–1089. [CrossRef]
7. O'Hara, A.M.; Shanahan, F. The gut flora as a forgotten organ. *EMBO Rep.* **2006**, *7*, 688–693.
8. Baquero, F.; Nombela, C. The microbiome as a human organ. *Clin. Microbiol. Infect.* **2012**, *18*, 2–4. [CrossRef]
9. Clarke, G.; Stilling, R.; Kennedy, P.J.; Stanton, C.; Cryan, J.F.; Dinan, T.G. Minireview: Gut Microbiota: The Neglected Endocrine Organ. *Mol. Endocrinol.* **2014**, *28*, 1221–1238. [CrossRef]
10. Schmidt, T.; Raes, J.; Bork, P. The Human Gut Microbiome: From Association to Modulation. *Cell* **2018**, *172*, 1198–1215. [CrossRef]
11. Claus, S.P.; Guillou, H.; Ellero-Simatos, S. The gut microbiota: A major player in the toxicity of environmental pollutants? *Npj Biofilms Microbiomes* **2016**, *2*, 1–11. [CrossRef] [PubMed]
12. Lu, K.; Abo, R.P.; Schlieper, K.A.; Graffam, M.E.; Levine, S.S.; Wishnok, J.S.; Swenberg, J.A.; Tannenbaum, S.R.; Fox, J.G. Arsenic Exposure Perturbs the Gut Microbiome and Its Metabolic Profile in Mice: An Integrated Metagenomics and Metabolomics Analysis. *Environ. Health Perspect.* **2014**, *122*, 284–291. [CrossRef] [PubMed]
13. Gao, B.; Bian, X.; Mahbub, R.; Lu, K. Sex-Specific Effects of Organophosphate Diazinon on the Gut Microbiome and Its Metabolic Functions. *Environ. Health Perspect.* **2017**, *125*, 198–206. [CrossRef] [PubMed]
14. Gao, B.; Tu, P.; Bian, X.; Chi, L.; Ru, H.; Lu, K. Profound perturbation induced by triclosan exposure in mouse gut microbiome: A less resilient microbial community with elevated antibiotic and metal resistomes. *BMC Pharmacol. Toxicol.* **2017**, *18*, 46. [CrossRef]
15. Bian, X.; Tu, P.; Chi, L.; Gao, B.; Ru, H.; Lu, K. Saccharin induced liver inflammation in mice by altering the gut microbiota and its metabolic functions. *Food Chem. Toxicol.* **2017**, *107*, 530–539. [CrossRef] [PubMed]
16. Cho, I.; Blaser, M. The human microbiome: At the interface of health and disease. *Nat. Rev. Genet.* **2012**, *13*, 260–270. [CrossRef]
17. Manichanh, C.; Borruel, N.; Casellas, F.; Guarner, F. The gut microbiota in IBD. *Nat. Rev. Gastroenterol. Hepatol.* **2012**, *9*, 599–608. [CrossRef]
18. Hartstra, A.V.; Bouter, K.E.; Bäckhed, F.; Nieuwdorp, M. Insights Into the Role of the Microbiome in Obesity and Type 2 Diabetes. *Diabetes Care* **2015**, *38*, 159–165. [CrossRef]

19. Wang, Z.; Klipfell, E.; Bennett, B.J.; Koeth, R.; Levison, B.S.; Dugar, B.; Feldstein, A.E.; Britt, E.B.; Fu, X.; Chung, Y.-M.; et al. Gut flora metabolism of phosphatidylcholine promotes cardiovascular disease. *Nature* **2011**, *472*, 57–63. [CrossRef]

20. Schnabl, B.; Brenner, D.A. Interactions between the intestinal microbiome and liver diseases. *Gastroenterology* **2014**, *146*, 1513–1524. [CrossRef]

21. Louis, P.; Hold, G.; Flint, H.J. The gut microbiota, bacterial metabolites and colorectal cancer. *Nat. Rev. Genet.* **2014**, *12*, 661–672. [CrossRef] [PubMed]

22. Collins, S.M.; Surette, M.; Bercik, P. The interplay between the intestinal microbiota and the brain. *Nat. Rev. Genet.* **2012**, *10*, 735–742. [CrossRef] [PubMed]

23. Tremlett, H.; Bauer, K.C.; Appel-Cresswell, S.; Finlay, B.B.; Waubant, E. The gut microbiome in human neurological disease: A review. *Ann. Neurol.* **2017**, *81*, 369–382. [CrossRef] [PubMed]

24. Ananthakrishnan, A.N. Epidemiology and risk factors for IBD. *Nat. Rev. Gastroenterol. Hepatol.* **2015**, *12*, 205–217. [CrossRef]

25. Thayer, K.; Heindel, J.J.; Bucher, J.R.; Gallo, M.A. Role of Environmental Chemicals in Diabetes and Obesity: A National Toxicology Program Workshop Review. *Environ. Health Perspect.* **2012**, *120*, 779–789. [CrossRef]

26. Alavanja, M.C.R.; Hoppin, J.A.; Kamel, F. Health Effects of Chronic Pesticide Exposure: Cancer and Neurotoxicity. *Annu. Rev. Public Health* **2004**, *25*, 155–197. [CrossRef]

27. Council, N.R. *Complex Mixtures: Methods for in Vivo Toxicity Testing*; National Academies Press: Washington, DC, USA, 1988.

28. Andersen, M.; Krewski, D. Toxicity Testing in the 21st Century: Bringing the Vision to Life. *Toxicol. Sci.* **2009**, *107*, 324–330. [CrossRef]

29. Koppel, N.; Rekdal, V.M.; Balskus, E.P. Chemical transformation of xenobiotics by the human gut microbiota. *Science* **2017**, *356*, 2770. [CrossRef]

30. Zhang, L.; Nichols, R.G.; Correll, J.; Murray, I.A.; Tanaka, N.; Smith, P.B.; Hubbard, T.D.; Sebastian, A.; Albert, I.; Hatzakis, E.; et al. Persistent Organic Pollutants Modify Gut Microbiota–Host Metabolic Homeostasis in Mice Through Aryl Hydrocarbon Receptor Activation. *Environ. Health Perspect.* **2015**, *123*, 679–688. [CrossRef]

31. Thompson, J.; Oliveira, R.; Djukovic, A.; Ubeda, C.; Xavier, K.B. Manipulation of the Quorum Sensing Signal AI-2 Affects the Antibiotic-Treated Gut Microbiota. *Cell Rep.* **2015**, *10*, 1861–1871. [CrossRef]

32. Becattini, S.; Taur, Y.; Pamer, E.G. Antibiotic-Induced Changes in the Intestinal Microbiota and Disease. *Trends Mol. Med.* **2016**, *22*, 458–478. [CrossRef] [PubMed]

33. Langdon, A.; Crook, N.; Dantas, G. The effects of antibiotics on the microbiome throughout development and alternative approaches for therapeutic modulation. *Genome Med.* **2016**, *8*, 39. [CrossRef] [PubMed]

34. Lozupone, C.A.; Stombaugh, J.I.; Gordon, J.I.; Jansson, J.; Knight, R. Diversity, stability and resilience of the human gut microbiota. *Nature* **2012**, *489*, 220–230. [CrossRef] [PubMed]

35. Ferrer, M.; Santos, V.A.P.M.D.; Ott, S.J.; Moya, A. Gut microbiota disturbance during antibiotic therapy: A multi-omic approach. *Gut Microbes* **2013**, *5*, 64–70. [CrossRef]

36. Cho, I.; Yamanishi, S.; Cox, L.M.; Methe, B.A.; Zavadil, J.; Li, K.; Gao, Z.; Mahana, U.; Raju, K.; Teitler, I.; et al. Antibiotics in early life alter the murine colonic microbiome and adiposity. *Nature* **2012**, *488*, 621–626. [CrossRef]

37. Livanos, A.E.; Greiner, T.U.; Vangay, P.; Pathmasiri, W.; Stewart, D.; McRitchie, S.; Li, H.; Chung, J.; Sohn, J.; Kim, S.; et al. Antibiotic-mediated gut microbiome perturbation accelerates development of type 1 diabetes in mice. *Nat. Microbiol.* **2016**, *1*, 16140. [CrossRef]

38. Le Bastard, Q.; Grégoire, M.; Chapelet, G.; Javaudin, F.; Dailly, E.; Batard, E.; Knights, D.; Montassier, E.; Al-Ghalith, G.A. Systematic review: Human gut dysbiosis induced by non-antibiotic prescription medications. *Aliment. Pharmacol. Ther.* **2018**, *47*, 332–345. [CrossRef]

39. Forslund, K.; Consortium, M.; Hildebrand, F.; Nielsen, T.; Falony, G.; Le Chatelier, E.; Sunagawa, S.; Prifti, E.; Vieira-Silva, S.; Gudmundsdottir, V.; et al. Disentangling type 2 diabetes and metformin treatment signatures in the human gut microbiota. *Nature* **2015**, *528*, 262–266. [CrossRef]

40. Rogers, M.; Aronoff, D.M. The influence of non-steroidal anti-inflammatory drugs on the gut microbiome. *Clin. Microbiol. Infect.* **2016**, *22*, e171–e179. [CrossRef]

41. Imhann, F.; Bonder, M.J.; Vila, A.V.; Fu, J.; Mujagic, Z.; Vork, L.; Tigchelaar, E.F.; Jankipersadsing, S.A.; Cenit, M.C.; Harmsen, H.J.M.; et al. Proton pump inhibitors affect the gut microbiome. *Gut* **2016**, *65*, 740–748. [CrossRef]

42. Flowers, S.A.; Evans, S.; Ward, K.M.; McInnis, M.G.; Ellingrod, V.L. Interaction Between Atypical Antipsychotics and the Gut Microbiome in a Bipolar Disease Cohort. *Pharmacother. J. Hum. Pharmacol. Drug Ther.* **2017**, *37*, 261–267. [CrossRef] [PubMed]

43. Maier, L.; Pruteanu, M.; Kuhn, M.; Zeller, G.; Telzerow, A.; Anderson, E.E.; Brochado, A.R.; Fernandez, K.C.; Dose, H.; Mori, H.; et al. Extensive impact of non-antibiotic drugs on human gut bacteria. *Nature* **2018**, *555*, 623–628. [CrossRef] [PubMed]

44. Van De Wiele, T.; Gallawa, C.M.; Kubachk, K.M.; Creed, J.T.; Basta, N.; Dayton, E.A.; Whitacre, S.; Du Laing, G.; Bradham, K. Arsenic Metabolism by Human Gut Microbiota upon in Vitro Digestion of Contaminated Soils. *Environ. Health Perspect.* **2010**, *118*, 1004–1009. [CrossRef] [PubMed]

45. Liebert, C.A.; Wireman, J.; Smith, T.; Summers, A. Phylogeny of mercury resistance (mer) operons of gram-negative bacteria isolated from the fecal flora of primates. *Appl. Environ. Microbiol.* **1997**, *63*, 1066–1076. [CrossRef] [PubMed]

46. Richardson, J.B.; Dancy, B.C.R.; Horton, C.L.; Lee, Y.S.; Madejczyk, M.; Xu, Z.Z.; Ackermann, G.; Humphrey, G.; Palacios, G.; Knight, R.; et al. Exposure to toxic metals triggers unique responses from the rat gut microbiota. *Sci. Rep.* **2018**, *8*, 6578. [CrossRef] [PubMed]

47. Chi, L.; Bian, X.; Gao, B.; Tu, P.; Ru, H.; Lu, K. The Effects of an Environmentally Relevant Level of Arsenic on the Gut Microbiome and Its Functional Metagenome. *Toxicol. Sci.* **2017**, *160*, 193–204. [CrossRef] [PubMed]

48. Lu, K.; Mahbub, R.; Cable, P.H.; Ru, H.; Parry, N.; Bodnar, W.M.; Wishnok, J.S.; Stýblo, M.; Swenberg, J.A.; Fox, J.G.; et al. Gut Microbiome Phenotypes Driven by Host Genetics Affect Arsenic Metabolism. *Chem. Res. Toxicol.* **2014**, *27*, 172–174. [CrossRef]

49. Chi, L.; Bian, X.; Gao, B.; Ru, H.; Tu, P.; Lu, K. Sex-Specific Effects of Arsenic Exposure on the Trajectory and Function of the Gut Microbiome. *Chem. Res. Toxicol.* **2016**, *29*, 949–951. [CrossRef]

50. Lu, K.; Cable, P.H.; Abo, R.P.; Ru, H.; Graffam, M.E.; Schlieper, K.A.; Parry, N.; Levine, S.S.; Bodnar, W.M.; Wishnok, J.S.; et al. Gut Microbiome Perturbations Induced by Bacterial Infection Affect Arsenic Biotransformation. *Chem. Res. Toxicol.* **2013**, *26*, 1893–1903. [CrossRef]

51. Chi, L.; Gao, B.; Bian, X.; Tu, P.; Ru, H.; Lu, K. Manganese-induced sex-specific gut microbiome perturbations in C57BL/6 mice. *Toxicol. Appl. Pharmacol.* **2017**, *331*, 142–153. [CrossRef]

52. Gao, B.; Chi, L.; Mahbub, R.; Bian, X.; Tu, P.; Ru, H.; Lu, K. Multi-Omics Reveals that Lead Exposure Disturbs Gut Microbiome Development, Key Metabolites, and Metabolic Pathways. *Chem. Res. Toxicol.* **2017**, *30*, 996–1005. [CrossRef] [PubMed]

53. Samsel, A.; Seneff, S. Glyphosate's Suppression of Cytochrome P450 Enzymes and Amino Acid Biosynthesis by the Gut Microbiome: Pathways to Modern Diseases. *Entropy* **2013**, *15*, 1416–1463. [CrossRef]

54. Costacurta, A.; Vanderleyden, J. Synthesis of Phytohormones by Plant-Associated Bacteria. *Crit. Rev. Microbiol.* **1995**, *21*, 1–18. [CrossRef] [PubMed]

55. Hashimoto, T.; Perlot, T.; Rehman, A.; Trichereau, J.; Ishiguro, H.; Paolino, M.; Sigl, V.; Hanada, T.; Hanada, R.; Lipinski, S.; et al. ACE2 links amino acid malnutrition to microbial ecology and intestinal inflammation. *Nature* **2012**, *487*, 477–481. [CrossRef] [PubMed]

56. Amrhein, N.; Deus, B.; Gehrke, P.; Steinrücken, H.C. The Site of the Inhibition of the Shikimate Pathway by Glyphosate. *Plant Physiol.* **1980**, *66*, 830–834. [CrossRef] [PubMed]

57. Jin, C.; Luo, T.; Zhu, Z.; Pan, Z.; Yang, J.; Wang, W.; Fu, Z.; Jin, Y. Imazalil exposure induces gut microbiota dysbiosis and hepatic metabolism disorder in zebrafish. *Comp. Biochem. Physiol. Part C Toxicol. Pharmacol.* **2017**, *202*, 85–93. [CrossRef] [PubMed]

58. Jin, C.; Xia, J.; Wu, S.; Tu, W.; Pan, Z.; Fu, Z.; Wang, Y. Insights Into a Possible Influence on Gut Microbiota and Intestinal Barrier Function During Chronic Exposure of Mice to Imazalil. *Toxicol. Sci.* **2018**, *162*, 113–123. [CrossRef]

59. Timofeeva, O.A.; Roegge, C.S.; Seidler, F.J.; Slotkin, T.A.; Levin, E.D. Persistent cognitive alterations in rats after early postnatal exposure to low doses of the organophosphate pesticide, diazinon. *Neurotoxicol. Teratol.* **2008**, *30*, 38–45. [CrossRef]

60. Slotkin, T.A.; Ryde, I.T.; Levin, E.D.; Seidler, F.J. Developmental neurotoxicity of low dose diazinon exposure of neonatal rats: Effects on serotonin systems in adolescence and adulthood. *Brain Res. Bull.* **2008**, *75*, 640–647. [CrossRef]

61. Bian, X.; Chi, L.; Gao, B.; Tu, P.; Ru, H.; Lu, K. The artificial sweetener acesulfame potassium affects the gut microbiome and body weight gain in CD-1 mice. *PLoS ONE* **2017**, *12*, e0178426. [CrossRef]

62. Gao, B.; Chi, L.; Tu, P.; Bian, X.; Thomas, J.; Ru, H.; Lu, K. The organophosphate malathion disturbs gut microbiome development and the quorum-Sensing system. *Toxicol. Lett.* **2018**, *283*, 52–57. [CrossRef] [PubMed]

63. Drasar, B.S.; Renwick, A.G.; Williams, R.T. The role of the gut flora in the metabolism of cyclamate. *Biochem. J.* **1972**, *129*, 881–890. [CrossRef] [PubMed]

64. Renwick, A.; Tarka, S. Microbial hydrolysis of steviol glycosides. *Food Chem. Toxicol.* **2008**, *46*, S70–S74. [CrossRef] [PubMed]

65. Krishnan, R.; Wilkinson, I.; Joyce, L.; Rofe, A.M.; Bais, R.; Conyers, R.A.; Edwards, J.B. The effect of dietary xylitol on the ability of rat caecal flora to metabolise xylitol. *Aust. J. Exp. Boil. Med Sci.* **1980**, *58*, 639–652. [CrossRef] [PubMed]

66. Suez, J.; Korem, T.; Zeevi, D.; Zilberman-Schapira, G.; Thaiss, C.A.; Maza, O.; Israeli, D.; Zmora, N.; Gilad, S.; Weinberger, A.; et al. Artificial sweeteners induce glucose intolerance by altering the gut microbiota. *Nature* **2014**, *514*, 181–186. [CrossRef]

67. Bian, X.; Chi, L.; Gao, B.; Tu, P.; Ru, H.; Lu, K. Gut Microbiome Response to Sucralose and Its Potential Role in Inducing Liver Inflammation in Mice. *Front. Physiol.* **2017**, *8*, 487. [CrossRef]

68. Palmnäs, M.S.A.; Cowan, T.E.; Bomhof, M.R.; Su, J.; Reimer, R.A.; Vogel, H.J.; Hittel, D.S.; Shearer, J. Low-Dose Aspartame Consumption Differentially Affects Gut Microbiota-Host Metabolic Interactions in the Diet-Induced Obese Rat. *PLoS ONE* **2014**, *9*, e109841.

69. Chi, L.; Bian, X.; Gao, B.; Tu, P.; Lai, Y.; Ru, H.; Lu, K. Effects of the Artificial Sweetener Neotame on the Gut Microbiome and Fecal Metabolites in Mice. *Molecules* **2018**, *23*, 367. [CrossRef] [PubMed]

70. Chassaing, B.; Koren, O.; Goodrich, J.K.; Poole, A.; Srinivasan, S.; Ley, R.E.; Gewirtz, A.T. Dietary emulsifiers impact the mouse gut microbiota promoting colitis and metabolic syndrome. *Nature* **2015**, *519*, 92–96. [CrossRef] [PubMed]

71. Roca-Saavedra, P.; Mendez-Vilabrille, V.; Miranda, J.M.; Nebot, C.; Cobas, A.C.; Franco, C.M.; Cepeda, A. Food additives, contaminants and other minor components: Effects on human gut microbiota—A review. *J. Physiol. Biochem.* **2018**, *74*, 69–83. [CrossRef] [PubMed]

72. Narrowe, A.B.; Albuthi-Lantz, M.; Smith, E.P.; Bower, K.J.; Roane, T.M.; Vajda, A.; Miller, C.S. Perturbation and restoration of the fathead minnow gut microbiome after low-level triclosan exposure. *Microbiome* **2015**, *3*, 6. [CrossRef] [PubMed]

73. Hu, J.; Raikhel, V.; Gopalakrishnan, K.; Fernandez-Hernandez, H.; Lambertini, L.; Manservisi, F.; Falcioni, L.; Bua, L.; Belpoggi, F.L.; Teitelbaum, S.; et al. Effect of postnatal low-dose exposure to environmental chemicals on the gut microbiome in a rodent model. *Microbiome* **2016**, *4*, 26. [CrossRef] [PubMed]

74. Gaulke, C.A.; Barton, C.L.; Proffitt, S.; Tanguay, R.L.; Sharpton, T.J. Triclosan Exposure Is Associated with Rapid Restructuring of the Microbiome in Adult Zebrafish. *PLoS ONE* **2016**, *11*, e0154632. [CrossRef] [PubMed]

75. Poole, A.; Pischel, L.; Ley, C.; Suh, G.; Goodrich, J.K.; Haggerty, T.D.; Ley, R.E.; Parsonnet, J. Crossover Control Study of the Effect of Personal Care Products Containing Triclosan on the Microbiome. *mSphere* **2016**, *1*, e00056-15. [CrossRef]

76. Chi, L.; Mahbub, R.; Gao, B.; Bian, X.; Tu, P.; Ru, H.; Lu, K. Nicotine Alters the Gut Microbiome and Metabolites of Gut–Brain Interactions in a Sex-Specific Manner. *Chem. Res. Toxicol.* **2017**, *30*, 2110–2119. [CrossRef]

77. Cani, P.D.; Amar, J.; Iglesias, M.A.; Poggi, M.; Knauf, C.; Bastelica, D.; Neyrinck, A.M.; Fava, F.; Tuohy, K.; Chabo, C.; et al. Metabolic Endotoxemia Initiates Obesity and Insulin Resistance. *Diabetes* **2007**, *56*, 1761–1772. [CrossRef]

78. Amar, J.; Chabo, C.; Waget, A.; Klopp, P.; Vachoux, C.; Bermúdez-Humarán, L.G.; Smirnova, N.; Berge, M.; Sulpice, T.; Lahtinen, S.; et al. Intestinal mucosal adherence and translocation of commensal bacteria at the early onset of type 2 diabetes: Molecular mechanisms and probiotic treatment. *EMBO Mol. Med.* **2011**, *3*, 559–572. [CrossRef]

79. Schertzer, J.D.; Tamrakar, A.K.; Magalhães, J.G.; Pereira, S.; Bilan, P.J.; Fullerton, M.D.; Liu, Z.; Steinberg, G.R.; Giacca, A.; Philpott, D.J.; et al. NOD1 Activators Link Innate Immunity to Insulin Resistance. *Diabetes* **2011**, *60*, 2206–2215. [CrossRef]

80. Tremaroli, V.; Bäckhed, F. Functional interactions between the gut microbiota and host metabolism. *Nature* **2012**, *489*, 242–249. [CrossRef]

81. Tolhurst, G.; Heffron, H.; Lam, Y.S.; Parker, H.E.; Habib, A.M.; Diakogiannaki, E.; Cameron, J.; Grosse, J.; Reimann, F.; Gribble, F.M. Short-Chain Fatty Acids Stimulate Glucagon-Like Peptide-1 Secretion via the G-Protein–Coupled Receptor FFAR2. *Diabetes* **2012**, *61*, 364–371. [CrossRef]

82. Thomas, C.; Gioiello, A.; Noriega, L.; Strehle, A.; Oury, J.; Rizzo, G.; Macchiarulo, A.; Yamamoto, H.; Mataki, C.; Pruzanski, M.; et al. TGR5-Mediated Bile Acid Sensing Controls Glucose Homeostasis. *Cell Metab.* **2009**, *10*, 167–177. [CrossRef] [PubMed]

83. Zelante, T.; Iannitti, R.G.; Cunha, C.; De Luca, A.; Giovannini, G.; Pieraccini, G.; Zecchi, R.; D'Angelo, C.; Massi-Benedetti, C.; Fallarino, F.; et al. Tryptophan Catabolites from Microbiota Engage Aryl Hydrocarbon Receptor and Balance Mucosal Reactivity via Interleukin-22. *Immunity* **2013**, *39*, 372–385. [CrossRef]

84. Venkatesh, M.; Mukherjee, S.; Wang, H.; Li, H.; Sun, K.; Benechet, A.; Qiu, Z.; Maher, L.; Redinbo, M.R.; Phillips, R.S.; et al. Symbiotic bacterial metabolites regulate gastrointestinal barrier function via the xenobiotic sensor PXR and Toll-like receptor 4. *Immunity* **2014**, *41*, 296–310. [CrossRef]

85. Lamas, B.; Richard, M.L.; Leducq, V.; Pham, H.-P.; Michel, M.-L.; Da Costa, G.; Bridonneau, C.; Jegou, S.; Hoffmann, T.W.; Natividad, J.M.; et al. CARD9 impacts colitis by altering gut microbiota metabolism of tryptophan into aryl hydrocarbon receptor ligands. *Nat. Med.* **2016**, *22*, 598–605. [CrossRef] [PubMed]

86. Biesalski, H.K. Nutrition meets the microbiome: Micronutrients and the microbiota. *Ann. N. Y. Acad. Sci.* **2016**, *1372*, 53–64. [CrossRef] [PubMed]

87. Dicksved, J.; Halfvarson, J.; Rosenquist, M.; Järnerot, G.; Tysk, C.; Apajalahti, J.; Engstrand, L.; Jansson, J. Molecular analysis of the gut microbiota of identical twins with Crohn's disease. *ISME J.* **2008**, *2*, 716–727. [CrossRef] [PubMed]

88. Frank, D.N.; Amand, A.L.S.; Feldman, R.A.; Boedeker, E.C.; Harpaz, N.; Pace, N.R. Molecular-phylogenetic characterization of microbial community imbalances in human inflammatory bowel diseases. *Proc. Natl. Acad. Sci. USA* **2007**, *104*, 13780–13785. [CrossRef]

89. Carroll, I.M.; Ringel-Kulka, T.; Keku, T.O.; Chang, Y.-H.; Packey, C.D.; Sartor, R.B.; Ringel, Y. Molecular analysis of the luminal- and mucosal-associated intestinal microbiota in diarrhea-predominant irritable bowel syndrome. *Am. J. Physiol. Liver Physiol.* **2011**, *301*, G799–G807. [CrossRef]

90. Young, V.B.; Schmidt, T.M. Antibiotic-Associated Diarrhea Accompanied by Large-Scale Alterations in the Composition of the Fecal Microbiota. *J. Clin. Microbiol.* **2004**, *42*, 1203–1206. [CrossRef]

91. Chang, J.Y.; Antonopoulos, D.A.; Kalra, A.; Tonelli, A.; Khalife, W.T.; Schmidt, T.M.; Young, V.B. Decreased Diversity of the Fecal Microbiome in Recurrent *Clostridium difficile*–Associated Diarrhea. *J. Infect. Dis.* **2008**, *197*, 435–438. [CrossRef]

92. Folke, C.; Carpenter, S.; Walker, B.; Scheffer, M.; Elmqvist, T.; Gunderson, L.; Holling, C.S. Regime Shifts, Resilience, and Biodiversity in Ecosystem Management. *Annu. Rev. Ecol. Evol. Syst.* **2004**, *35*, 557–581. [CrossRef]

93. Van Der Waaij, D.; Berghuis, J.M.; Lekkerkerk, J.E.C. Colonization resistance of the digestive tract of mice during systemic antibiotic treatment. *Epidemiology Infect.* **1972**, *70*, 605–610. [CrossRef] [PubMed]

94. Levine, J.M.; D'Antonio, C.M. Elton Revisited: A Review of Evidence Linking Diversity and Invasibility. *Oikos* **1999**, *87*, 15. [CrossRef]

95. Elmqvist, T.; Folke, C.; Nyström, M.; Peterson, G.; Bengtsson, J.; Walker, B.; Norberg, J. Response diversity, ecosystem change, and resilience. *Front. Ecol. Environ.* **2003**, *1*, 488–494. [CrossRef]

96. Theriot, C.M.; Koenigsknecht, M.J.; Carlson, P.E.; Hatton, G.E.; Nelson, A.M.; Li, B.; Huffnagle, G.B.; Li, J.; Young, V.B. Antibiotic-induced shifts in the mouse gut microbiome and metabolome increase susceptibility to *Clostridium difficile* infection. *Nat. Commun.* **2014**, *5*, 3114. [CrossRef]

97. Turnbaugh, P.J.; Hamady, M.; Yatsunenko, T.; Cantarel, B.L.; Duncan, A.; Ley, R.E.; Sogin, M.L.; Jones, W.J.; Roe, B.A.; Affourtit, J.P.; et al. A core gut microbiome in obese and lean twins. *Nature* **2009**, *457*, 480–484. [CrossRef]

98. Høverstad, T.; Midtvedt, T. Short-Chain Fatty Acids in Germfree Mice and Rats. *J. Nutr.* **1986**, *116*, 1772–1776.

99. Wostmann, B.S.; Larkin, C.; Moriarty, A.; Bruckner-Kardoss, E. Dietary intake, energy metabolism, and excretory losses of adult male germfree Wistar rats. *Lab. Anim. Sci.* **1983**, *33*, 46–50.

100. Turnbaugh, P.J.; Ley, R.E.; Mahowald, M.A.; Magrini, V.; Mardis, E.R.; Gordon, J.I. An obesity-associated gut microbiome with increased capacity for energy harvest. *Nature* **2006**, *444*, 1027–1031. [CrossRef]

101. Turnbaugh, P.J.; Ley, R.E.; Hamady, M.; Fraser, C.M.; Knight, R.; Gordon, J.I. The Human Microbiome Project. *Nature* **2007**, *449*, 804–810. [CrossRef]

102. Smith, M.I.; Yatsunenko, T.; Manary, M.; Trehan, I.; Mkakosya, R.; Cheng, J.; Kau, A.; Rich, S.S.; Concannon, P.; Mychaleckyj, J.C.; et al. Gut Microbiomes of Malawian Twin Pairs Discordant for Kwashiorkor. *Science* **2013**, *339*, 548–554. [CrossRef] [PubMed]

103. Nguyen, T.L.A.; Vieira-Silva, S.; Liston, A.; Raes, J. How informative is the mouse for human gut microbiota research? *Dis. Model. Mech.* **2015**, *8*, 1–16. [CrossRef] [PubMed]

104. Strimbu, K.; Tavel, J.A. What are biomarkers? *Curr. Opin. HIV AIDS* **2010**, *5*, 463–466. [CrossRef] [PubMed]

105. Dietert, R.R.; Silbergeld, E.K. Biomarkers for the 21st Century: Listening to the Microbiome. *Toxicol. Sci.* **2015**, *144*, 208–216. [CrossRef]

106. Zhang, S.; Jin, Y.; Zeng, Z.; Liu, Z.; Fu, Z. Subchronic Exposure of Mice to Cadmium Perturbs Their Hepatic Energy Metabolism and Gut Microbiome. *Chem. Res. Toxicol.* **2015**, *28*, 2000–2009. [CrossRef]

107. Jin, C.; Zeng, Z.; Fu, Z.; Jin, Y. Oral imazalil exposure induces gut microbiota dysbiosis and colonic inflammation in mice. *Chemosphere* **2016**, *160*, 349–358. [CrossRef]

108. Jin, Y.; Zeng, Z.; Wu, Y.; Zhang, S.; Fu, Z. Oral Exposure of Mice to Carbendazim Induces Hepatic Lipid Metabolism Disorder and Gut Microbiota Dysbiosis. *Toxicol. Sci.* **2015**, *147*, 116–126. [CrossRef]

109. Gao, B.; Chi, L.; Tu, P.; Gao, N.; Lu, K. The Carbamate Aldicarb Altered the Gut Microbiome, Metabolome, and Lipidome of C57BL/6J Mice. *Chem. Res. Toxicol.* **2018**, *32*, 67–79. [CrossRef]

110. Zeng, M.Y.; Inohara, N.; Núñez, G. Mechanisms of inflammation-driven bacterial dysbiosis in the gut. *Mucosal Immunol.* **2017**, *10*, 18–26. [CrossRef]

111. Lloyd-Price, J.; Abu-Ali, G.; Huttenhower, C. The healthy human microbiome. *Genome Med.* **2016**, *8*, 51. [CrossRef]

112. Montrose, D.C.; Horelik, N.A.; Madigan, J.P.; Stoner, G.D.; Wang, L.-S.; Bruno, R.S.; Park, H.J.; Giardina, C.; Rosenberg, D.W. Anti-inflammatory effects of freeze-dried black raspberry powder in ulcerative colitis. *Carcinogenesis* **2011**, *32*, 343–350. [CrossRef] [PubMed]

113. Muir, J.; Shepherd, S.; Rosella, O.; Gibs, P. Fructans exacerbate gastrointestinal symptoms in irritable bowel syndrome and Crohn's disease. In Proceedings of the 55th Australian Cereal Chemistry Conference: Biomolecular Aspects of Analysis, Food and Health, Charles Sturt University, Wagga Wagga, NSW, Australia, 3–7 July 2005.

114. Miles, J.P.; Zou, J.; Kumar, M.-V.; Pellizzon, M.; Ulman, E.; Ricci, M.; Gewirtz, A.T.; Chassaing, B. Supplementation of Low- and High-fat Diets with Fermentable Fiber Exacerbates Severity of DSS-induced Acute Colitis. *Inflamm. Bowel Dis.* **2017**, *23*, 1133–1143. [CrossRef] [PubMed]

115. Everard, A.; Belzer, C.; Geurts, L.; Ouwerkerk, J.P.; Druart, C.; Bindels, L.B.; Guiot, Y.; Derrien, M.; Muccioli, G.G.; Delzenne, N.M.; et al. Cross-talk between *Akkermansia muciniphila* and intestinal epithelium controls diet-induced obesity. *Proc. Natl. Acad. Sci. USA* **2013**, *110*, 9066–9071. [CrossRef] [PubMed]

116. Greer, R.L.; Dong, X.; De Moraes, A.; Zielke, R.A.; Fernandes, G.D.R.; Peremyslova, E.; Vasquez-Perez, S.; Schoenborn, A.A.; Gomes, E.P.; Pereira, A.C.; et al. *Akkermansia muciniphila* mediates negative effects of IFNγ on glucose metabolism. *Nat. Commun.* **2016**, *7*, 13329. [CrossRef]

117. Li, J.; Lin, S.; Vanhoutte, P.M.; Woo, C.W.H.; Xu, A. *Akkermansia Muciniphila* Protects Against Atherosclerosis by Preventing Metabolic Endotoxemia-Induced Inflammation in *Apoe*$^{-/-}$ Mice. *Circulation* **2016**, *133*, 2434–2446. [CrossRef]

118. Schneeberger, M.; Everard, A.; Gómez-Valadés, A.G.; Matamoros, S.; Ramírez, S.; Delzenne, N.M.; Gomis, R.; Claret, M.; Cani, P.D. *Akkermansia muciniphila* inversely correlates with the onset of inflammation, altered adipose tissue metabolism and metabolic disorders during obesity in mice. *Sci. Rep.* **2015**, *5*, 16643. [CrossRef]

119. Derrien, M.; Belzer, C.; De Vos, W.M. *Akkermansia muciniphila* and its role in regulating host functions. *Microb. Pathog.* **2017**, *106*, 171–181. [CrossRef]

120. Anhê, F.F.; Roy, D.; Pilon, G.; Dudonné, S.; Matamoros, S.; Varin, T.V.; Garofalo, C.; Moine, Q.; Desjardins, Y.; Levy, E.; et al. A polyphenol-rich cranberry extract protects from diet-induced obesity, insulin resistance and intestinal inflammation in association with increased *Akkermansia* spp. population in the gut microbiota of mice. *Gut* **2015**, *64*, 872–883.

121. Tu, P.; Bian, X.; Chi, L.; Gao, B.; Ru, H.; Knobloch, T.J.; Weghorst, C.; Lu, K. Characterization of the Functional Changes in Mouse Gut Microbiome Associated with Increased *Akkermansia muciniphila* Population Modulated by Dietary Black Raspberries. *ACS Omega* **2018**, *3*, 10927–10937. [CrossRef]

122. Tu, P.; Bian, X.; Chi, L.; Xue, J.; Gao, B.; Lai, Y.; Ru, H.; Lu, K. Metabolite Profiling of the Gut Microbiome in Mice with Dietary Administration of Black Raspberries. *ACS Omega* **2020**, *5*, 1318–1325. [CrossRef]

123. Tu, P.; Xue, J.; Bian, X.; Chi, L.; Gao, B.; Leng, J.; Ru, H.; Knobloch, T.J.; Weghorst, C.M.; Lu, K.; et al. Dietary administration of black raspberries modulates arsenic biotransformation and reduces urinary 8-oxo-2′-deoxyguanosine in mice. *Toxicol. Appl. Pharmacol.* **2019**, *377*, 114633. [CrossRef] [PubMed]

124. Roopchand, D.E.; Carmody, R.N.; Kuhn, P.; Moskal, K.; Rojas-Silva, P.; Turnbaugh, P.J.; Raskin, I. Dietary Polyphenols Promote Growth of the Gut Bacterium *Akkermansia muciniphila* and Attenuate High-Fat Diet–Induced Metabolic Syndrome. *Diabetes* **2015**, *64*, 2847–2858. [CrossRef] [PubMed]

125. Derrien, M.; Vaughan, E.E.; Plugge, C.M.; De Vos, W.M. *Akkermansia muciniphila gen. nov., sp. nov.*, a human intestinal mucin-degrading bacterium. *Int. J. Syst. Evol. Microbiol.* **2004**, *54*, 1469–1476. [CrossRef] [PubMed]

126. Lindén, S.K.; Sutton, P.; Karlsson, N.G.; Korolik, V.; McGuckin, M.A. Mucins in the mucosal barrier to infection. *Mucosal Immunol.* **2008**, *1*, 183–197.

127. Kleessen, B.; Hartmann, L.; Blaut, M. Fructans in the diet cause alterations of intestinal mucosal architecture, released mucins and mucosa-associated bifidobacteria in gnotobiotic rats. *Br. J. Nutr.* **2003**, *89*, 597–606. [CrossRef]

128. Everard, A.; Lazarevic, V.; Derrien, M.; Girard, M.; Muccioli, G.G.; Neyrinck, A.M.; Possemiers, S.; Van Holle, A.; François, P.; De Vos, W.M.; et al. Responses of Gut Microbiota and Glucose and Lipid Metabolism to Prebiotics in Genetic Obese and Diet-Induced Leptin-Resistant Mice. *Diabetes* **2011**, *60*, 2775–2786. [CrossRef]

129. De La Cuesta-Zuluaga, J.; Mueller, N.; Corrales-Agudelo, V.; Velásquez-Mejía, E.P.; Carmona, J.A.; Abad, J.M.; Escobar, J.S. Metformin Is Associated With Higher Relative Abundance of Mucin-Degrading *Akkermansia muciniphila* and Several Short-Chain Fatty Acid–Producing Microbiota in the Gut. *Diabetes Care* **2017**, *40*, 54–62. [CrossRef]

130. Wu, H.; Esteve, E.; Tremaroli, V.; Khan, M.T.; Caesar, R.; Mannerås-Holm, L.; Ståhlman, M.; Olsson, L.M.; Serino, M.; Planas-Fèlix, M.; et al. Metformin alters the gut microbiome of individuals with treatment-naive type 2 diabetes, contributing to the therapeutic effects of the drug. *Nat. Med.* **2017**, *23*, 850–858. [CrossRef]

Review

The Versatile Roles of the tRNA Epitranscriptome during Cellular Responses to Toxic Exposures and Environmental Stress

Sabrina M. Huber [1,†]**, Andrea Leonardi** [2,3]**, Peter C. Dedon** [1,4] **and Thomas J. Begley** [2,5,*]

[1] Department of Biological Engineering, Massachusetts Institute of Technology, Cambridge, MA 02139, USA; smhuber@mit.edu or sabrina.huber@hest.ethz.ch (S.M.H.); pcdedon@mit.edu (P.C.D.)
[2] RNA Institute, State University of New York, Albany, NY 12222, USA; aleonardi2@albany.edu
[3] College of Nanoscale Science and Engineering, State University of New York, Albany, NY 12203, USA
[4] Antimicrobial Resistance Interdisciplinary Research Group, Singapore-MIT Alliance for Research and Technology, Singapore 138602, Singapore
[5] Department of Biological Sciences, State University of New York, Albany, NY 12222, USA
* Correspondence: tbegley@albany.edu; Tel.: +1-518-437-4486
† Present address: Laboratory of Toxicology, ETH Zürich, Zürich 8092, Switzerland.

Received: 4 March 2019; Accepted: 21 March 2019; Published: 25 March 2019

Abstract: Living organisms respond to environmental changes and xenobiotic exposures by regulating gene expression. While heat shock, unfolded protein, and DNA damage stress responses are well-studied at the levels of the transcriptome and proteome, tRNA-mediated mechanisms are only recently emerging as important modulators of cellular stress responses. Regulation of the stress response by tRNA shows a high functional diversity, ranging from the control of tRNA maturation and translation initiation, to translational enhancement through modification-mediated codon-biased translation of mRNAs encoding stress response proteins, and translational repression by stress-induced tRNA fragments. tRNAs need to be heavily modified post-transcriptionally for full activity, and it is becoming increasingly clear that many aspects of tRNA metabolism and function are regulated through the dynamic introduction and removal of modifications. This review will discuss the many ways that nucleoside modifications confer high functional diversity to tRNAs, with a focus on tRNA modification-mediated regulation of the eukaryotic response to environmental stress and toxicant exposures. Additionally, the potential applications of tRNA modification biology in the development of early biomarkers of pathology will be highlighted.

Keywords: epitranscriptomics; tRNA modifications; stress response mechanisms; codon-biased translation

1. Introduction

In an ever-changing environment, living systems are subjected to stresses such as temperature fluctuations, nutrient limitations, and exposures that damage intracellular biomolecules. Cells respond to these stresses by activation of response, repair, and adaptation pathways or, if the damage is too severe, by initiation of cell death systems. Among the well-studied stress response mechanisms are the heat shock, unfolded protein, DNA damage, oxidative, and nutrient stress responses, which are reviewed in detail elsewhere [1–3]. The molecular underpinnings of these mechanisms have been extensively explored at the level of signaling pathways, changes in transcription, and the proteome. Here, we focus on recently emerging mechanisms of cellular stress response involving the dynamic regulation of ribonucleoside modifications that control transfer RNA (tRNA) metabolism, structure, and function, which, in turn, modulate protein translation and a cell's ability to cope with stress. We discuss how tRNA modifications can affect tRNA stability, maturation, and codon recognition in mRNA, all of which regulate cell survival and adaptation during environmental stress and xenobiotic exposures. Aberrant modification of tRNAs is also directly linked to human diseases; including neurological disorders, metabolic diseases, and cancer [4–6]. As such, tRNA-based stress response mechanisms have the potential for diagnostic applications and could be useful biomarkers, as they have been shown to be linked to increased levels of reactive oxygen species (ROS), mitochondrial dysfunction, DNA damage, and changes in metabolism [7–12].

2. tRNA-Based Regulation of Cellular Processes

tRNAs are best known as the adaptor molecules of the translation machinery, in which they deliver the appropriate amino acids to the ribosomes according to the interaction of their anticodons with the codons of messenger RNAs (mRNAs). tRNAs are extensively post-transcriptionally modified by epitranscriptomic writer enzymes, with these RNA modifications having structural and functional roles, as well as playing downstream roles in different biological processes. A detailed description of these RNA modifications is given in Section 3. Beyond their role as adaptors in protein synthesis, tRNAs have been shown to be critically involved in many other cellular processes. Aminoacylated-tRNAs (aa-tRNAs) act as amino acid donors not only to nascent peptide chains during translation, but also to membrane lipids, to peptidoglycan precursors, and to the amino-terminus of proteins [13]. In addition, tRNAs act as biosynthetic precursors of antimicrobial molecules and tetrapyrroles [13,14]. Uncharged tRNAs are used by retroviruses to prime DNA synthesis and are thus essential for viral replication [15–17]. During the priming of DNA synthesis in human immunodeficiency virus type 1 (HIV-1) infections, the 3′-end of the primer tRNA is complementary to a region of the viral RNA called the primer binding site [18].

tRNAs also function as stress sensors and are key for initiating stress responses [19]. For example, eukaryotic cells respond to nutrient deprivation by global inhibition of protein synthesis, which can be mediated by tRNAs acting as signaling molecules. The limited availability of extracellular amino acids during starvation leads to the accumulation of uncharged cytosolic tRNAs that directly bind to the protein kinase GCN2 [20,21]. This tRNA binding activates the GCN2 kinase activity and results in the phosphorylation of eIF2, which in turn regulates translation in amino acid-starved cells and promotes increased synthesis of the transcription factor GCN4 that activates amino acid biosynthetic genes. Interestingly, the GCN4-regulated transcripts exhibit a codon bias that can be linked to regulation by modified wobble uridines in tRNA [22], as discussed shortly. Furthermore, cytosolic and mitochondrial tRNAs can inhibit apoptosis by binding to cytochrome c, thereby blocking the formation of the apoptosome [23,24]. Furthermore, tRNAs serve as a source of small noncoding RNAs called tRNA-derived fragments (tRF), which vary in length, biogenesis, sequence, and function [25,26]. Examples of functions of tRFs in the literature include the repression of gene expression in a miRNA-like fashion, inhibition of translation through ribosome binding, and modulation of cell proliferation [25,27–29]. Interestingly, the formation of tRFs can be controlled by ribonucleoside modifications which is discussed in more detail in Section 4.3.

Here we explore how tRNA modifications play a role in this tRNA-based regulation of cellular processes in the face of stressful conditions.

3. Ribonucleoside Modifications

Post-transcriptional RNA modifications were first described in 1957, when the presence of pseudouridine was demonstrated in yeast RNA extracts [30]. Since then, more than 140 chemically distinct ribonucleosides have been characterized, some of which are conserved throughout all domains of life. Modified ribonucleosides occur in all RNA classes, with tRNAs containing the most numerous and chemically diverse modifications [31,32]. On average, ~17% of the 80–90 ribonucleosides in tRNA are modified, ranging from relatively small structural changes, such as methylation or acetylation, to highly intricate chemical alterations that give rise to so-called hypermodified bases such as queuosine (Q) (Figure 1A). The type, location, and abundance of modifications not only vary between different tRNAs, but also differ within similar isoacceptors (tRNAs bearing the same amino acid but different anticodons). Ribonucleoside modifications are located throughout tRNA molecules and can affect tRNA stability, folding, localization, transport, processing, and function [33]. The anticodon loops of nearly all tRNAs are heavily modified, predominantly at positions 34 and 37 [31,32]. The ribonucleotide at the wobble position 34 pairs with the third mRNA codon base in the aminoacyl-tRNA binding site (A-site) of the ribosome during decoding and thus is crucial for accurate reading of the genetic code. Modifications at the wobble position of the tRNA anticodon have been shown to allow decoding of multiple mRNA codons differing by the third nucleoside (synonymous codons) or restrict pairing with noncognate codons [34]. Position 37 in tRNA is adjacent to the 3′-side of the anticodon and has been shown to prevent frame shifting during translocation by stabilizing codon–anticodon interactions [35]. It is important to note that tRNA modifications play an important role during HIV-1 infections. There is an additional interaction between an A-rich loop located upstream of the primer binding site region and the lysine tRNA anticodon loop dependent on a sulfur in the wobble base modification 5-methoxycarbonylmethyl-2-thiouridine (mcm^{5}s^{2}U). This demonstrates the importance of modified ribonucleosides in tRNA function [18]. There is also evidence that tRNA modifications play important roles in organelle function. For example, decreased tRNA modification has been directly linked to the mitochondrial diseases myoclonus epilepsy associated with ragged-red fibers (MERRF) and mitochondrial encephalopathy, lactic acidosis, and stroke-like episodes (MELAS) [8,36,37]. These diseases are caused by a mutation in mitochondrial tRNA genes, with the change in sequence preventing formation of the tRNA modifications s^2 and 5-taurinomethyluridine (tm^{5}U), both disturbing codon–anticodon interactions to disrupt protein synthesis [37,38]. As mitochondrial translation is used to synthesize key enzymes involved in metabolism, the resulting mistranslation defects can alter proteins involved in the electron transport chain and ATP synthesis. The resulting defect in energy production can be linked to the muscle weakness and neurological dysfunction associated with MERRF and MELAS, and it provides a mechanistic link between tRNA modifications and energy metabolism.

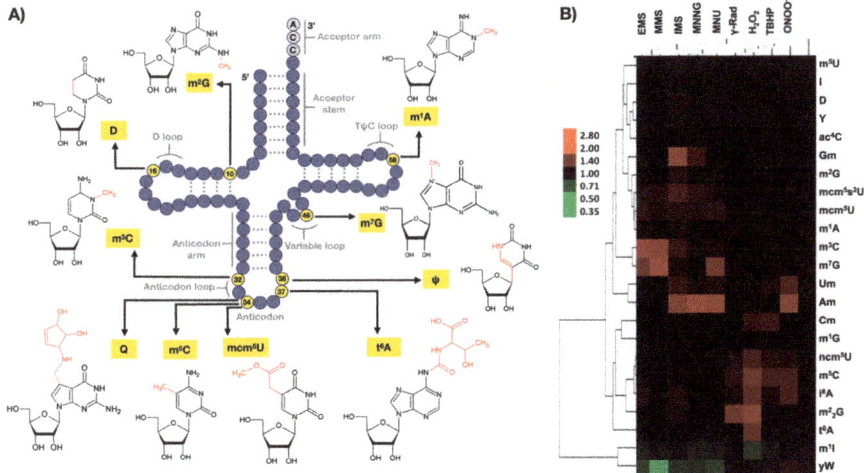

Figure 1. (**A**) Clover-leaf structure of eukaryotic tRNA formed through base pairing in the acceptor stem, D-loop, anticodon arm, and TψC-loop. Structures and positions of some modifications discussed in the review are indicated. (**B**) Hierarchical cluster analysis of average fold-change values for tRNA modifications in total tRNA from *S. cerevisiae* exposed to equitoxic (LD$_{80}$) doses of various alkylating (EMS, ethyl methansulfonate; MMS, methyl methansulfonate; IMS, isopropyl methanesulfonate; MNNG, *N*-methyl-*N'*-nitro-*N*-nitrosoguanidine; MNU, *N*-nitro-*N*-methylurea) and oxidizing agents (γ-Rad, γ-radiation; H$_2$O$_2$, hydrogen peroxide; TBHP, tert-butyl hydroperoxide; ONOO$^-$, peroxynitrite). Yeast cells were exposed to LD$_{80}$ doses of the agents and tRNA modifications were quantified by liquid chromatography—tandem mass spectroscopy (LC-MS/MS). The fold-change values were derived from the average of normalized MS signal intensity data from five biological replicates relative to unexposed controls, and hierarchical clustering analysis was performed in log space (log$_2$) and visualized as a heat map. Reproduced from [11], ACS publications, 2015.

3.1. tRNA Modification as Dynamic Marks

RNA modifications were long considered to be static and stable marks after their post-transcriptional enzymatic introduction by "writer" proteins and their removal from the transcriptome was thought to occur passively via degradation of the modified RNA followed by transcription of new unmodified RNA. However, the findings that exposure of wild type yeast to various alkylating and oxidizing agents caused signature patterns of changes in the relative quantities of numerous modified ribonucleosides in tRNAs (Figure 1B) [9,11], demonstrated the responsiveness of modified ribonucleosides to environmental cues and highlighted the potential for reversibility. The pattern of up- and downregulation of these modifications were unique to each stressor and predictive of exposure. For example, in response to the alkylating agents, mcm^{5}U, mcm^{5}s^{2}U, m^{3}C, and m^{7}G were all increased, but were relatively unchanged in response to oxidizing agents. In contrast m^{5}C, i^{6}A, ncm^{5}U and, to some extent, m$^2_2$G, were increased in response to oxidizing agents, but show little change in response to alkylating agents.

Technological advances throughout recent years lead to the identification of a number of "eraser" enzymes that are able to catalyze the active removal of modified residues, in particular methylation marks, and opened the door for dynamic regulation of modified ribonucleosides in tRNA (Figure 2). The reversibility of tRNA modifications, via the eraser ALKBH1, was recently shown to mediate the demethylation of m^{1}A in tRNAs [39]. Notably though, reversibility was first shown in 2011, when the fat mass and obesity-associated protein (FTO) and ALKBH5 were both shown to robustly remove N^6-methyladenosine (m^{6}A) on polyadenylated RNAs in vitro through an oxidative demethylation mechanism and contribute to m^{6}A levels in cellular mRNA (Figure 2) [40–42]. Since then, reversible N^6-adenosine methylation has been demonstrated to play key roles in a number of biological processes, including mRNA nuclear export, the association of nuclear speckle proteins, splicing, cap-independent translation, UV-induced DNA damage response, leukemogenesis, and drug response [42–45].

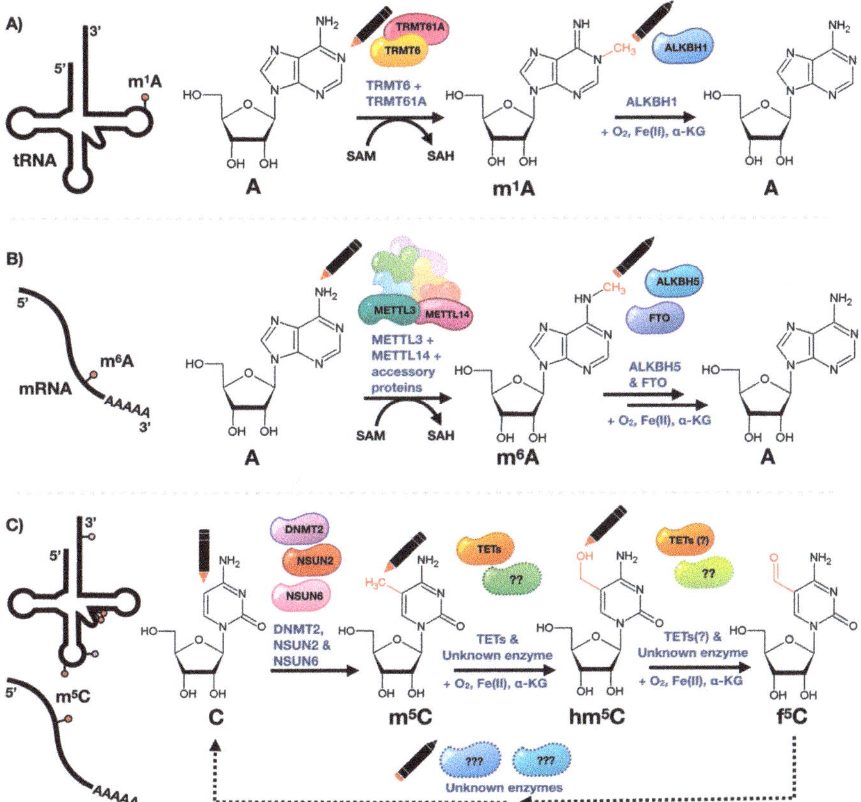

Figure 2. Reversibility and dynamics of RNA modifications. (**A**) The epitranscriptomic writer for the tRNA m^1A modification is a TRMT6/TRMT61A complex, which uses S-adenosyl methionine (SAM) as a methyl donor. ALKBH1 removes the methyl group and requires oxygen, iron and α-ketoglutarate cofactors for demethylase activity. (**B**) N^6-Methyladenosine (m^6A) is added to mRNA via the METTL3-METTL14 heterodimer, which along with accessory proteins forms the N^6-methyltransferase complex using SAM as methyl donor. The modification is removed by ALKBH5 or FTO requiring oxygen, iron, and α-ketoglutarate cofactors. (**C**) 5-Methylcytidine (m^5C) on cytosolic tRNAs is added by DNMT2, NSUN2, and NSUN6, while m^5C on mRNA is introduced by NSUN2 only. It undergoes further oxidative metabolism mediated by TET enzymes (and possible other unknown enzymes) to 5-hydroxymethylcytidine (hm^5C), requiring oxygen, iron, and α-ketoglutarate. Further TET activity results in formation of 5-formylcytidine (f^5C) using hm^5C as precursor. These modifications have not been shown to be fully reversible although it is predicted that there are erasers for m^5C, or its metabolites [46–48].

Interestingly, recent studies show that in vivo FTO preferentially targets N^6-2′-O-dimethyladenosine (m^6A_m), a highly prevalent mRNA modification found adjacent to the N^7-methylguanosine cap at the first encoded nucleotide position, with nearly 100-times greater catalytic activity compared to m^6A [49,50]. Using a transcriptome-wide map of m^6Am, it was shown that the presence of this modification in the extended cap confers increased mRNA stability by reducing the susceptibility to DCP2-mediated decapping, thus influencing mRNA abundance and protein synthesis [49]. The application of stable isotope metabolic tracing demonstrated that oxidative processing is not limited to methylated adenosine residues and that 5-methylcytosine in RNA undergoes similar oxidative metabolism via 5-hydroxy- and 5-formylcytosine [47]. While the ten-eleven translocation (Tet) family of Fe(II)- and α-ketoglutarate-dependent dioxygenases, well known for their oxidation of m^5C in DNA, have been shown to also promote the formation of hm^5C in RNA [46], there is substantial evidence that TETs are not the only/main family of enzymes able to catalyze the oxidation of m^5C in RNA. For instance, significant amounts of hm^5C are not only found in TET triple knockout mouse embryonic stem cells but also in organisms that do not express any TET enzymes, such as *C. elegans* [46,47]. In support of this hypothesis, the formation of 2′-O-methyl-5-hydroxymethylcytidine (hm^5Cm), a modification closely related to hm^5C, has been shown to be TET-independent [48]. It remains unclear whether hm^5C is a metabolic intermediate in the demethylation pathway and dynamic regulation of m^5C, or whether it is an epitranscriptomic mark with its own function. The involvement of hm^5C in an active demethylation pathway is supported by metabolic labeling studies in human embryonic kidney 293 cells that demonstrated that in small RNAs (<200 nt) hm^5C itself or hm^5C-containing transcripts are subject to enhanced turnover [48]. However, sucrose gradient fractionation followed by dot blotting in *Drosophila* S2 cells revealed that mRNAs heavily loaded with ribosomes have a high hm^5C content, suggesting that the function of RNA hydroxymethylation is to promote mRNA translation in vivo [51]. While RNA methylations have been at the forefront of dynamic examples of ribonucleoside modifications, it is likely that other examples of dynamically-regulated or reversible modifications will follow. These initial examples of RNA demethylases point towards the responsiveness of RNA modifications to environmental stimuli that allow organisms to react and adapt to changing environments.

3.2. tRNA Modifications Prevent Translational Infidelity and Proteotoxic stress

Proper anticodon–codon pairing and maintaining the correct reading frame on translated mRNA are key functions linked to tRNA modifications. Studies specific to wobble uridine U34-based modifications and their corresponding writers have been published highlighting how the C5 and C2 position on U34 play key roles in preventing protein synthesis errors. For example, the writer tRNA methyltransferase 9 (Trm9) from yeast completes the formation 5-methoxycarbonylmethyluridine (mcm^5U) and mcm^5s^2U by adding the terminal methyl group. The mcm^5U and mcm^5s^2U modifications are found at U34 on tRNAs that decode arginine, glutamine, glutamic acid, and lysine. The arginine codons AGA and AGG are found in a split codon box with the codons AGU and AGC for serine. The mcm^5U modification is needed to prevent pairing of $tRNA^{Arg}$, which normally decodes AGA and AGC, with the AGU and AGC codons for serine. As such, cells deficient in Trm9 and mcm^5-based modifications show increased levels of arginine misincoporation at serine codons [52]. In addition, *trm9Δ* cells show increased −1 frameshifts and activation of heat shock and unfolded protein response (UPR) pathways. Studies in yeast using cells deficient in the wobble uridine writers for s^2 (Ncs2) and c^5 (elp6) have employed a novel reporter system and ribosome profiling to demonstrate that there is increased proteotoxic stress due to tRNA modification defects and perturbed translation [53]. Additional studies in yeast have also demonstrated that wobble U tRNA modifications play important roles in maintaining reading frame, with unmodified or under modified tRNA not entering the ribosome A-site efficiently [54,55]. As the levels of mcm^5U and mcm^5s^2U change during the cell cycle and in response to specific exposures [52], there could be dynamic changes in translational fidelity during stress responses.

3.3. tRNA Modification Enzymes as Essential Features of the Cell Stress Response

Many genes encoding tRNA-modifying enzymes are essential for cell function, with losses causing severe defects in growth and development [56,57]. However, despite their conservation throughout evolution, many genes encoding tRNA modification enzymes are not essential under normal growth conditions. tRNA normally contains many modifications and is considered to be a stable RNA. Notably though, the loss of key epitranscriptomic marks can destabilize tRNA and lead to rapid degradation. For example, mature tRNA for valine missing m^5C and m^7G, specific to tRNA methyltransferase (Trm) 4 and Trm8, is surveyed and targeted by the rapid tRNA decay (RTD) pathway [58,59]. The ability to generally function in the absence of a tRNA modification is illustrated by the fact that tRNAs lacking modifications (e.g., in vitro transcribed tRNAs) still function in translation and that cells can compensate for lost modifications by increasing tRNA copy numbers to drive translation [60,61]. The absence of a modification often causes subtle phenotypic effects in cells, so it has been challenging to elucidate the exact biological functions of more than a few modifications. However, the deletion of writer proteins often increases the cellular sensitivity to specific stresses, demonstrating the importance of RNA modifications in adaptation to environmental changes. For instance, depletion of N^1-methyladenosine (m^1A) in the hyperthermophilic bacterium *Thermus thermophilus* resulted in a thermosensitive phenotype, suggesting a role of m^1A in temperature adaptation [62]. In the same organism, loss of pseudouridine at tRNA position 55 ($\psi55$) caused abnormal increases in the levels of other modified nucleosides (Gm, m^5s^2U, and m^1A) and led to growth retardation at lower temperatures [63]. An emerging literature now documents the critical roles of tRNA modifications in the cellular response to physiological changes, environmental changes, and stressful exposures [12,64–66].

4. Mechanisms by which tRNA Modifications Function in the Cell Stress Response

4.1. m^1A Affects Translation Initiation during Cell Stress

Initiation of protein synthesis in eukaryotes is mediated by eukaryotic initiation factors (eIFs) and involves the assembly of the initiator tRNA, the 40S, and 60S ribosomal subunits into an 80S ribosome at the start codon of mRNAs, typically AUG coding for methionine [67]. The initiator methionyl-tRNA (tRNA^{iMet}) is used exclusively during initiation of protein synthesis and is different from the elongator methionyl-tRNA, which is solely used for insertion of methionine into a growing polypeptide chain [68]. Initiator tRNAs carry a highly conserved N^1-methyladenosine residue at position 58 (m^1A58) that is key for the formation of a tertiary substructure not seen in elongator tRNAs (Figures 1A and 3A) [69]. In yeast, m^1A58 is essential for viability. Impaired function of m^1A methyltransferases resulted in growth arrest and ultimately cell death, which could be attributed to rapid and specific degradation of mature tRNA^{iMet} in the absence of m^1A58 by the use of pulse-chase experiments [70,71]. Hence, m^1A58 provides direct means of regulating the intracellular levels of tRNA^{iMet} and thus the initiation of protein synthesis without affecting the elongation step of translation. The eraser ALKBH1 was recently shown to mediate the demethylation of m^1A in tRNAs, providing evidence of reversible tRNA methylation [39]. Interestingly, glucose deprivation of HeLa cells resulted in increased expression of ALKBH1 which correlated with decreased levels of m^1A and attenuated protein synthesis [39]. The levels of the m^1A58 methyltransferase heterodimer Trmt6/Trmt61 showed no significant changes [39], suggesting the presence of an active demethylation pathway of m^1A under glucose starvation. These studies provide the first evidence that reversible N^1-adenosine methylation is involved in the control of protein synthesis in response to nutrient availability.

4.2. Wobble tRNA Modifications Regulate Codon-Biased Translation of Stress Response Proteins

Post-transcriptional modifications at the wobble position in the anticodon loop of transfer RNAs have direct means to influence the decoding of the genetic code by mediating codon–anticodon interactions. Depending on the type of chemical modification, the interaction with certain codons is preferred due to increased stabilization of specific base pairs. tRNA wobble modifications should be able to regulate the decoding rates of synonymous codons which differ by the third nucleoside, with dynamic changes in modification levels regulating translation. Using a unique bioanalytical platform, we have shown that stressors cause the reprogramming of dozens of modified ribonucleosides in tRNA that regulate the selective translation of codon-biased mRNAs of critical stress response proteins required for cell survival [11,72]. For example, exposure of yeast to the oxidizing agent hydrogen peroxide (H_2O_2) caused an increase in 5-methylcytosine (m^5C) found in the wobble position of leucine tRNA that reads UUG (1 of six leucine codons) (Figure 1). Codon reporter systems and proteomic studies have linked m^5C to the enhanced translation of UUG-enriched mRNAs for oxidative stress response genes (Figure 3B). The role of m^5C-based translation of UUG codons in cell survival was demonstrated when loss of the writer tRNA methyltransferase 4 (Trm4), catalyzing the formation of wobble m^5C, rendered the cells sensitive to H2O2 exposure. Similar results have been shown for writers of mcm^5U and mcm^5Um modifications in yeast and mouse models, as deficiencies lead to sensitivity to alkylating and oxidizing agents, respectfully [12,73,74]. In both cases, codon specific reporter constructs, transcript- and protein-based studies, and increased levels of wobble uridine modifications in response to stress have supported the idea that stress promotes the translation of codon specific transcripts, which is coordinated by epitranscriptomic reprogramming. A similar phenomenon has been shown to occur in mycobacteria exposed to hypoxia [75]. This stress led to increased steady-state levels of proteins derived from ACG-enriched genes, with the increase dependent upon stress-induced conversion of mo^5U to cmo^5U at the wobble position of the UGU anticodon on the threonine tRNA that reads the ACG codon (Figure 3B).

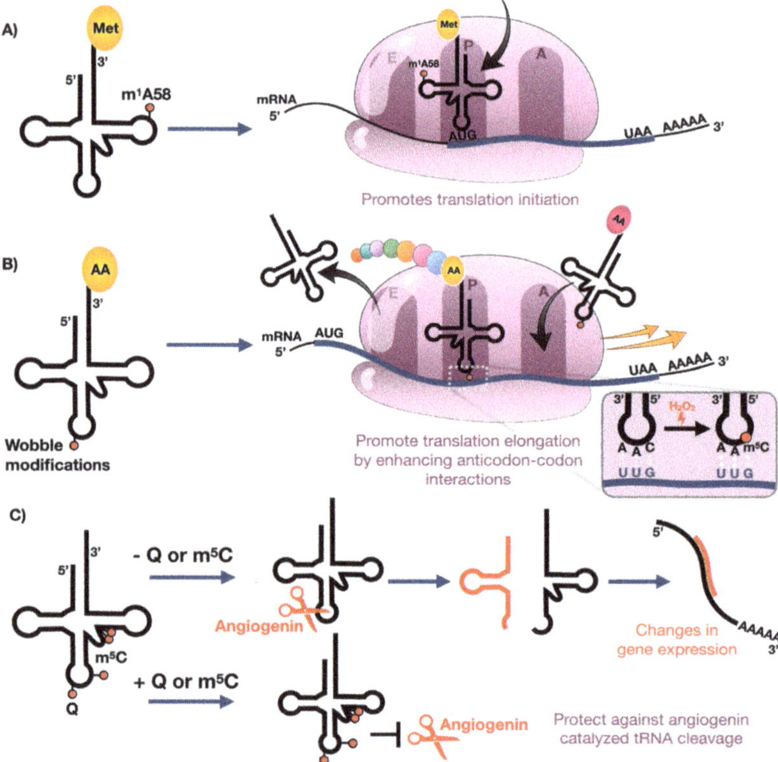

Figure 3. Stress-induced changes in tRNA modification levels can regulate (**A**) translation initiation, (**B**) translation elongation, and (**C**) tRNA cleavage.

4.3. tRNA Modifications Restrict Stress-Induced tRNA Cleavage

A third role for modified nucleosides in tRNA involves regulation of tRNA degradation and cleavage of tRNAs into small regulatory RNA fragments (Figure 3C). The latter is illustrated by angiogenin-mediated endonucleolytic cleavage of tRNAs in the anticodon loop, which is a widely conserved oxidative stress response in eukaryotes [76–78]. One example of modification-dependent angiogenin cleavage of tRNA under stress involves 5-methylcytidine (m^5C). In nuclear-encoded eukaryotic tRNAs, m^5C commonly occurs at six cytidine positions, namely, C34 and C38 in the anticodon loop; C48, C49, and C50 in the variable region; and C72 in the acceptor stem [79]. Cytosine-C5 methylation in mitochondrial encoded tRNAs is restricted to the variable loop and the acceptor stem and appears at C48, C49, and C72 [80]. Introduction of methyl groups to these sites is mediated by several members of a large protein family of conserved RNA m^5C-methyltransferases, namely tRNA aspartic acid MTase 1 (TRDMT1, also known as DNMT2) and the NOP2/Sun domain proteins (NSUN) 2 and 6 [79,81]. While TRDMT1 specifically methylates position 38 in glycine, aspartic acid, and valine tRNAs [76,82], NSUN6 is responsible for the methylation of position 72 in threonine and cysteine tRNAs [81]. NSUN2 has wider substrate specificity and methylates the vast majority of tRNAs at positions 34, 48, 49, and 50 [83,84]. Studies by Blanco et al. revealed that the absence of NSUN2-dependent m^5C sites in the variable loop leads to increased tRNA cleavage by the endonuclease angiogenin and the accumulation of 5′ tRNA-derived small RNA fragments in mouse models and dermal fibroblasts obtained from patients with Dubowitz-like syndrome [85]. These 5′ tRNA fragments induce cellular stress responses that lead to reduced protein translation rates, decreased cell size, and increased cell death in vitro and in vivo causing a syndromic disorder characterized by growth and neurodevelopmental deficiencies [85]. Similar to NSUN2-mediated methylation in the variable region, DNMT2-mediated cytosine-C5 methylation in the anticodon loop can also protect tRNAs from endonucleolytic cleavage by angiogenin under stress in *Drosophila* [76]. The prominent role of Dnmt2 in the stress response is further confirmed by the fact that *Drosophila* Dnmt2 mutants show significantly reduced viability under oxidative or heat stress [76]. While the presence of m^5C limits the fragmentation of tRNA at various locations under stress, it is currently unclear how cytidine-C5 methylation modulates the activity of stress-induced endonucleases. The lack of m^5C could result in a more flexible tRNA structure in which the anticodon loop becomes more exposed to tRNA cleavage enzymes. Alternatively, the modification could mask the sequence recognition motif of these enzymes.

m^5C undergoes oxidative processing to hm^5C [46–48]. It has been shown by quantitative isotope dilution-mass spectrometry that hm^5C is enriched in tRNA fractions of HEK293T cells [48]. This provides evidence that m^5C is dynamically controlled in tRNAs. While it is currently unclear whether hm^5C also protects tRNAs against stress-induced cleavage, metabolic labeling studies have shown that hm^5C is subject to enhanced turnover in RNA either due to specific tRNA degradation/cleavage or reversibility to unmodified C [48]. The oxidation of m^5C to hm^5C could provide means of dynamically regulating the cleavage potential of tRNAs under changing environmental conditions. Further studies should address the exact positions of hm^5C and how this oxidative derivative of m^5C affects angiogenin-mediated cleavage of tRNAs.

Another example of a link between tRNA modifications and stress-induced tRNA cleavage involves queuosine restriction of tRNA cleavage during oxidative stress. Queuosine (Q) is a hypermodified residue found at the wobble position of tRNAs with GUN anticodons, namely histidine, asparagine, tyrosine, and aspartic acid (Figure 1A) [86]. While bacteria can synthesize Q de novo by a complex biosynthetic pathway [87], eukaryotes lack its synthesis pathways and rely on the uptake of the micronutrient queuine from dietary sources and the gut microbes, for subsequent enzymatic incorporation into tRNA [88,89]. Since Q-deficient mice do not exhibit any pathological symptoms in a stress-less environment, it has been suggested that the role of Q may be to protect the organisms against stress. A link between Q and the oxidative stress response in mice was established when it was shown that exogenous administration of queuine to mice with Dalton's lymphoma ascites transplanted (DLAT) tumors improved the activities of antioxidant enzymes, such as catalase, superoxide dismutase, and glutathione peroxidase [90]. While this promotion of the antioxidant defense system could be evoked by queuine itself or by Q-modified tRNA, Wang et al. recently showed that Q-deficient HEK293T and HeLa cells produce significantly more tRNA halves from tRNA[His] and tRNA[Asn] upon arsenite stress and angiogenin treatment (Figure 3), suggesting that Q directly protects its cognate tRNAs against ribonuclease cleavage [91]. The total tRNA pool is not altered. In mammals, Q can be further modified by an unknown glycosyltransferase by addition of a mannosyl or galactosyl group to yield manQ or galQ, respectively [92]. The current lack of high-throughput methods for the simultaneous detection and quantification of Q, manQ ,and galQ from limited starting material, has hampered the studies of the Q-derivatives and thus their exact physiological roles remain poorly understood.

Modifications do not always restrict the cleavage of tRNAs. For instance, wobble mcm^5s^2U is a target for the eukaryotic γ-toxin secreted by Kluyveromyces lactis killer strains and promotes the cleavage of tRNAs, causing irreversible growth arrest of sensitive yeast cells [93]. The importance of modified ribonucleosides is not limited to the eukaryotic host defense response. Colicin E5 and PrrC are *E. coli* endoribonucleases that specifically cleave Q- and 5-methylaminomethyl-2-thiouridine (mnm^5s^2U)-modified tRNAs, respectively [94,95]. These examples demonstrate that tRNA modifications can be critical determinants in defending host cells from the invasion of viruses or biotic stresses.

4.4. tRNA Modifications Affect tRNA Maturation during Stress

Eukaryotic tRNAs are initially transcribed as larger precursors (pre-tRNAs) that require a variety of post-transcriptional alterations to become fully mature and functional tRNAs. These processing steps include the removal of the $5'$-leader and $3'$-trailer sequences, addition of the nucleotides CCA to the $3'$-end, intron splicing, and introduction of a large number of ribonucleoside modifications [96,97]. Using Northern blot analysis and RNA sequencing, it was shown that tRNA maturation is differentially regulated during temperature and nonfermentable carbon source stress in yeast. Accumulation of aberrant tRNA precursors was observed upon shifting yeast to elevated temperatures and/or to glycerol-containing medium [98]. Interestingly, several tRNA modifications are added at the pre-tRNA stage [97,99], ensuring proper folding. For instance, studies in *Xenopus oocytes* showed sequential addition of base modifications during tRNA tyrosine maturation [100]. While in this particular organism, m^1A, ψ, and m^5C already occur in the pre-tRNA with immature $5'$-leader and $3'$-trailer sequences, m^2_2G, m^2G, and D are introduced after maturation of the $5'$- and $3'$-termini but before intron splicing [100]. The order and location of incorporation of some modifications can be species-specific. For instance, wobble inosine modifications are incorporated into pre-tRNAs in the nucleus in human tRNAs and into mature tRNAs in the cytosol in *Trypanosoma*, respectively [101,102]. These findings demonstrate that tRNA modifications are stringently coupled with tRNA processing and maturation and suggest that environmental cues can affect tRNA precursor forms by controlling tRNA modification levels in a tRNA- and condition-specific manner.

5. RNA Modifications as Potential Biomarkers of Exposure and Disease Pathology

Based on these diverse roles for tRNA modifications in the cell stress response, it is reasonable to propose that ribonucleoside modifications can serve as biomarkers of specific stresses and environmental changes. Support for this idea comes from the observed role of tRNA modifications as sensors for changes in environmental and intracellular conditions. For instance, mitochondrial t^6A is sensitive to intracellular CO_2 [103]. The growth of HEK293T cells in sodium bicarbonate-free medium in the absence of CO_2, caused a significant decrease in the frequency of t^6A in mitochondrial tRNAs (mt-tRNAs), which could be rescued by the addition of sodium bicarbonate to the cell culture medium [103]. As a result, Lin et al. speculated that hypoxic conditions in solid tumors could affect t^6A formation as mitochondrial CO_2 is predominantly provided by the TCA cycle. In support of this, they found hypomodification of t^6A37 in mitochondrial tRNA serine isolated from solid tumor xenografts [103].

Similarly, agents and exposures that cause macromolecular damage lead to the predictable reprogramming of RNA modifications. The predictive power in tRNA modifications is illustrated with the response of yeast exposed to four oxidants and five alkylating agents [11]. tRNA modification patterns accurately distinguished between the two types of toxicant, with 14 modified ribonucleosides forming the basis for a data-driven model that predicted toxicant chemistry with >80% sensitivity and specificity [11]. tRNA modification subpatterns also distinguished among chemically similar toxicants such as SN1 and SN2 alkylating agents [11]. This distinction further linked to codon-biased translation: SN2-induced increases in m^{3}C in tRNA led to selective translation of threonine-rich membrane proteins from genes enriched with ACC and ACT degenerate codons for threonine [11]. tRNA modifications thus serve as predictive biomarkers of exposure.

However, other types of stress response can also promote characteristic epitranscriptomic changes. Looking beyond tRNA, analysis of all cellular RNAs (including mRNA, rRNA, tRNA, and snoRNA) at the nucleoside level has been used to demonstrate that there is broad reprogramming of mRNA-, rRNA- and, tRNA-based modifications in response to osmotic stress. For example, using yeast, it was observed that there were dramatic increases in monomethylated C, representing m^{3}C, m^{5}C, Cm, and m^{4}C in mRNA, rRNA, tRNA, and snoRNA during the osmotic stress response [64]. Also, the tRNA-based wobble U34 modifications mcm^{5}U and mcm^{5}s^{2}U, as well as i^{6}A at position 37, were increased in response to osmotic stress. Mechanistically the reason for osmotic stress induced changes in the epitranscriptome could be to promote RNA stabilization, RNA localization as well as translational regulation. Regardless of the mechanistic details, the observation that there are stress-induced changes in modification levels specific to mRNA- and rRNA-based species supports the idea that global epitranscriptomic changes are ingrained in regulatory responses.

The preceding examples with cultured cells and specific stresses illustrate the potential for tRNA modifications to serve as biomarkers of disease and pathology. Defects in RNA modification or their corresponding writers have been linked to cancer, neurodegenerative and neurological defects, and diabetes [4–6]. For example, defects in Q levels have been observed in ovarian and lung tumors [104,105]. Defects in the mitochondrial and nuclear writer of dimethylguanosine (m$^2{}_{,2}$G), tRNA methyltransferase 1 (TRMT1), have been shown to promote intellectual disabilities, with this epitranscriptomic system linked to redox metabolism [106]. Type II diabetes has also been linked to defects in tRNA modifications, with sequence variants and a mouse model defective in the 2-methylthio-N6-threonylcarbamoyladenosine (ms^{2}t^{6}A) writer CDKAL1 demonstrating how modification of tRNA for lysine plays an important role in maintaining pancreatic islet function and controlling glucose levels [107,108]. Expression of RNA modification systems have also been linked to cancer survival, as the wobble U modification writer enzyme ALKBH8 has been shown to be required for growth of bladder cancers [109], while the ALKBH3 eraser of m^{6}A in RNA is required for survival of non-small cell lung cancer and other cancers [110,111].

6. Conclusions and Perspectives

The dynamic regulation of RNA modifications plays an important role in the response of cells to environmental fluctuations and xenobiotic exposures. On average there are 13 modifications in each tRNA and many distinct tRNA isoacceptors in each cell. Global analyses have demonstrated that there are coordinated changes in tRNA modifications, for example, the alkylation-induced increases in mcm^{5}U and mcm^{5}s^{2}U occurring as there are increases in m^{3}C and m^{7}G [11]. These coordinated changes in epitranscriptomic marks suggest that a program of translational regulation specific to many codon–anticodon pairs is driving the translational response to stress. While almost all tRNA modifications occur in multiple tRNAs at several positions, toxicant-induced changes are affecting very specific positions in individual tRNAs. It is currently not understood, how multiple tRNAs bearing the same modification can be differentially regulated when the sites are targeted by a single enzyme. RNA modification enzymes may rely on the presence of other modifications for the introduction of a modification, as observed for the Dnmt2-mediated introduction of m^{5}C, which is stimulated by the presence of Q at position 34 [112]. The investigation whether such crosstalk between modifications is a general phenomenon requires the development of novel techniques allowing the single-base resolution mapping of individual modifications, such as bisulfite sequencing [113]. The presence of certain modifications could lead to structural changes in the three-dimensional structure of tRNA, making other sites more accessible for modification enzymes that were previously hidden. Furthermore, it will not only be essential to unveil how tRNA modifications are coordinately regulated, but also to start integrating the tRNA modification landscape with other epitranscriptomic marks on messenger RNA, such as m^{6}A (Figure 4). So far, tRNA, mRNA, and modifications of other RNA types have all been considered as separate entities and their interplay remains completely unstudied. However, some enzymes, like NSUN2, target multiple RNA species for modification, pointing towards the existence of interaction systems between multiple RNA modification systems. It is also important to place translational regulation in the context of a larger program of stress-induced epitranscriptomic changes, which should be regulating tRNA stability, translation initiation, and microRNA-based regulation of transcripts. The observation that there are stress-induced changes in modification levels specific to mRNA- and rRNA-based species supports the idea that global epitranscriptomic changes are ingrained in regulatory responses [64]. Linking tRNA modification-based translational regulation in the context of mRNA modification-based regulation should be an active area of research in the future. There is abundant data to show that the most prevalent mRNA modification—m^{6}A—plays dynamic roles in regulating fertility and development, and we suggest that tRNA-based and other epitranscriptomic marks should be important for human development. Further, the identification of m^{6}A erasers in the form of demethylase enzymes highlights the dynamic potential of epitranscriptomic marks and suggests that a wide array of tRNA-specific erasers should be present in human cells.

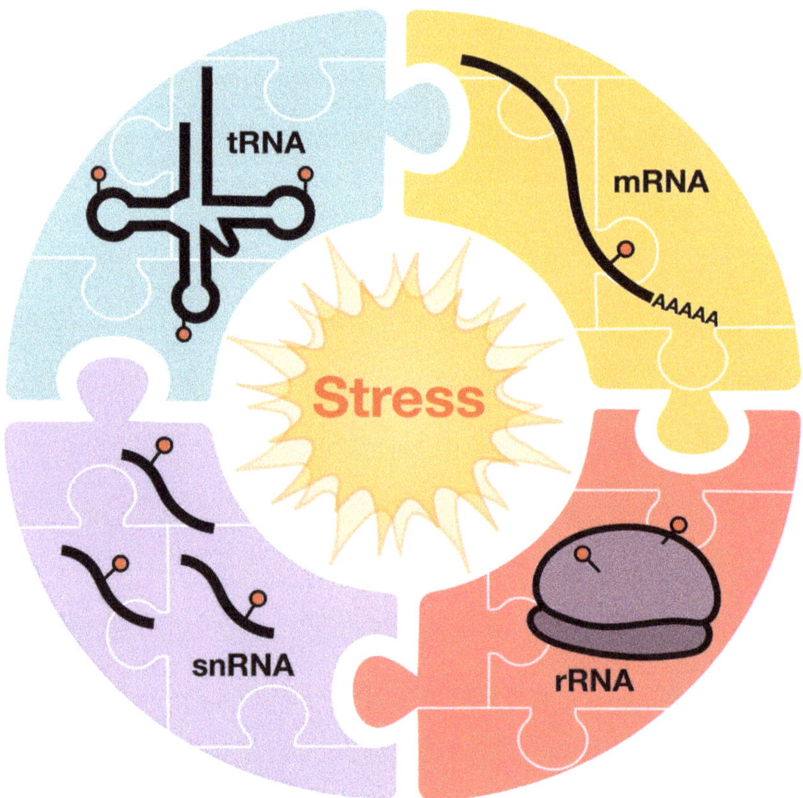

Figure 4. Coordinated changes in epitranscriptomic marks on tRNA, mRNA, rRNA, and snRNA are theorized to drive the translational response to stress, and other physiological responses.

Author Contributions: The manuscript was written by S.M.H, P.C.D, and T.J.B. The figures and legends were produced and written by A.L. All authors have given approval to the final version of the manuscript.

Funding: This research was funded by grants from the National Institutes of Health, grant numbers R01ES026856 and R01ES024615, and by a grant from the National Science Foundation of Singapore to support the Singapore-MIT Alliance for Research and Technology Antimicrobial Resistance IRG. S.M.H was supported by an Early Postdoc.Mobility Fellowship from the Swiss National Science Foundation (P2SKP_174681).

Acknowledgments: We would like to thank members of the Dedon and Begley labs for constructive comments.

Conflicts of Interest: The authors declare no conflicts of interest.

References

1. Fulda, S.; Gorman, A.M.; Hori, O.; Samali, A. Cellular Stress Responses: Cell Survival and Cell Death. *Int. J. Cell Biol.* **2010**. [CrossRef] [PubMed]
2. Ciccia, A.; Elledge, S.J. The DNA damage response: Making it safe to play with knives. *Mol. Cell* **2010**, *40*, 179–204. [CrossRef] [PubMed]
3. Schröder, M.; Kaufman, R.J. The mammalian unfolded protein response. *Annu. Rev. Biochem.* **2005**, *74*, 739–789. [CrossRef]
4. Torres, A.G.; Batlle, E.; Ribas de Pouplana, L. Role of tRNA modifications in human diseases. *Trends Mol. Med.* **2014**, *20*, 306–314. [CrossRef]
5. Zhang, X.; Cozen, A.E.; Liu, Y.; Chen, Q.; Lowe, T.M. Small RNA Modifications: Integral to Function and Disease. *Trends Mol. Med.* **2016**, *22*, 1025–1034. [CrossRef]
6. Sarin, L.P.; Leidel, S.A. Modify or die?—RNA modification defects in metazoans. *RNA Biol.* **2015**, *11*, 1555–1567. [CrossRef]
7. Mishima, E.; Inoue, C.; Saigusa, D.; Inoue, R.; Ito, K.; Suzuki, Y.; Jinno, D.; Tsukui, Y.; Akamatsu, Y.; Araki, M.; et al. Conformational change in transfer RNA is an early indicator of acute cellular damage. *J. Am. Soc. Nephrol.* **2014**, *25*, 2316–2326. [CrossRef] [PubMed]
8. Kirino, Y.; Yasukawa, T.; Ohta, S.; Akira, S.; Ishihara, K.; Watanabe, K.; Suzuki, T. Codon-specific translational defect caused by a wobble modification deficiency in mutant tRNA from a human mitochondrial disease. *Proc. Natl. Acad. Sci. USA* **2004**, *101*, 15070–15075. [CrossRef]
9. Chan, C.T.Y.; Dyavaiah, M.; DeMott, M.S.; Taghizadeh, K.; Dedon, P.C.; Begley, T.J. A Quantitative Systems Approach Reveals Dynamic Control of tRNA Modifications during Cellular Stress. *PLoS Genet.* **2010**, *6*, e1001247. [CrossRef] [PubMed]
10. Patil, A.; Dyavaiah, M.; Joseph, F.; Rooney, J.P.; Chan, C.T.Y.; Dedon, P.C.; Begley, T.J. Increased tRNA modification and gene-specific codon usage regulate cell cycle progression during the DNA damage response. *Cell Cycle* **2012**, *11*, 3656–3665. [CrossRef] [PubMed]
11. Chan, C.T.; Deng, W.; Li, F.; DeMott, M.S.; Babu, I.R.; Begley, T.J.; Dedon, P.C. Highly Predictive Reprogramming of tRNA Modifications Is Linked to Selective Expression of Codon-Biased Genes. *Chem. Res. Toxicol.* **2015**, *28*, 978–988. [CrossRef]
12. Endres, L.; Begley, U.; Clark, R.; Gu, C.; Dziergowska, A.; Małkiewicz, A.; Melendez, J.A.; Dedon, P.C.; Begley, T.J. Alkbh8 Regulates Selenocysteine-Protein Expression to Protect against Reactive Oxygen Species Damage. *PLoS ONE* **2015**, *10*, e0131335. [CrossRef]
13. Banerjee, R.; Chen, S.; Dare, K.; Gilreath, M.; Praetorius-Ibba, M.; Raina, M.; Reynolds, N.M.; Rogers, T.; Roy, H.; Yadavalli, S.S.; et al. tRNAs: Cellular barcodes for amino acids. *FEBS Lett.* **2010**, *584*, 387–395. [CrossRef]
14. Francklyn, C.S.; Minajigi, A. tRNA as an active chemical scaffold for diverse chemical transformations. *FEBS Lett.* **2010**, *584*, 366–375. [CrossRef]
15. Harada, F.; Sawyer, R.C.; Dahlberg, J.E. A primer ribonucleic acid for initiation of in vitro Rous sarcoma virus deoxyribonucleic acid synthesis. *J. Biol. Chem.* **1975**, *250*, 3487–3497. [PubMed]
16. Harada, F.; Peters, G.G.; Dahlberg, J.E. The primer tRNA for Moloney murine leukemia virus DNA synthesis. Nucleotide sequence and aminoacylation of tRNAPro. *J. Biol. Chem.* **1979**, *254*, 10979–10985. [PubMed]
17. Mak, J.; Kleiman, L. Primer tRNAs for reverse transcription. *J. Virol.* **1997**, *71*, 8087–8095.

18. Isel, C.; Marquet, R.; Keith, G.; Ehresmann, C.; Ehresmann, B. Modified nucleotides of tRNA(3Lys) modulate primer/template loop-loop interaction in the initiation complex of HIV-1 reverse transcription. *J. Biol. Chem.* **1993**, *268*, 25269–25272. [PubMed]

19. Zhong, J.; Xiao, C.; Gu, W.; Du, G.; Sun, X.; He, Q.-Y.; Zhang, G. Transfer RNAs Mediate the Rapid Adaptation of *Escherichia coli* to Oxidative Stress. *PLoS Genet.* **2015**, *11*, e1005302. [CrossRef]

20. Wek, S.A.; Zhu, S.; Wek, R.C. The histidyl-tRNA synthetase-related sequence in the eIF-2 alpha protein kinase GCN2 interacts with tRNA and is required for activation in response to starvation for different amino acids. *Mol. Cell. Biol.* **1995**, *15*, 4497–4506. [CrossRef]

21. Dong, J.; Qiu, H.; Garcia-Barrio, M.; Anderson, J.; Hinnebusch, A.G. Uncharged tRNA Activates GCN2 by Displacing the Protein Kinase Moiety from a Bipartite tRNA-Binding Domain. *Mol. Cell* **2000**, *6*, 269–279. [CrossRef]

22. Doyle, F.; Leonardi, A.; Endres, L.; Tenenbaum, S.A.; Dedon, P.C.; Begley, T.J. Gene- and genome-based analysis of significant codon patterns in yeast, rat and mice genomes with the CUT Codon UTilization tool. *Methods* **2016**, *107*, 98–109. [CrossRef]

23. Mei, Y.; Yong, J.; Liu, H.; Shi, Y.; Meinkoth, J.; Dreyfuss, G.; Yang, X. tRNA binds to cytochrome *c* and inhibits caspase activation. *Mol. Cell* **2010**, *37*, 668–678. [CrossRef]

24. Mei, Y.; Yong, J.; Stonestrom, A.; Yang, X. tRNA and cytochrome *c* in cell death and beyond. *Cell Cycle* **2010**, *9*, 2936–2939. [CrossRef]

25. Lee, Y.S.; Shibata, Y.; Malhotra, A.; Dutta, A. A novel class of small RNAs: tRNA-derived RNA fragments (tRFs). *Genes Dev.* **2009**, *23*, 2639–2649. [CrossRef]

26. Anderson, P.; Ivanov, P. tRNA fragments in human health and disease. *FEBS Lett.* **2014**, *588*, 4297–4304. [CrossRef]

27. Maute, R.L.; Schneider, C.; Sumazin, P.; Holmes, A.; Califano, A.; Basso, K.; Dalla-Favera, R. tRNA-derived microRNA modulates proliferation and the DNA damage response and is down-regulated in B cell lymphoma. *Proc. Natl. Acad. Sci. USA* **2013**, *110*, 1404–1409. [CrossRef]

28. Gebetsberger, J.; Zywicki, M.; Künzi, A.; Polacek, N. tRNA-derived fragments target the ribosome and function as regulatory non-coding RNA in Haloferax volcanii. *Archaea* **2012**, *2012*, 260909. [CrossRef] [PubMed]

29. Sobala, A.; Hutvagner, G. Small RNAs derived from the 5′ end of tRNA can inhibit protein translation in human cells. *RNA Biol.* **2013**, *10*, 553–563. [CrossRef]

30. Davis, F.F.; Allen, F.W. Ribonucleic acids from yeast which contain a fifth nucleoside. *J. Biol. Chem.* **1957**, *227*, 907–915. [PubMed]

31. Boccaletto, P.; Machnicka, M.A.; Purta, E.; Piątkowski, P.; Bagiński, B.; Wirecki, T.K.; de Crécy-Lagard, V.; Ross, R.; Limbach, P.A.; Kotter, A.; et al. MODOMICS: A database of RNA modification pathways. 2017 update. *Nucleic Acids Res.* **2018**, *46*, D303–D307. [CrossRef] [PubMed]

32. Cantara, W.A.; Crain, P.F.; Rozenski, J.; McCloskey, J.A.; Harris, K.A.; Zhang, X.; Vendeix, F.A.P.; Fabris, D.; Agris, P.F. The RNA modification database, RNAMDB: 2011 update. *Nucleic Acids Res.* **2011**, *39*, D195–D201. [CrossRef]

33. Jackman, J.E.; Alfonzo, J.D. Transfer RNA modifications: Nature's combinatorial chemistry playground. *Wiley Interdiscip. Rev. RNA* **2013**, *4*, 35–48. [CrossRef] [PubMed]

34. Agris, P.F.; Vendeix, F.A.P.; Graham, W.D. tRNA's Wobble Decoding of the Genome: 40 Years of Modification. *J. Mol. Biol.* **2007**, *366*, 1–13. [CrossRef]

35. Bjork, G.R.; Wikstrom, P.M.; Bystrom, A.S. Prevention of translational frameshifting by the modified nucleoside 1-methylguanosine. *Science* **1989**, *244*, 986. [CrossRef] [PubMed]

36. Shoffner, J.M.; Lott, M.T.; Lezza, A.M.S.; Seibel, P.; Ballinger, S.W.; Wallace, D.C. Myoclonic epilepsy and ragged-red fiber disease (MERRF) is associated with a mitochondrial DNA tRNALys mutation. *Cell* **1990**, *61*, 931–937. [CrossRef]

37. Yasukawa, T.; Suzuki, T.; Ishii, N.; Ohta, S.; Watanabe, K. Wobble modification defect in tRNA disturbs codon-anticodon interaction in a mitochondrial disease. *EMBO J.* **2001**, *20*, 4794–4802. [CrossRef]

38. Yasukawa, T.; Suzuki, T.; Suzuki, T.; Ueda, T.; Ohta, S.; Watanabe, K. Modification Defect at Anticodon Wobble Nucleotide of Mitochondrial tRNAsLeu(UUR) with Pathogenic Mutations of Mitochondrial Myopathy, Encephalopathy, Lactic Acidosis, and Stroke-like Episodes. *J. Biol. Chem.* **2000**, *275*, 4251–4257. [CrossRef] [PubMed]

39. Liu, F.; Clark, W.; Luo, G.; Wang, X.; Fu, Y.; Wei, J.; Wang, X.; Hao, Z.; Dai, Q.; Zheng, G.; et al. ALKBH1-Mediated tRNA Demethylation Regulates Translation. *Cell* **2016**, *167*, 816–828.e816. [CrossRef] [PubMed]

40. Fu, Y.; Jia, G.; Pang, X.; Wang, R.N.; Wang, X.; Li, C.J.; Smemo, S.; Dai, Q.; Bailey, K.A.; Nobrega, M.A.; et al. FTO-mediated formation of N^6-hydroxymethyladenosine and N^6-formyladenosine in mammalian RNA. *Nat. Commun.* **2013**, *4*, 1798. [CrossRef]

41. Jia, G.; Fu, Y.; Zhao, X.; Dai, Q.; Zheng, G.; Yang, Y.; Yi, C.; Lindahl, T.; Pan, T.; Yang, Y.-G.; et al. N6-methyladenosine in nuclear RNA is a major substrate of the obesity-associated FTO. *Nat. Chem. Biol.* **2011**, *7*, 885–887. [CrossRef]

42. Zheng, G.; Dahl, J.A.; Niu, Y.; Fedorcsak, P.; Huang, C.-M.; Li, C.J.; Vågbø, C.B.; Shi, Y.; Wang, W.-L.; Song, S.-H.; et al. ALKBH5 is a mammalian RNA demethylase that impacts RNA metabolism and mouse fertility. *Mol. Cell* **2013**, *49*, 18–29. [CrossRef] [PubMed]

43. Zhou, J.; Wan, J.; Gao, X.; Zhang, X.; Jaffrey, S.R.; Qian, S.-B. Dynamic m^6A mRNA methylation directs translational control of heat shock response. *Nature* **2015**, *526*, 591–594. [CrossRef]

44. Xiang, Y.; Laurent, B.; Hsu, C.-H.; Nachtergaele, S.; Lu, Z.; Sheng, W.; Xu, C.; Chen, H.; Ouyang, J.; Wang, S.; et al. RNA m^6A methylation regulates the ultraviolet-induced DNA damage response. *Nature* **2017**, *543*, 573–576. [CrossRef] [PubMed]

45. Li, Z.; Weng, H.; Su, R.; Weng, X.; Zuo, Z.; Li, C.; Huang, H.; Nachtergaele, S.; Dong, L.; Hu, C.; et al. FTO Plays an Oncogenic Role in Acute Myeloid Leukemia as a N^6-Methyladenosine RNA Demethylase. *Cancer Cell* **2017**, *31*, 127–141. [CrossRef]

46. Fu, L.; Guerrero, C.R.; Zhong, N.; Amato, N.J.; Liu, Y.; Liu, S.; Cai, Q.; Ji, D.; Jin, S.G.; Niedernhofer, L.J.; et al. Tet-mediated formation of 5-hydroxymethylcytosine in RNA. *J. Am. Chem. Soc.* **2014**, *136*, 11582–11585. [CrossRef] [PubMed]

47. Huber, S.M.; van Delft, P.; Mendil, L.; Bachman, M.; Smollett, K.; Werner, F.; Miska, E.A.; Balasubramanian, S. Formation and abundance of 5-hydroxymethylcytosine in RNA. *Chembiochem Eur. J. Chem. Biol.* **2015**, *16*, 752–755. [CrossRef] [PubMed]

48. Huber, S.M.; van Delft, P.; Tanpure, A.; Miska, E.A.; Balasubramanian, S. 2′-O-Methyl-5-hydroxymethylcytidine: A Second Oxidative Derivative of 5-Methylcytidine in RNA. *J. Am. Chem. Soc.* **2017**, *139*, 1766–1769. [CrossRef]

49. Mauer, J.; Luo, X.; Blanjoie, A.; Jiao, X.; Grozhik, A.V.; Patil, D.P.; Linder, B.; Pickering, B.F.; Vasseur, J.-J.; Chen, Q.; et al. Reversible methylation of m^6A$_m$ in the 5′ cap controls mRNA stability. *Nature* **2017**, *541*, 371–375. [CrossRef]

50. Meyer, K.D.; Jaffrey, S.R. Rethinking m^6A Readers, Writers, and Erasers. *Annu. Rev. Cell Dev. Biol.* **2017**, *33*, 319–342. [CrossRef] [PubMed]

51. Delatte, B.; Wang, F.; Ngoc, L.V.; Collignon, E.; Bonvin, E.; Deplus, R.; Calonne, E.; Hassabi, B.; Putmans, P.; Awe, S.; et al. Transcriptome-wide distribution and function of RNA hydroxymethylcytosine. *Science* **2016**, *351*, 282. [CrossRef] [PubMed]

52. Patil, A.; Chan, C.T.; Dyavaiah, M.; Rooney, J.P.; Dedon, P.C.; Begley, T.J. Translational infidelity-induced protein stress results from a deficiency in Trm9-catalyzed tRNA modifications. *RNA Biol.* **2012**, *9*, 990–1001. [CrossRef] [PubMed]

53. Nedialkova, D.D.; Leidel, S.A. Optimization of Codon Translation Rates via tRNA Modifications Maintains Proteome Integrity. *Cell* **2015**, *161*, 1606–1618. [CrossRef]

54. Tukenmez, H.; Xu, H.; Esberg, A.; Bystrom, A.S. The role of wobble uridine modifications in +1 translational frameshifting in eukaryotes. *Nucleic Acids Res.* **2015**, *43*, 9489–9499. [CrossRef] [PubMed]

55. Paredes, J.A.; Carreto, L.; Simões, J.; Bezerra, A.R.; Gomes, A.C.; Santamaria, R.; Kapushesky, M.; Moura, G.R.; Santos, M.A.S. Low level genome mistranslations deregulate the transcriptome and translatome and generate proteotoxic stress in yeast. *BMC Biol.* **2012**, *10*, 55. [CrossRef] [PubMed]

56. Tuorto, F.; Lyko, F. Genome recoding by tRNA modifications. *Open Biol.* **2016**, *6*, 160287. [CrossRef] [PubMed]

57. Kirchner, S.; Ignatova, Z. Emerging roles of tRNA in adaptive translation, signalling dynamics and disease. *Nat. Rev. Genet.* **2014**, *16*, 98. [CrossRef] [PubMed]

58. Chernyakov, I.; Whipple, J.M.; Kotelawala, L.; Grayhack, E.J.; Phizicky, E.M. Degradation of several hypomodified mature tRNA species in *Saccharomyces cerevisiae* is mediated by Met22 and the 5′-3′ exonucleases Rat1 and Xrn1. *Genes Dev.* **2008**, *22*, 1369–1380. [CrossRef] [PubMed]

59. Alexandrov, A.; Chernyakov, I.; Gu, W.; Hiley, S.L.; Hughes, T.R.; Grayhack, E.J.; Phizicky, E.M. Rapid tRNA Decay Can Result from Lack of Nonessential Modifications. *Mol. Cell* **2006**, *21*, 87–96. [CrossRef] [PubMed]

60. Cload, S.T.; Liu, D.R.; Froland, W.A.; Schultz, P.G. Development of improved tRNAs for in vitro biosynthesis of proteins containing unnatural amino acids. *Chem. Biol.* **1996**, *3*, 1033–1038. [CrossRef]

61. Esberg, A.; Huang, B.; Johansson, M.J.O.; Byström, A.S. Elevated Levels of Two tRNA Species Bypass the Requirement for Elongator Complex in Transcription and Exocytosis. *Mol. Cell* **2006**, *24*, 139–148. [CrossRef] [PubMed]

62. Droogmans, L.; Roovers, M.; Bujnicki, J.M.; Tricot, C.; Hartsch, T.; Stalon, V.; Grosjean, H. Cloning and characterization of tRNA (m1A58) methyltransferase (TrmI) from Thermus thermophilus HB27, a protein required for cell growth at extreme temperatures. *Nucleic Acids Res.* **2003**, *31*, 2148–2156. [CrossRef] [PubMed]

63. Ishida, K.; Kunibayashi, T.; Tomikawa, C.; Ochi, A.; Kanai, T.; Hirata, A.; Iwashita, C.; Hori, H. Pseudouridine at position 55 in tRNA controls the contents of other modified nucleotides for low-temperature adaptation in the extreme-thermophilic eubacterium *Thermus thermophilus*. *Nucleic Acids Res.* **2011**, *39*, 2304–2318. [CrossRef]

64. Rose, R.E.; Pazos, M.A., 2nd; Curcio, M.J.; Fabris, D. Global Epitranscriptomics Profiling of RNA Post-Transcriptional Modifications as an Effective Tool for Investigating the Epitranscriptomics of Stress Response. *Mol. Cell. Proteom.* **2016**, *15*, 932–944. [CrossRef]

65. Basanta-Sanchez, M.; Temple, S.; Ansari, S.A.; D'Amico, A.; Agris, P.F. Attomole quantification and global profile of RNA modifications: Epitranscriptome of human neural stem cells. *Nucleic Acids Res.* **2016**, *44*, e26. [CrossRef]

66. Chan, C.T.Y.; Pang, Y.L.J.; Deng, W.; Babu, I.R.; Dyavaiah, M.; Begley, T.J.; Dedon, P.C. Reprogramming of tRNA modifications controls the oxidative stress response by codon-biased translation of proteins. *Nat. Commun.* **2012**, *3*, 937. [CrossRef]

67. Pestova, T.V.; Kolupaeva, V.G.; Lomakin, I.B.; Pilipenko, E.V.; Shatsky, I.N.; Agol, V.I.; Hellen, C.U.T. Molecular mechanisms of translation initiation in eukaryotes. *Proc. Natl. Acad. Sci. USA* **2001**, *98*, 7029. [CrossRef] [PubMed]

68. Kozak, M. Comparison of initiation of protein synthesis in procaryotes, eucaryotes, and organelles. *Microbiol. Rev.* **1983**, *47*, 1–45.

69. Basavappa, R.; Sigler, P.B. The 3 A crystal structure of yeast initiator tRNA: Functional implications in initiator/elongator discrimination. *EMBO J.* **1991**, *10*, 3105–3111. [CrossRef]

70. Anderson, J.; Phan, L.; Cuesta, R.; Carlson, B.A.; Pak, M.; Asano, K.; Björk, G.R.; Tamame, M.; Hinnebusch, A.G. The essential Gcd10p-Gcd14p nuclear complex is required for 1-methyladenosine modification and maturation of initiator methionyl-tRNA. *Genes Dev.* **1998**, *12*, 3650–3662. [CrossRef]

71. Kadaba, S.; Krueger, A.; Trice, T.; Krecic, A.M.; Hinnebusch, A.G.; Anderson, J. Nuclear surveillance and degradation of hypomodified initiator tRNAMet in S. cerevisiae. *Genes Dev.* **2004**, *18*, 1227–1240. [CrossRef] [PubMed]

72. Su, D.; Chan, C.T.Y.; Gu, C.; Lim, K.S.; Chionh, Y.H.; McBee, M.E.; Russell, B.S.; Babu, I.R.; Begley, T.J.; Dedon, P.C. Quantitative analysis of ribonucleoside modifications in tRNA by HPLC-coupled mass spectrometry. *Nat. Protoc.* **2014**, *9*, 828–841. [CrossRef] [PubMed]

73. Deng, W.; Babu, I.R.; Su, D.; Yin, S.; Begley, T.J.; Dedon, P.C. Trm9-Catalyzed tRNA Modifications Regulate Global Protein Expression by Codon-Biased Translation. *PLoS Genet.* **2015**, *11*, e1005706. [CrossRef] [PubMed]

74. Begley, U.; Dyavaiah, M.; Patil, A.; Rooney, J.P.; DiRenzo, D.; Young, C.M.; Conklin, D.S.; Zitomer, R.S.; Begley, T.J. Trm9-catalyzed tRNA modifications link translation to the DNA damage response. *Mol. Cell* **2007**, *28*, 860–870. [CrossRef] [PubMed]

75. Chionh, Y.H.; McBee, M.; Babu, I.R.; Hia, F.; Lin, W.; Zhao, W.; Cao, J.; Dziergowska, A.; Malkiewicz, A.; Begley, T.J.; et al. tRNA-mediated codon-biased translation in mycobacterial hypoxic persistence. *Nat. Commun.* **2016**, *7*, 13302. [CrossRef]

76. Schaefer, M.; Pollex, T.; Hanna, K.; Tuorto, F.; Meusburger, M.; Helm, M.; Lyko, F. RNA methylation by Dnmt2 protects transfer RNAs against stress-induced cleavage. *Genes Dev.* **2010**, *24*, 1590–1595. [CrossRef] [PubMed]

77. Fu, H.; Feng, J.; Liu, Q.; Sun, F.; Tie, Y.; Zhu, J.; Xing, R.; Sun, Z.; Zheng, X. Stress induces tRNA cleavage by angiogenin in mammalian cells. *FEBS Lett.* **2008**, *583*, 437–442. [CrossRef] [PubMed]

78. Thompson, D.M.; Lu, C.; Green, P.J.; Parker, R. tRNA cleavage is a conserved response to oxidative stress in eukaryotes. *RNA* **2008**, *14*, 2095–2103. [CrossRef]

79. Motorin, Y.; Lyko, F.; Helm, M. 5-methylcytosine in RNA: Detection, enzymatic formation and biological functions. *Nucleic Acids Res.* **2010**, *38*, 1415–1430. [CrossRef] [PubMed]

80. Suzuki, T.; Suzuki, T. A complete landscape of post-transcriptional modifications in mammalian mitochondrial tRNAs. *Nucleic Acids Res.* **2014**, *42*, 7346–7357. [CrossRef]

81. Haag, S.; Warda, A.S.; Kretschmer, J.; Günnigmann, M.A.; Höbartner, C.; Bohnsack, M.T. NSUN6 is a human RNA methyltransferase that catalyzes formation of m^5C72 in specific tRNAs. *RNA* **2015**, *21*, 1532–1543. [CrossRef]

82. Goll, M.G.; Kirpekar, F.; Maggert, K.A.; Yoder, J.A.; Hsieh, C.-L.; Zhang, X.; Golic, K.G.; Jacobsen, S.E.; Bestor, T.H. Methylation of tRNAAsp by the DNA Methyltransferase Homolog Dnmt2. *Science* **2006**, *311*, 395. [CrossRef]

83. Khoddami, V.; Cairns, B.R. Identification of direct targets and modified bases of RNA cytosine methyltransferases. *Nat. Biotechnol.* **2013**, *31*, 458. [CrossRef] [PubMed]

84. Hussain, S.; Sajini, A.A.; Blanco, S.; Dietmann, S.; Lombard, P.; Sugimoto, Y.; Paramor, M.; Gleeson, J.G.; Odom, D.T.; Ule, J.; et al. NSun2-Mediated Cytosine-5 Methylation of Vault Noncoding RNA Determines Its Processing into Regulatory Small RNAs. *Cell Rep.* **2013**, *4*, 255–261. [CrossRef]

85. Blanco, S.; Dietmann, S.; Flores, J.V.; Hussain, S.; Kutter, C.; Humphreys, P.; Lukk, M.; Lombard, P.; Treps, L.; Popis, M.; et al. Aberrant methylation of tRNAs links cellular stress to neuro-developmental disorders. *Embo J.* **2014**, *33*, 2020–2039. [CrossRef] [PubMed]

86. Harada, F.; Nishimura, S. Possible anticodon sequences of tRNAHis, tRNAAsn, and tRNAAsp from *Escherichia coli*. Universal presence of nucleoside O in the first position of the anticodons of these transfer ribonucleic acid. *Biochemistry* **1972**, *11*, 301–308. [CrossRef]

87. El Yacoubi, B.; Bailly, M.; de Crécy-Lagard, V. Biosynthesis and Function of Posttranscriptional Modifications of Transfer RNAs. *Annu. Rev. Genet.* **2012**, *46*, 69–95. [CrossRef] [PubMed]

88. Reyniers, J.P.; Pleasants, J.R.; Wostmann, B.S.; Katze, J.R.; Farkas, W.R. Administration of exogenous queuine is essential for the biosynthesis of the queuosine-containing transfer RNAs in the mouse. *J. Biol. Chem.* **1981**, *256*, 11591–11594.

89. Katze, J.R.; Gunduz, U.; Smith, D.L.; Cheng, C.S.; McCloskey, J.A. Evidence that the nucleic acid base queuine is incorporated intact into tRNA by animal cells. *Biochemistry* **1984**, *23*, 1171–1176. [CrossRef] [PubMed]

90. Pathak, C.; Jaiswal, Y.K.; Vinayak, M. Queuine promotes antioxidant defence system by activating cellular antioxidant enzyme activities in cancer. *Biosci. Rep.* **2008**, *28*, 73. [CrossRef]

91. Wang, X.; Matuszek, Z.; Huang, Y.; Parisien, M.; Dai, Q.; Clark, W.; Schwartz, M.H.; Pan, T. Queuosine modification protects cognate tRNAs against ribonuclease cleavage. *RNA* **2018**, *24*, 1305–1313. [CrossRef] [PubMed]

92. Kasai, H.; Nakanishi, K.; Macfarlane, R.D.; Torgerson, D.F.; Ohashi, Z.; McCloskey, J.A.; Gross, H.J.; Nishimura, S. The structure of Q* nucleoside isolated from rabbit liver transfer ribonucleic acid. *J. Am. Chem. Soc.* **1976**, *98*, 5044–5046. [CrossRef] [PubMed]

93. Lu, J.; Huang, B.; Esberg, A.; Johansson, M.J.O.; Byström, A.S. The Kluyveromyces lactis gamma-toxin targets tRNA anticodons. *RNA* **2005**, *11*, 1648–1654. [CrossRef]

94. Ogawa, T.; Inoue, S.; Yajima, S.; Hidaka, M.; Masaki, H. Sequence-specific recognition of colicin E5, a tRNA-targeting ribonuclease. *Nucleic Acids Res.* **2006**, *34*, 6065–6073. [CrossRef] [PubMed]

95. Jiang, Y.; Meidler, R.; Amitsur, M.; Kaufmann, G. Specific interaction between anticodon nuclease and the tRNALys wobble base11Edited by D. Draper. *J. Mol. Biol.* **2001**, *305*, 377–388. [CrossRef]

96. Hopper, A.K.; Phizicky, E.M. tRNA transfers to the limelight. *Genes Dev.* **2003**, *17*, 162–180. [CrossRef]

97. Hopper, A.K. Transfer RNA post-transcriptional processing, turnover, and subcellular dynamics in the yeast Saccharomyces cerevisiae. *Genetics* **2013**, *194*, 43–67. [CrossRef]

98. Foretek, D.; Wu, J.; Hopper, A.K.; Boguta, M. Control of Saccharomyces cerevisiae pre-tRNA processing by environmental conditions. *RNA* **2016**, *22*, 339–349. [CrossRef]

99. Ohira, T.; Miyauchi, K.; Sakaguchi, Y.; Suzuki, T.; Suzuki, T. Precise analysis of modification status at various stage of tRNA maturation in *Saccharomyces cerevisiae*. *Nucleic Acids Symp. Ser.* **2009**, *53*, 301–302. [CrossRef] [PubMed]

100. Nishikura, K.; De Robertis, E.M. RNA processing in microinjected Xenopus oocytes: Sequential addition of base modifications in a spliced transfer RNA. *J. Mol. Biol.* **1981**, *145*, 405–420. [CrossRef]
101. Torres, A.G.; Piñeyro, D.; Rodríguez-Escribà, M.; Camacho, N.; Reina, O.; Saint-Léger, A.; Filonava, L.; Batlle, E.; Ribas de Pouplana, L. Inosine modifications in human tRNAs are incorporated at the precursor tRNA level. *Nucleic Acids Res.* **2015**, *43*, 5145–5157. [CrossRef] [PubMed]
102. Gaston, K.W.; Rubio, M.A.T.; Spears, J.L.; Pastar, I.; Papavasiliou, F.N.; Alfonzo, J.D. C to U editing at position 32 of the anticodon loop precedes tRNA 5′ leader removal in trypanosomatids. *Nucleic Acids Res.* **2007**, *35*, 6740–6749. [CrossRef] [PubMed]
103. Lin, H.; Miyauchi, K.; Harada, T.; Okita, R.; Takeshita, E.; Komaki, H.; Fujioka, K.; Yagasaki, H.; Goto, Y.-I.; Yanaka, K.; et al. CO_2-sensitive tRNA modification associated with human mitochondrial disease. *Nat. Commun.* **2018**, *9*, 1875. [CrossRef] [PubMed]
104. Baranowski, W.; Dirheimer, G.; Jakowicki, J.A.; Keith, G. Deficiency of Queuine, a Highly Modified Purine Base, in Transfer RNAs from Primary and Metastatic Ovarian Malignant Tumors in Women. *Cancer Res.* **1994**, *54*, 4468.
105. Huang, B.-S.; Wu, R.-T.; Chien, K.-Y. Relationship of the Queuine Content of Transfer Ribonucleic Acids to Histopathological Grading and Survival in Human Lung Cancer. *Cancer Res.* **1992**, *52*, 4696.
106. Dewe, J.M.; Fuller, B.L.; Lentini, J.M.; Kellner, S.M.; Fu, D. TRMT1-Catalyzed tRNA Modifications Are Required for Redox Homeostasis to Ensure Proper Cellular Proliferation and Oxidative Stress Survival. *Mol. Cell. Biol.* **2017**, *37*, e00214–e00217. [CrossRef]
107. Steinthorsdottir, V.; Thorleifsson, G.; Reynisdottir, I.; Benediktsson, R.; Jonsdottir, T.; Walters, G.B.; Styrkarsdottir, U.; Gretarsdottir, S.; Emilsson, V.; Ghosh, S.; et al. A variant in CDKAL1 influences insulin response and risk of type 2 diabetes. *Nat. Genet.* **2007**, *39*, 770. [CrossRef]
108. Wei, F.-Y.; Suzuki, T.; Watanabe, S.; Kimura, S.; Kaitsuka, T.; Fujimura, A.; Matsui, H.; Atta, M.; Michiue, H.; Fontecave, M.; et al. Deficit of tRNA(Lys) modification by Cdkal1 causes the development of type 2 diabetes in mice. *J. Clin. Investig.* **2011**, *121*, 3598–3608. [CrossRef] [PubMed]
109. Shimada, K.; Nakamura, M.; Anai, S.; De Velasco, M.; Tanaka, M.; Tsujikawa, K.; Ouji, Y.; Konishi, N. A Novel Human AlkB Homologue, ALKBH8, Contributes to Human Bladder Cancer Progression. *Cancer Res.* **2009**, *69*, 3157. [CrossRef]
110. Ueda, Y.; Ooshio, I.; Fusamae, Y.; Kitae, K.; Kawaguchi, M.; Jingushi, K.; Hase, H.; Harada, K.; Hirata, K.; Tsujikawa, K. AlkB homolog 3-mediated tRNA demethylation promotes protein synthesis in cancer cells. *Sci. Rep.* **2017**, *7*, 42271. [CrossRef]
111. Tasaki, M.; Shimada, K.; Kimura, H.; Tsujikawa, K.; Konishi, N. ALKBH3, a human AlkB homologue, contributes to cell survival in human non-small-cell lung cancer. *Br. J. Cancer* **2011**, *104*, 700–706. [CrossRef] [PubMed]
112. Ehrenhofer-Murray, A.E. Cross-Talk between Dnmt2-Dependent tRNA Methylation and Queuosine Modification. *Biomolecules* **2017**, *7*, 14. [CrossRef] [PubMed]
113. Schaefer, M.; Pollex, T.; Hanna, K.; Lyko, F. RNA cytosine methylation analysis by bisulfite sequencing. *Nucleic Acids Res.* **2009**, *37*, e12. [CrossRef] [PubMed]

Article

Determinants of Hair Manganese, Lead, Cadmium and Arsenic Levels in Environmentally Exposed Children

Thomas Jursa [1], Cheryl R. Stein [2] and Donald R. Smith [1],*

[1] Department of Microbiology and Environmental Toxicology, University of California, Santa Cruz, CA 95064, USA; tpjursa@ucsc.edu

[2] Department of Child and Adolescent Psychiatry, Hassenfeld Children's Hospital at NYU Langone, Child Study Center, New York University, New York, NY 10016, USA; Cheryl.Stein@nyumc.org

* Correspondence: drsmith@ucsc.edu; Tel.: +1-831-459-5041

Received: 5 February 2018; Accepted: 20 March 2018; Published: 22 March 2018

Abstract: Biomarkers of environmental metal exposure in children are important for elucidating exposure and health risk. While exposure biomarkers for As, Cd, and Pb are relatively well defined, there are not yet well-validated biomarkers of Mn exposure. Here, we measured hair Mn, Pb, Cd, and As levels in children from the Mid-Ohio Valley to determine within and between-subject predictors of hair metal levels. Occipital scalp hair was collected in 2009–2010 from 222 children aged 6–12 years (169 female, 53 male) participating in a study of chemical exposure and neurodevelopment in an industrial region of the Mid-Ohio Valley. Hair samples from females were divided into three two centimeter segments, while males provided a single segment. Hair was cleaned and processed in a trace metal clean laboratory, and analyzed for As, Cd, Mn, and Pb by magnetic sector inductively coupled plasma mass spectrometry. Hair Mn and Pb levels were comparable (median 0.11 and 0.15 µg/g, respectively) and were ~10-fold higher than hair Cd and As levels (0.007 and 0.018 µg/g, respectively). Hair metal levels were higher in males compared to females, and varied by ~100–1000-fold between all subjects, and substantially less (<40–70%) between segments within female subjects. Hair Mn, Pb, and Cd, but not As levels systematically increased by ~40–70% from the proximal to distal hair segments of females. There was a significant effect of season of hair sample collection on hair Mn, Pb, and Cd, but not As levels. Finally, hair metal levels reported here are ~2 to >10-fold lower than levels reported in other studies in children, most likely because of more rigorous hair cleaning methodology used in the present study, leading to lower levels of unresolved exogenous metal contamination of hair.

Keywords: manganese; lead; cadmium; arsenic; hair; children; environment

---

1. Introduction

Exposure biomarkers play an important role in estimating the internal dose of a person exposed to an environmental contaminant, and they are often essential in determining exposure–health effect relationships in epidemiological studies [1–9]. For example, lead (Pb) levels in blood and bone are accepted as well-validated Pb exposure biomarkers, and they helped establish the association between Pb exposure and health risk in children and adults [7,10,11]. Similarly, blood cadmium (Cd) levels have been shown to reflect Cd exposure from environmental sources [8,9]. However, for metals such as manganese (Mn), studies are mixed on whether blood Mn levels are an indicator of exposure and risk of health effects, presenting a need to develop and validate alternative exposure biomarkers [1–3]. Recent studies have suggested that hair Mn levels may help fill this need [2,3,12–14].

A number of studies have reported associations between levels of Mn in hair and Mn exposure and associated health effects in children and adolescents [12–15], including a recent review showing that hair was the most consistent and valid biomarker of Mn-associated health effects in children [3]. Similarly, levels of some other metals in hair, including mercury and arsenic, have been reported as both meaningful exposure biomarkers and indicators of health risks from exposure [16,17]. Hair may provide some practical advantages over other tissues as an exposure biomarker; scalp hair grows at a rate of roughly one centimeter per month, providing a possible indicator of exposure integrated over periods of one to five months or more, depending on the length of collected hair [18,19]. Moreover, analysis of sequential sections of hair may provide useful information on the temporal variability of exposure, although few studies have investigated whether sequential hair segments are useful for retrospective exposure assessment over the duration of hair growth [17].

However, the potential utility of hair as an exposure/effect biomarker is not without some challenges. Most notably, hair is susceptible to contamination from exogenous sources such as dust, water, and use of hair products [14,17,20]. Studies reporting hair metal levels as a biomarker of exposure vary widely in the type of method used to remove exogenous contamination from the hair, with methods varying from a simple water rinse to detergent and acid sonication [2,12,14,17,21–25]. Not surprisingly, reported hair metal levels vary widely by study, though it is unclear if this reflects differences in exposure versus differences in effectiveness of cleaning exogenous contamination. Studies have shown that metal levels in hair are derived largely from exogenous contamination, with the rigor of hair cleaning prior to analyses affecting the contribution of exogenous contamination to the measured hair metal levels [14,20,26].

Here, we determined levels of Mn, Pb, Cd and arsenic (As) in scalp hair samples from 222 male and female children aged 6–12 years living in the Mid-Ohio Valley. For 169 female subjects, hair samples were cut into sequential segments to determine the reproducibility of hair metal levels within the same subject, and the variation in hair metal levels over different seasons of growth. Hair samples were cleaned using a rigorous cleaning method shown to effectively remove exogenous metal contamination [14] and processed for analyses by inductively coupled plasma–mass spectrometry (ICP-MS).

2. Methods

2.1. Subjects

Hair samples were collected and processed from 222 subjects age 6–12 years (169 female, 53 male) recruited through the C8 Health Project Neurobehavioral Development Follow-up, which was investigating the neurodevelopmental health effects of perfluoroocatnoate (PFOA) exposure in southeastern Ohio and northwestern West Virginia. A detailed description of subject recruitment, as well as information on subject demographics, residence, and medical histories collected via maternal report at the time of the neurodevelopment follow-up study in 2009–2010 is provided elsewhere [27]. Mothers provided informed consent and children provided verbal assent; child and mother each received $50 for participation. The Mount Sinai Program for the Protection of Human Subjects and the Battelle Centers for Public Health Research & Evaluation Institutional Review Board approved all study procedures. Investigations were carried out following the rules of the Declaration of Helsinki of 1975 (https://www.wma.net/what-we-do/medical-ethics/declaration-of-helsinki/), revised in 2008. Relevant to the present study, this Ohio River Valley region also hosts the longest operating ferromanganese refinery in North America in Marietta, OH (Eramet Marietta, Inc., Marietta, OH, USA), and studies by others have reported elevated Mn exposures and associated health effects in children in the region [2,28].

Hair samples were collected proximal to the occipital lobe scalp and stored in zip top plastic bags at room temperature until analysis. Male subjects provided a single 2 cm segment of hair proximal to the scalp, while female subjects provided longer hair samples that were cut into sequential 2 cm segments (0–2, 2–4, 4–6 cm from the scalp), based on the overall length of the sample. A single 2 cm hair sample was analyzed for all male subjects ($n = 53$), since most males had hair too short to provide multiple segments. For females (n = 169), three sequential 2 cm hair segments were available for 159 subjects (referred to as proximal, medial, and distal 2 cm segments, relative to the scalp), two hair segments were available for n = 4 females (n = 3 for proximal and medial, n = 1 for proximal and distal), and one hair segment was available for n = 6 females (all proximal).

2.2. Experimental

All cleaning and processing of hair samples was conducted in a HEPA filtered-air trace metal clean room, using acid-cleaned labware and ultrapure trace metal grade reagents. Individual hair segments/samples weighing 5–30 mg each were cleaned of exogenous metal contamination as described previously [14]. Briefly, samples were placed in acid-cleaned 5 mL polypropylene syringe tubes and sonicated (20 min) in 0.5% Triton, rinsed five-times with ultrapure Milli-Q water, sonicated (10 min) in 1 N trace metal grade nitric acid (Fisher Scientific, Santa Clara, CA, USA), rinsed with 1 N nitric acid, and rinsed five-times with Milli-Q water, and then dried at 65 °C for 48 h. Subsequently, samples were digested in 0.5 mL 15.7 N quartz-distilled nitric acid (Fisher Scientific, optima grade) at 80 °C for 6 h in a Class-100 HEPA filtered-air fume hood, and then diluted with 5 mL Milli-Q water. For analyses, 0.25 mL of digestate was transferred to an acid-cleaned polyethylene microfuge tube, diluted with 0.25 mL Milli-Q water, and centrifuged at 13,000× g for analysis. Rhodium and thallium were added to samples as internal standards, and samples analyzed for Mn, Pb, Cd, and As by magnetic sector inductively coupled plasma mass spectrometry (Thermo Element XR ICP-MS, Waltham, MA, USA), as described elsewhere [1,14]. Methane was added to the argon (Ar) carrier gas to minimize ArCl formation. ^{208}Pb, ^{111}Cd, and ^{113}Cd were measured in low resolution, while ^{55}Mn and ^{75}Ar were measured in medium resolution. Typical analytical limits of detection (LOD's) over five analytical runs were 0.0077, 0.0038, 0.0004, and 0.0018 ng/mL for Mn, Pb, Cd, and As, respectively. For metal levels below the analytical LOD, the LOD was multiplied by 0.5 and adjusted using the sample dilution factor and sample weight of processed hair to derive a value for half the procedural detection limit, and that value included in the data set for statistical analyses. Standard reference material (SRM) NIES 13 (human hair) was used to assess analytical accuracy; mean SRM recoveries (% recovery ± % RSD) based on 17 replicates over five analytical runs averaged 101 ± 10 for Pb and 98 ± 6 for Cd (both certified values), and 75 ± 14 for Mn and 98 ± 9 for As (both non-certified reference values).

2.3. Data Analyses

Summary data are expressed as median or mean ± standard error (SE), or mean ± standard deviation (SD). If necessary, data were square root- or log-transformed to achieve normality and variance equality. To examine within and between subject variation in hair metal levels, data were analyzed using mixed models with metal concentration as dependent variable, hair segment as independent variable, and a repeated statement identifying subject (within subject referring to separate segments of the same strand of hair). Models were adjusted for subject age (continuous) and season of collection, with four seasonal periods (August 2009–October 2009; November 2009–February 2010; March 2010–May 2010; June 2010–August 2010, selected to reflect different climatological seasons that were balanced by the number of subjects). A within-subject effect of hair segment on metal concentration was identified by a significant Type 3 effect. To determine whether the within-subject variation differed by tertile of hair metal concentration, we stratified models by tertile of hair metal concentrations and calculated least square means with a Tukey adjustment. We qualitatively examined the upper and lower confidence bounds of the difference in hair metal levels across tertiles. If the difference and bounds were comparable across tertiles then we concluded that the segment effect on hair metal levels did not differ by tertile. Lastly, to assess whether the segment effect differed by season we added a segment–season interaction term to the unstratified models and identified differences by a significant Type 3 effect. A p-value $\leq$ 0.05 for the various outcomes was considered statistically significant. All data were analyzed using SAS (Version 14.1) or JMP (Version 13.0) software (SAS Institute Inc., Cary, NC, USA, 2016).

3. Results

3.1. Hair Metal Levels in Children Vary by Several Orders of Magnitude between Subjects and Were Highly Correlated

In this study of 222 subjects (53 male, 169 female), median levels of Mn and Pb in the proximal segment of children's hair were comparable at 0.109 µg/g (range 0.005–4.10 µg/g) and 0.152 µg/g (range 0.008–7.73 µg/g), respectively (Table 1, Figure 1). Hair levels of Cd and As were both about 10-fold lower than Mn and Pb levels, with median Cd levels of 0.007 µg/g (range 0.0005–0.463) and median As levels of 0.018 µg/g (range 0.004–0.438). Notably, for all four metals there was a ~>100-fold range in proximal segment metal levels between subjects, suggesting substantial differences in the exposure levels and/or incorporation of metal levels into hair between subjects. Moreover, levels of all four metals were higher in the proximal segment of males compared to the proximal segment of females (p's < 0.0001), based on mixed model analysis on log10 transformed data with sex (fixed) and subject (random) factors (n = 157 female, 45–53 male, depending on metal). The large majority of hair samples possessed metal levels that were above the analytical detection limits (>97% for Mn, Pb, and Cd, 92% for As).

Table 1. Hair metal concentrations (µg metal/g hair) of two centimeter hair segments from males and females *.

Sex	Metal	Segment	N	Mean	Median	Range	S.D.
Males	Mn	proximal	53	0.406	0.232	0.0248–4.10	0.616
	Pb		52	0.829	0.414	0.0415–7.73	1.38
	Cd		51	0.0540	0.0188	0.0039–0.463	0.0861
	As		45	0.0630	0.0366	0.0061–0.438	0.0893
Females	Mn	proximal	169	0.184	0.0948	0.0045–3.40	0.356
		medial	157	0.226	0.126	0.0067–4.63	0.463
		distal	155	0.253	0.135	0.0184–3.12	0.388
	Pb	proximal	169	0.242	0.134	0.0080–5.66	0.493
		medial	156	0.229	0.163	0.0042–1.95	0.229
		distal	153	0.284	0.208	0.0078–2.49	0.298
	Cd	proximal	169	0.0119	0.0050	0.0005–0.182	0.0242
		medial	157	0.0202	0.0088	0.0005–0.335	0.0491
		distal	155	0.0215	0.0128	0.0004–0.401	0.0369
	As	proximal	169	0.0227	0.0165	0.0044–0.214	0.0259
		medial	157	0.0197	0.0155	0.0011–0.0789	0.0143
		distal	155	0.0238	0.0170	0.0024–0.310	0.0307

* The reported 'n' and the ranges in metal levels are for measured values above the limit of detection (see text).

Associations between the levels of different metals in hair could suggest similar exposure sources. Spearman's correlation analysis of metal levels in all hair samples of male (proximal segment) and female (proximal, medial, and distal segments) subjects shows that all four metals are highly correlated (p < 0.0001), with the correlation between Cd and Pb being strongest (Spearman's ρ = 0.5780), followed by the correlation between Mn and Cd (ρ = 0.4475), and Mn and Pb (ρ = 0.3513). Correlations between As and the other three metals were weaker (ρ < 0.21) (Table 2).

Table 2. Spearman's correlation (*n* = number of hair samples) between hair metal levels in all hair samples from males (proximal segment) and females (proximal, medial, and distal segments); all Spearman's ρ values are highly significant, $p < 0.0001$.

	Mn		
Pb	0.3513 (530)	**Pb**	
Cd	0.4475 (532)	0.5780 (529)	**Cd**
As	0.1872 (526)	0.1814 (523)	0.2081 (526)

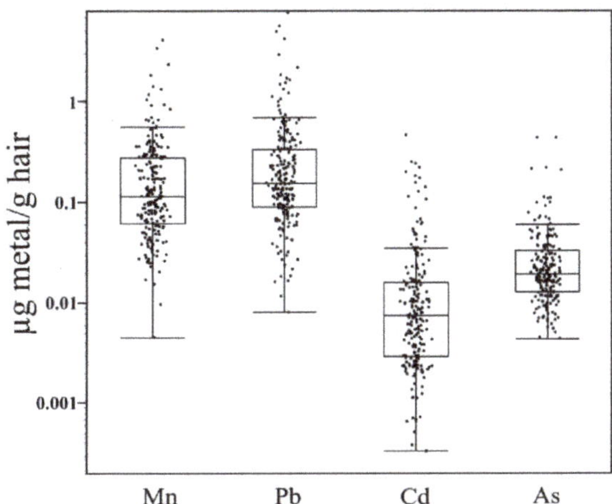

Figure 1. Hair metal concentrations (μg/g, note log scale) in the proximal 2 cm hair segment for male (*n* = 53) and female (*n* = 169) subjects. The horizontal line within the box represents the median, while the upper and lower margins of the boxes represent the 75th and 25th percentiles; whiskers are drawn to the furthest data point within 1.5-times the interquartile range. *N* = 214–222; hair metal values below the limit of detection are excluded.

3.2. Variance in Hair Metal Levels between-Subjects is Much Greater than within-Subjects, and Hair Mn, Pb, and Cd, but Not As Concentrations Increase from Proximal to Distal Segments

To determine whether metal levels in hair segments grown over a period of one to two months were more variable between subjects than between adjacent hair segments within a subject, hair samples from female subjects, who typically provided hair samples sufficiently long for segmentation, were cut into sequential two centimeter segments (designated proximal, medial, and distal segments relative to the scalp) for analyses; each two centimeter segment is assumed to reflect roughly two months of hair growth and metal exposure [18,19]. Variance component analysis was used to determine the contribution of subject and hair segment (as variance components) to the variation in hair metal levels between and within subjects. The subject variance component accounted for 65–73% of the variance in hair metal levels, and was substantially greater than the within-subject (i.e., hair segment) variance component, which accounted for 0.1–4% of the variance in hair Mn, Pb, and As, and 12% of the variance in Cd (Table 3). The remainder of the variance in hair metal levels (i.e., 17–35%) was accounted for by the subject–hair segment interaction (Table 3).

Table 3. Percent of variance in hair metal levels attributed to between-subject, within-subject (i.e., hair segment), and subject–segment interaction components according to interclass correlation analysis.

Component	Mn	Pb	Cd	As
Between-subject	72.8	71.5	70.3	65.1
Within-subject	4.1	3.8	12.4	0.1
Interaction	23.1	24.7	17.3	34.8

Given the modest contribution of the hair segment (i.e., within-subject) factor to the variability in hair metal levels in female subjects (i.e., 4–12% for Mn, Pb, and Cd), data for the proximal, medial, and distal two centimeter hair segments of female subjects were analyzed to determine if the metal concentrations *systematically* varied between segments of hair within subjects. To facilitate this, since hair metal concentrations varied substantially between subjects, hair metal concentrations for each segment were expressed as a percentage of the average of the three segments for each female subject, and the geometric mean across subjects of the normalized (%) value for each segment was then calculated for the subjects. Results show that Mn, Pb, and Cd, but not As hair metal concentrations systematically *increased* from the proximal to distal segment (Figure 2a). The relative increase in metal levels from proximal to medial to distal segments was comparable for Mn (72%, 94%, 111% of the three-segment average, respectively) and Pb (74%, 90%, 114% of the three-segment average), and slightly greater for Cd (57%, 92%, 133% of the three-segment average). In contrast, As concentrations were relatively invariant across the proximal, medial, and distal hair segments (98%, 91%, 96% of the three-segment average, respectively) (Figure 2a). Consistent with this, in mixed model analysis with a repeated statement identifying subject, a Type 3 test for fixed effects revealed a significant effect of hair segment on Mn, Pb, and Cd levels ($p < 0.001$, $p = 0.0015$, and $p < 0.001$, respectively), but no significant variation between hair segments in As concentration ($p = 0.658$).

To explore whether this systematic relative increase in hair Mn, Pb, and Cd levels from proximal to distal segments was comparable for subjects with low versus high hair metal levels, we stratified subjects into tertiles based on their three-segment average hair metal concentrations and performed mixed model analysis with a repeated statement identifying subject, as above. For perspective, the systematic ~40–70% relative increase in hair Mn, Pb, and Cd concentrations from the proximal to distal hair segments noted above, while significant, is small compared to the ~100–1000-fold difference in hair metal levels between subjects (Table 1), or the $\geq$5-fold difference in mean hair metal levels of the lowest and highest tertiles of hair metal levels (Figure 2b). Interestingly, mixed model results show that the relative increase in metal levels from proximal to distal hair segments does not vary by tertile of hair metal concentration for any of the metals (p's = 0.2–>0.9). To visualize this, we stratified subjects into tertiles by their three-segment average hair metal concentrations as above, and normalized the metal concentration for each segment to the three-segment average per subject (expressed as a percent), and then calculated the geometric mean for these normalized (%) values for each tertile of hair metal concentrations. These data reflect the mixed model null results noted above by showing that the *relative* increase in hair Mn, Pb, and Cd from proximal to distal hair segments within female subjects does not differ by tertile of hair metal levels (Figure 2c).

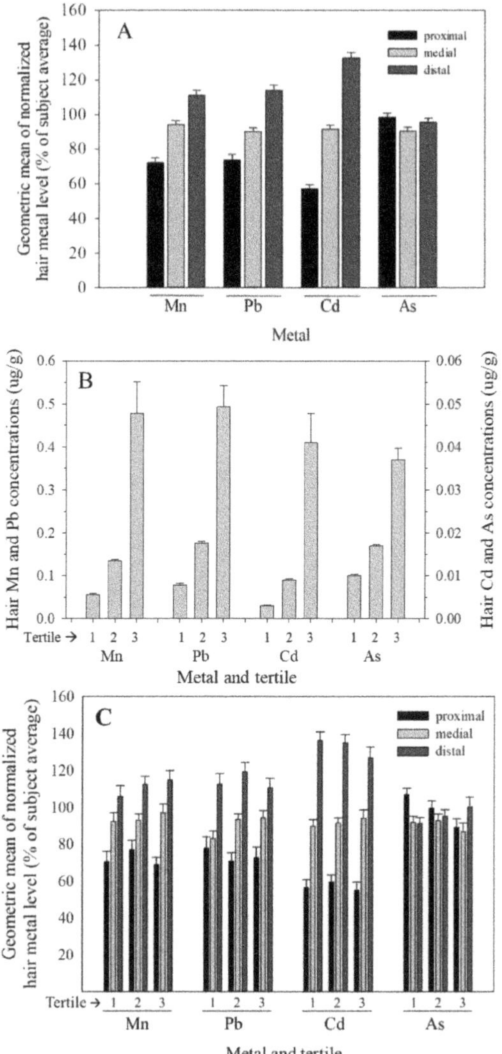

Figure 2. (**A**) Normalized hair Mn, Pb, and Cd, but not As levels systematically increase from proximal to distal two centimeter hair segments from female subjects. Normalized hair metal levels (%) for each segment were calculated by dividing the hair segment metal concentration by the average of all three segments (proximal, medial, distal) for each individual subject. Data are geometric mean (±SE) for all female subjects (n = 153–155 per segment and metal); (**B**) Hair Mn, Pb, Cd, and As concentrations differ by ≥5-fold between the lowest and highest tertiles of hair metal levels. Data are mean (±SE) three-segment average of female subjects segregated into tertiles (n = 56–57 per tertile and metal); (**C**) The relative increase in hair Mn, Pb, and Cd levels from proximal to distal segments is comparable for subjects with lower (first tertile) versus higher (third tertile) hair metal levels (see text for details).

3.3. Hair Metal Levels Vary Seasonally

To explore whether the season of hair sample collection, as a surrogate of possible seasonal differences in exposure or residual exogenous contamination, could explain the systematic increase in hair Mn, Pb, and Cd concentrations from the proximal to distal hair segments, we performed analysis in which a season factor and a segment–season interaction were added to the mixed-model noted above. The four season intervals of August–October 2009, November 2009–February 2010, March–May 2010, and June–August 2010 were selected to align with climate seasons in the Mid-Ohio Valley and to achieve reasonable balance in the number of female subjects across the four season intervals. Results show a significant effect of season on hair Mn, Pb, and Cd levels (p's < 0.001, 0.023, and <0.001, respectively), but not As (p = 0.60). There was no season–segment interaction for Mn or Pb (p's > 0.70), although for Cd the interaction was trending towards significance (p = 0.073).

This effect of season on hair Mn, Pb, and Cd levels is illustrated by generally lower metal concentrations in samples collected in the late fall to spring seasons, and higher concentrations in hair samples collected in summer to early fall seasons (Figure 3). For example, the lowest hair Mn, Pb, and Cd concentrations (seasonal medians of 0.081, 0.146, and 0.0059 μg/g, respectively; all three segments per subject combined) were for samples collected in the November–February (Mn and Pb) or March–May (Cd) seasons. In contrast, the highest hair Mn, Pb, and Cd levels (seasonal medians of 0.157, 0.235, and 0.0142 μg/g, respectively) were for samples collected in the June–August (Mn) or August–October (Pb, Cd) seasons. Across subjects, median levels of hair Mn increased by ~90% between the two seasons, whereas Pb increased by ~60% and Cd by ~240%.

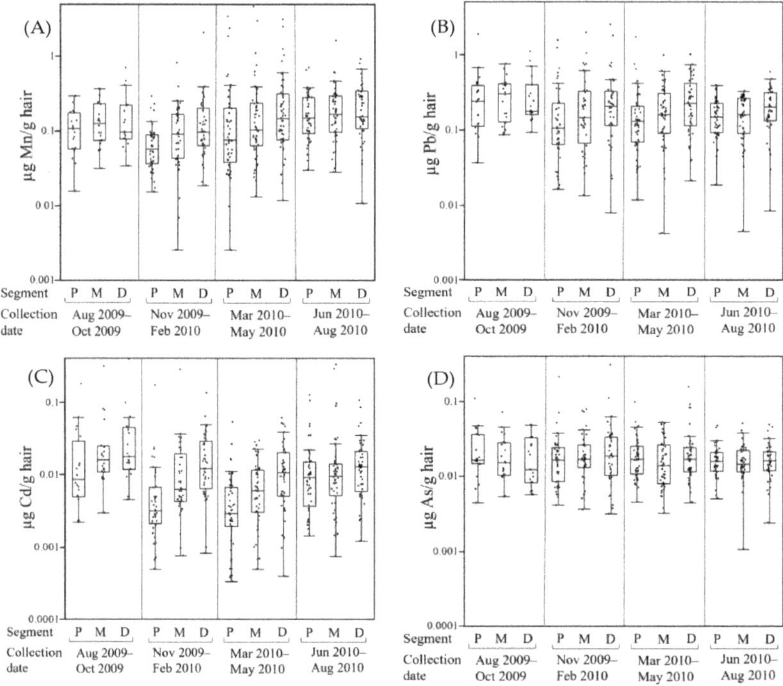

Figure 3. Hair segment Mn (**A**), Pb (**B**), Cd (**C**), and As (**D**) levels in females (μg/g, note log scale) vary with season of collection. Hair metal levels are plotted by three to four months season of collection intervals. The horizontal line within the box represents the median, while the upper and lower margins of the boxes represent the 75th and 25th percentiles; the whiskers are drawn to the furthest data point within 1.5 times the interquartile range. Only female subjects with proximal (P), medial (M), and distal (D) hair segments are shown (n = 153–155 subjects per metal).

We similarly performed mixed model analysis restricted to metal levels in the proximal segments of males and females, with sex and season of collection as fixed effects and subject as a random effect. Results show a significant effect of sex (p's < 0.0001) and season (p's ≤ 0.004) on proximal segment Mn, Pb, and Cd levels, with higher metal levels in males and higher levels in hair collected in summer/early fall versus winter/spring. For proximal segment As levels, there was a significant effect of sex ($p < 0.0001$, males higher), but no effect of season of hair collection ($p = 0.137$). There was no sex–season interaction for any of the metals (p's > 0.43), indicating that the season of collection did not differently affect male and female hair metal levels.

4. Discussion

Hair offers several advantages over other biological tissues/matrices as an exposure biomarker, most notably the potential to retrospectively reconstruct exposures over sequential integrated periods of weeks to months, depending on the length of hair. Here we report Mn, Pb, Cd, and As levels in hair samples from male and female children/adolescents age 6–12 years living in the Mid-Ohio Valley, a region noted for its industrial activity, including the longest operating ferromanganese refinery in North America in Marietta, OH (Eramet Marietta, Inc., Marietta, OH, USA) [2]. This study is unique in that hair samples from females ($n = 159$) were divided into sequential two centimeter segments, and all samples were cleaned prior to analyses using a rigorous cleaning method previously shown to effectively remove exogenous metal contamination [14].

4.1. Correlations between Metals Suggests Some Shared Exposure Sources/Pathways

The significant correlations among hair levels of all four metals, with correlations of hair Mn, Pb, and Cd being strongest (Spearman's ρ's ~0.35–0.58 for all hair samples, Table 2) suggests some shared environmental sources/pathways for the incorporation of these metals into hair, and possibly also similar chemistries of interaction of several of these metals with hair keratin. The Mid-Ohio valley region has a history of industrial activity, including ferromanganese alloy and perfluoroocatnoate (PFOA) chemical manufacturing, and air monitoring in 2007–2008 measured levels of Cd, As, and particularly Mn that exceeded ATSDR and EPA health-based comparison values [29]. Further, human hair is a complex biological matrix composed predominantly of proteins (65–95%), water (up to 32% by weight depending on its moisture content), lipids, pigment, and trace elements that are coordinated with the functional groups of protein amino acids or with fatty-acid groups of lipids [19]. Though the protein composition of hair may vary across individuals, it is generally rich in polar and charged amino acids, including hydroxyls, amides, acidic and basic amino acids, and disulfides, and these are the components of hair that may readily coordinate with endogenously and exogenously incorporated metals [19,30]. Thus, the stronger correlations between hair Pb and Cd could also reflect similar chemistries of interaction of these two metals with hair keratin; Cd and Pb have similar affinities to sulfur and nitrogen ligands, while Mn coordinates strongly with oxygen ligands [19,30]. Since As is assumed to exist mainly as oxyanion species, it likely chemically coordinates with different functional groups in hair keratin fibers than the cationic metals.

4.2. Hair Metal Levels Vary Substantially More between Subjects than within Subjects, and Hair Mn, Pb, and Cd, but Not As Concentrations Increase from Proximal to Distal Segments

We found that hair Mn, Pb, Cd, and As levels were higher in males compared to females, and that levels varied by ~100–1000-fold between all subjects (Table 1, Figure 1), but varied comparatively little within subjects, with relative changes of ~40–60% in hair Mn, Pb, and Cd, and <10% in As levels across the proximal, medial, and distal hair segments of female subjects (Figure 3). Consistent with this, variance component analysis showed that the between subject factor accounted for the majority (65–73%) of variance in hair metal levels, while the within subject component (i.e., variation between hair segments within female subjects) accounted for ~4% of the variance in Mn and Pb, 12% in Cd, but very little of the within subject variance in As (0.1%). Together, these findings suggest that hair metal levels reflect important between subject differences in metal exposure and incorporation of metals into hair.

Notably, concentrations of hair Mn, Pb, and Cd, but not As systematically increased from the proximal to distal two centimeter hair segment in female subjects. Since hair samples were collected from subjects throughout the Mid-Ohio Valley over a 13 months period, it is unlikely that this increase in metal concentrations from proximal to distal hair segments reflects temporal differences in metal exposure common to all subjects. Alternatively, we considered whether residual exogenous metal contamination that remained after rigorous cleaning could account for the increase in hair metal levels across segments. Given that subjects likely inhabited environments with inherently different environmental exposure burdens, as suggested by the 100–1000-fold difference in hair metal levels between subjects, we reasoned that subjects in higher metal exposure environments would have experienced both higher endogenous metal exposures, leading to greater metal incorporation into growing hair, and higher exogenous metal contamination of hair compared to subjects living in lower metal burden environments. Moreover, we considered that the older, distal hair segments likely acquired more exogenous metal contamination than the younger proximal segments, because the older distal segments were in contact with the environment roughly four months longer than the proximal segments, and thus were exposed to a greater cumulative environmental exposure burden. Following this logic with the assumption that environmental exposures from water, air, dust, etc. are likely the primary exposure source(s) for both the endogenous and exogenous components of hair metal levels, the *relative* contribution of residual exogenous contamination to total hair metal levels would scale with (i) the magnitude of environmental metal contamination and the duration of time the hair was in contact with the environment, and (ii) the endogenous (metabolically incorporated) component of hair metal levels. In this case the *relative* (percent) increase in hair Mn, Pb, and Cd levels from the proximal to distal hair segments would be comparable for subjects in the lowest and highest tertiles of hair metal levels, which is consistent with our findings (Figure 2c). Collectively, these findings further suggest that the proximal segment hair metal levels reflect predominantly endogenously-incorporated metals, while distal segment metal levels reflect both endogenously-incorporated metals and a relatively small but notable exogenously-added (contamination) component of hair metals (i.e., Mn, Pb, and Cd), the latter in spite of the rigorous hair cleaning methodology used here [14].

Skröder et al. [17] similarly reported systematic increases in hair Mn, Pb, and Cd, but not As in sequential hair segments over eight-centimeter of hair length in a small number of Bangladeshi children ($n = 19$). In that study the relative increase in hair metal levels with distance from the scalp was 4.6-fold for Mn, and roughly two to three-fold for Pb and Cd-relative increases that are much greater than the ~40–70% relative increase from proximal to distal hair segments observed in the present study. Skröder et al. interpreted their findings as evidence of exogenous metal contamination that was most pronounced for hair Mn levels. In light of (i) the very elevated groundwater Mn levels in the Bangladeshi subjects' environment; and (ii) our studies showing that hair is readily and significantly contaminated from direct contact with Mn-contaminated water and that exogenous Mn contamination from water is incompletely removed even with rigorous cleaning [14], it is likely that the hair Mn levels reported by Skröder et al. are dominated by unresolved exogenous contamination.

4.3. Hair Metal Levels Vary by Season of Collection

We found that the season of hair collection was associated with hair levels of Mn, Pb, and Cd, but not As, and that there was no season–sex (proximal segments only) or season–hair segment (females only) interaction in the mixed model analyses. This suggests that the main effect of season of hair collection may reflect a contribution of seasonal differences in residual exogenous metal contamination that slightly but measurably contributed to hair metal levels. Since hair samples were collected over a 13 month period, the proximal and distal hair segments of females would have grown over different seasons depending on the season of hair collection [18,19]. Male and female subjects whose hair was collected in late summer/early fall, when outdoor activity and some routes of exposure might be greatest, had higher levels of Mn, Cd, and Pb in proximal hair segments (males and females) and across all three segments of females, compared to hair samples collected in winter and spring. If hair contained only endogenously-incorporated metals, then seasonal differences in exposure would be associated with the season of hair growth, not the season of hair collection as observed.

4.4. Hair Metal Concentrations Reported Here Are Generally Lower than other Studies in Children

To facilitate inter-study comparison of reported hair metal levels in children, we summarized reported findings from 15 studies of similarly aged children (Table 4). Median or mean hair Mn levels differ by ~120-fold across studies, while hair levels of Pb, Cd, and As differ across studies by ~15-fold, 4-fold, and 180-fold, respectively. These differences between studies may reflect, at least in part, differences in endogenous metal exposure and incorporation of metals from the circulation into hair. However, because of the susceptibility of hair to environmental metal contamination, differences in hair metal levels between studies likely also reflect differences in unresolved exogenous contamination. The listed studies used a variety of different hair cleaning methodologies, from no cleaning to multi-stage cleaning procedures employing various combinations of detergents and/or solvents, weak acid, and sonication (Table 4). For example, Skröder et al. [17] reported median hair Mn levels of 5.0 µg/g in Bangladeshi children exposed to elevated Mn in drinking water, while Hernandez-Bonilla et al. [22] and Menezes-Filho et al. [21] reported hair Mn levels greater than 10 µg/g in Mexican and Brazilian children, respectively, living in the vicinity of ferromanganese alloy plants. These hair Mn levels are nearly two orders of magnitude or more higher than levels reported in the present study, or levels reported by Torrente et al. [31] and Lucas et al. [12] for Spanish and Italian children, respectively, living in areas impacted by industrial emissions. Skröder et al. [17] and Hernandez-Bonilla et al. [22] reported cleaning hair prior to analysis with a Triton detergent wash, while Menezes-Filho et al. [21] and Torrente et al. [31] used a Triton wash with ultrasound sonication. The present study and Lucas et al. [12] used Triton sonication followed by sonication in a 1 N nitric acid solution. While it is difficult to separate the influence of environmental exposure from the efficacy of hair cleaning methods to reduce exogenous hair contamination, these data suggest that studies that used more rigorous cleaning procedures reported lower hair Mn (and generally other metals) concentrations, consistent with studies showing that the rigor of hair cleaning prior to analyses significantly influences hair metal levels from exogenous contamination [14,20,26].

There are several studies from different geographical regions that used the same cleaning method, as well as studies from the same geographical region that used different cleaning methods that can be readily compared to estimate the extent that differences in hair metal levels between studies reflect differences in exposure versus differences in exogenous contamination due to different hair cleaning methods. For example, prior studies from our lab in Italian children exposed to environmental metals from industrial ferroalloy emissions [12,14] used the same Triton sonication followed by dilute nitric acid sonication hair cleaning method as the present study. Hair Mn and Pb levels are very comparable between these studies (Table 4), allowing us to conclude that the subjects in the present study had ~10–30% higher Mn exposure and ~10% lower Pb exposure levels compared to the Italian subjects, based on hair metal levels. We can also compare hair Mn levels between two studies from the same region that used different hair cleaning methods. Haynes et al. [2] used a Triton detergent hair cleaning method (without sonication) and reported geometric mean hair Mn levels of 0.417 µg/g from children in the same Ohio Valley region as the present study, which are greater than three-fold higher than levels in the present study (geometric mean Mn of 0.119 µg/g, median 0.109 µg/g, Table 4). This difference in hair Mn levels between the two studies may be due to differences in Mn exposure between cohorts. However, it may also be that they are due to differences in hair cleaning methods, given that our prior study [14] found that a Triton sonication hair cleaning method similar to that used by Haynes et al. [2] yielded hair Mn and Pb levels that were ~2.5–4-fold higher than hair metal levels following the Triton + weak nitric acid sonication method used in the present study. Collectively, these findings suggest that the extent that hair metal levels reflect endogenous exposure will vary substantially depending on the hair cleaning method and the extent that cleaning reduces exogenous hair contamination, the latter of which may also vary between subjects and hair type [14,26].

Table 4. Comparison of hair metal concentrations in children from the present study with levels reported in children (ages 4–14) in other studies that used a variety of different hair cleaning methods prior to analysis.

Location (Study) *	Cleaning Method [#]	Sub-Population (N) [&]	Mn µg/g	Pb µg/g	Cd µg/g	As µg/g
U.S., Ohio (this study) [a]	T/S, N/S	all (222)	0.109 (0.441)	0.152 (0.829)	0.007 (0.050)	0.018 (0.049)
Italy [12], [a]	T/S, N/S	all (501)	0.098 (0.139)	NA	NA	NA
Spain [32], [b]	T/S, E/S	males (96) females (124) combined (220)	NA NA 0.137 (NA)	NA NA 0.14 (NA)	0.003 (0.003-0.004) 0.006 (0.004-0.007) NA	NA NA 0.017 (NA)
Brazil [21], [a]	T/S	exposed males (34) exposed females (36)	12.1 (9.9) 12.4 (13.4)	NA NA	NA NA	NA NA
Brazil [33], [a]	T/S	referents (44) exposed (88)	NA NA	2.09 (2.06) 1.26 (3.70)	NA NA	NA NA
Spain [31], [c]	T/S	urban area (45) industrial area (54)	0.26 (0.90) 0.18 (0.28)	0.32 (0.30) 1.59 (3.01)	<0.03 <0.03	NA NA
Tibet [34], [b]	T	exposed (22) unexposed 1 (24) unexposed 2 (24)	4.28 (5.36) 2.87 (3.05) 2.44 (3.00)	NA NA NA	NA NA NA	NA NA NA
Bangladesh [17], [d]	T	all (207)	5.0 (1.4–23)	1.6 (0.50–6.4)	0.029 (0.0008–0.150)	0.53 (0.14–2.9)
U.S., Ohio [2], [b]	T	all (370)	0.417 (0.002)	NA	NA	NA
Greece [35], [b]	T	urban area 1 (11) urban area 2 (21) suburban area (19)	NA NA NA	0.78 (1.47) 1.29 (6.86) 0.60 (0.67)	0.014 (0.028) 0.023 (0.021) 0.015 (0.037)	0.020 (0.029) 0.036 (0.011) 0.026 (0.009)
Mexico [22], [b]	T	unexposed (93) exposed (79)	0.57 (0.49–0.66) 12 (10.7–13.8)	NA NA	NA NA	NA NA
Spain [36], [d]	A/S	all (648)	0.33 (0.12–0.94)	0.70 (0.17–4.28)	0.018 (0.004–0.079)	0.07 (<0.05–0.26)
Italy [24], [a]	A/S	males (130) females (94)	0.27 (0.25) 0.31 (0.27)	0.78 (0.76) 0.79 (0.80)	0.03 (0.05) 0.03 (0.05)	<0.01 <0.01
Russia [37], [a]	A	unexposed (84) exposed (82)	2.25 (3.77) 1.60 (3.51)	1.55 (2.98) 2.48 (4.20)	0.12 (0.18) 0.11 (0.14)	0.030 (0.034) 0.020 (0.042)
Vietnam [25], [c]	A	control males (5) control females (4) exposed males (22) exposed female (44)	NA NA NA NA	NA NA NA NA	NA NA NA NA	0.31 (0.07) 0.36 (0.17) 2.76 (2.54) 5.59 (7.90)
Quebec [23], [b]	U	males (148) females (164)	0.75 (NA) 0.8 (NA)	NA NA	NA NA	NA NA

* Study citation number and data type; [a] = data are median (SD); [b] = data are geometric mean (SD, geometric SD, or 95% CI); [c] = data are mean (SD); [d] = data are median (5th–95th percentile). NA = not analyzed or reported. [#] Hair cleaning methods, T = Triton, N = nitric acid, /S = sonicated, E = ethanol, A = acetone, U = uncleaned. [&] Study sub-population by gender and study sub-site location when reported (all = both males and females across study sub-sites).

4.5. Hair Metal Levels as a Biomarker of Exposure and Associated Health Effects

Studies have reported mixed results regarding the extent that hair metal levels are associated with metal levels in environmental media (e.g., water, soil, dust, and airborne particles, etc.), or biomarkers of endogenous metal exposure (blood, urine, nails). Lucas et al. [12] reported low but statistically significant correlations between children's hair Mn and Mn levels in household dust (ρ's ~ 0.27, $p < 0.001$) and airborne particles ($\rho = 0.126$, $p < 0.05$). Bouchard et al. [13] and Oulhote et al. [23] reported that hair Mn levels were higher in Canadian children exposed to Mn-contaminated water compared to children living in homes with a private well with lower water Mn levels, while Skröder et al. [17] reported no correlation between Mn levels in water and hair Mn levels in Bangladeshi children.

Studies have also reported mixed results on the associations between metal levels in hair and other exposure biomarkers. Hair As levels have been shown to reflect the internal body burden of As, based on strong correlations between hair As levels with As levels in erythrocytes ($\rho = 0.73$, $p < 0.001$) and urine ($\rho = 0.66$, $p < 0.001$) [17], while hair Cd and Pb levels are not generally recognized as reliable predictors of exposure and internal dose [7,11,17]. A number of studies have reported no association between hair Mn and Mn levels in blood [2,12,38] or erythrocytes [17], while others reported low but significant correlations between hair and blood Mn levels [22,39], and associations between hair Mn and Mn levels in fingernails ($\rho = 0.247$, $p < 0.001$) [12]. In their recent study, Skröder et al. [17] concluded that levels of Mn in hair do not reflect the actual internal Mn dose in Bangladeshi children, but the authors acknowledged that their findings strongly pointed to significant external contamination of hair from Mn-contaminated water, which would preclude the ability to even test whether hair Mn reflects the internal Mn burden. Finally, there is substantial evidence showing the hair Mn levels are associated with a number of neurodevelopmental health effects, including reduced IQ, learning, memory, and perceptual reasoning, and greater hyperactive and oppositional behaviors [2,13,23,38–41], leading Coetzee et al. [3] to conclude in their recent review that hair was the most consistent and valid biomarker of manganese exposure and associated neurodevelopmental health effects in children.

The present study had several limitations. First, the parent C8 Health Project Neurobehavioral Development Follow-up study, which was investigating the neurodevelopmental health effects of perfluoroocatnoate (PFOA) exposure in southeastern Ohio and northwestern West Virginia [27], did not assess metal exposures in the subjects' environment (e.g., air, dust, water) or in other biomarker tissues (e.g., blood), thereby limiting our ability to interpret the hair metal levels reported here as a biomarker of environmental metal exposures or internalized body burden. Second, sequential hair segments were available only for female subjects, and not males, precluding assessment of a sex–hair segment interaction in our statistical models.

Acknowledgments: We thank Patricia Leung for analytical assistance, and Ashley M. Pajak for assistance with the statistical analysis plan. This study was supported in part by grants R21 ES019643 and R01 ES019222 from the National Institute of Environmental Health Sciences.

Author Contributions: T.J. performed hair metal analyses and assisted in data analysis and interpretation, and drafting the manuscript. C.R.S. conceived of and supervised the parent study on chemical exposure and neurodevelopment, including hair sample collection, and participated in data analysis and interpretation for the present study. D.R.S. participated in statistical analyses, data interpretation, and drafting the manuscript.

Conflicts of Interest: The authors declare no conflict of interest.

References

1. Smith, D.; Gwiazda, R.; Bowler, R.; Roels, H.; Park, R.; Taicher, C.; Lucchini, R. Biomarkers of Mn Exposure in Humans. *Am. J. Ind. Med.* **2007**, *50*, 801–811. [CrossRef] [PubMed]
2. Haynes, E.N.; Sucharew, H.; Kuhnell, P.; Alden, J.; Barnas, M.; Wright, R.O.; Parsons, P.J.; Aldous, K.M.; Praamsma, M.L.; Beidler, C.; et al. Manganese Exposure and Neurocognitive Outcomes in Rural School-Age Children: The Communities Actively Researching Exposure Study (Ohio, USA). *Environ. Health Perspect.* **2015**, *123*. [CrossRef] [PubMed]
3. Coetzee, D.J.; McGovern, P.M.; Rao, R.; Harnack, L.J.; Georgieff, M.K.; Stepanov, I. Measuring the Impact of Manganese Exposure on Children's Neurodevelopment: Advances and Research Gaps in Biomarker-Based Approaches. *Environ. Health* **2016**, *15*, 91. [CrossRef] [PubMed]
4. Arora, M.; Austin, C.; Sarrafpour, B.; Hernańdez-Ávila, M.; Hu, H.; Wright, R.O.; Tellez-Rojo, M.M. Determining Prenatal, Early Childhood and Cumulative Long-Term Lead Exposure Using Micro-Spatial Deciduous Dentine Levels. *PLoS ONE* **2014**, *9*. [CrossRef] [PubMed]
5. Sanders, A.P.; Claus Henn, B.; Wright, R.O. Perinatal and Childhood Exposure to Cadmium, Manganese, and Metal Mixtures and Effects on Cognition and Behavior: A Review of Recent Literature. *Curr. Environ. Health Rep.* **2015**, *2*, 284–294. [CrossRef] [PubMed]
6. Zoni, S.; Lucchini, R.G. Manganese Exposure: Cognitive, Motor and Behavioral Effects on Children: A Review of Recent Findings. *Curr. Opin. Pediatr.* **2013**, *25*, 255–260. [CrossRef] [PubMed]

7. Bergdahl, I.A.; Skerfving, S. Biomonitoring of Lead Exposure—Alternatives to Blood. *J. Toxicol. Environ. Health Part A* **2008**, *71*, 1235–1243. [CrossRef] [PubMed]
8. Fowler, B.A. Monitoring of Human Populations for Early Markers of Cadmium Toxicity: A Review. *Toxicol. Appl. Pharmacol.* **2009**, *238*, 294–300. [CrossRef] [PubMed]
9. Roels, H.A.; Hoet, P.; Lison, D. Usefulness of Biomarkers of Exposure to Inorganic Mercury, Lead, or Cadmium in Controlling Occupational and Environmental Risks of Nephrotoxicity. *Ren. Fail.* **1999**, *21*, 251–262. [CrossRef] [PubMed]
10. Lanphear, B.P.; Hornung, R.; Khoury, J.; Yolton, K.; Baghurst, P.; Bellinger, D.C.; Canfield, R.L.; Dietrich, K.N.; Bornschein, R.; Greene, T.; et al. Low-Level Environmental Lead Exposure and Children's Intellectual Function: An International Pooled Analysis. *Environ. Health Perspect.* **2005**, *113*, 894–899. [CrossRef] [PubMed]
11. *Measuring Lead Exposure in Infants, Children, and other Sensitive Populations*; National Academies Press: Washington, DC, USA, 1993.
12. Lucas, E.L.; Bertrand, P.; Guazzetti, S.; Donna, F.; Peli, M.; Jursa, T.P.; Lucchini, R.; Smith, D.R. Impact of Ferromanganese Alloy Plants on Household Dust Manganese Levels: Implications for Childhood Exposure. *Environ. Res.* **2015**, *138*, 279–290. [CrossRef] [PubMed]
13. Bouchard, M.F.; Sauvé, S.; Barbeau, B.; Legrand, M.; Brodeur, M.È.; Bouffard, T.; Limoges, E.; Bellinger, D.C.; Mergler, D. Intellectual Impairment in School-Age Children Exposed to Manganese from Drinking Water. *Environ. Health Perspect.* **2011**, *119*, 138–143. [CrossRef] [PubMed]
14. Eastman, R.R.; Jursa, T.P.; Benedetti, C.; Lucchini, R.G.; Smith, D.R. Hair as a Biomarker of Environmental Manganese Exposure. *Environ. Sci. Technol.* **2013**, *47*, 1629–1637. [CrossRef] [PubMed]
15. Crinella, F.M. Does Soy-Based Infant Formula Cause ADHD? Update and Public Policy Considerations. *Expert Rev. Neurother.* **2012**, *12*, 395–407. [CrossRef] [PubMed]
16. Branco, V.; Caito, S.; Farina, M.; Teixeira da Rocha, J.; Aschner, M.; Carvalho, C. Biomarkers of Mercury Toxicity: Past, Present, and Future Trends. *J. Toxicol. Environ. Health Part B* **2017**, *20*, 119–154. [CrossRef] [PubMed]
17. Skröder, H.; Kippler, M.; Nermell, B.; Tofail, F.; Levi, M.; Rahman, S.M.; Raqib, R.; Vahter, M. Major Limitations in Using Element Concentrations in Hair as Biomarkers of Exposure to Toxic and Essential Trace Elements in Children. *Environ. Health Perspect.* **2017**, *125*, 67021. [CrossRef] [PubMed]
18. Harkey, M.R. Anatomy and Physiology of Hair. *Forensic Sci. Int.* **1993**, *63*, 9–18. [CrossRef]
19. Robbins, C.R. *Chemical and Physical Behavior of Human Hair*, 5th ed.; Springer: Berlin/Heidelberg, Germany, 2012.
20. Razagui, I.B.-A. A Comparative Evaluation of Three Washing Procedures for Minimizing Exogenous Trace Element Contamination in Fetal Scalp Hair of Various Obstetric Outcomes. *Biol. Trace Elem. Res.* **2008**, *123*, 47–57. [CrossRef] [PubMed]
21. Menezes-Filho, J.A.; de Carvalho-Vivas, C.F.; Viana, G.F.S.; Ferreira, J.R.D.; Nunes, L.S.; Mergler, D.; Abreu, N. Elevated Manganese Exposure and School-Aged Children's Behavior: A Gender-Stratified Analysis. *Neurotoxicology* **2014**, *45*, 293–300. [CrossRef] [PubMed]
22. Hernández-Bonilla, D.; Schilmann, A.; Montes, S.; Rodríguez-Agudelo, Y.; Rodríguez-Dozal, S.; Solís-Vivanco, R.; Ríos, C.; Riojas-Rodríguez, H. Environmental Exposure to Manganese and Motor Function of Children in Mexico. *Neurotoxicology* **2011**, *32*, 615–621. [CrossRef] [PubMed]
23. Oulhote, Y.; Mergler, D.; Barbeau, B.; Bellinger, D.C.; Bouffard, T.; Brodeur, M.-È.; Saint-Amour, D.; Legrand, M.; Sauvé, S.; Bouchard, M.F. Neurobehavioral Function in School-Age Children Exposed to Manganese in Drinking Water. *Environ. Health Perspect.* **2014**, *122*, 1343–1350. [CrossRef] [PubMed]
24. Dongarrà, G.; Lombardo, M.; Tamburo, E.; Varrica, D.; Cibella, F.; Cuttitta, G. Concentration and Reference Interval of Trace Elements in Human Hair from Students Living in Palermo, Sicily (Italy). *Environ. Toxicol. Pharmacol.* **2011**, *32*, 27–34. [CrossRef] [PubMed]
25. Hanh, H.T.; Kim, K.-W.; Bang, S.; Hoa, N.M. Community Exposure to Arsenic in the Mekong River Delta, Southern Vietnam. *J. Environ. Monit.* **2011**, *13*, 2025. [CrossRef] [PubMed]
26. Stauber, J.L.; Florence, T.M. Manganese in Scalp Hair: Problems of Exogenous Manganese and Implications for Manganese Monitoring in Groote Eylandt Aborigines. *Sci. Total Environ.* **1989**, *83*, 85–98. [CrossRef]
27. Stein, C.R.; Savitz, D.A.; Bellinger, D.C. Perfluorooctanoate and Neuropsychological Outcomes in Children. *Epidemiology* **2013**, *24*, 590–599. [CrossRef] [PubMed]

28. Rugless, F.; Bhattacharya, A.; Succop, P.; Dietrich, K.N.; Cox, C.; Alden, J.; Kuhnell, P.; Barnas, M.; Wright, R.; Parsons, P.J.; et al. Childhood Exposure to Manganese and Postural Instability in Children Living near a Ferromanganese Refinery in Southeastern Ohio. *Neurotoxicol. Teratol.* **2014**, *41*, 71–79. [CrossRef] [PubMed]

29. Agency for Toxic Substances and Disease Registry (ATSDR). *Health Consultation: Marietta Area Air Investigation Marietta, Ohio*; ATSDR: Atlanta, GA, USA, 2009.

30. Da Silva, J.J.R.F.; Williams, R.J.P. *The Biological Chemistry of the Elements: The Inorganic Chemistry of Life*; Oxford University Press: Oxford, UK, 2001.

31. Torrente, M.; Colomina, M.T.; Domingo, J.L. Metal Concentrations in Hair and Cognitive Assessment in an Adolescent Population. *Biol. Trace Elem. Res.* **2005**, *104*, 215–222. [CrossRef]

32. Molina-Villalba, I.; Lacasaña, M.; Rodríguez-Barranco, M.; Hernández, A.F.; Gonzalez-Alzaga, B.; Aguilar-Garduño, C.; Gil, F. Biomonitoring of Arsenic, Cadmium, Lead, Manganese and Mercury in Urine and Hair of Children Living near Mining and Industrial Areas. *Chemosphere* **2015**, *124*, 83–91. [CrossRef] [PubMed]

33. Menezes-Filho, J.A.; de Sousa Viana, G.F.; Paes, C.R. Determinants of Lead Exposure in Children on the Outskirts of Salvador, Brazil. *Environ. Monit. Assess.* **2012**, *184*, 2593–2603. [CrossRef] [PubMed]

34. Guo, Y.; Li, H.; Yang, L.; Li, Y.; Wei, B.; Wang, W.; Gong, H.; Guo, M.; Nima, C.; Zhao, S.; et al. Trace Element Levels in Scalp Hair of School Children in Shigatse, Tibet, an Endemic Area for Kaschin-Beck Disease (KBD). *Biol. Trace Elem. Res.* **2017**, *180*, 15–22. [CrossRef] [PubMed]

35. Evrenoglou, L.; Partsinevelou, S.A.; Stamatis, P.; Lazaris, A.; Patsouris, E.; Kotampasi, C.; Nicolopoulou-Stamati, P. Children Exposure to Trace Levels of Heavy Metals at the North Zone of Kifissos River. *Sci. Total Environ.* **2013**, *443*, 650–661. [CrossRef] [PubMed]

36. Llorente Ballesteros, M.T.; Navarro Serrano, I.; Izquierdo Álvarez, S. Reference Levels of Trace Elements in Hair Samples from Children and Adolescents in Madrid, Spain. *J. Trace Elem. Med. Biol.* **2017**, *43*, 113–120. [CrossRef] [PubMed]

37. Drobyshev, E.J.; Solovyev, N.D.; Ivanenko, N.B.; Kombarova, M.Y.; Ganeev, A.A. Trace Element Biomonitoring in Hair of School Children from a Polluted Area by Sector Field Inductively Coupled Plasma Mass Spectrometry. *J. Trace Elem. Med. Biol.* **2017**, *39*, 14–20. [CrossRef] [PubMed]

38. Menezes-Filho, J.A.; Novaes, C.d.O.; Moreira, J.C.; Sarcinelli, P.N.; Mergler, D. Elevated Manganese and Cognitive Performance in School-Aged Children and Their Mothers. *Environ. Res.* **2011**, *111*, 156–163. [CrossRef] [PubMed]

39. Torres-Agustín, R.; Rodríguez-Agudelo, Y.; Schilmann, A.; Solís-Vivanco, R.; Montes, S.; Riojas-Rodríguez, H.; Cortez-Lugo, M.; Ríos, C. Effect of Environmental Manganese Exposure on Verbal Learning and Memory in Mexican Children. *Environ. Res.* **2013**, *121*, 39–44. [CrossRef] [PubMed]

40. Bouchard, M.; Mergler, D.; Baldwin, M.; Panisset, M.; Bowler, R.; Roels, H.A. Neurobehavioral Functioning after Cessation of Manganese Exposure: A Follow-up after 14 Years. *Am. J. Ind. Med.* **2007**, *50*, 831–840. [CrossRef] [PubMed]

41. Wright, R.O.; Amarasiriwardena, C.; Woolf, A.D.; Jim, R.; Bellinger, D.C. Neuropsychological Correlates of Hair Arsenic, Manganese, and Cadmium Levels in School-Age Children Residing near a Hazardous Waste Site. *Neurotoxicology* **2006**, *27*, 210–216. [CrossRef] [PubMed]

Article

Biomonitoring of Urinary Benzene Metabolite SPMA in the General Population in Central Italy

Giovanna Tranfo [1], Daniela Pigini [1], Enrico Paci [1], Lisa Bauleo [2,*], Francesco Forastiere [2] and Carla Ancona [2]

[1] INAIL Research, Department of Occupational and Environmental Medicine, Epidemiology and Hygiene, Via di Fontana Candida 1, 00078 Monte Porzio Catone, Italy; g.tranfo@inail.it (G.T.); d.pigini@inail.it (D.P.); e.paci@inail.it (E.P.)

[2] Lazio Regional Health Service, Department of Epidemiology, Via Cristoforo Colombo, 112, 00147 Rome, Italy; fran.forastiere@gmail.com (F.F.); c.ancona@deplazio.it (C.A.)

* Correspondence: l.bauleo@deplazio.it; Tel.: +39-06-997-221-78; Fax: +39-06-997-211-1

Received: 7 June 2018; Accepted: 10 July 2018; Published: 11 July 2018

Abstract: Background: Benzene is an important component of cigarette smoke and car exhaust. Products containing benzene in concentrations greater than 0.1% are prohibited in Europe, but 1% of benzene is still allowed in gasoline. The purpose of the study was to assess the levels of urine benzene biomarkers in a sample of the general population not occupationally exposed to benzene, resident in the period 2013–2014 in Central Italy, compared to other groups. Methods: The urinary levels of the benzene metabolites S-phenyl-mercapturic acid (SPMA) and cotinine (nicotine metabolite) were determined by means of HPLC with mass spectrometric detection in 1076 subjects. Results: The median SPMA value in smokers was 1.132 μg/g of creatinine while in non-smokers it was 0.097 μg/g of creatinine, and the 95th percentile results were seven times higher. Conclusion: The main source of benzene exposure in the studied population was active smoking, however, non-smokers were also exposed to airborne benzene concentrations. The concentration ranges found in this study can be used as a background reference for occupational exposure assessment to benzene by means of SPMA biomonitoring.

Keywords: human biomonitoring; urine; non-occupational exposure; S-phenyl-mercapturic acid; HPLC-MS/MS

---

1. Introduction

Benzene is an important chemical compound used in the manufacturing of polymers, plastics, rubber, dyes, detergents, and other products, and it is a ubiquitous pollutant of indoor and outdoor air. In fact, even though in Europe the sale to or use by consumers of products that contain benzene in concentrations greater than 0.1% by weight is restricted by the European regulation on Registration and Evaluation of Chemicals (REACH), human exposure to this substance can still be due to industrial use, combustions of organics and natural gas, and motor fuels, as 1% in volume of benzene is allowed in gasoline according to Directive 98/70/EC [1].

The Directive 2008/50/EC on ambient air quality and cleaner air for Europe sets objectives for ambient air quality in order to protect human health and the environment as a whole; these objectives relate to sulphur dioxide, nitrogen dioxide, particulate matter, lead, benzene, and carbon monoxide. For benzene, the limit value has been set at 5 μg/m^3 as the annual average since 1 January 2010 [2].

Benzene exposure for industrial sectors can be easily assessed if the quantity of material used and the working environment are well defined, while for the general population, it is harder to quantify because individual lifestyles are extremely variable, ambient weather conditions can impact exposure, and living environments are different [3].

The air concentrations for different employment sectors can range from 1 µg/m^3 to over 1000 µg/m^3 (aviation workers, service station workers, bus drivers, police, urban workers, fishermen, and shoe production workers), but in the last decades, in developed countries, airborne benzene has progressively decreased as a consequence of preventive actions [4]; furthermore, outdoor ambient air concentrations of benzene are dependent on geographical location (i.e., rural versus urban). Benzene is also an important component of cigarette smoke. The contribution of environmental tobacco smoke is a major cause of indoor benzene exposure and depends upon local restrictions on smoking. Other significant sources of indoor benzene exposure are incense burning and traffic emissions.

Benzene is a known carcinogen and causes hematotoxicity at exposure levels below 1 ppm (3.25 mg/m^3) [5], which is the occupational limit value in the Recommendation from the Scientific Committee on Occupational Exposure Limits for benzene [6]. Exposure to a time-weighted average (TWA) of 1–5 parts per million (ppm) benzene in ambient air for 40 years is associated with an increased risk of acute myeloid leukemia [7]. The capacity of individuals to metabolize benzene is modulated by genetic factors [8]. Benzene uptake occurs mainly by inhalation in both occupational and non-occupational exposure. Benzene is oxidized to benzene oxide (BO) from cytochrome P450 (CYP) oxidase. A small fraction of BO (about 1% of total benzene uptake at 0.1–10 ppm of exposure) is detoxified through conjugation with glutathione (GSH) by glutathione-S-transferases (GSTs), and excreted in urine as S-phenyl-mercapturic acid (SPMA) [9]. S-phenyl-mercapturic acid (SPMA) is a very specific urinary biomarker of benzene. Its half-life is in a range of 9–13 h, and its accumulation is not probable, therefore, it is a biomarker of recent exposure [3].

Human biomonitoring data integrates all sources of possible exposure to a chemical, but it does not provide information on a single route of exposure. Biomonitoring permits an assessment of the exposure by quantitating a dose biomarker in biological fluids. The knowledge of the concentration of a chemical substances in biological fluids (biomarkers) measured in subjects without occupational exposure, accounts for the biological variability of the examined population and of other factors like residence and lifestyle [3,10]; it is also important for the evaluation of action levels or biological exposure limit (both environmental and occupational).

Cigarette smoking is a major source of exposure to benzene in active smokers [4] and is also able to affect the levels of biological markers of exposure to benzene in non-smokers exposed to passive smoking [11,12].

It has been estimated that smokers receive about 90% of their benzene intake from smoking [7].

An Italian surveillance study showed that during the period from 2014–2017 in the population of 18 and 69-years-olds, non-smokers were the majority (56.4%), ex-smokers (17.6%) were the minority, and 26% of the population were active smokers [13].

Exposure to benzene of the general population has been reduced significantly in Europe both in the outdoors and indoors by lowering the benzene content in gasoline and prohibiting smoking in many public places; therefore, population biomarker values are changing over time, and must be periodically updated.

The objective of the present study was to provide a measure of the environmental benzene exposure, according to smoking status and occupation, during the period from 2013–2014 in a sample of the general population living in Central Italy with no declared occupational exposure to benzene.

2. Materials and Methods

2.1. Study Population

The present study is part of a larger biomonitoring study, in which samples were collected between May 2013 and December 2014 from a population randomly selected from the municipality registers of the area of Civitavecchia (Italy) from the about 130,000 inhabitants [14]. The study protocol was approved by the local ethics committee. The subjects who agreed to participate to the study gave a written informed consent and filled in a questionnaire for collecting information on age, lifestyle, and food habits, cigarette, cigar, or tobacco smoking, the starting age for smoking, the end age for smoking for ex-smokers, electronic cigarette smoking, passive smoking, drug use, working activities, hobbies, use of chemical products, and in particular the possible occupational exposure to benzene. Part of this information was collected for the purpose of studying exposure to different chemical pollutants and was not used in the present study. The group studied included 1076 subjects aged 35–69 years. Information about the occupation status (employed, nor employed, housewives, retired) of participants was assessed during the interview. Housewife was considered a separate occupation. The smoking status was assessed using the urinary concentration of cotinine, and the cutoff value for the definition of smoker was set at urinary cotinine $\geq$100 µg/g of creatinine [15].

2.2. Urine Sample Collection and Preparation

Fasting subjects collected the first urine of the morning in empty plastic sterile containers on the same day as the medical visit: 30 mL of each sample were transferred into 50 mL Teflon tubes, identified with the subject code, frozen at $-20\,^{\circ}$C, and later transported to the laboratory where they were stored at $-20\,^{\circ}$C until analysis. Urinary creatinine was determined by the method of Jaffè using alkaline picrate test with UV/Vis detection at 490 nm [16]. The samples having a creatinine concentration higher than 3 g/L or lower than 0.3 g/L were discarded and the corresponding volunteers were excluded from the study in accordance with the American Conference of Governmental Industrial Hygienists (ACGIH) recommendation (ACGIH, Cincinnati 2014) [17].

2.3. Analytical Method

The concentration of the benzene and nicotine metabolites, SPMA and cotinine, was determined by isotopic dilution HPLC-MS/MS according to an analytical method previously validated in our laboratories [15]. Briefly, 3 mL of urine were acidified at pH 2, in order to hydrolyze the precursor of SPMA, with the deuterium labeled internal standards. Solid phase extraction (SPE) purification was carried out on Sep-pack C18 cartridges in two steps in order to elute a first fraction containing the acidic metabolite (SPMA) and subsequently a second one at pH 8 containing the cotinine: the two fractions were then injected separately into the API 4000 HPLC-MS/MS system. The HPLC-MS/MS is a Series 200 LC quaternary pump (PerkinElmer, Norwalk, CT, USA) coupled with an AB/Sciex API 4000 triple-quadrupole mass spectrometry detector equipped with a Turbo Ion Spray (TIS) probe. A Sinergi Fusion C18 analytical column (150 $\times$ 4.6-mm, 4-µm) was used for the analysis of urine samples and for the calibration standards for SPMA and cotinine. The mobile phase was a linear gradient of acetonitrile and acetic acid 1.0% *v/v* in water, flow rate 600 µL/min. The total run time was 10 min for the SPMA and 5 min for the cotinine. The precursor$\rightarrow$product ionic transitions monitored were 238.1$\rightarrow$109.1 for SPMA and 240.1$\rightarrow$109.1 for SPMAd$_2$ (in the negative ion mode) and 177.3$\rightarrow$80.10 for cotinine and 180.3$\rightarrow$80.10 for cotinine-d$_3$ (in the positive ion mode).

The precision for SPMA at the calibration level of 2 µg/L is 3%, and at the lowest level it is 10%. The limits of detection (LOD) calculated using the approach based on the standard deviation of the response and the slope, and expressed as 3.3 σ/S, were 0.026 µg/L for SPMA and 12.41 µg/L for cotinine. The final concentrations of both analytes were expressed in µg/g of creatinine to normalize values with respect to urine dilution variability. SPMA data below the LOD accounted for 20% of the population, and they have been substituted with the value of $\frac{1}{2}$ LOD in order to calculate the geometric mean (GM) and perform other statistical analysis.

2.4. Statistical Analysis

Descriptive statistics were carried out, and the SPMA concentration is presented as geometric mean (GM) with its geometric standard deviation (GSD), 5th, 50th, and 95th percentile. The relation among SPMA urinary concentrations and demographic characteristics (gender, age, occupation) in non-smokers was assessed using a linear regression model in which the dependent variable was the log-transformed SPMA. In this case, the measure of risks is expressed in term of Geometric mean ratio (GMR). Confidence interval of 95% were calculated. SAS (SAS Institute Inc., Cary, NC, USA) and STATA version 13 (StataCorp, College Station, TX, USA) software programs were used for the statistical analyses.

3. Results

Table 1 describes the characteristic of the population sample by smoking status.

Table 1. Characteristics of the study population by smoking status.

Population Group		Total		Smokers		Non Smokers	
		1076		296		780	
Gender	Males	461	42.8%	117	39.5%	344	44.1%
	Females	615	57.2%	179	60.5%	436	55.9%
Age group (years)	35–44	233	21.7%	78	26.4%	155	19.9%
	45–54	335	31.1%	93	31.4%	242	31.0%
	55–64	323	30.0%	90	30.4%	233	29.9%
	35–69	185	17.2%	35	11.8%	150	19.2%
Occupation	Employed	576	53.5%	173	58.4%	403	51.7%
	Unemployed	53	4.9%	19	6.4%	34	4.4%
	Housewives	217	20.2%	58	19.6%	159	20.4%
	Retired/Disabled	229	21.3%	46	15.5%	184	23.6%

Active smokers were 27.5% of the sample, females 57.2% of the sample, and 53.5% were employed. In the study sample, active smokers were younger, and there were no subjects with declared occupational exposure to benzene as assessed from the ABC study interview.

Table 2 reports the distribution of SPMA concentrations expressed in µg/g creatinine (Geometric mean and <25th, 25th–50th, 50th–75th, >75th percentile cut-offs) by the main characteristics of the ABC sample according to smoking status, assessed by means of the cotinine concentration of the same urine samples. Values expressed in µg/L are reported in the Supplemental Material (Table S1). Reference values for the biological monitoring of occupational exposures are generally normalized on the basis of creatinine concentration or specific gravity to account for fluctuations in urine dilution. We preferred the use of creatinine as there are many more published results for comparison, and because this is the parameter used by the ACGIH.

Table 2. Urinary S-phenyl-mercapturic acid (SPMA) concentrations (μg/g creatinine).

Group		N.	Geometric Mean (GSD)	5th Percentile	50th Percentile	95th Percentile	Min–Max
All subjects		1076	0.139 (7.409)	<LOD	0.151	3.403	<LOD–15.487
Smokers (cotinine ≥ 100 μg/g creatinine)							
All		296	0.926 (4.708)	0.058	1.132	6.612	<LOD–15.487
Gender	Males	117	0.691 (4.669)	0.053	0.710	6.02	<LOD–15.487
	Females	179	1.120 (4.612)	0.078	1.505	7.464	<LOD–11.917
Age group (years)	35–44	78	0.800 (4.699)	0.052	1.101	5.739	<LOD–10.972
	45–54	93	0.806 (4.729)	0.069	0.967	6.088	<LOD–9.887
	55–64	90	1.262 (4.392)	0.107	1.701	9.150	<LOD–15.487
	35–69	35	0.841 (5.283)	0.081	1.259	6.166	<LOD–11.917
Occupation	Employed	173	0.767 (4.628)	0.06	0.940	5.981	<LOD–15.487
	Unemployed	19	0.846 (7.116)	<LOD	0.979	7.180	<LOD–7.948
	Housewives	58	1.533 (4.827)	0.095	2.447	8.355	<LOD–11.917
	Retired	46	0.785 (4.245)	0.102	1.103	3.858	<LOD–5.702
Non Smokers (cotinine < 100 μg/g creatinine)							
All		780	0.068 (5.236)	<LOD	0.097	0.699	<LOD–2.667
Gender	Males	344	0.059 (5.522)	<LOD	0.088	0.633	<LOD–2.667
	Females	436	0.075 (4.985)	<LOD	0.103	0.706	<LOD–2.380
Age group (years)	35–44	155	0.059 (5.475)	<LOD	0.079	0.597	<LOD–1.480
	45–54	242	0.058 (5.479)	<LOD	0.089	0.562	<LOD–1.283
	55–64	233	0.081 (4.668)	<LOD	0.115	0.671	<LOD–2.380
	35–69	150	0.075 (5.400)	<LOD	0.098	1.000	<LOD–2.667
Occupation	Employed	403	0.061 (5.399)	<LOD	0.094	0.658	<LOD–1.283
	Unemployed	34	0.080 (5.224)	<LOD	0.117	0.710	<LOD–1.480
	Housewives	159	0.077 (4.471)	<LOD	0.099	0.625	<LOD–1.463
	Retired	184	0.080 (5.149)	<LOD	0.105	0.907	<LOD–2.667

GSD: geometric standard deviation; LOD: Limit of detection, for SPMA 0.026 μg/L.

Among smokers, we found higher SPMA concentrations (μg/g creatinine) in females (GSD (SD) 1.120 (4.612)) and people aged 55–64 years (1.262 (4.392)); housewives showed the highest values ((1.533 (4.827) when compared with the other occupational categories. In non-smokers, the median SPMA concentration was 0.1 μg/g of creatinine, with the 95th percentile equal to 0.7 μg/g of creatinine. No particular differences in gender, age, and occupation categories were observed, although the highest value was found in the oldest /retired subjects (aged ≥ 55), while retired and unemployed showed the highest median.

Table 3 shows results from the multivariate approach; when adjusting for participants' gender, age, and occupation, we found that SPMA concentrations were higher in females compared to males (p = 0.053) while no differences were found between age groups and occupational categories.

Table 3. Association between urinary concentration of SPMA and gender, age, and occupation. GMR = geometric mean ratio.

Group		GMR	C.I. 95%			p-Value
Gender	Males	1.00				
	Females	1.28	1.00	-	1.65	0.053
Age group (years)	35–44	1.00				
	45–54	1.03	0.75	-	1.41	0.848
	55–64	1.29	0.92	-	1.80	0.144
	>65	1.03	0.66	-	1.61	0.898
Occupation	Employed	1.00				
	Unemployed	1.15	0.66	-	1.98	0.622
	Housewives	0.95	0.69	-	1.32	0.782
	Retired	1.27	0.88	-	1.85	0.208

4. Discussion

Human biomonitoring is a powerful tool used in national and international surveys to assess the integrated exposure of the population to xenobiotics from different sources. In the population studied, the main source of exposure to benzene was active smoking: the data show that the mean SPMA value in smokers is about ten times that of non-smokers. The highest SPMA level was found in female smokers: women present a higher concentration of metabolites both when normalized for the creatinine and when not (μg/L), even when considering that creatinine concentrations are usually lower than in men.

The assessment of benzene exposure by means of biological monitoring does not permit source apportionment as it is integrated information. In the study sample, there were no subjects with declared occupational exposure to benzene and, therefore, benzene metabolites found in non-smokers could derive from passive smoking, use of incense in the home, or exposure to traffic. The median SPMA concentration of non-smokers was 0.1 μg/g of creatinine, that refers only to 50% of the sample, while the 95th percentile was 0.7 μg/g of creatinine, that is seven times higher than the median. This means that efforts to avoid or reduce benzene exposure should be made at the individual level by subjects having the higher biomarker (exposure) values to reach the lower ones.

Non-smokers' values were further stratified by age and occupational classes to be available as reference or "controls" for studies on subjects with similar characteristics. The highest value was found in the oldest/retired subjects (aged $\geq$ 65), while unemployed show the highest median. A possible explanation for these differences can be found in life habits (all these subjects have more free time) or in metabolic parameters (slower metabolism, more fat tissue). The same exercise cannot be done for the smokers, as any difference would mainly be due to the number of cigarettes.

We also examined other published data regarding urinary concentrations of SPMA in subjects without occupational exposure, all analytically determined by means of HPLC-MS/MS and published from 2011 to 2015. Only seven papers reported such values, four of which referred to population studies [18–21] and three to controls in studies on occupational exposures [22–24].

Comparison of our results with data from the literature shows that other studies included significantly fewer subjects than this one. Mean and median SPMA urinary concentrations are comparable to those found in the five studies on subjects who were resident in Italy, while the geometric mean found in this study for non-smokers is lower than those reported for African [20] and Chinese subjects [19], where airborne benzene concentrations are apparently higher. Moreover, none of the recent studies explored different age or job classes. Results summarized in Tables are reported as Supplemental Material (Tables S2 and S3).

5. Conclusions

Tobacco smoke (active or passive) is confirmed to be the main source of benzene exposure for the general population, and it is the most important confounding factor in the biological monitoring of occupational and environmental exposure to benzene, especially when exposure levels are low and very low. However, benzene exposure can still be high in non-smokers, and individual exposure to benzene can be reduced.

Biological values determined in the general population can be used to understand whether the levels found in workers are indicative of a professional exposure. This is particularly important for substances for which health-based exposure-limit values have not been defined, and it is particularly useful to aid in the interpretation of biological monitoring for genotoxic carcinogens. The comparison should be made taking into account the possible confounding factors, like smoking, that can produce significantly higher levels of urinary SPMA than does the occupational exposure to benzene. The values presented for urinary SPMA in this paper are a sound basis for the definition of occupational exposure to benzene in non-smoking subjects who reside and work in Italy. For smokers, due to the low benzene exposure values reached because of the European legislation on occupational and environmental safety and health, the amount of SPMA produced by smoking could completely mask the airborne benzene contribution. However, SPMA levels higher than the 95th percentile of the smokers group found in this paper should be further investigated while also considering the cotinine concentration of the subjects.

It must be stressed that environmental levels of airborne benzene should further decrease in time and, therefore, biological values, especially for non-smokers, should be reassessed periodically.

Supplementary Materials: The following are available online at http://www.mdpi.com/2305-6304/6/3/37/s1, Table S1. Urinary SPMA concentrations (μg/L); Table S2. Summary of the population studies published from year 2011; Table S3. Summary of the occupational exposure studies published from year 2011 (data of controls).

Author Contributions: G.T., F.F., L.B., and C.A. conceived and designed the experiments. E.P. performed the experiments. L.B., D.P., and E.P. analyzed the data. L.B., C.A., D.P., and G.T. wrote the paper.

Funding: This research received no external funding.

Acknowledgments: The analyses described were carried out in the laboratories of the National Institute for Insurance against Accidents at Work (INAIL) Research, Department of Occupational and Environmental Medicine, Epidemiology and Hygiene, Italy, using the research funds of the Institute itself.

Conflicts of Interest: The authors declare no conflict of interest.

References

1. DIRETTIVA 98/70/CE del Parlamento Europeo e del Consiglio del 13 Ottobre 1998 relativa Alla Qualità Della Benzina e del Combustibile Diesel e Recante Modificazione Della Direttiva 93/12/CEE del Consiglio. GUCE, 28.12.1998. Available online: https://eur-lex.europa.eu/legal-content/IT/TXT/?uri=CELEX%3A31998L0070 (accessed on 5 July 2018).
2. Directive 2008/50/EC of the European Parliament and of the Council May 21, 2008 on Ambient Air Quality and Cleaner Air for Europe. GUCE, 11.6.2008. Available online: https://eur-lex.europa.eu/legalcontent/IT/TXT/?uri=uriserv%3AOJ.L_.2016.344.01.0001.01.ENG& toc=OJ%3ALcontent/IT/TXT/?uri=uriserv%3AOJ.L_.2016.344.01.0001.01.ENG&toc=OJ%3AL (accessed on 5 July 2018).
3. Arnold, S.M.; Angerer, J.; Boogaard, P.J.; Hughes, M.F.; O'Lone, R.B.; Robison, S.H.; Schnatter, A.R. The use of biomonitoring data in exposure and human health risk assessment: Benzene case study. *Crit. Rev. Toxicol.* **2013**, *43*, 119–153. [CrossRef] [PubMed]
4. Fustinoni, S.; Consonni, D.; Campo, L.; Buratti, M.; Colombi, A.; Pesatori, A.C.; Bonzini, M.; Bertazzi, P.A.; Vito Foà, V.; Garte, S.; et al. Monitoring Low Benzene Exposure: Comparative Evaluation of Urinary Biomarkers, Influence of Cigarette Smoking, and Genetic Polymorphisms. *Cancer Epidemiol. Prev. Biomark.* **2005**, *14*, 2237–2244. [CrossRef] [PubMed]
5. Fustinoni, S.; Campo, L.; Mercadante, R.; Manini, P. Methodological issues in the biological monitoring of urinary benzene and S-phenyl-mercapturic acid at low exposure levels. *J. Chromatogr. B* **2010**, *878*, 2534–2540. [CrossRef] [PubMed]
6. Recommendation from the Scientific Committee on Occupational Exposure Limits for Benzene. SCOEL/SUM/140, December 1991. Available online: http://ec.europa.eu/social/keyDocuments.jsp? advSearchKey=recommendation&mode=advancedSubmit&langId=en&policyArea=&type=0&country= 0&year=0 (accessed on 5 July 2018).
7. Johnson, E.S.; Langård, S.; Lin, Y.-S. A critique of benzene exposure in the general population. *Sci. Total Environ.* **2007**, *374*, 183–198. [CrossRef] [PubMed]
8. Carbonari, D.; Proietto, A.; Fioretti, M.; Tranfo, G.; Paci, E.; Papacchini, M.; Mansi, A. Influence of genetic polymorphism on t,t-MA/S-PMA ratio in 301 benzene exposed subjects. *Toxicol. Lett.* **2014**, *231*, 205–212. [CrossRef] [PubMed]
9. Carbonari, D.; Chiarella, P.; Mansi, A.; Pigini, D.; Iavicoli, S.; Tranfo, G. Biomarkers of susceptibility following benzene exposure: Influence of genetic polymorphisms on benzene metabolism and health effects. *Biomark. Med.* **2016**, *10*, 145–163. [CrossRef] [PubMed]
10. Aprea, C.; Sciarra, G.; Bozzi, N.; Pagliantini, M.; Perico, A.; Bavazzano, P.; Leandri, A.; Carrieri, M.; Scapellato, M.L.; Bettinelli, M.; et al. Reference Values of Urinary trans,trans-muconic Acid: Italian Multicentric Study. *Arch. Environ. Contam. Toxicol.* **2008**, *55*, 329–340. [CrossRef] [PubMed]

11. Protano, C.; Guidotti, M.; Manini, P.; Petyx, M.; La Torre, G.; Vitali, M. Benzene exposure in childhood: Role of living environments and assessment of available tools. *Environ. Int.* **2010**, *36*, 779–787. [CrossRef] [PubMed]

12. Fustinoni, S.; Campo, L.; Satta, G.; Campagna, M.; Ibba, A.; Tocco, M.G.; Atzeri, S.; Avataneo, G.; Flore, C.; Meloni, M.; et al. Environmental and lifestyle factors affect benzene uptake biomonitoringof residents near a petrochemical plant. *Environ. Int.* **2012**, *39*, 2–7. [CrossRef] [PubMed]

13. PASSI. Available online: http://www.epicentro.iss.it/passi/default.asp (accessed on 5 July 2018).

14. Ancona, C.; Bauleo, L.; Biscotti, G.; Bocca, B.; Caimi, S.; Cruciani, F.; Di Lorenzo, S.; Petrolati, M.; Pino, A.; Piras, G.; et al. On behalf of the ABC Study Group. A survey on lifestyle and level of biomarkers of environmental exposure in residents in Civitavecchia (Italy). *Ann. Ist. Super. Sanità* **2017**, *52*, 488–494.

15. Tranfo, G.; Pigini, D.; Paci, E.; Marini, F.; Bonanni, R.C. Association of exposure to benzene and smoking with oxidative damage to nucleic acids by means of biological monitoring of general population volunteers. *Environ. Sci. Pollut. Res.* **2016**, 1–10. [CrossRef] [PubMed]

16. Kroll, M.H.; Chesler, R.; Hagengruber, C. Automated determination of urinary creatinine without sample dilution: Theory and practice. *Clin. Chem.* **1986**, *32*, 446–452. [PubMed]

17. American Conference of Governmental Industrial Hygienists (ACGIH) recommendation (ACGIH, Cincinnati 2014). *Am. J. Ind. Med.* **1994**, *26*, 133–143.

18. Protano, C.; Andreoli, R.; Manini, P.; Vitali, M. Urinary trans, trans-muconic acid and S-phenyl-mercapturic acid are indicative of exposure to urban benzene pollution during childhood. *Sci. Total Environ.* **2012**, *435–436*, 115–123. [CrossRef] [PubMed]

19. Hecht, S.S.; Koh, W.-P.; Wang, R.; Chen, M.; Carmella, S.G.; Murphy, S.E.; Yuan, J.-M. Elevated Levels of Mercapturic Acids of Acrolein and Crotonaldehyde in the Urine of Chinese Women in Singapore Who Regularly Cook at Home. *PLoS ONE* **2015**, *10*, e0120023. [CrossRef] [PubMed]

20. Tuakuila, J. S-phenyl-mercapturic acid (S-PMA) levels in urine as an indicator of exposure to benzene in the Kinshasa population. *Int. J. Hyg. Environ. Health* **2013**, *216*, 494–498. [CrossRef] [PubMed]

21. Ranzi, A.; Fustinoni, S.; Erspamer, L.; Campo, L.; Gatti, M.G.; Bechtold, P.; Bonassi, S.; Trenti, T.; Goldoni, C.A.; Bertazzi, P.A.; et al. Biomonitoring of the general population living near a modern solid waste incinerator: A pilot study in Modena, Italy. *Environ. Int.* **2013**, *61*, 88–97. [CrossRef] [PubMed]

22. Fustinoni, S.; Campo, L.; Mercadante, R.; Consonni, D.; Mielzynska, D.; Bertazzi, P.A. A quantitative approach to evaluate urinary benzene and S-phenyl-mercapturic acid as biomarkers of low benzene exposure. *Biomarkers* **2011**, 1–12. [CrossRef]

23. Lovreglio, P.; Barbieri, A.; Carrieri, M.; Sabatini, L.; Fracasso, M.E.; Doria, D.; Drago, I.; Basso, A.; D'Errico, M.N.; Bartolucci, G.B.; et al. Minore validità del benzene urinario rispetto all'acido S-fenilmercapturico nel rilevare l'esposizione occupazionale ed ambientale a concentrazioni molto basse di benzene. *G. Ital. Med. Lav. Ergon.* **2011**, *33*, 117–124. [PubMed]

24. Campagna, M.; Satta, G.; Campo, L.; Flore, V.; Ibba, A.; Meloni, M.; Tocco, M.G.; Avataneo, G.; Flore, C.; Fustinoni, S.; Cocco, P. Biological monitoring of low-level exposure to benzene. *Med. Lav.* **2012**, *103*, 338–346. [PubMed]

 *toxics*

Review

A Review of Biomonitoring of Phthalate Exposures

Yu Wang [1], Hongkai Zhu [1] and Kurunthachalam Kannan [1,2,*]

[1] Wadsworth Center, New York State Department of Health, Albany, New York, NY 12201, USA; wangyu@mail.nankai.edu.cn (Y.W.); Hongkai.Zhu@health.ny.gov (H.Z.)

[2] Department of Environmental Health Sciences, School of Public Health, State University of New York at Albany, Albany, NY 12201, USA

[*] Correspondence: kurunthachalam.kannan@health.ny.gov; Tel.: +1-518-474-0015; Fax: +1-518-473-2895

Received: 31 January 2019; Accepted: 29 March 2019; Published: 5 April 2019

Abstract: Phthalates (diesters of phthalic acid) are widely used as plasticizers and additives in many consumer products. Laboratory animal studies have reported the endocrine-disrupting and reproductive effects of phthalates, and human exposure to this class of chemicals is a concern. Several phthalates have been recognized as substances of high concern. Human exposure to phthalates occurs mainly via dietary sources, dermal absorption, and air inhalation. Phthalates are excreted as conjugated monoesters in urine, and some phthalates, such as di-2-ethylhexyl phthalate (DEHP), undergo secondary metabolism, including oxidative transformation, prior to urinary excretion. The occurrence of phthalates and their metabolites in urine, serum, breast milk, and semen has been widely reported. Urine has been the preferred matrix in human biomonitoring studies, and concentrations on the order of several tens to hundreds of nanograms per milliliter have been reported for several phthalate metabolites. Metabolites of diethyl phthalate (DEP), dibutyl- (DBP) and diisobutyl- (DiBP) phthalates, and DEHP were the most abundant compounds measured in urine. Temporal trends in phthalate exposures varied among countries. In the United States (US), DEHP exposure has declined since 2005, whereas DiNP exposure has increased. In China, DEHP exposure has increased since 2000. For many phthalates, exposures in children are higher than those in adults. Human epidemiological studies have shown a significant association between phthalate exposures and adverse reproductive outcomes in women and men, type II diabetes and insulin resistance, overweight/obesity, allergy, and asthma. This review compiles biomonitoring studies of phthalates and exposure doses to assess health risks from phthalate exposures in populations across the globe.

Keywords: phthalate; DEHP; biomonitoring; human exposure; toxicity; reproductive

---

1. Introduction

Phthalates are diesters of phthalic acid (1,2-benzenedicarboxylic acid) and are synthetic organic chemicals used in industries as solvents, plasticizers, and additives in polyvinyl chloride (PVC) plastics or personal care products (PCPs) [1]. More than 25 phthalates are used in commercial applications, with each adding unique qualities to the product into which it is incorporated. Ten commonly used phthalates (Figure 1, Table 1) are dimethyl phthalate (DMP), diethyl phthalate (DEP), dibutyl phthalate (DBP), diisobutyl phthalate (DiBP), benzylbutyl phthalate (BzBP), dicyclohexyl phthalate (DCHP), di(2-ethylhexyl) phthalate (DEHP), di-n-octyl phthalate (DnOP), di-isononyl phthalate (DiNP), and di-isodecyl phthalate (DiDP). DEHP, one of the major phthalates in commerce, was first synthesized for use as a plasticizer in 1933 [2]. The application of DEHP as an additive in polyvinyl chloride (PVC) to impart the flexibility of plastic has made phthalates popular around the world. The addition of phthalates to PVC makes it not only flexible but also malleable and durable. PVC products may contain up to 50% (by weight) phthalates [1].

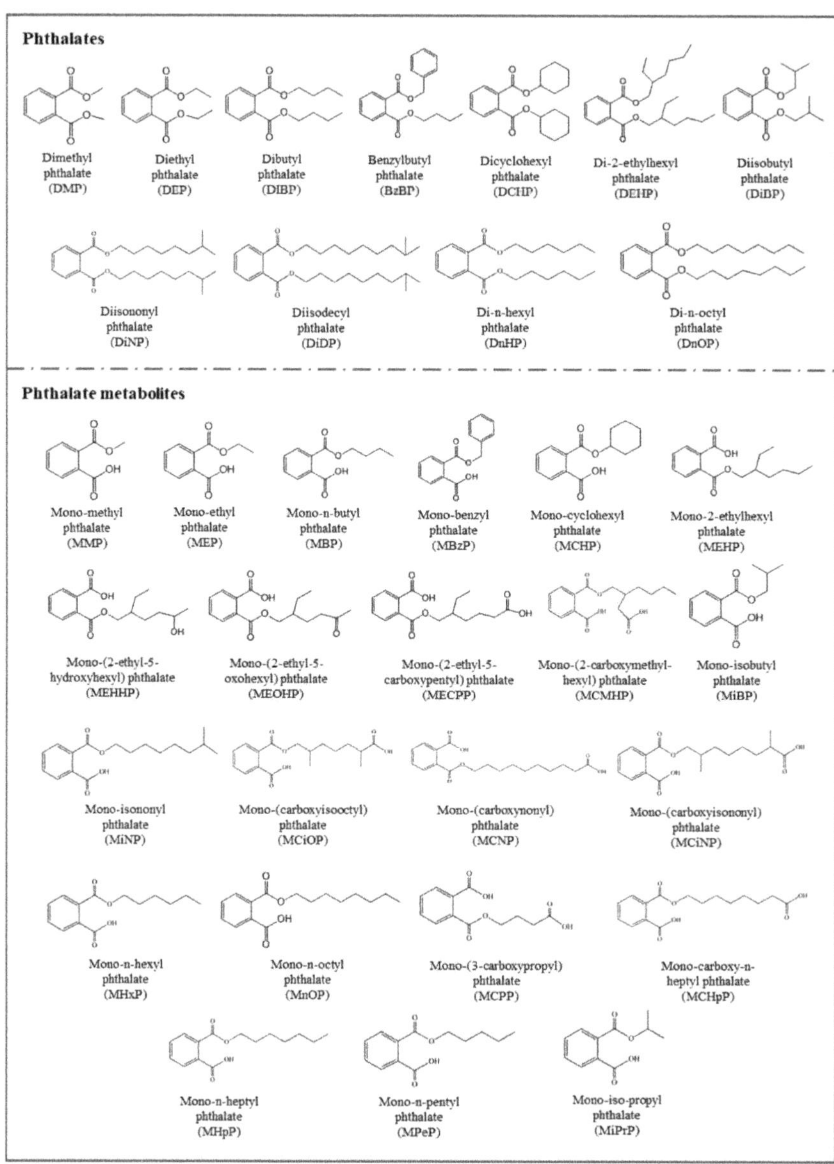

Figure 1. Chemical structures of major phthalates and their metabolites studied in the literature.

Table 1. Major phthalate diesters and their corresponding metabolites studied in the literature.

Parent Compounds	Abb.	Major Metabolites	Abb.
Dimethyl phthalate	DMP	Mono-methyl phthalate	MMP
Diethyl phthalate	DEP	Mono-ethyl phthalate	MEP
Dibutyl phthalate	DBP	Mono-n-butyl phthalate	MBP
Benzylbutyl phthalate	BzBP	Mono-benzyl phthalate (some mono-n-butyl phthalate)	MBzP
Dicyclohexyl phthalate	DCHP	Mono-cyclohexyl phthalate	MCHP
Di-2-ethylhexyl phthalate	DEHP	Mono-2-ethylhexyl phthalate	MEHP
		Mono-(2-ethyl-5-hydroxyhexyl) phthalate	MEHHP (5OH-MEHP)
		Mono-(2-ethyl-5-oxohexyl) phthalate	MEOHP (5oxo-MEHP)
		Mono-(2-ethyl-5-carboxypentyl) phthalate	MECPP (5cx-MEPP)
		Mono-(2-carboxymethyl-hexyl) phthalate	MCMHP (2cx-MMHP)
Diisobutyl phthalate	DiBP	Mono-isobutyl phthalate	MiBP
Diisononyl phthalate	DiNP	Mono-isononyl phthalate	MiNP
		Mono-(carboxyisooctyl) phthalate	MCiOP
Diisodecyl phthalate	DiDP	Mono-(carboxynonyl) phthalate	MCNP
		Mono-(carboxyisononyl) phthalate	MCiNP
Di-n-hexyl phthalate	DnHP	Mono-n-hexyl phthalate	MHxP
Di-*n*-octyl phthalate	DnOP	Mono-n-octyl phthalate	MnOP
		Mono-(3-carboxypropyl) phthalate	MCPP
		Mono-carboxy-n-heptyl phthalate	MCHpP
		Mono-n-heptyl phthalate	MHpP
		Mono-n-pentyl phthalate	MPeP
		Mono-iso-propyl phthalate	MiPrP

Currently, phthalates are used in many types of products, including building materials, automotive parts, medical devices, food packaging, cosmetics, perfumes, toys, teethers, adhesives, paints, floorings, lubricants, hair sprays, shampoos, soaps, nail polishes, and detergents [3–5]. The annual global production of phthalate was 4.7 million metric tons in 2006 [6,7] and ~8 million metric tons in 2015 [8]. In most commercial products, DEHP, DiNP, and BzBP are used as additives, and they easily migrate from those products into the environment through evaporation, leaching, and abrasion [9]. Phthalates have been measured in a range of environmental matrices, including sludge, dust, soil, air, and water [4], and are ubiquitous contaminants in the environment.

Phthalates are reproductive and developmental toxicants [10]. In laboratory animal studies, DEHP has been reported to affect the reproductive system and development [11,12]. Further, changes in hepatic structure and function and kidney function as well as disruption of thyroid signaling, immune function, and metabolic homeostasis were reported [13–16]. The US Environmental Protection Agency (EPA) classified DEHP and BzBP as probable and possible human carcinogens, respectively. European authorities have classified phthalates with three to six carbons in their backbone as Repr 1B Agents (i.e., presumed human reproductive toxicants) (https://echa.europa.eu/substance-information/-/substanceinfo/100.239.213).

Human exposure to phthalates arises mainly from ingestion, inhalation, and dermal absorption [17,18]. Human biomonitoring studies have measured parent phthalate in serum [19] and their metabolites in human urine [20,21], semen [22,23], and breast milk [24,25]. Studies have demonstrated that phthalate exposure is associated with oxidative stress in humans [26,27]. Some studies have linked phthalate exposure to premature thelarche [28,29], endometriosis [30,31], low semen quality [32], reduced testosterone levels [33], obesity, diabetes, and breast cancer [34,35]. Phthalates are regarded as endocrine-disrupting compounds [36]. One of the most significant effects of phthalates is in terms of fetal development and reproductive anomalies and is referred to as "phthalate syndrome" (e.g., developmental or testicular effects, insulin like factor 3 production) [37,38]. In addition, phthalate exposure might be linked to insulin resistance and obesity in human populations [39,40].

In 1999, the European Union (EU) temporarily banned the use of six phthalates in children's toys: DiNP, DEHP, DBP, BzBP, DiDP, and DnOP (http://europa.eu/rapid/press-release_IP-05-838_en. htm). Further, in 2009, these phthalates were restricted in toys in Europe (https://eur-lex.europa.eu/ legal-content/EN/TXT/?uri=CELEX%3A32009L0048). The US followed suit in 2008 by passing the Consumer Products Safety Improvement Act, which banned the same six phthalates in children's toys (https://www.cpsc.gov/Regulations-Laws--Standards/Statutes/The-Consumer-Product-Safety-Improvement-Act). Many industries began substituting alternative chemicals for phthalates in their products, and several substitutive phthalate and non-phthalate plasticizers are currently used in many products [41,42]. Although six phthalates are now restricted in children's products in the US and EU, they are unregulated and continue to be used in toys in many other parts of the world, including China and India. In addition, children continue to be exposed to phthalates in cosmetics and PCPs as well as in school supplies made of PVC, including notebooks and binders, art supplies, backpacks, lunchboxes, paperclips, and umbrellas (https://www.sustainableproduction.org/downloads/PhthalateAlternatives-January2011.pdf). Raincoats, boots, handbags, and soft plastic shoes also may contain phthalates.

A search on the basis of Web of Science Core Collection, BIOSIS Previews, Derwent Innovations Index, MEDLINE, and ScieELO Citation Index was carried out to identify studies relevant to biomonitoring and epidemiology on phthalates and phthalate metabolites. Topics of interest included studies on phthalates in urine, serum, and other biofluids. The search terms used were: phthalic acid/phthalates OR phthalate metabolites AND biomonitoring OR epidemiological studies. Publications between 2000 and 2018 were extracted. This review provides a summary of human biomonitoring studies of phthalate diesters and their monoester (primary) and oxidative (secondary) metabolites as well as select epidemiological studies that link phthalate exposure to health outcomes in human populations.

2. Sources of Phthalates

Owing to their widespread use in consumer products, phthalates are ubiquitous in the environment, and a variety of sources have been reported to contribute to human exposure. For the purpose of exposure analysis, phthalates have often been grouped as lower molecular weight (ester side-chain lengths, one to four carbons; DMP, DEP, and DBP), and higher molecular weight (ester side-chain lengths, five or more carbons; DEHP, DiNP, DiDP, and BzBP) phthalates [43]. The high molecular weight phthalates are used primarily in PVC polymers and plastisol applications, plastics, food packaging, and food processing materials, vinyl toys and vinyl floor coverings, and building products. The low molecular weight phthalates are often used in non-PVC applications, such as personal care products, paints, adhesives, and enteric-coated tablets [44]. BzBP, DEHP, DiNP, DBP, and DiBP are used in toys, bags, gloves, and plastic tubing for improving flexibility and making the polymeric products soft and malleable [4]. DMP and DEP are widely used in cosmetics, such as perfumes, aftershaves, shampoos, makeup, and nail care products [4]. Cosmetics and personal care products are the major sources of human exposure to low molecular weight phthalates. Food packaging plastic film contains phthalates (such as DBP and DEP) at levels of up to 10% by weight. Plasticizer migration occurs when food packaging films come in direct contact with foods, and fatty foods and high temperatures increase the migration [45]. Diet has been a major source of exposure to high molecular weight phthalates, especially DEHP. In particular, foods packaged in plastic/PVC materials contribute to exposure to DEHP in humans [46].

The major source of exposure to DEP—one of the major phthalates found in human urine—is cosmetics and personal care products [17]. Studies have reported elevated concentrations of phthalates in indoor air and dust [47]. In fact, among various contaminants measured in indoor dust, phthalates, especially DEHP and DEP, are the major contaminants in indoor dust and air [46]. Phthalates also were reported to occur in pharmaceuticals, especially in over-the-counter medications/syrups and in pills with enteric coatings [48,49]. Medical devices that are suspected to contain DEHP include intravenous (IV) storage bags, ventilator tubing, IV infusion sets, endotracheal tubes, IV infusion catheters, nasogastric tubes, blood storage bags, enteral and parenteral nutrition storage bags, blood administration sets, urinary catheters, PVC exam gloves, suction catheters, chest tubes, nasal cannula tubing, hemodialysis tubing, syringes, extracorporeal membrane oxygenation tubing, and cardiopulmonary bypass tubing [50].

Exposure doses to phthalates have been calculated through the ingestion of foods, air inhalation, and dust ingestion for the general population in the US (sampled during 2011–2014) (Table 2) [46]. Dust ingestion is a major source of exposure to phthalates in infants and toddlers, whereas diet is the major source for children and adults. The exposure doses are in the range of a sub to low µg/kg bw/d. Further details of exposure doses calculated through biomonitoring data are provided below.

Table 2. Human exposure doses to total phthalates for the US population through various pathways (µg/kg bw/d).

Exposure Route	Dust Ingestion	Dust Dermal Absorption	Personal Care Products (Dermal)	Diet	Indoor Air Inhalation
Infants (<1 y *)	1.12	0.001	0.0095	-	0.845
Toddlers (1–3 y)	1.7	0.0008	0.0059	-	0.423
Children (3–11 y)	0.468	0.0006	-	4.68	0.203
Teenagers (11–18 y)	0.291	0.0005	-	-	0.089
Adults (>18 y)	0.233	0.0002	0.013–0.49	1.03	0.07

* y = years old; "-" means not reported; data source: Tran and Kannan, 2015 [46].

3. Biomonitoring of Phthalates

Due to the ubiquitous occurrence and widespread exposure of phthalates, their metabolites are one of the most examined environmental chemicals in human biomonitoring studies. The reported half-life of phthalates diesters in blood plasma or urine of humans and rodents was less than 24 h. Several studies have reviewed pharmacokinetics of phthalate esters, and these studies have found rapid hydrolysis of diesters to monoesters in the gastrointestinal tract [1,2]. Binding of DEHP metabolites to blood plasma proteins, existence of biliary excretion, and enterohepatic circulation in humans have been suggested [2]. Nevertheless, urinary excretion has been the major elimination pathway of phthalates [2]. Urinary concentrations of phthalate metabolites are generally 5–20 times higher than that in lipid-rich compartments. For example, urinary concentrations of mono-2-ethylhexyl phthalate (MEHP), mono-isobutyl phthalate (MIBP), mono-ethyl phthalate (MEP), and mono-n-butyl phthalate (MBP) were 20–100 times those in blood or milk [24]. Phthalate metabolites have been measured in various body fluids, including urine [47,51], serum [52,53], semen [32,54], breast milk [55,56], and saliva [57] (Table 3). Phthalates can cross the placental barrier [58] and have been measured in amniotic fluid in human studies [59]. To date, studies that report partitioning of phthalates among various tissues and organs in an organism, at state-state exposure conditions, are not available. It is worth noting that a few earlier reviews have described biomonitoring of phthalates in humans [60]. Biomonitoring studies that report concentrations of phthalates metabolites are presented in Table 3.

Table 3. Reported concentrations of major phthalate metabolites in human specimens collected from various countries.

Matrix	Country/Region	Studied Population	Concentration						Reference
			MMP	MEP	MBP	MiBP	MDEHP	Unit	
Urine	Australia	30 non-occupational exposure		18.5	11.8	7.3	25.2	µg/L; median	[61]
Urine	Austria	251 children/adolescents; 272 adults; 72 senior citizens		25	10	28	15.5	µg/L; median	[62]
Urine	Belgium	261 persons		34.3	33.3	24.3	11.7	µg/L; median	[63]
Urine	Belgium	210 adolescents			38.5		52.7	µg/L; median	[64]
Urine	Belgium	123 men 138 women		37.6	31.3	26.2	17.1	µg/L; median	[65]
Urine	Belgium	25 persons		20.4	15.6	15.9	12.01	µg/L; median	[66]
Urine	Brazil	300 children (6–14 years old).	8.3	57.3	42.4	43.8	109	µg/L; median	[67]
Urine	Canada	3236 persons (6–49 years old)		49.1	23.8		40.9	µg/L; median	[68]
Urine	Canada	2000 women (first trimester)		32.02	11.59			µg/L; GM	[69]
Urine	Canada	80 infants				7.01	10.63	µg/L; median	[70]
Urine	China	108 young adults	31.8	37.5	67	57.2	65.3	µg/L; median	[71]
Urine	China	21 women 19 men	16.5	20.7	49.6	44	44.2	µg/L; median	[72]
Urine	China	430 children (208 girls and 222 boys)	15.7	4.14	21.9		14.3	µg/L; median	[73]
Urine	China	183 samples	14.6	22.1	63.5	57.1	76.1	µg/L; median	[51]
Urine	China	364 males (19–44 years old)		28.2	47.1		42	µg/L; median	[74]
Urine	China	39 children (5–9 years)		28.5	232	81.3	79.1	µg/L; median	[75]
Urine	Czech	117 women	ND	56.7			32.2	µg/L; median	[76]
Urine	Czech	120 children	ND	31.6			61.9	µg/L; median	[76]
Urine	Denmark	60 men		54.5	36.8	47.3	68.1	µg/L; median	[77]
Urine	Denmark	145 women		74	26	48	67	µg/L; GM	[78]
Urine	Denmark	143 children		28	39	74	99	µg/L; GM	[78]
Urine	Denmark	129 children		29	111		107	µg/L; median	[79]
Urine	Denmark	441 children		16.6	80.1	72.2	89.8	µg/L; median	[80]
Urine	Europe	171 individuals		49.9	0		4.5	µg/g CR; median	[42]
Urine	Europe	1335 children		34.4	38.4	45.4	47.6	µg/L; median	[81]
Urine	Europe	1347 mother		48.2	23.9	30.1	29.2	µg/L; median	[81]
Urine	Europe	1301 mother		72	18.3	23.3	22.4	µg/L; median	[82]
Urine	France	279 mothers		43.5	35.7	53.7	84.6	µg/L; median	[83]
Urine	Germany	634 individuals			109	35.4	45.3	µg/L; median	[84]
Urine	Germany	254 children					99.9	µg/L; median	[85]
Urine	Germany	53 women 32 men		90.2	181		83.3	µg/L; median	[86]
Urine	Germany	399 individuals			49.6	44.9	38.8	µg/L; median	[87]
Urine	Germany	120 females and 120 males			19.6	25.5	19.3	µg/L; median	[41]
Urine	Germany	30 males and 30 females (2015)	2.8	13.5	8.0	9.8	12.3	µg/L; median	[88]
Urine	Germany	30 males and 30 females (2007)	8.0	53.6	16.4	19.3	33.4	µg/L; median	[88]
Urine	Germany	111 children (48 girls and 63 boys)			53.6	74.9	130.1	µg/L; median	[89]
Urine	Germany	465 children (8–10 years old)			52.5	62.8	75.7	µg/L; median	[90]
Urine	Germany	599 children			95.6	94.3	174.6	µg/L; median	[91]
Urine	Germany	600 children (3–14 years old)			96		85	µg/L; median	[91]
Urine	Germany	207 infants (1–5 month)		12.1			1.1	µg/L; median	[92]
Urine	Germany	104 mothers		50.5		66.6	28.9	µg/L; median	[93]
Urine	Germany	104 children		39.1	56.5	103.9	55.7	µg/L; median	[93]
Urine	Greece	239 women		142	32.1	36.7	44.6	µg/L; median	[94]

Table 3. *Cont.*

Matrix	Country/Region	Studied Population	Concentration						Reference
			MMP	MEP	MBP	MiBP	MDEHP	Unit	
Urine	Greece	239 children		35.3	23.3	36	45.6	µg/L; median	[94]
Urine	Hungary	115 women	ND	55			32.4	µg/L; median	[76]
Urine	Hungary	117 children	ND	47			56.7	µg/L; median	[76]
Urine	India	15 women 7 men	8.6	150	13	18.3	77.9	µg/L; median	[72]
Urine	Iran	56 children and adolescent (6–18 years)	17.4	28.2	42.9		44.9	µg/L; median	[95]
Urine	Ireland	120 mothers		50.2	18.5	23.8	17	µg/g CR; GM	[96]
Urine	Ireland	120 children		38.7	26.1	41.4	32.8	µg/g CR; GM	[96]
Urine	Israel	205 adults (20–74 years old)			27.9	37.6	81.7	µg/L; median	[97]
Urine	Italy	83 women (2011)		73.1	38.8		15.6	µg/g CR; median	[42]
Urine	Italy	111 women (2016)		49.9	0		4.5	µg/g CR; median	[42]
Urine	Italy	83 females		61.0	32.5		10.5	µg/L; median	[98]
Urine	Italy	74 males		73.2	41.2		15.2	µg/L; median	[98]
Urine	Japan	8 women 27 men	18.2	16.4	17.7	7.5	35.1	µg/L; median	[72]
Urine	Japan	80 women (controls)		21.4	84.3		72.7	µg/L; median	[99]
Urine	Japan	57 women (cases)		39.6	87.2		89.3	µg/L; median	[99]
Urine	Japan	35 adults 1 children	33	18	36		5	µg/L; median	[100]
Urine	Japan	111 pregnant women	5.70	7.75	46.6		18.5	µg/L; median	[101]
Urine	Korea	39 children		19.2	107	53.4	145.6	µg/L; median	[102]
Urine	Korea	60 individuals	10	13.4	16.7	4.5	43.6	µg/L; median	[72]
Urine	Korea	25 adults		80	134	40.4	125.8	µg/L; median	[103]
Urine	Korea	305 women			41		23.7	µg/g CR; median	[104]
Urine	Korea	1646 elderly people			39.5		44.8	µg/L; median	[105]
Urine	Korea	6478 adults			44.2		88.2	µg/L; median	[106]
Urine	Korea	6003 adults			24.2		52.2	µg/L; median	[107]
Urine	Korea	171 children		2.71	12.4	5.25	12.3	µg/L; median	[108]
Urine	Korea	392 children					185	µg/L; median	[109]
Urine	Korea	265 mothers					67.4	µg/L; median	[109]
Urine	Korea	297 adults					55.7	µg/L; median	[109]
Urine	Kuwait	22 women 24 men	10.1	411	113	54.1	180.4	µg/L; median	[72]
Urine	Malaysia	19 women 10 men	6.3	18.6	10.5	10.8	27.5	µg/L; median	[72]
Urine	Netherlands	100 women	ND	112	43.2	41.3	61.8	µg/L; median	[110]
Urine	Norway	10 women	2	310	41.1	57	112.3	µg/L; median	[111]
Urine	Norway	61 adults		24.2	13.4	12.8		µg/L; median	[112]
Urine	Norway	116 pregnant women		55	25	20	26	µg/L; median	[113]
Urine	Portugal	112 children (4–18 years)		59.4	12.7	16.9	40.4	µg/L; median	[114]
Urine	Saudi Arabia	130 individuals	8.65	47.5	38.5	38.5	117.1	µg/L; median	[26]
Urine	Slovakia	129 occupational exposure			110	39.2	55.9	µg/L; median	[115]
Urine	Slovakia	68 occupational exposure population		201	103	61.4	82.7	µg/L; median	[116]
Urine	Slovakia	125 women	ND	54.8			36.7	µg/L; median	[76]
Urine	Slovakia	127 children	ND	39.6			82.8	µg/L; median	[76]
Urine	Slovakia	85 occupational exposure		78.5	85.6		21.5	µg/L; median	[117]
Urine	Slovakia	70 general population		78.1	96		14.7	µg/L; median	[117]

Table 3. *Cont.*

Matrix	Country/Region	Studied Population	Concentration						Reference
			MMP	MEP	MBP	MiBP	MDEHP	Unit	
Urine	Spain	391 pregnant women		246	27.1	28.4	87.8	µg/L; median	[118]
Urine	Spain	120 children		198.9			63	µg/g CR; GM	[119]
Urine	Spain	120 mothers		150.8			33.3	µg/g CR; GM	[119]
Urine	Sweden	314 men		41	47		48.4	µg/L; median	[120]
Urine	Sweden	38 women	1.2	35	46	16	35	µg/L; median	[24]
Urine	Taiwan	41 women and 19 men (21–67 years)	32.3		36.5		15.9	µg/L; median	
Urine	Taiwan	155 women	5.7	25.3	80		22.6	µg/L; median	
Urine	Taiwan	30 (children, age: 2)			100.4	17.2	195.8	µg/L; median	[121]
Urine	Taiwan	59 (children, age: 5)			75.2	25.2	148.9	µg/L; median	
Urine	Taiwan	100 women			72.3	12.5	96.8	µg/L; median	
Urine	U.S.	45 males (subfertile couples)		108	24.7		91.4	µg/L; median	[122]
Urine	U.S.	35 children		177.7	52.4	16.6	1025.9	µg/L; median	[123]
Urine	U.S.	7600–10,031 individuals	1.4	167	18.9	3.6	73.1	µg/g CR; median	[124]
Urine	U.S.	12–18 months toddlers		13.2-1388	6.6–2540		<1.7–47.3	µg/L; median	[125]
Urine	U.S.	186 persons in Northern Manhattan		199	36			µg/L; median	[126]
Urine	U.S.	446 pregnant women		41.1				µg/g CR; GM	[127]
Urine	U.S.	378 pregnant women		47	13.7	9.47	14	µg/L; median	[128]
Urine	U.S.	482 individuals		141	17.8	7.6	106.6	µg/L; median	[27]
Urine	U.S.	2772 adults		167			35.4	µg/g CR; median	[129]
Urine	U.S.	392 children of 6–11 years old		96.9			69.9	µg/g CR; median	[129]
Urine	U.S.	2350 individuals	1.8	194.4	20.7	3.7	73	µg/L; median	[130]
Urine	U.S.	33 lactating women					35.7–45.9	µg/L; median	[131]
Urine	U.S.	50 pregnant women (18–38)		61.5	18.2		31.1	µg/L; median	[132]
Urine	U.S.	406 men	4.5	145	14.5		5.2	µg/L; median	[133]
Urine	Vietnam	16 women 14 men	8.4	7.2	19.1	13.6	56.7	µg/L; median	[72]
Serum	Denmark	60 men		<LOD	ND	<LOD	8.4	µg/L; median	[77]
Serum	Sweden	36 women		0.5	0.5	0.5	0.5	µg/L; median	[24]
Seminal plasma	Denmark	60 men		<LOD	<LOD	ND	<LOD	µg/L; median	[77]
Breast milk	Denmark	65 women	0.1	0.9	4.3		9.5	µg/L; median	[25]
Breast milk	Finland	65 women	0.1	1.0	12.0		13.0	µg/L; median	[25]
Breast milk	Sweden	42 women		ND	0.5	ND	0.49)	µg/L; median	[24]
Milk	Switzerland	54 women			6.0	24.3	26.2	µg/L; median	[134]
Milk	U.S.	33 lactating women					0.3–0.7	µg/L; median	[131]
Nail	Belgian	10 individuals		64	74		138	µg/g CR; median	[135]
Nail	Norway	61 adults	89.7	104.8	89.3			µg/g CR; GM	[112]

MDEHP: Sum of five DEHP metabolites (MEHP, MEHHP, MEOHP, MECPP, and MCMHP); ND: Not detected; LOD: Limit of detection; LOQ: Limit of quantification; GM = Geometric mean; CR = Creatinine.

Phthalate diesters (parent compounds) were measured in blood plasma of women with endometriosis in India, and a significant association was found between phthalate exposure and the risk of developing endometriosis [136]. Similarly, studies have determined phthalates in serum samples of couples from Greenland, Poland, and Ukraine that showed that the DEHP levels were associated with reduced time to achieve pregnancy [137]. Phthalate diesters and their metabolites also have been measured in breast milk, serum, and urine from Swedish women [24]. In milk and serum samples, the concentrations of phthalate diesters and their metabolites were below the method limit of detection (0.12–3.0 µg/L). Detectable concentrations of phthalate metabolites, however, were found in urine (0.1–1000 µg/L). Measurements of phthalate diesters in breast milk and serum are prone to false positives due to background contamination. Medical devices, including blood collection devices and plastic containers that are used to collect and store samples, can contain phthalate diesters [49]. If the samples were to be analyzed for phthalate diesters, caution should be taken with the screening devices used to collect and store samples. A comprehensive review of challenges associated with low-level phthalate analysis in biological specimens has been published [17].

3.1. Phthalate Metabolites in Urine

Although microbial degradation of DEHP to MEHP in soils through lipase and esterase enzymes has been shown, environmental degradation/transformation of parent phthalates is slow [25,138]. Because phthalates have a short half-life in human bodies and are excreted quickly in urine as monoester metabolites, the metabolites are suitable biomarkers for human exposure to parent compounds. The half-life of phthalates in human bodies (in plasma and urine) is less than 24 h, and following metabolism, monoesters of phthalates are conjugated with glucuronide or sulfate and excreted in urine [139]. Analysis of metabolites in urine involves enzymatic deconjugation followed by purification. Assessment of human exposure to phthalates is based mainly on the measurement of their urinary monoester metabolites, although several secondary and oxidative metabolites have been reported to occur in human specimens [139]. For instance, DMP, DEP, and DBP undergo degradation/hydrolysis and form their corresponding monoesters, i.e., MMP, MEP and MBP, respectively. Both hydrolysis and oxidation products are formed from the metabolism of DEHP. MEHP, the hydrolysis product of DEHP, is not a major metabolite. The oxidative metabolites, MEOHP, MEHHP, MECPP, and MCMHP, however, are the major metabolites of DEHP and are appropriate biomarkers of exposure to this compound [21]. Some studies suggest, however, that MEHP is more toxic than are other oxidative metabolites. The general metabolic pathways of phthalate esters in humans are shown in Figure 2.

General Population Adults: A large number of studies have reported measurements of phthalate metabolites in human specimens collected from European (Germany, Netherlands, Denmark, Norway, Sweden, Greece, the Czech Republic, Hungary, Slovakia, and Spain) and Asian countries (Japan, China, South Korea, India, Taiwan, Vietnam, Saudi Arabia, Malaysia, and Kuwait) as well as from North American countries. The number of phthalate metabolites measured in urine samples varied considerably; as new analytical standards become made available commercially, more metabolites were added to the list of compounds measured in urine. Although a majority of the recent studies measure close to 20 phthalate metabolites, studies conducted a decade ago measured 10 or fewer metabolites of phthalates.

In general, the concentrations of the sum of 22 phthalate metabolites measured in human urine were on the order of several to hundreds of parts-per-billion (µg/L) [21]. In a majority of the biomonitoring studies, metabolites of DEHP, DEP, and DBP were the major compounds identified in urine, and the profile varied depending on the country. Urine samples collected from 32 men and 53 women (age: 7–64 years) from northern Bavaria (Germany) contained MBP (median: 181 µg/L), MEP (90.2 µg/L), and major DEHP metabolites, such as MEHHP (46.8 µg/L) and MEOHP (36.5 µg/L) [86]. The median concentrations of DEHP metabolites, namely, MEHP, MEOHP, and MEHHP, were 4.5, 28.3, and 35.9 µg/L, respectively, and these three metabolites were highly intercorrelated. The concentration ratios, MEHHP/MEHP, and MEOHP/MEHP, were calculated to be 8.2, and 5.9, respectively. These ratios suggest that MEHP is further oxidized to form MEHHP and MEOHP [86].

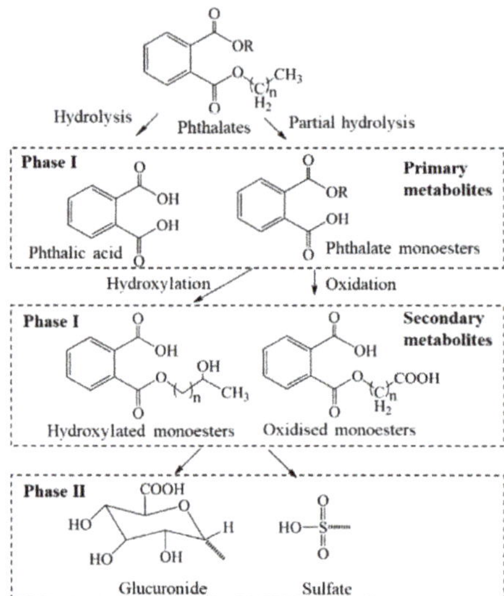

Figure 2. Metabolic pathways of phthalate esters in humans.

The urinary concentrations of phthalate metabolites in general populations vary among countries. Some of the highest concentrations of total phthalate metabolites were found in urine collected in 2006–2007 from Kuwaitis, with a maximum value of 19,300 µg/L and a median value for of 1050 µg/L [72]. The occurrence of phthalate metabolites was investigated in urine from Germans, and MBP was found at high concentrations (median: 49 µg/L) [87]. The median concentrations of phthalate metabolites in urine samples from Germany decreased significantly from 2002 to 2008 [41]. Similarly, urinary phthalate metabolites measured in 2015 were significantly lower than those in 2007 in Germany [88].

Biomonitoring studies in other European countries, including France [83], Belgium [65,66], Slovakia [117], and Norway [112], report phthalate metabolite concentrations in the range of 1–100 µg/L in urine from adults. MBP and DEHP metabolites were the predominant compounds found in those studies. Further, a comparative analysis of biomonitoring data in Europe suggested a significant decline in phthalate metabolite concentrations (especially MEP, MBP, MBzP, and DEHP metabolites) from 2011 to 2016 [42]. Several alternative plasticizers, however, are used as replacements for DEHP in European countries. Common alternatives include Hexamoll DINCH (DINCH), acetyl tributyl citrate (ATBC), dioctyl terephthalate (DOTP), 2,2,4-trimethyl 1,3-pentanediol diisobutyrate (TXIB), trioctyl trimellitate (TOTM), and di-(2-ethylhexyl) adipate (DEHA).

In North America, the distribution of phthalate metabolites in urine has been summarized in nationwide monitoring surveys. For example, the US National Health and Nutrition Examination Survey (NHANES) of the Centers for Disease Control and Prevention (CDC) showed that MEP, MEHP, MEHHP, and MEOHP concentrations in urine from adults >20 years of age were 167, 3.99, 18.8, and 12.6 µg/g creatinine (CR), respectively [129]. The NHANES program has measured 15 phthalate metabolites in urine. The weighted geometric mean concentration of 15 phthalate metabolites in the US general population was 125 µg/L for the samples collected in the period of 2007–2008. MEP was the major compound found in urine, accounting for >70% of the total concentrations, which was followed by mono-(2-ethyl-5-hydroxyhexyl) phthalate (MEHHP; ~18% of the total phthalate concentrations). The NHANES data for the US general population in the period of 2005–2006 showed that MCNP, a metabolite of DiDP, was found at a median concentration of 2.70 µg/L [140]. The updated NHANES report for 2013–2014 are available (https://wwwn.cdc.gov/Nchs/Nhanes/2013-2014/SSPHTE_H.htm). A 67% decline in DEHP exposure in the US population between 2005/6 and 2011/12 has been reported [141]. Several factors have been shown to affect exposures. The NHANES data showed that several phthalate urinary metabolites were higher in males, Hispanics, and African Americans [142]. The Human Biomonitoring Program of Health Canada measured 11 phthalate metabolites in urine samples of 3236 Canadians and found median MEP and MEHHP concentrations at 49.1 and 23.4 µg/L, respectively [68]. Since 2001, there has been clear evidence of a decline in DEP, DBP, and DEHP exposure in the US [115]. In contrast, urinary DiNP concentrations in the US population increased significantly during the period 2005/6–2011/12 (www.cdc.gov/exposurereport).

Urinary concentrations of phthalate metabolites have been reported for several Asian countries [97]. The measured concentrations in Asian countries were similar to those reported in Europe and North America, although the profiles were distinct. For instance, MBP and MiBP were the major metabolites found in urine from China, and their respective median concentrations were 61.2 and 51.7 µg/L [51]. Similar concentrations of MBP and the sum of DEHP metabolites were reported in urine from Nanjing city (47.1 and 42.0 µg/L) [74] and Taiwan (47.1 and 42.0 µg/L) [121]. In contrast, DEHP metabolites were predominant in urine from Japan, Malaysia, and Vietnam [72]. A nationwide survey of urine samples from 6478 adults during the period of 2012–2014 in Korea showed median urinary concentrations of DEHP metabolites (88.2 µg/L) that were twofold higher than that of MBP (44.2 µg/L) [106,107]. In Israel, phthalate metabolites were found in urine samples collected from 250 adults (ages 20–74), with median concentrations that ranged from 17.1 µg/L (MEOHP) to 37.6 µg/L (MiBP) [97]. DEHP exposure in the Chinese population has increased since 2001 [143].

The global distribution of major phthalate metabolites measured in urine from general populations is presented in Figure 3. Urine samples collected from Kuwait during 2006-2007 contained the highest median concentrations of MEP (411 µg/L), MBP (113 µg/L), and DEHP metabolites (180 µg/L), with a sum of median phthalate metabolite concentrations (median) at 1,050 µg/L [72]. This value is the highest among all countries studied. The profiles of phthalate metabolites varied, with MEP as the predominant metabolite in Indian and Kuwaiti urine samples (49% of the total), which were similar to those found in the US. In China (52%), MBP was the major metabolite found in urine. In Korea (46%), Japan (31%), and Vietnam (52%), DEHP metabolites were the dominant ones. MMP accounted for <8% of the total phthalate metabolite concentrations in all Asian countries, except for Japan, where it was 20%. Overall, MEP and DEHP metabolites were the major phthalate metabolites found in urine from most Asian countries, a pattern similar to that found in the US [130]. The reported urinary concentrations of phthalate metabolites among several European countries were similar whereas information for African countries and Australia/Oceanian countries is limited.

Pregnant Women: Phthalates have been widely studied for exposure levels in pregnant women. MEP (222 µg/g CR) was the predominant phthalate metabolite found in urine samples of pregnant women from the Netherlands (Generation R study) [110]. Similar exposure levels were reported for pregnant women from the US [131,132], Canada [69], and Norway [113], with MEP median concentrations exceeding 30 µg/L. In a study of urinary phthalate metabolite concentrations in Spanish pregnant women (*n* = 391), the median concentration of MEP was reported at 246 µg/g CR [118].

Several studies have examined phthalate metabolite concentrations in matched urine samples of newborns and mothers. Maternal urinary concentrations of MEHHP and MEOHP in Korea were 17.7 and 14.7 µg/L, respectively, which were two- to threefold higher than those found in newborns (5.79 and 3.27 µg/L) [144]. Another study, however, showed similar urinary concentrations of phthalate metabolites between 120 mother-and-child pairs [96]. Occurrence of phthalate metabolites in pregnant women suggests potential exposure in the fetus.

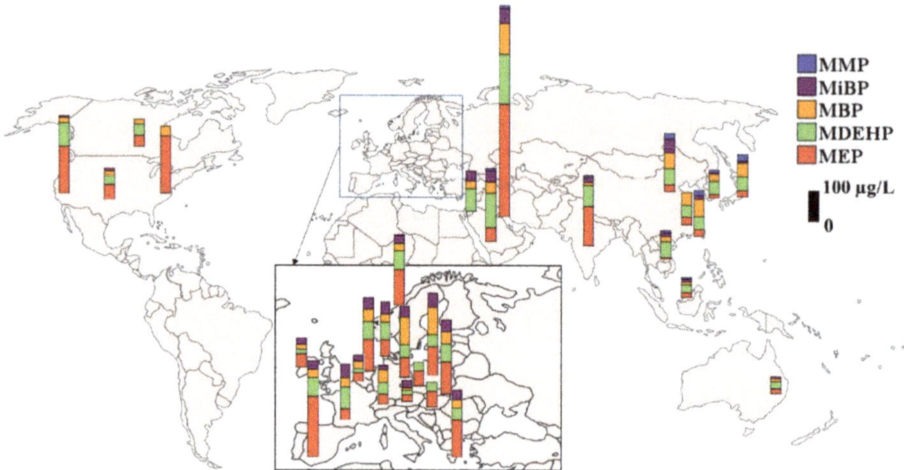

Figure 3. Urinary concentrations of phthalate metabolites reported in adults (general population) from several countries (MDEHP: Sum of DEHP metabolites; biomonitoring data published after 2000; median concentration is presented).

Children: The NHANES data showed that the concentrations of urinary phthalate metabolites in children 6–11 years old were higher than those in adolescents and adults [142]. Several studies support the CDC's findings that children have higher urinary concentrations than do adults of DBP, BzBP, and DEHP [41,145]. Differences in urinary concentrations of phthalates among infants, children, and adults may reflect different sources and routes of intake. Ingestion is thought to be a primary pathway of exposure to some phthalates, especially those in food packaging [146]. The mouthing behavior of infants and toddlers could potentially increase their exposures to phthalates in toys and other products made with plasticized polymers. The global distribution of reported urinary phthalate metabolite concentrations in children is shown in Figure 4.

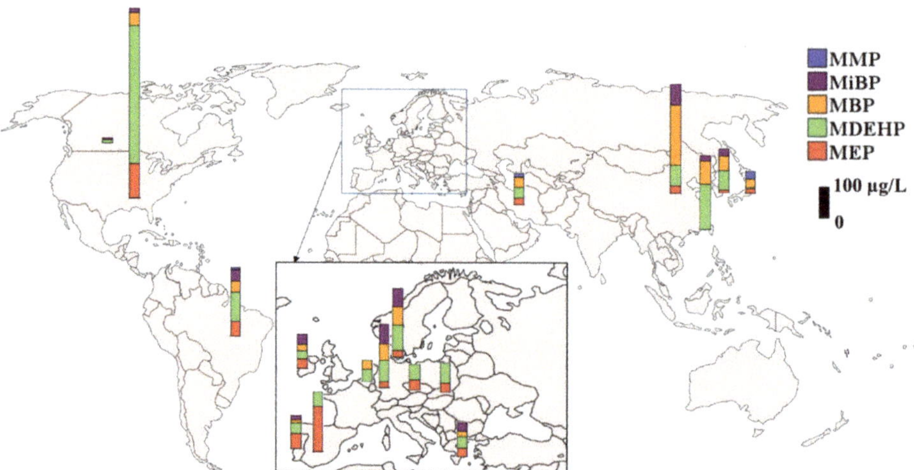

Figure 4. Urinary concentrations of phthalate metabolites reported in children (general population) from several countries (MDEHP: Sum of DEHP metabolites; biomonitoring data published after 2000, median concentration is presented).

MEP, MBP, and DEHP metabolites were the dominant compounds detected in urine from children. Spot urine samples from 5- to 7-year-old German children contained a median phthalate metabolite concentration (sum of 5 metabolites) of 76.9 µg/L, with DEHP metabolites as major compounds [89]. A similar concentration of DEHP metabolites at 75.7 µg/L was found in urine samples from 8- to 10-year-old German children (n = 465) [90]. Several biomonitoring studies reported comparable concentrations of DEHP metabolites and MBP in urine from children in China [75], Korea [108], Canada [70], Brazil [67] and Portugal [114].

In urine samples collected from children in Beijing, China, MBP was the most abundant metabolite (median: 232 µg/L), followed by MiBP (81.3 µg/L), MECPP (79.1 µg/L), and MEP (28.5 µg/L). A significant association between the concentrations of parent phthalate diesters in handwipes and the corresponding monoester metabolites in urine were observed in urine from children, which suggested that dermal absorption is an important exposure pathway for phthalates in children [75]. Mean urinary concentrations of MBP decreased as the children aged [91]. Among children, urinary DBP and DEHP metabolites in boys were higher than those in girls, whereas urinary MEP concentrations were positively correlated with age in both genders [79]. Urinary concentrations of MEP in adolescents were higher than those in children, which was associated with high cosmetic usage among teenagers [79,95].

Urinary phthalate metabolite concentrations have been reported for children and adults from 17 European countries, namely, Belgium, Cyprus, Czech Republic, Denmark, Germany, Hungary, Ireland, Luxembourg, Poland, Portugal, Romania, Slovenia, Slovak Republic, Spain, Sweden, Switzerland, and the United Kingdom (DEMOCOPHES); the geometric mean concentrations of MEP, MBZP, MBP, MiBP, and ΣDEHP metabolites were 34.4, 7.15, 34.8, 45.4, and 47.6 µg/L for children (n = 1355) and were 48.2, 4.51, 23.9, 30.1, and 29.2 µg/L for mothers [81], which suggested that children in those countries were more highly exposed to several phthalates than were their mothers. Nevertheless, some studies reported higher urinary MEP concentrations in mothers (45.1–72.0 µg/L) than in children (12.1–16.4 µg/L) [82,92]. The concentrations of DEHP metabolites were reported to be similar between mothers and children [82,109,119]. A significant positive correlation existed in urinary phthalate metabolite concentrations between children and their parents. MECPP, an oxidative metabolite of DEHP, was predominant in urine from children (92.7%) relative to that found in adults (56.7–57.6%). Studies have found that children possess enhanced oxidative metabolism for DEHP [91,109,147]. Another study of urinary phthalate metabolites in 104 paired mothers and school-aged children reported higher concentrations of secondary DEHP metabolites in children than in mothers [93]. A study from Austria showed higher urinary concentrations of phthalate metabolites in children than adults [62]. Overall, these studies suggest higher exposures to phthalates in children than adults.

Highly Exposed Populations: Highly exposed individuals have urinary phthalate metabolite concentrations that often exceed those at the 95th percentile of the general population (https://www.ncbi.nlm.nih.gov/books/NBK215044/). Neonates who receive medical treatments such as transfusions are widely recognized as potentially highlexposed [148]. A study from Slovakia showed that the urinary concentrations of DEHP metabolites, MiBP, and MBP in occupationally exposed individuals from plastic industry were 55.9, 39.2, and 110 µg/L, respectively [115], which were higher than those in urine from women of no known occupational exposures [61]. The median concentrations of MEP, MBP, MiBP, and DEHP metabolites in urine from hairdressing apprentices who attended vocational training schools in Slovakia were 201, 103, 61.4, and 82.7 µg/L, respectively [116]. Some medications contain phthalates in their coatings or delivery systems [49] and may contribute to the high exposures of children, pregnant women, and others who take these medications.

Exposure Assessment: The concentrations of phthalate metabolites measured in urine can be used to assess the amount of parent phthalate to which humans are exposed, when the fraction of the metabolite excreted in urine is known, as presented in the equation below [147]:

$$\text{Estimated parent phthalate concentration} = \frac{\text{Metabolite concentration}}{\text{Excretion fraction}} \tag{1}$$

The estimated daily intake (*EDI*) of parent phthalates is then calculated by taking the average weight of an individual with the average urinary excretion rate, as shown in the equation below:

$$\text{Estimated daily intake (EDI)} = \frac{\text{Estimated parent phthalate concentration} \times \text{Daily urine excretion volume}}{\text{Average body weight}} \quad (2)$$

Several studies have estimated exposure doses to phthalates in populations, which allowed for comparison against a reference dose (*RfD*), the maximum acceptable oral dose of a toxic substance, of the US EPA. The estimated mean daily exposure doses to DEP and DBP in Asian countries and the US were one to two orders of magnitude below the EPA *RfD* (DEP = 800, DBP = 100, and DEHP = 20 µg/kg body weight (bw)/day). The estimated daily exposure doses to DEHP in Kuwait and India, however, were close to the RfD of the US EPA [72]. Similarly, high concentrations of DEHP metabolites (mean concentration = 338 µg/L) were reported in urine from the Saudi population [26].

The calculated EDI_{max} values for DEHP and DBP were 8 and 0.08 µg/kg bw/day, respectively, for the population in Taiwan, which were one to two orders of magnitude lower than the tolerable daily intake (*TDI*) values (the daily intake amount of a chemical that has been assessed to be safe for human being on a long-term basis) suggested for DEHP (50 µg/kg bw) and DBP (10 µg/kg bw) by the European Food Safety Authority (EFSA) [147].

The 95th percentile for DEHP exposure doses calculated for the general population ($n = 85$) and children ($n = 254$) from Germany were 21 and 25 µg/kg bw/day, respectively, which exceeded the RfD (20 µg/kg bw/day) and the TDI (20–48 µg/kg bw/day) [149]. Further, elevated exposure to phthalates, especially DEHP, in neonates admitted to intensive care units was reported (median: 42 µg/kg bw/day; 95th percentile: 1780 µg/kg bw/day) [149], and the exposure dose was higher than the RfD.

Some studies defined "Biomonitoring Equivalents (BEs)" as the concentration or range of concentrations of a chemical or its metabolite in a biological medium (blood, urine, or other medium) that is consistent with an existing health-based exposure guideline (e.g., RfD and TDI) [150,151]. BE values for MBP, MBzP, and MEP were reported at 18000, 3800 and 2700 µg/L, respectively [150], and the BE values range from 1500 to 3600 µg/L for MiNP [151]. These values may be used as screening tools for evaluation of biomonitoring data for phthalate metabolites in the context of existing risk assessments and for prioritization of the potential need for additional risk assessment efforts for each of these compounds relative to other chemicals [150,151].

Although current exposure doses in the general population are below the tolerance limits reported by environmental agencies, certainly population groups, especially children, are exposed to high levels of phthalates. Studies of the effects of phthalates from early life stage exposures are warranted.

3.2. Phthalate Metabolites in Serum

The biomonitoring studies of human phthalate exposure have been based on urinary concentrations of phthalate metabolites. However, when only serum was available for analysis, MEP and MiBP representing low molecular weight phthalates, and MECPP and MCiOP representing high molecular weight phthalates, have been used as indicators of phthalate exposure [77]. A study reported the correlations of phthalate metabolite concentrations among urine, serum, and seminal plasma of young Danish men [77]. The mean concentrations of MEP, MBP, and DEHP metabolites were one to two orders of magnitude lower in serum (MEP: 4.2, MBP: 0.4, and DEHP: 7.6 µg/L) and seminal plasma (1.0, 0.8, and 0.6 µg/L) than in urine (326, 42.5, and 115 µg/L). Another study, however, showed that the distribution pattern of monoester metabolites in serum was similar to that of urine [152], especially for MEHP (the metabolite of DEHP) [152]. Nevertheless, MEHHP, MEOHP, MECPP, and MCMHP were found at much higher concentrations in urine than in serum [153]. The presence of MEHP in serum was more likely related to contamination that arises from sampling devices.

Whole blood and cord blood samples from 128 healthy pregnant women and their newborns were analyzed for phthalate metabolites. Median concentrations of MEHHP and MEOHP were 0.31 and <LOD µg/L in maternal blood and 0.32 and <LOD µg/L in cord blood, respectively. MEHHP and

MEOHP also were reported to occur in the placenta at concentrations of 0.09 and <LOD ng/g [144]. MBP, MEHP, MEP, and MiBP were detected in blood serum at median concentrations of 0.54 0.49, 0.50, and 0.5 µg/L, respectively [24], and these concentrations were at least an order of magnitude lower than those measured in urine.

In the serum of patients who were undergoing dialysis [53,154–156], phthalate acid (PA) was found as a metabolite of phthalates at remarkably high concentrations of 5.22 ± 3.94 mg/L [155]. Another study also reported the occurrence of PA in serum (0.205 ± 0.067 mg/L) of patients who were undergoing dialysis [154]. Accumulation of PA in patients who are undergoing dialysis has been suggested [156]. Serum concentrations of MEHP and DEHP were reported in autistic children [157].

3.3. Phthalate Metabolites in Amniotic Fluid, Breast Milk, Semen, and Saliva

MBP was found in >93% of amniotic fluid samples collected from the US [59] at concentrations two- to threefold lower than those of serum and four- to sevenfold lower than those of urine [59]. Studies have reported the occurrence of phthalates in breast milk [158]; the reported concentrations in breast milk were much lower than those in urine but similar to those in amniotic fluid. Monoester metabolites of phthalates were measured in breast milk from 33 lactating mothers in North Carolina. MCPP (0.2 µg/L) and MEOHP (0.3 µg/L), MECPP (0.1–0.4 µg/L), and MEHHP (0.2–0.3 µg/L) were detected in some samples [131]. MiNP was the major metabolite found in breast milk collected from mothers in Denmark (101 µg/L) and Finland (89 µg/L) [25,159]. Median concentrations of MBP, MBzP, and MEHP in breast milk were 0.54, 0.50, and 0.49 µg/L, respectively [24].

Human saliva samples (n = 39) also contained phthalate metabolites [57,160]. Salivary concentrations of phthalate metabolites in 39 adult volunteers were in the ranges of <1 to 10.6 µg/L for PA, 91.4 µg/L for MEP, 65.8 µg/L for MBP, and 354 µg/L for MBzP. MBP was the most (85%) frequently detected compound in saliva [57]. Two phthalate metabolites (2.2 µg/L MCPP and 2.3 µg/L MECPP) were detected in a saliva sample from a US woman [131].

MBP and MBzP were found in semen from US men [32,54]. High concentrations of DEHP and its metabolites ($\sum$40.6 µg/L) were found in semen from German men [161]. Studies have also indicated that semen quality can be affected by environmentally relevant phthalate exposures [121]. Further, DEHP (4.20 µg/L) and DBP (2.06 µg/L) were reported at high concentrations in male seminal plasma from men in the US. The metabolites of DEHP ($\sum$0.98 µg/L) and MBP (2.97 µg/L) also were present in considerable concentrations in seminal plasma in the same study [162]. These results suggested that phthalate metabolites can partition in seminal plasma. Similarly, DEHP (2.09 µg/L) and DBP (1.75 µg/L), as well as their metabolites, were found as the predominant phthalates/phthalate metabolites in seminal plasma from male partners who were planning for pregnancy. This study showed adverse associations between seminal phthalate metabolite concentrations and semen quality [163].

Phthalate metabolites were measured in nail samples from Belgium, and the total concentrations ranged between <12 and 7980 ng/g. It should be noted, however, that some phthalates, especially DBP, are used in nail polishes and that care should be exercised in interpreting such measurements. MEHP, MBP, and MEP were the major metabolites detected in every nail sample, with a median concentration of 138, 74, and 64 ng/g, respectively [135]. Another study of nail samples from Oslo, Norway, showed the presence of monoesters, such as MMP (geometric mean 89.7 ng/g), MEP (104.8 ng/g), and MBP (89.3 ng/g) [112]. The utility of other biologic matrices, such as blood, breast milk, semen, and nails, for assessing human exposure to phthalates remains largely unknown due to the limited data.

4. Select Epidemiological Studies Linking Phthalate Exposure and Health Outcomes

Controlled laboratory animal studies on the toxic effects of phthalates have enabled understanding of biological plausibility and potential mechanisms of actions of this class of chemicals. Thus far, the majority of the laboratory animal exposure/toxicity studies have focused on DEHP and DBP/DiBP, with limited studies examining the toxicities of other phthalates [164–191]. The reproductive and developmental effects of phthalates are among the most studied and well-described toxic endpoints in those studies. The toxic endpoints determined in animal studies, following phthalate exposure, include retention of nipples, anogenital distance, pathological changes in testes and male reproductive accessory glands, hypospadias, cryptorchidism, and semen parameters. Phthalates have well-documented anti-androgenic activity in rodent studies that result in reduced circulating testosterone. Several reviews have been published on the toxicity of phthalates [10,14,168–170]. As a class of well-studied endocrine disrupting chemicals, exposure to phthalates has been linked to sex anomalies, endometriosis, altered reproductive development, early puberty and fertility, breast and skin cancer, allergy and asthma, overweight and obesity, insulin resistance, and type II diabetes.

4.1. Diabetes

Diabetes is a metabolic disease that results in elevated blood glucose levels. Epidemiological studies in the US [192,193] reported that women with higher urinary concentrations of MBP, MiBP, MBzP, and MCPP and those of DEHP metabolites showed increased risk of diagnosis for diabetes in comparison with those who had lower concentrations of phthalates. Phthalate exposures have been shown to result in insulin resistance [166,194].

4.2. Overweight and Obesity

Overweight and obesity can be associated with many chronic diseases, including diabetes. Phthalate exposure was associated with increased body mass and waist circumference [195]. Some phthalate metabolites (MEP, MBP, and MiBP) were associated with obesity in children, whereas MEHP, MECPP, MEHHP, MEOHP, MBzP, and MCNP were associated with obesity in adults. Further, DEHP metabolites were found to be significantly associated with obesity in adult females and older males [196]. Urinary concentration of MBP was associated with fat deposition in boys in China [197].

Several studies have shown a significant association between obesity and phthalate exposure [193,196,198]. MEP, MEHP, MBzP, MEHHP, and MEOHP were associated with obesity in the US population [198]. MBzP, MEHHP, MEOHP, and MEP were associated with increased waist circumference and BMI [193] In contrast, higher concentrations of MEP and DEHP were found in the serum and urine of individuals who were undergoing weight loss [199]. Food intake is the main source of phthalate exposure (for high molecular weight phthalates). Therefore, overweight population with high food intake might have high phthalate exposures.

4.3. Allergy and Asthma

Exposure to high molecular weight phthalates are is associated with allergies and asthma [200,201]. Studies indicated that children are prone to exposure to DEHP, BzBP, DBP, and DEP and that exposure was associated with allergic rhinitis, atopic dermatitis, and conjunctivitis [202]. DEHP and BzBP and their monoesters are regarded as allergens, and exposure to them has been associated with asthma and wheezing in adults [200,201]. Exposure of DEHP, BzBP, DBP, and DEP during gestation has been associated with allergic responses in infants and toddlers [200]. Urinary MEHP concentrations are correlated with asthma in children [203]. Prenatal exposure to DEHP metabolites and BzBP has been associated with the risk of developing asthma at the age of 7 years and older [204].

4.4. Reproductive Health

Urinary MEP and MBP and the metabolites of DEHP and DiNP are associated with anomalies in pubertal development in girls [205]. A significant association between urinary concentrations of MBzP, MEHP, and MEP and increased risk of endometriosis was found in women [206]. Exposure to MEP, MiBP, and MBP pose an increased risk of pregnancy loss in Chinese women.

Poor semen quality was associated with exposure to phthalate metabolites. MBP and MBzP were strongly associated with spermatotoxicity and subfertility in males [32,54]. Significantly higher concentrations of DEHP (4.66 µg/mL) and MEHP (3.19 µg/mL) were found in the urine of 40 Turkish boys with gynecomastia as compared to that of control groups [207]. Several reviews have appeared on the reproductive and developmental toxicities of phthalates [208]. Whereas some inconsistencies exist across phthalates for specific health outcomes associated with exposures, moderate to strong evidence of male reproductive effects have been demonstrated in the literature [208]. Because humans are exposed to thousands of harmful chemicals, establishing the link between exposure to a single substance class and adverse health outcomes is fraught with uncertainties.

5. Conclusions and Perspectives

Human biomonitoring studies are useful in elucidating exposures and body burdens of phthalates at a population level. Although the sources of exposure to phthalates are well described, several questions about cumulative exposures to phthalates throughout the life span, relative contributions of various sources to cumulative exposures, and mixed exposures that may include phthalates or other chemicals that may elicit common adverse outcomes remain unanswered. Biomonitoring studies clearly demonstrate that human exposures are almost ubiquitous, and, in most cases, children have higher exposures than do adults. The existing studies indicate that the observed associations between phthalate exposure and disease outcomes are exploratory and preliminary, the health effects of phthalate exposure warrant further study. Robust analytical methods exist to measure more than 20 phthalate metabolites in urine, a preferred matrix of choice for biomonitoring studies. Although studies have reported the occurrence of phthalate metabolites in other human specimens, including serum, seminal plasma, and amniotic fluid, the relevance of these matrices in understanding toxic effects needs further investigation. Although biomonitoring studies select major biomarkers/metabolites of phthalates, several other intermediate and transformation products of phthalates appear to exist in human specimens. These intermediates may have more pronounced effects on health. Lack of analytical standards hinders the identification of those intermediate biological transformation products of phthalates. Further, the interaction of phthalate metabolites with other contaminants should be considered in future investigations.

There is a lack of biomonitoring data on phthalate exposures in developing countries in Africa and South America. Studies are needed in those regions with regard to exposures and associated health outcomes in populations. Further, epigenetic effects of phthalate exposures warrant further investigation.

Funding: Research reported in this publication was supported, in part, by the National Institute of Environmental Health Sciences of the National Institutes of Health under Award No. U2CES026542-01. The content is solely the responsibility of the authors and does not necessarily represent the official views of the National Institutes of Health.

Conflicts of Interest: The authors declare no conflict of interest.

References

1. Latini, G. Monitoring phthalate exposure in humans. *Clin. Chim. Acta* **2005**, *361*, 20–29. [CrossRef] [PubMed]
2. Frederiksen, H.; Skakkebaek, N.E.; Andersson, A.M. Metabolism of phthalates in humans. *Mol. Nutr. Food Res.* **2007**, *51*, 899–911. [CrossRef] [PubMed]
3. Petersen, J.H.; Breindahl, T. Plasticizers in total diet samples, baby food and infant formulae. *Food Addit. Contam.* **2000**, *17*, 133–141. [CrossRef] [PubMed]
4. Wormuth, M.; Scheringer, M.; Vollenweider, M.; Hungerbuhler, K. What are the sources of exposure to eight frequently used phthalic acid esters in Europeans? *Risk Anal.* **2006**, *26*, 803–824. [CrossRef] [PubMed]
5. Graham, P.R. Phthalate ester plasticizers—Why and how they are used. *Environ. Health Perspect.* **1973**, *3*, 3–12. [PubMed]
6. Mackintosh, C.E.; Maldonado, J.A.; Ikonomou, M.G.; Gobas, F.A.P.C. Sorption of phthalate esters and PCBs in a marine ecosystem. *Environ. Sci. Technol.* **2006**, *40*, 3481–3488. [CrossRef]
7. Sirivarasai, J.; Wananukul, W.; Kaojarern, S.; Chanprasertyothin, S.; Thongmung, N.; Ratanachaiwong, W.; Sura, T.; Sritara, P. Association between inflammatory marker, environmental lead exposure and glutathione *S*-transferase gene. *Toxicol. Lett.* **2013**, *221*, 61. [CrossRef]

8. Net, S.; Sempere, R.; Delmont, A.; Paluselli, A.; Ouddane, B. Occurrence, fate, behavior and ecotoxicological state of phthalates in different environmental matrices. *Environ. Sci. Technol.* **2015**, *49*, 4019–4035. [CrossRef]

9. Gimeno, P.; Thomas, S.; Bousquet, C.; Maggio, A.F.; Civade, C.; Brenier, C.; Bonnet, P.A. Identification and quantification of 14 phthalates and 5 non-phthalate plasticizers in PVC medical devices by GC–MS. *J. Chromatogr. B* **2014**, *949–950*, 99–108. [CrossRef] [PubMed]

10. Kay, V.R.; Bloom, M.S.; Foster, W.G. Reproductive and developmental effects of phthalate diesters in males. *Crit. Rev. Toxicol.* **2014**, *44*, 467–498. [CrossRef]

11. Talsness, C.E.; Andrade, A.J.M.; Kuriyama, S.N.; Taylor, J.A.; vom Saal, F.S. Components of plastic: Experimental studies in animals and relevance for human health. *Philos. Trans. R. Soc. Lond. B Biol. Sci.* **2009**, *364*, 2079–2096. [CrossRef] [PubMed]

12. Gray, J.L.E.; Ostby, J.; Furr, J.; Wolf, C.J.; Lambright, C.; Parks, L.; Veeramachaneni, D.N.; Wilson, V.; Price, M.; Hotchkiss, A.; et al. Effects of environmental antiandrogens on reproductive development in experimental animals. *Hum. Reprod. Update* **2001**, *7*, 248–264. [CrossRef] [PubMed]

13. Shehata, A.; Mohamed, Z.; El-Haleem, M.; Samak, M. Effects of exposure to plasticizers di-(2-ethylhexyl) phthalate and trioctyltrimellitate on the histological structure of adult male albino rats' liver. *J. Clin. Toxicol.* **2013**, *3*, 169–178.

14. Rusyn, I.; Peters, J.M.; Cunningham, M.L. Modes of action and species-specific effects of di-(2-ethylhexyl)phthalate in the liver. *Crit. Rev. Toxicol.* **2006**, *36*, 459–479. [CrossRef] [PubMed]

15. Wei, Z.; Song, L.; Wei, J.; Chen, T.; Chen, J.; Lin, Y.; Xia, W.; Xu, B.; Li, X.; Chen, X.; et al. Maternal exposure to di-(2-ethylhexyl)phthalate alters kidney development through the renin–angiotensin system in offspring. *Toxicol. Lett.* **2012**, *212*, 212–221. [CrossRef] [PubMed]

16. Crocker, J.F.S.; Safe, S.H.; Acott, P. Effects of chronic phthalate exposure on the kidney. *J. Toxicol. Environ. Health* **1988**, *23*, 433–444. [CrossRef]

17. Guo, Y.; Kannan, K. A survey of phthalates and parabens in personal care products from the United States and its implications for human exposure. *Environ. Sci. Technol.* **2013**, *47*, 14442–14449. [CrossRef]

18. Guo, Y.; Wang, L.; Kannan, K. Phthalates and parabens in personal care products from China: Concentrations and human exposure. *Arch. Environ. Contam. Toxicol.* **2014**, *66*, 113–119. [CrossRef] [PubMed]

19. Specht, I.O.; Toft, G.; Hougaard, K.S.; Lindh, C.H.; Lenters, V.; Jonsson, B.A.G.; Heederik, D.; Giwercman, A.; Bonde, J.P.E. Associations between serum phthalates and biomarkers of reproductive function in 589 adult men. *Environ. Int.* **2014**, *66*, 146–156. [CrossRef] [PubMed]

20. Dong, R.H.; Zhou, T.; Zhao, S.Z.; Zhang, H.; Zhang, M.R.; Chen, J.S.; Wang, M.; Wu, M.; Li, S.G.; Chen, B. Food consumption survey of Shanghai adults in 2012 and its associations with phthalate metabolites in urine. *Environ. Int.* **2017**, *101*, 80–88. [CrossRef]

21. Silva, M.J.; Samandar, E.; Preau, J.L., Jr.; Reidy, J.A.; Needham, L.L.; Calafat, A.M. Quantification of 22 phthalate metabolites in human urine. *J. Chromatogr. B Anal. Technol. Biomed. Life Sci.* **2007**, *860*, 106–112. [CrossRef]

22. Chen, Q.; Yang, H.; Zhou, N.Y.; Sun, L.; Bao, H.Q.; Tan, L.; Chen, H.Q.; Ling, X.; Zhang, G.W.; Huang, L.P.; et al. Phthalate exposure, even below US EPA reference doses, was associated with semen quality and reproductive hormones: Prospective MARHCS study in general population. *Environ. Int.* **2017**, *104*, 58–68. [CrossRef] [PubMed]

23. Nassan, F.L.; Coull, B.A.; Skakkebaek, N.E.; Williams, M.A.; Dadd, R.; Minguez-Alarcon, L.; Krawetz, S.A.; Hait, E.J.; Korzenik, J.R.; Moss, A.C.; et al. A crossover-crossback prospective study of dibutyl-phthalate exposure from mesalamine medications and semen quality in men with inflammatory bowel disease. *Environ. Int.* **2016**, *95*, 120–130. [CrossRef]

24. Högberg, J.; Hanberg, A.; Berglund, M.; Skerfving, S.; Remberger, M.; Calafat, A.M.; Filipsson, A.F.; Jansson, B.; Johansson, N.; Appelgren, M.; et al. Phthalate diesters and their metabolites in human breast milk, blood or serum, and urine as biomarkers of exposure in vulnerable populations. *Environ. Health Perspect.* **2008**, *116*, 334–339. [CrossRef] [PubMed]

25. Main, K.M.; Mortensen, G.K.; Kaleva, M.M.; Boisen, K.A.; Damgaard, I.N.; Chellakooty, M.; Schmidt, I.M.; Suomi, A.-M.; Virtanen, H.E.; Petersen, J.H.; et al. Human breast milk contamination with phthalates and alterations of endogenous reproductive hormones in infants three months of age. *Environ. Health Perspect.* **2006**, *114*, 270–276. [CrossRef]

26. Asimakopoulos, A.G.; Xue, J.; De Carvalho, B.P.; Iyer, A.; Abualnaja, K.O.; Yaghmoor, S.S.; Kumosani, T.A.; Kannan, K. Urinary biomarkers of exposure to 57 xenobiotics and its association with oxidative stress in a population in Jeddah, Saudi Arabia. *Environ. Res.* **2016**, *150*, 573–581. [CrossRef] [PubMed]

27. Ferguson, K.K.; McElrath, T.F.; Chen, Y.-H.; Mukherjee, B.; Meeker, J.D. Urinary phthalate metabolites and biomarkers of oxidative stress in pregnant women: A repeated measures analysis. *Environ. Health Perspect.* **2015**, *123*, 210–216. [CrossRef]

28. Colón, I.; Caro, D.; Bourdony, C.J.; Rosario, O. Identification of phthalate esters in the serum of young Puerto Rican girls with premature breast development. *Environ. Health Perspect.* **2000**, *108*, 895–900. [PubMed]

29. Buck Louis, G.M.; Gray, L.E.; Marcus, M.; Ojeda, S.R.; Pescovitz, O.H.; Witchel, S.F.; Sippell, W.; Abbott, D.H.; Soto, A.; Tyl, R.W.; et al. Environmental factors and puberty timing: Expert panel research needs. *Pediatrics* **2008**, *121*, 192–207. [CrossRef] [PubMed]

30. Cobellis, L.; Latini, G.; Felice, C.D.; Razzi, S.; Paris, I.; Ruggieri, F.; Mazzeo, P.; Petraglia, F. High plasma concentrations of di-(2-ethylhexyl)-phthalate in women with endometriosis. *Hum. Reprod.* **2003**, *18*, 1512–1515. [CrossRef] [PubMed]

31. Reddy, B.S.; Rozati, R.; Reddy, B.V.R.; Raman, N. General gynaecology: Association of phthalate esters with endometriosis in Indian women. *Int. J. Gynaecol. Obstet.* **2006**, *113*, 515–520. [CrossRef] [PubMed]

32. Duty, S.M.; Silva, M.J.; Barr, D.B.; Brock, J.W.; Ryan, L.; Chen, Z.; Herrick, R.F.; Christiani, D.C.; Hauser, R. Phthalate exposure and human semen parameters. *Epidemiology* **2003**, *14*, 269–277. [CrossRef]

33. Joensen Ulla, N.; Frederiksen, H.; Jensen Martin, B.; Lauritsen Mette, P.; Olesen Inge, A.; Lassen Tina, H.; Andersson, A.M.; Jørgensen, N. Phthalate excretion pattern and testicular function: A study of 881 healthy danish men. *Environ. Health Perspect.* **2012**, *120*, 1397–1403. [CrossRef] [PubMed]

34. Yaghjyan, L.; Sites, S.; Ruan, Y.; Chang, S.H. Associations of urinary phthalates with body mass index, waist circumference and serum lipids among females: National Health and Nutrition Examination Survey 1999–2004. *Int. J. Obstet.* **2015**, *36*, 994–1000. [CrossRef] [PubMed]

35. López-Carrillo, L.; Hernández-Ramírez, R.U.; Calafat, A.M.; Torres-Sánchez, L.; Galván-Portillo, M.; Needham, L.L.; Ruiz-Ramos, R.; Cebrián, M.E. Exposure to phthalates and breast cancer risk in Northern Mexico. *Environ. Health Perspect.* **2010**, *118*, 539–544. [CrossRef] [PubMed]

36. Monographs on the Evaluation of Carcinogenic Risks to Humans. Available online: https://monographs.iarc. fr/wp-content/uploads/2018/06/mono77.pdf (accessed on 5 April 2019).

37. Le Moal, J.; Sharpe, R.M.; Jørgensen, N.; Levine, H.; Jurewicz, J.; Mendiola, J.; Swan, S.H.; Virtanen, H.; Christin-Maître, S.; Cordier, S.; et al. Toward a multi-country monitoring system of reproductive health in the context of endocrine disrupting chemical exposure. *Eur. J. Public Health.* **2016**, *26*, 76–83. [CrossRef]

38. Sharpe, R.M.; Skakkebaek, N.E. Testicular dysgenesis syndrome: Mechanistic insights and potential new downstream effects. *Fertil. Steril.* **2008**, *89*, 33–38. [CrossRef] [PubMed]

39. Kuo, C.C.; Moon, K.; Thayer, K.A.; Navas-Acien, A. Environmental chemicals and type 2 diabetes: An updated systematic review of the epidemiologic evidence. *Curr. Diabetes Rep.* **2013**, *13*, 831–849. [CrossRef]

40. Swan, S.H. Environmental phthalate exposure in relation to reproductive outcomes and other health endpoints in humans. *Environ. Res.* **2008**, *108*, 177–184. [CrossRef]

41. Goen, T.; Dobler, L.; Koschorreck, J.; Muller, J.; Wiesmuller, G.A.; Drexler, H.; Kolossa-Gehring, M. Trends of the internal phthalate exposure of young adults in Germany—Follow-up of a retrospective human biomonitoring study. *Int. J. Hyg. Environ. Health* **2011**, *215*, 36–45. [CrossRef] [PubMed]

42. Tranfo, G.; Caporossi, L.; Pigini, D.; Capanna, S.; Papaleo, B.; Paci, E. Temporal trends of urinary phthalate concentrations in two populations: Effects of REACH authorization after five years. *Int. J. Environ. Res. Public Health* **2018**, *15*, 1950. [CrossRef]

43. North, M.L.; Takaro, T.K.; Diamond, M.L.; Ellis, A.K. Effects of phthalates on the development and expression of allergic disease and asthma. *Ann. Allergy Asthma Immunol.* **2014**, *112*, 496–502. [CrossRef] [PubMed]

44. Wittassek, M.; Koch, H.M.; Angerer, J.; Bruning, T. Assessing exposure to phthalates—The human biomonitoring approach. *Mol. Nutr. Food Res.* **2011**, *55*, 7–31. [CrossRef] [PubMed]

45. Sapozhnikova, Y.; Hoh, E. Suspect screening of chemicals in food packaging film by comprehensive two-dimensional gas chromatography coupled to time-of-flight mass spectrometry. *LCGC N. Am.* **2019**, *37*, 52–60.

46. Tran, T.M.; Kannan, K. Occurrence of phthalate diesters in particulate and vapor phases in indoor air and implications for human exposure in Albany, New York, USA. *Arch. Environ. Contam. Toxicol.* **2015**, *68*, 489–499. [CrossRef]

47. Guo, Y.; Kannan, K. Comparative assessment of human exposure to phthalate esters from house dust in China and the United States. *Environ. Sci. Technol.* **2011**, *45*, 3788–3794. [CrossRef] [PubMed]

48. Moreta, C.; Tena, M.T.; Kannan, K. Analytical method for the determination and a survey of parabens and its derivatives in pharmaceuticals. *Environ. Res.* **2015**, *142*, 452–460. [CrossRef] [PubMed]

49. Hauser, R.; Duty, S.; Godfrey-Bailey, L.; Calafat, A.M. Medications as a source of human exposure to phthalates. *Environ. Health Perspect.* **2004**, *112*, 751–753. [CrossRef] [PubMed]

50. U.S. FDA. Safety Assessment of Di(2-ethylhexyl)phthalate (DEHP) Released from PVC Medical Devices, Rockville, MD 20852. Available online: https://noharm-global.org/documents/safety-assessment-dehp-released-pvc-medical-devices (accessed on 5 April 2019).

51. Guo, Y.; Wu, Q.; Kannan, K. Phthalate metabolites in urine from China, and implications for human exposures. *Environ. Int.* **2011**, *37*, 893–898. [CrossRef] [PubMed]

52. Lind, P.M.; Roos, V.; Ronn, M.; Johansson, L.; Ahlstrom, H.; Kullberg, J.; Lind, L. Serum concentrations of phthalate metabolites are related to abdominal fat distribution two years later in elderly women. *Environ. Health* **2012**, *11*, 21–29. [CrossRef] [PubMed]

53. Kato, K.; Silva, M.J.; Brock, J.W.; Reidy, J.A.; Malek, N.A.; Hodge, C.C.; Nakazawa, H.; Needham, L.L.; Barr, D.B. Quantitative Detection of Nine Phthalate Metabolites in Human Serum Using Reversed-Phase High-Performance Liquid Chromatography-Electrospray Ionization-Tandem Mass Spectrometry. *J. Anal. Toxicol.* **2003**, *27*, 284–289. [CrossRef]

54. Hauser, R.; Meeker, J.D.; Duty, S.; Silva, M.J.; Calafat, A.M. Altered semen quality in relation to urinary concentrations of phthalate monoester and oxidative metabolites. *Epidemiology* **2006**, *17*, 682–691. [CrossRef]

55. Zhu, J.P.; Phillips, S.P.; Feng, Y.L.; Yang, X.F. Phthalate esters in human milk: Concentration variations over a 6-month postpartum time. *Environ. Sci. Technol.* **2006**, *40*, 5276–5281. [CrossRef] [PubMed]

56. Damgaard, I.N.; Skakkebæk, N.E.; Toppari, J.; Virtanen, H.E.; Shen, H.; Schramm, K.W.; Petersen, J.H.; Jensen, T.K.; Main, K.M.; the Nordic Cryptorchidism Study, G. Persistent pesticides in human breast milk and cryptorchidism. *Environ. Health Perspect.* **2006**, *114*, 1133–1138. [CrossRef] [PubMed]

57. Silva, M.J.; Reidy, J.A.; Samandar, E.; Herbert, A.R.; Needham, L.L.; Calafat, A.M. Detection of phthalate metabolites in human saliva. *Arch. Toxicol.* **2005**, *79*, 647–652. [CrossRef]

58. Fennell, T.R.; Krol, W.L.; Sumner, S.C.J.; Snyder, R.W. Pharmacokinetics of dibutylphthalate in pregnant rats. *Toxicol. Sci.* **2004**, *82*, 407–418. [CrossRef]

59. Silva, M.J.; Reidy, J.A.; Herbert, A.R.; Preau, J.L.; Needham, L.L.; Calafat, A.M. Detection of phthalate metabolites in human amniotic fluid. *Bull. Environ. Contam. Toxicol.* **2004**, *72*, 1226–1231. [CrossRef] [PubMed]

60. Choi, J.; Knudsen, L.E.; Mizrak, S.; Joas, A. Identification of exposure to environmental chemicals in children and older adults using human biomonitoring data sorted by age: Results from a literature review. *Int. J. Hyg. Environ. Health* **2017**, *220*, 282–298. [CrossRef]

61. Heffernan, A.L.; Thompson, K.; Eaglesham, G.; Vijayasarathy, S.; Mueller, J.F.; Sly, P.D.; Gomez, M.J. Rapid, automated online SPE-LC-QTRAP-MS/MS method for the simultaneous analysis of 14 phthalate metabolites and 5 bisphenol analogues in human urine. *Talanta* **2016**, *151*, 224–233. [CrossRef]

62. Hartmann, C.; Uhl, M.; Weiss, S.; Koch, H.M.; Scharf, S.; Konig, J. Human biomonitoring of phthalate exposure in Austrian children and adults and cumulative risk assessment. *Int. J. Hyg. Environ. Health* **2015**, *218*, 489–499. [CrossRef]

63. Dewalque, L.; Charlier, C.; Pirard, C. Estimated daily intake and cumulative risk assessment of phthalate diesters in a Belgian general population. *Toxicol. Lett.* **2014**, *231*, 161–168. [CrossRef]

64. Geens, T.; Bruckers, L.; Covaci, A.; Schoeters, G.; Fierens, T.; Sioen, I.; Vanermen, G.; Baeyens, W.; Morrens, B.; Loots, I.; et al. Determinants of bisphenol A and phthalate metabolites in urine of Flemish adolescents. *Environ. Res.* **2014**, *134*, 110–117. [CrossRef]

65. Dewalque, L.; Pirard, C.; Charlier, C. Measurement of urinary biomarkers of parabens, benzophenone-3, and phthalates in a Belgian population. *Biomed. Res. Int.* **2014**, *2014*, 649314–649327. [CrossRef]

66. Dewalque, L.; Pirard, C.; Dubois, N.; Charlier, C. Simultaneous determination of some phthalate metabolites, parabens and benzophenone-3 in urine by ultra high pressure liquid chromatography tandem mass spectrometry. *J. Chromatogr. B Anal. Technol. Biomed. Life Sci.* **2014**, *949–950*, 37–47. [CrossRef]

67. Rocha, B.A.; Asimakopoulos, A.G.; Barbosa, F., Jr.; Kannan, K. Urinary concentrations of 25 phthalate metabolites in Brazilian children and their association with oxidative DNA damage. *Sci. Total Environ.* **2017**, *586*, 152–162. [CrossRef]

68. Saravanabhavan, G.; Guay, M.; Langlois, E.; Giroux, S.; Murray, J.; Haines, D. Biomonitoring of phthalate metabolites in the Canadian population through the Canadian Health Measures Survey (2007–2009). *Int. J. Hyg. Environ. Health* **2013**, *216*, 652–661. [CrossRef]

69. Arbuckle, T.E.; Davis, K.; Marro, L.; Fisher, M.; Legrand, M.; LeBlanc, A.; Gaudreau, E.; Foster, W.G.; Choeurng, V.; Fraser, W.D.; et al. Phthalate and bisphenol A exposure among pregnant women in Canada—Results from the MIREC study. *Environ. Int.* **2014**, *68*, 55–65. [CrossRef]

70. Arbuckle, T.E.; Fisher, M.; MacPherson, S.; Lang, C.; Provencher, G.; LeBlanc, A.; Hauser, R.; Feeley, M.; Ayotte, P.; Neisa, A.; et al. Maternal and early life exposure to phthalates: The plastics and personal-care products use in pregnancy (P4) study. *Sci. Total Environ.* **2016**, *551–552*, 344–356. [CrossRef]

71. Gao, C.-J.; Liu, L.-Y.; Ma, W.-L.; Ren, N.-Q.; Guo, Y.; Zhu, N.-Z.; Jiang, L.; Li, Y.-F.; Kannan, K. Phthalate metabolites in urine of Chinese young adults: Concentration, profile, exposure and cumulative risk assessment. *Sci. Total Environ.* **2016**, *543*, 19–27. [CrossRef]

72. Guo, Y.; Alomirah, H.; Cho, H.-S.; Minh, T.B.; Mohd, M.A.; Nakata, H.; Kannan, K. Occurrence of phthalate metabolites in human urine from several Asian countries. *Environ. Sci. Technol.* **2011**, *45*, 3138–3144. [CrossRef]

73. Shen, Q.; Shi, H.J.; Zhang, Y.H.; Cao, Y. Dietary intake and phthalates body burden in boys and girls. *Arch. Public Health* **2015**, *73*, 5–10. [CrossRef]

74. Zhang, J.; Liu, L.; Wang, X.; Huang, Q.; Tian, M.; Shen, H. Low-level environmental phthalate exposure associates with urine metabolome alteration in a Chinese male cohort. *Environ. Sci. Technol.* **2016**, *50*, 5953–5960. [CrossRef]

75. Gong, M.; Weschler, C.J.; Liu, L.; Shen, H.; Huang, L.; Sundell, J.; Zhang, Y. Phthalate metabolites in urine samples from Beijing children and correlations with phthalate levels in their handwipes. *Indoor Air* **2015**, *25*, 572–581. [CrossRef]

76. Černá, M.; Malý, M.; Rudnai, P.; Középesy, S.; Náray, M.; Halzlová, K.; Jajcaj, M.; Grafnetterová, A.; Krsková, A.; Antošová, D.; et al. Case study: Possible differences in phthalates exposure among the Czech, Hungarian, and Slovak populations identified based on the DEMOCOPHES pilot study results. *Environ. Res.* **2015**, *141*, 118–124. [CrossRef] [PubMed]

77. Frederiksen, H.; Jorgensen, N.; Andersson, A.M. Correlations between phthalate metabolites in urine, serum, and seminal plasma from young Danish men determined by isotope dilution liquid chromatography tandem mass spectrometry. *J. Anal. Toxicol.* **2010**, *34*, 400–410. [CrossRef] [PubMed]

78. Frederiksen, H.; Nielsen, J.K.S.; Mørck, T.A.; Hansen, P.W.; Jensen, J.F.; Nielsen, O.; Andersson, A.-M.; Knudsen, L.E. Urinary excretion of phthalate metabolites, phenols and parabens in rural and urban Danish mother–child pairs. *Int. J. Hyg. Environ. Health* **2013**, *216*, 772–783. [CrossRef]

79. Frederiksen, H.; Aksglaede, L.; Sorensen, K.; Skakkebaek, N.E.; Juul, A.; Andersson, A.-M. Urinary excretion of phthalate metabolites in 129 healthy Danish children and adolescents: Estimation of daily phthalate intake. *Environ. Res.* **2011**, *111*, 656–663. [CrossRef] [PubMed]

80. Langer, S.; Beko, G.; Weschler, C.J.; Brive, L.M.; Toftum, J.; Callesen, M.; Clausen, G. Phthalate metabolites in urine samples from Danish children and correlations with phthalates in dust samples from their homes and daycare centers. *Int. J. Hyg. Environ. Health* **2014**, *217*, 78–87. [CrossRef] [PubMed]

81. Schwedler, G.; Seiwert, M.; Fiddicke, U.; Issleb, S.; Holzer, J.; Nendza, J.; Wilhelm, M.; Wittsiepe, J.; Koch, H.M.; Schindler, B.K.; et al. Human biomonitoring pilot study DEMOCOPHES in Germany: Contribution to a harmonized European approach. *Int. J. Hyg. Environ. Health* **2017**, *220*, 686–696. [CrossRef]

82. Haug, L.S.; Sakhi, A.K.; Cequier, E.; Casas, M.; Maitre, L.; Basagana, X.; Andrusaityte, S.; Chalkiadaki, G.; Chatzi, L.; Coen, M.; et al. In-utero and childhood chemical exposome in six European mother-child cohorts. *Environ. Int.* **2018**, *121*, 751–763. [CrossRef]

83. Zeman, F.A.; Boudet, C.; Tack, K.; Floch Barneaud, A.; Brochot, C.; Pery, A.R.; Oleko, A.; Vandentorren, S. Exposure assessment of phthalates in French pregnant women: Results of the ELFE pilot study. *Int. J. Hyg. Environ. Health* **2013**, *216*, 271–279. [CrossRef]

84. Wittassek, M.; Wiesmuller, G.A.; Koch, H.M.; Eckard, R.; Dobler, L.; Muller, J.; Angerer, J.; Schluter, C. Internal phthalate exposure over the last two decades—A retrospective human biomonitoring study. *Int. J. Hyg. Environ. Health* **2007**, *210*, 319–333. [CrossRef]

85. Becker, K.; Seiwert, M.; Angerer, J.; Heger, W.; Koch, H.M.; Nagorka, R.; Roßkamp, E.; Schlüter, C.; Seifert, B.; Ullrich, D. DEHP metabolites in urine of children and DEHP in house dust. *Int. J. Hyg. Environ. Health* **2004**, *207*, 409–417. [CrossRef]

86. Koch, H.M.; Drexler, H.; Angerer, J. An estimation of the daily intake of di(2-ethylhexyl)phthalate (DEHP) and other phthalates in the general population. *Int. J. Hyg. Environ. Health* **2003**, *206*, 77–83. [CrossRef]

87. Fromme, H.; Bolte, G.; Koch, H.M.; Angerer, J.; Boehmer, S.; Drexler, H.; Mayer, R.; Liebl, B. Occurrence and daily variation of phthalate metabolites in the urine of an adult population. *Int. J. Hyg. Environ. Health* **2007**, *210*, 21–33. [CrossRef]

88. Koch, H.M.; Ruther, M.; Schutze, A.; Conrad, A.; Palmke, C.; Apel, P.; Bruning, T.; Kolossa-Gehring, M. Phthalate metabolites in 24-h urine samples of the German Environmental Specimen Bank (ESB) from 1988 to 2015 and a comparison with US NHANES data from 1999 to 2012. *Int. J. Hyg. Environ. Health* **2017**, *220*, 130–141. [CrossRef]

89. Koch, H.M.; Wittassek, M.; Bruning, T.; Angerer, J.; Heudorf, U. Exposure to phthalates in 5–6 years old primary school starters in Germany—A human biomonitoring study and a cumulative risk assessment. *Int. J. Hyg. Environ. Health* **2011**, *214*, 188–195. [CrossRef]

90. Kasper-Sonnenberg, M.; Koch, H.M.; Wittsiepe, J.; Bruning, T.; Wilhelm, M. Phthalate metabolites and bisphenol A in urines from German school-aged children: Results of the Duisburg birth cohort and Bochum cohort studies. *Int. J. Hyg. Environ. Health* **2014**, *217*, 830–838. [CrossRef]

91. Becker, K.; Goen, T.; Seiwert, M.; Conrad, A.; Pick-Fuss, H.; Muller, J.; Wittassek, M.; Schulz, C.; Kolossa-Gehring, M. GerES IV: Phthalate metabolites and bisphenol A in urine of German children. *Int. J. Hyg. Environ. Health* **2009**, *212*, 685–692. [CrossRef]

92. Volkel, W.; Kiranoglu, M.; Schuster, R.; Fromme, H. Hbmnet Phthalate intake by infants calculated from biomonitoring data. *Toxicol. Lett.* **2014**, *225*, 222–229. [CrossRef]

93. Kasper-Sonnenberg, M.; Koch, H.M.; Wittsiepe, J.; Wilhelm, M. Levels of phthalate metabolites in urine among mother-child-pairs—Results from the Duisburg birth cohort study, Germany. *Int. J. Hyg. Environ. Health* **2012**, *215*, 373–382. [CrossRef]

94. Myridakis, A.; Fthenou, E.; Balaska, E.; Vakinti, M.; Kogevinas, M.; Stephanou, E.G. Phthalate esters, parabens and bisphenol-A exposure among mothers and their children in Greece (Rhea cohort). *Environ. Int.* **2015**, *83*, 1–10. [CrossRef] [PubMed]

95. Jeddi, M.Z.; Gorji, M.E.; Rietjens, I.M.C.M.; Louisse, J.; Bruinen de Bruin, Y.; Liska, R. Biomonitoring and subsequent risk assessment of combined exposure to phthalates in Iranian children and adolescents. *Int. J. Environ. Res. Public Health* **2018**, *15*, 2336. [CrossRef]

96. Cullen, E.; Evans, D.; Griffin, C.; Burke, P.; Mannion, R.; Burns, D.; Flanagan, A.; Kellegher, A.; Schoeters, G.; Govarts, E.; et al. Urinary phthalate concentrations in mothers and their children in Ireland: Results of the DEMOCOPHES human biomonitoring study. *Int. J. Environ. Res. Public Health* **2017**, *14*, 1456. [CrossRef] [PubMed]

97. Berman, T.; Goldsmith, R.; Goen, T.; Spungen, J.; Novack, L.; Levine, H.; Amitai, Y.; Shohat, T.; Grotto, I. Urinary concentrations of environmental contaminants and phytoestrogens in adults in Israel. *Environ. Int.* **2013**, *59*, 478–484. [CrossRef] [PubMed]

98. Tranfo, G.; Papaleo, B.; Caporossi, L.; Capanna, S.; De Rosa, M.; Pigini, D.; Corsetti, F.; Paci, E. Urinary metabolite concentrations of phthalate metabolites in Central Italy healthy volunteers determined by a validated HPLC/MS/MS analytical method. *Int. J. Hyg. Environ. Health* **2013**, *216*, 481–485. [CrossRef]

99. Itoh, H.; Iwasaki, M.; Hanaoka, T.; Sasaki, H.; Tanaka, T.; Tsugane, S. Urinary phthalate monoesters and endometriosis in infertile Japanese women. *Sci. Total Environ.* **2009**, *408*, 37–42. [CrossRef] [PubMed]

100. Itoh, H.; Yoshida, K.; Masunaga, S. Quantitative identification of unknown exposure pathways of phthalates based on measuring their metabolites in human urine. *Environ. Sci. Technol.* **2007**, *41*, 4542–4547. [CrossRef]

101. Suzuki, Y.; Yoshinaga, J.; Mizumoto, Y.; Serizawa, S.; Shiraishi, H. Foetal exposure to phthalate esters and anogenital distance in male newborns. *Int. J. Androl.* **2012**, *35*, 236–244. [CrossRef]

102. Kim, S.; Kang, S.; Lee, G.; Lee, S.; Jo, A.; Kwak, K.; Kim, D.; Koh, D.; Kho, Y.L.; Kim, S.; et al. Urinary phthalate metabolites among elementary school children of Korea: Sources, risks, and their association with oxidative stress marker. *Sci. Total Environ.* **2014**, *472*, 49–55. [CrossRef]

103. Ji, K.; Lim Kho, Y.; Park, Y.; Choi, K. Influence of a five-day vegetarian diet on urinary levels of antibiotics and phthalate metabolites: A pilot study with "Temple Stay" participants. *Environ. Res.* **2010**, *110*, 375–382. [CrossRef]

104. Jo, A.; Kim, H.; Chung, H.; Chang, N. Associations between dietary intake and urinary bisphenol A and phthalates levels in Korean women of reproductive age. *Int. J. Environ. Res. Public Health* **2016**, *13*, 680. [CrossRef] [PubMed]

105. Kim, J.H.; Lee, S.; Shin, M.Y.; Kim, K.N.; Hong, Y.C. Risk assessment for phthalate exposures in the elderly: A repeated biomonitoring study. *Sci. Total Environ.* **2018**, *618*, 690–696. [CrossRef]

106. Choi, W.; Kim, S.; Baek, Y.W.; Choi, K.; Lee, K.; Kim, S.; Yu, S.D.; Choi, K. Exposure to environmental chemicals among Korean adults-updates from the second Korean National Environmental Health Survey (2012–2014). *Int. J. Hyg. Environ. Health* **2017**, *220*, 29–35. [CrossRef] [PubMed]

107. Park, C.; Choi, W.; Hwang, M.; Lee, Y.; Kim, S.; Yu, S.; Lee, I.; Paek, D.; Choi, K. Associations between urinary phthalate metabolites and bisphenol A levels, and serum thyroid hormones among the Korean adult population—Korean National Environmental Health Survey (KoNEHS) 2012–2014. *Sci. Total Environ.* **2017**, *584–585*, 950–957. [CrossRef]

108. Kim, S.; Lee, J.; Park, J.; Kim, H.J.; Cho, G.J.; Kim, G.H.; Eun, S.H.; Lee, J.J.; Choi, G.; Suh, E.; et al. Urinary phthalate metabolites over the first 15months of life and risk assessment—CHECK cohort study. *Sci. Total Environ.* **2017**, *607–608*, 881–887. [CrossRef] [PubMed]

109. Song, N.R.; On, J.W.; Lee, J.; Park, J.D.; Kwon, H.J.; Yoon, H.J.; Pyo, H. Biomonitoring of urinary di(2-ethylhexyl) phthalate metabolites of mother and child pairs in South Korea. *Environ. Int.* **2013**, *54*, 65–73. [CrossRef] [PubMed]

110. Ye, X.; Pierik, F.H.; Hauser, R.; Duty, S.; Angerer, J.; Park, M.M.; Burdorf, A.; Hofman, A.; Jaddoe, V.W.; Mackenbach, J.P.; et al. Urinary metabolite concentrations of organophosphorous pesticides, bisphenol A, and phthalates among pregnant women in Rotterdam, the Netherlands: The Generation R study. *Environ. Res.* **2008**, *108*, 260–267. [CrossRef] [PubMed]

111. Ye, X.; Pierik, F.H.; Angerer, J.; Meltzer, H.M.; Jaddoe, V.W.V.; Tiemeier, H.; Hoppin, J.A.; Longnecker, M.P. Levels of metabolites of organophosphate pesticides, phthalates, and bisphenol A in pooled urine specimens from pregnant women participating in the Norwegian Mother and Child Cohort Study (MoBa). *Int. J. Hyg. Environ. Health* **2009**, *212*, 481–491. [CrossRef]

112. Giovanoulis, G.; Alves, A.; Papadopoulou, E.; Cousins, A.P.; Schutze, A.; Koch, H.M.; Haug, L.S.; Covaci, A.; Magner, J.; Voorspoels, S. Evaluation of exposure to phthalate esters and DINCH in urine and nails from a Norwegian study population. *Environ. Res.* **2016**, *151*, 80–90. [CrossRef] [PubMed]

113. Sabaredzovic, A.; Sakhi, A.K.; Brantsaeter, A.L.; Thomsen, C. Determination of 12 urinary phthalate metabolites in Norwegian pregnant women by core-shell high performance liquid chromatography with on-line solid-phase extraction, column switching and tandem mass spectrometry. *J. Chromatogr. B Anal. Technol. Biomed. Life Sci.* **2015**, *1002*, 343–352. [CrossRef]

114. Correia-Sa, L.; Kasper-Sonnenberg, M.; Palmke, C.; Schutze, A.; Norberto, S.; Calhau, C.; Domingues, V.F.; Koch, H.M. Obesity or diet? Levels and determinants of phthalate body burden—A case study on Portuguese children. *Int. J. Hyg. Environ. Health* **2018**, *221*, 519–530. [CrossRef] [PubMed]

115. Petrovicova, I.; Kolena, B.; Sidlovska, M.; Pilka, T.; Wimmerova, S.; Trnovec, T. Occupational exposure to phthalates in relation to gender, consumer practices and body composition. *Environ. Sci. Pollut. Res. Int.* **2016**, *23*, 24125–24134. [CrossRef] [PubMed]

116. Kolena, B.; Petrovicova, I.; Sidlovska, M.; Pilka, T.; Neuschlova, M.; Valentova, I.; Rybansky, L.; Trnovec, T. Occupational phthalate exposure and health outcomes among hairdressing apprentices. *Hum. Exp. Toxicol.* **2017**, *36*, 1100–1112. [CrossRef]

117. Pilka, T.; Petrovicova, I.; Kolena, B.; Zatko, T.; Trnovec, T. Relationship between variation of seasonal temperature and extent of occupational exposure to phthalates. *Environ. Sci. Pollut. Res. Int.* **2015**, *22*, 434–440. [CrossRef]

118. Valvi, D.; Monfort, N.; Ventura, R.; Casas, M.; Casas, L.; Sunyer, J.; Vrijheid, M. Variability and predictors of urinary phthalate metabolites in Spanish pregnant women. *Int. J. Hyg. Environ. Health* **2015**, *218*, 220–231. [CrossRef]

119. Cutanda, F.; Koch, H.M.; Esteban, M.; Sanchez, J.; Angerer, J.; Castano, A. Urinary levels of eight phthalate metabolites and bisphenol A in mother-child pairs from two Spanish locations. *Int. J. Hyg. Environ. Health* **2015**, *218*, 47–57. [CrossRef]

120. Axelsson, J.; Rylander, L.; Rignell-Hydbom, A.; Jönsson, B.A.G.; Lindh, C.H.; Giwercman, A. Phthalate exposure and reproductive parameters in young men from the general Swedish population. *Environ. Int.* **2015**, *85*, 54–60. [CrossRef]

121. Chen, M.L.; Chen, J.S.; Tang, C.L.; Mao, I.F. The internal exposure of Taiwanese to phthalate—An evidence of intensive use of plastic materials. *Environ. Int.* **2008**, *34*, 79–85. [CrossRef]

122. Wirth, J.J.; Rossano, M.G.; Potter, R.; Puscheck, E.; Daly, D.C.; Paneth, N.; Krawetz, S.A.; Protas, B.M.; Diamond, M.P. A pilot study associating urinary concentrations of phthalate metabolites and semen quality. *Syst. Biol. Reprod. Med.* **2008**, *54*, 143–154. [CrossRef]

123. Teitelbaum, S.L.; Britton, J.A.; Calafat, A.M.; Ye, X.; Silva, M.J.; Reidy, J.A.; Galvez, M.P.; Brenner, B.L.; Wolff, M.S. Temporal variability in urinary concentrations of phthalate metabolites, phytoestrogens and phenols among minority children in the United States. *Environ. Res.* **2008**, *106*, 257–269. [CrossRef] [PubMed]

124. Ferguson, K.K.; Loch-Caruso, R.; Meeker, J.D. Urinary phthalate metabolites in relation to biomarkers of inflammation and oxidative stress: NHANES 1999–2006. *Environ. Res.* **2011**, *111*, 718–726. [CrossRef]

125. Brock, J.W.; Caudill, S.P.; Silva, M.J.; Needham, L.L.; Hilborn, E.D. Phthalate monoesters levels in the urine of young children. *Bull. Environ. Contam. Toxicol.* **2002**, *68*, 309–314. [CrossRef] [PubMed]

126. Just, A.C.; Adibi, J.J.; Rundle, A.G.; Calafat, A.M.; Camann, D.E.; Hauser, R.; Silva, M.J.; Whyatt, R.M. Urinary and air phthalate concentrations and self-reported use of personal care products among minority pregnant women in New York city. *J. Expo. Sci. Environ. Epidemiol.* **2010**, *20*, 625–633. [CrossRef] [PubMed]

127. Polinski, K.J.; Dabelea, D.; Hamman, R.F.; Adgate, J.L.; Calafat, A.M.; Ye, X.; Starling, A.P. Distribution and predictors of urinary concentrations of phthalate metabolites and phenols among pregnant women in the Healthy Start Study. *Environ. Res.* **2018**, *162*, 308–317. [CrossRef]

128. Wenzel, A.G.; Brock, J.W.; Cruze, L.; Newman, R.B.; Unal, E.R.; Wolf, B.J.; Somerville, S.E.; Kucklick, J.R. Prevalence and predictors of phthalate exposure in pregnant women in Charleston, SC. *Chemosphere* **2018**, *193*, 394–402. [CrossRef]

129. Calafat, A.M.; McKee, R.H. Integrating biomonitoring exposure data into the risk assessment process: Phthalates [diethyl phthalate and di(2-ethylhexyl) phthalate] as a case study. *Environ. Health Perspect.* **2006**, *114*, 1783–1789. [CrossRef] [PubMed]

130. Colacino Justin, A.; Harris, T.R.; Schecter, A. Dietary intake Is associated with phthalate body burden in a nationally representative sample. *Environ. Health Perspect.* **2010**, *118*, 998–1003. [CrossRef]

131. Hines, E.P.; Calafat, A.M.; Silva, M.J.; Mendola, P.; Fenton, S.E. Concentrations of phthalate metabolites in milk, urine, saliva, and Serum of lactating North Carolina women. *Environ. Health Perspect.* **2009**, *117*, 86–92. [CrossRef]

132. Buckley, J.P.; Palmieri, R.T.; Matuszewski, J.M.; Herring, A.H.; Baird, D.D.; Hartmann, K.E.; Hoppin, J.A. Consumer product exposures associated with urinary phthalate levels in pregnant women. *J. Expo. Sci. Environ. Epidemiol.* **2012**, *22*, 468–475. [CrossRef]

133. Duty, S.M.; Ackerman, R.M.; Calafat, A.M.; Hauser, R. Personal care product use predicts urinary concentrations of some phthalate monoesters. *Environ. Health Perspect.* **2005**, *113*, 1530–1535. [CrossRef]

134. Schlumpf, M.; Kypke, K.; Wittassek, M.; Angerer, J.; Mascher, H.; Mascher, D.; Vökt, C.; Birchler, M.; Lichtensteiger, W. Exposure patterns of UV filters, fragrances, parabens, phthalates, organochlor pesticides, PBDEs, and PCBs in human milk: Correlation of UV filters with use of cosmetics. *Chemosphere* **2010**, *81*, 1171–1183. [CrossRef] [PubMed]

135. Alves, A.; Vanermen, G.; Covaci, A.; Voorspoels, S. Ultrasound assisted extraction combined with dispersive liquid-liquid microextraction (US-DLLME)—A fast new approach to measure phthalate metabolites in nails. *Anal. Bioanal. Chem.* **2016**, *408*, 6169–6180. [CrossRef] [PubMed]

136. Reddy, B.S.; Rozati, R.; Reddy, S.; Kodampur, S.; Reddy, P.; Reddy, R. High plasma concentrations of polychlorinated biphenyls and phthalate esters in women with endometriosis: A prospective case control study. *Fertil. Steril.* **2006**, *85*, 775–779. [CrossRef] [PubMed]

137. Specht, I.O.; Bonde, J.P.; Toft, G.; Lindh, C.H.; Jonsson, B.A.G.; Jorgensen, K.T. Serum phthalate levels and time to pregnancy in couples from Greenland, Poland and Ukraine. *PLoS ONE* **2015**, *10*, 1371–1385. [CrossRef] [PubMed]
138. Calafat, A.M.; Slakman, A.R.; Silva, M.J.; Herbert, A.R.; Needham, L.L. Automated solid phase extraction and quantitative analysis of human milk for 13 phthalate metabolites. *J. Chromatogr. B Anal. Technol. Biomed. Life Sci.* **2004**, *805*, 49–56. [CrossRef] [PubMed]
139. Silva, M.J.; Malek, N.A.; Hodge, C.C.; Reidy, J.A.; Kato, K.; Barr, D.B.; Needham, L.L.; Brock, J.W. Improved quantitative detection of 11 urinary phthalate metabolites in humans using liquid chromatography–atmospheric pressure chemical ionization tandem mass spectrometry. *J. Chromatogr. B* **2003**, *789*, 393–404. [CrossRef]
140. Calafat, A.M.; Wong, L.Y.; Silva, M.J.; Samandar, E.; Preau, J.L., Jr.; Jia, L.T.; Needham, L.L. Selecting adequate exposure biomarkers of diisononyl and diisodecyl phthalates: Data from the 2005-2006 National Health and Nutrition Examination Survey. *Environ. Health Perspect.* **2011**, *119*, 50–55. [CrossRef]
141. Zota, A.R.; Calafat, A.M.; Woodruff, T.J. Temporal trends in phthalate exposures: Findings from the National Health and Nutrition Examination Survey, 2001–2010. *Environ. Health Perspect.* **2014**, *122*, 235–241. [CrossRef]
142. CDC. NHANES Fourth Annual Report. 2012. Available online: https://www.cdc.gov/exposurereport/pdf/fourthreport.pdf (accessed on 5 April 2019).
143. Johns, L.E.; Cooper, G.S.; Galizia, A.; Meeker, J.D. Exposure assessment issues in epidemiology studies of phthalates. *Environ. Int.* **2015**, *85*, 27–39. [CrossRef]
144. Kim, J.H.; Park, H.; Lee, J.; Cho, G.; Choi, S.; Choi, G.; Kim, S.Y.; Eun, S.H.; Suh, E.; Kim, S.K.; et al. Association of diethylhexyl phthalate with obesity-related markers and body mass change from birth to 3 months of age. *J. Epidemiol. Community Health* **2016**, *70*, 466–472. [CrossRef]
145. Koch, H.M.; Drexler, H.; Angerer, J. Internal exposure of nursery-school children and their parents and teachers to di(2-ethylhexyl) phthalate (DEHP). *Int. J. Hyg. Environ. Health* **2004**, *207*, 15–22. [CrossRef] [PubMed]
146. Shea, K.M.; The AAP Committee on Environmental Health. Pediatric exposure and potential toxicity of phthalate plasticizers. *Pediatrics* **2003**, *111*, 1467–1474. [CrossRef] [PubMed]
147. Lin, S.; Ku, H.-Y.; Su, P.-H.; Chen, J.-W.; Huang, P.-C.; Angerer, J.; Wang, S.-L. Phthalate exposure in pregnant women and their children in central Taiwan. *Chemosphere* **2011**, *82*, 947–955. [CrossRef]
148. Calafat, A.M.; Weuve, J.; Ye, X.; Jia, L.T.; Hu, H.; Ringer, S.; Huttner, K.; Hauser, R. Exposure to bisphenol A and other phenols in neonatal intensive care unit premature infants. *Environ. Health Perspect.* **2009**, *117*, 639–644. [CrossRef]
149. Koch, H.M.; Preuss, R.; Angerer, J. Di(2-ethylhexyl)phthalate (DEHP): Human metabolism and internal exposure—An update and latest results. *Int. J. Androl.* **2006**, *29*, 155–165, discussion 181–185. [CrossRef] [PubMed]
150. Aylward, L.L.; Hays, S.M.; Gagne, M.; Krishnan, K. Derivation of Biomonitoring Equivalents for di-n-butyl phthalate (DBP), benzylbutyl phthalate (BzBP), and diethyl phthalate (DEP). *Regul. Toxicol. Pharmacol.* **2009**, *55*, 259–267. [CrossRef]
151. Hays, S.M.; Aylward, L.L.; Kirman, C.R.; Krishnan, K.; Nong, A. Biomonitoring Equivalents for di-isononyl phthalate (DINP). *Regul. Toxicol. Pharmacol.* **2011**, *60*, 181–188. [CrossRef] [PubMed]
152. Silva, M.J.; Barr, D.B.; Reidy, J.A.; Kato, K.; Malek, N.A.; Hodge, C.C.; Hurtz, D.; Calafat, A.M.; Needham, L.L.; Brock, J.W. Glucuronidation patterns of common urinary and serum monoester phthalate metabolites. *Arch. Toxicol.* **2003**, *77*, 561–567. [CrossRef]
153. Koch, H.M.; Bolt, H.M.; Preuss, R.; Angerer, J. New metabolites of di(2-ethylhexyl)phthalate (DEHP) in human urine and serum after single oral doses of deuterium-labelled DEHP. *Arch. Toxicol.* **2005**, *79*, 367–376. [CrossRef]
154. Mettang, T.; Alscher, D.M.; Pauli-Magnus, C.; Dunst, R.; Kuhlmann, U.; Rettenmeier, A.W. Phthalic acid is the main metabolite of the plasticizer di(2-ethylhexyl) phthalate in peritoneal dialysis patients. *Adv. Perit. Dial.* **1999**, *15*, 229–233. [PubMed]
155. Pollack, G.M.; Buchanan, J.F.; Slaughter, R.L.; Kohli, R.K.; Shen, D.D. Circulating concentrations of di(2-ethylhexyl) phthalate and its de-esterified phthalic acid products following plasticizer exposure in patients receiving hemodialysis. *Toxicol. Appl. Pharmacol.* **1985**, *79*, 257–267. [CrossRef]

156. Choi, J.; Eom, J.; Kim, J.; Lee, S.; Kim, Y. Association between some endocrine-disrupting chemicals and childhood obesity in biological samples of young girls: A cross-sectional study. *Environ. Toxicol. Pharmacol.* **2014**, *38*, 51–57. [CrossRef] [PubMed]

157. Kardas, F.; Bayram, A.K.; Demirci, E.; Akin, L.; Ozmen, S.; Kendirci, M.; Canpolat, M.; Oztop, D.B.; Narin, F.; Gumus, H.; et al. Increased serum phthalates (MEHP, DEHP) and bisphenol A concentrations in children with autism spectrum disorder: The role of endocrine disruptors in autism etiopathogenesis. *J. Child Neurol.* **2016**, *31*, 629–635. [CrossRef]

158. Kim, S.; Lee, J.; Park, J.; Kim, H.J.; Cho, G.; Kim, G.H.; Eun, S.H.; Lee, J.J.; Choi, G.; Suh, E.; et al. Concentrations of phthalate metabolites in breast milk in Korea: Estimating exposure to phthalates and potential risks among breast-fed infants. *Sci. Total Environ.* **2015**, *508*, 13–19. [CrossRef] [PubMed]

159. Mortensen, G.K.; Main, K.M.; Andersson, A.-M.; Leffers, H.; Skakkebæk, N.E. Determination of phthalate monoesters in human milk, consumer milk, and infant formula by tandem mass spectrometry (LC–MS–MS). *Anal. Bioanal. Chem.* **2005**, *382*, 1084–1092. [CrossRef]

160. Kato, K.; Silva, M.J.; Needham, L.L.; Calafat, A.M. Quantifying phthalate metabolites in human meconium and semen using automated off-line solid-phase extraction coupled with on-line SPE and isotope-dilution high-performance liquid chromatography-tandem mass spectrometry. *Anal. Chem.* **2006**, *78*, 6651–6655. [CrossRef] [PubMed]

161. Herr, C.; zur Nieden, A.; Koch, H.M.; Schuppe, H.-C.; Fieber, C.; Angerer, J.; Eikmann, T.; Stilianakis, N.I. Urinary di(2-ethylhexyl)phthalate (DEHP)—Metabolites and male human markers of reproductive function. *Int. J. Hyg. Environ. Health* **2009**, *212*, 648–653. [CrossRef]

162. Buck Louis, G.M.; Smarr, M.M.; Sun, L.; Chen, Z.; Honda, M.; Wang, W.; Karthikraj, R.; Weck, J.; Kannan, K. Endocrine disrupting chemicals in seminal plasma and couple fecundity. *Environ. Res.* **2018**, *163*, 64–70. [CrossRef]

163. Smarr, M.M.; Kannan, K.; Sun, L.; Honda, M.; Wang, W.; Karthikraj, R.; Chen, Z.; Weck, J.; Buck Louis, G.M. Preconception seminal plasma concentrations of endocrine disrupting chemicals in relation to semen quality parameters among male partners planning for pregnancy. *Environ. Res.* **2018**, *167*, 78–86. [CrossRef]

164. Fabjan, E.; Hulzebos, E.; Mennes, W.; Piersma, A.H. A category approach for reproductive effects of phthalates. *Crit. Rev. Toxicol.* **2006**, *36*, 695–726. [CrossRef] [PubMed]

165. Lyche, J.L.; Gutleb, A.C.; Bergman, A.; Eriksen, G.S.; Murk, A.J.; Ropstad, E.; Saunders, M.; Skaare, J.U. Reproductive and developmental toxicity of phthalates. *J. Toxicol. Environ. Health B Crit. Rev.* **2009**, *12*, 225–249. [CrossRef]

166. Casals-Casas, C.; Desvergne, B. Endocrine disruptors: From endocrine to metabolic disruption. *Annu. Rev. Physiol.* **2011**, *73*, 135–162. [CrossRef] [PubMed]

167. Christiansen, S.; Boberg, J.; Axelstad, M.; Dalgaard, M.; Vinggaard, A.M.; Metzdorff, S.B.; Hass, U. Low-dose perinatal exposure to di(2-ethylhexyl) phthalate induces anti-androgenic effects in male rats. *Reprod. Toxicol.* **2010**, *30*, 313–321. [CrossRef] [PubMed]

168. Andrade, A.J.M.; Grande, S.W.; Talsness, C.E.; Grote, K.; Golombiewski, A.; Sterner-Kock, A.; Chahoud, I. A dose-response study following in utero and lactational exposure to di-(2-ethylhexyl) phthalate (DEHP): Effects on androgenic status, developmental landmarks and testicular histology in male offspring rats. *Toxicology* **2006**, *225*, 64–74. [CrossRef] [PubMed]

169. Culty, M.; Thuillier, R.; Li, W.P.; Wang, Y.; Martinez-Arguelles, D.B.; Benjamin, C.G.; Triantafilou, K.M.; Zirkin, B.R.; Papadopoulos, V. In utero exposure to di-(2-ethylhexyl) phthalate exerts both short-term and long-lasting suppressive effects on testosterone production in the rat. *Biol. Reprod.* **2008**, *78*, 1018–1028. [CrossRef] [PubMed]

170. Gray, L.E.; Barlow, N.J.; Howdeshell, K.L.; Ostby, J.S.; Furr, J.R.; Gray, C.L. Transgenerational Effects of Di (2-Ethylhexyl) Phthalate in the Male CRL:CD(SD) Rat: Added Value of Assessing Multiple Offspring per Litter. *Toxicol. Sci.* **2009**, *110*, 411–425. [CrossRef]

171. Jarfelt, K.; Dalgaard, M.; Hass, U.; Borch, J.; Jacobsen, H.; Ladefoged, O. Antiandrogenic effects in male rats perinatally exposed to a mixture of di(2-ethylhexyl) phthalate and di(2-ethylhexyl) adipate. *Reprod. Toxicol.* **2005**, *19*, 505–515. [CrossRef]

172. Vo, T.T.B.; Jung, E.M.; Dang, V.H.; Jung, K.; Baek, J.; Choi, K.C.; Jeung, E.B. Differential effects of flutamide and di-(2-ethylhexyl) phthalate on male reproductive organs in a rat model. *J. Reprod. Dev.* **2009**, *55*, 400–411. [CrossRef]

173. Wilson, V.S.; Howdeshell, K.L.; Lambright, C.S.; Furr, J.; Gray, L.E. Differential expression of the phthalate syndrome in male Sprague-Dawley and Wistar rats after in utero DEHP exposure. *Toxicol. Lett.* **2007**, *170*, 177–184. [CrossRef]

174. Boberg, J.; Metzdorff, S.; Wortziger, R.; Axelstad, M.; Brokken, L.; Vinggaard, A.M.; Dalgaard, M.; Nellemann, C. Impact of diisobutyl phthalate and other PPAR agonists on steroidogenesis and plasma insulin and leptin levels in fetal rats. *Toxicology* **2008**, *250*, 75–81. [CrossRef] [PubMed]

175. Borch, J.; Axelstad, M.; Vinggaard, A.M.; Dalgaard, M. Diisobutyl phthalate has comparable anti-androgenic effects to di-n-butyl phthalate in fetal rat testis. *Toxicol. Lett.* **2006**, *163*, 183–190. [CrossRef] [PubMed]

176. Saillenfait, A.M.; Sabate, J.P.; Gallissot, F. Developmental toxic effects of diisobutyl phthalate, the methyl-branched analogue of di-n-butyl phthalate, administered by gavage to rats. *Toxicol. Lett.* **2006**, *165*, 39–46. [CrossRef]

177. Saillenfait, A.M.; Sabate, J.P.; Gallissot, F. Diisobutyl phthalate impairs the androgen-dependent reproductive development of the male rat. *Reprod. Toxicol.* **2008**, *26*, 107–115. [CrossRef] [PubMed]

178. Scott, H.M.; Hutchison, G.R.; Jobling, M.S.; McKinnell, C.; Drake, A.J.; Sharpe, R.M. Relationship between androgen action in the "Male Programming Window," fetal sertoli cell number, and adult testis size in the rat. *Endocrinology* **2008**, *149*, 5280–5287. [CrossRef] [PubMed]

179. Struve, M.F.; Gaido, K.W.; Hensley, J.B.; Lehmann, K.P.; Ross, S.M.; Sochaski, M.A.; Willson, G.A.; Dorman, D.C. Reproductive toxicity and pharmacokinetics of di-n-butyl phthalate (DBP) following dietary exposure of pregnant rats. *Birth Defects Res. B Dev. Reprod. Toxicol.* **2009**, *86*, 345–354. [CrossRef] [PubMed]

180. Kim, T.S.; Jung, K.K.; Kim, S.S.; Kang, I.H.; Baek, J.H.; Nam, H.S.; Hong, S.K.; Lee, B.M.; Hong, J.T.; Oh, K.W.; et al. Effects of in utero exposure to di(n-butyl) phthalate on development of male reproductive tracts in Sprague-Dawley rats. *J. Toxicol. Environ. Health A Curr. Issues* **2010**, *73*, 1544–1559. [CrossRef]

181. MacLeod, D.J.; Sharpe, R.M.; Welsh, M.; Fisken, M.; Scott, H.M.; Hutchison, G.R.; Drake, A.J.; van den Driesche, S. Androgen action in the masculinization programming window and development of male reproductive organs. *Int. J. Androl.* **2010**, *33*, 279–286. [CrossRef]

182. Jiang, J.T.; Sun, W.L.; Jing, Y.F.; Liu, S.B.; Ma, Z.; Hong, Y.; Ma, L.; Qin, C.; Liu, Q.; Stratton, H.J.; et al. Prenatal exposure to di-n-butyl phthalate induces anorectal malformations in male rat offspring. *Toxicology* **2011**, *290*, 322–326. [CrossRef]

183. Howarth, J.A.; Price, S.C.; Dobrota, M.; Kentish, P.A.; Hinton, R.H. Effects on male rats of di-(2-ethylhexyl) phthalate and di-n-hexylphthalate administered alone or in combination. *Toxicol. Lett.* **2001**, *121*, 35–43. [CrossRef]

184. Zhai, W.H.; Huang, Z.G.; Chen, L.; Feng, C.; Li, B.; Li, T.S. Thyroid endocrine disruption in Zebrafish larvae after exposure to mono-(2-ethylhexyl) phthalate (MEHP). *PLoS ONE* **2014**, *9*, 92465–92471. [CrossRef]

185. O'Connor, J.C.; Frame, S.R.; Ladics, G.S. Evaluation of a 15-day screening assay using intact male rats for identifying antiandrogens. *Toxicol. Sci.* **2002**, *69*, 92–108. [CrossRef] [PubMed]

186. Larsen, S.T.; Hansen, J.S.; Hansen, E.W.; Clausen, P.A.; Nielsen, G.D. Airway inflammation and adjuvant effect after repeated airborne exposures to di-(2-ethylhexyl)phthalate and ovalbumin in BALB/c mice. *Toxicology* **2007**, *235*, 119–129. [CrossRef] [PubMed]

187. Dearman, R.J.; Beresford, L.; Bailey, L.; Caddick, H.T.; Betts, C.J.; Kimber, I. Di-(2-ethylhexyl) phthalate is without adjuvant effect in mice on ovalbumin. *Toxicology* **2008**, *244*, 231–241. [CrossRef] [PubMed]

188. Li, Y.F.; Zhuang, M.Z.; Li, T.; Shi, N. Neurobehavioral toxicity study of dibutyl phthalate on rats following in utero and lactational exposure. *J. Appl. Toxicol.* **2009**, *29*, 603–611. [CrossRef]

189. Li, X.J.; Jiang, L.; Chen, L.; Chen, H.S.; Li, X. Neurotoxicity of dibutyl phthalate in brain development following perinatal exposure: A study in rats. *Environ. Toxicol. Pharmacol.* **2013**, *36*, 392–402. [CrossRef]

190. Tanaka, T. Reproductive and neurobehavioural toxicity study of bis(2-ethylhexyl) phthalate (DEHP) administered to mice in the diet. *Food Chem. Toxicol.* **2002**, *40*, 1499–1506. [CrossRef]

191. Hoshi, H.; Ohtsuka, T. Adult rats exposed to low-doses of di-n-butyl phthalate during gestation exhibit decreased grooming behavior. *Bull. Environ. Contam. Toxicol.* **2009**, *83*, 62–66. [CrossRef]

192. Sun, Q.; Cornelis, M.C.; Townsend, M.K.; Tobias, D.K.; Heather Eliassen, A.; Franke, A.A.; Hauser, R.; Hu, F.B. Association of urinary concentrations of bisphenol A and phthalate metabolites with risk of type 2 diabetes: A prospective investigation in the nurses' health study (NHS) and NHSII cohorts. *Environ. Health Perspect.* **2014**, *122*, 616–623. [CrossRef]

193. Stahlhut Richard, W.; van Wijngaarden, E.; Dye Timothy, D.; Cook, S.; Swan Shanna, H. Concentrations of urinary phthalate metabolites are associated with increased waist circumference and insulin resistance in adult U.S. males. *Environ. Health Perspect.* **2007**, *115*, 876–882. [CrossRef]

194. Trasande, L.; Sathyanarayana, S.; Spanier, A.J.; Trachtman, H.; Attina, T.M.; Urbina, E.M. Urinary phthalates are associated with higher blood pressure in childhood. *J. Pediatr.* **2013**, *163*, 747–753. [CrossRef]

195. Wang, H.; Zhou, Y.; Tang, C.; He, Y.; Wu, J.; Chen, Y.; Jiang, Q. Urinary phthalate metabolites are associated with body mass index and waist circumference in Chinese school children. *PLoS ONE* **2013**, *8*, 56800–56809. [CrossRef] [PubMed]

196. Buser, M.C.; Murray, H.E.; Scinicariello, F. Age and sex differences in childhood and adulthood obesity association with phthalates: Analyses of NHANES 2007–2010. *Int. J. Hyg. Environ. Health* **2014**, *217*, 687–694. [CrossRef] [PubMed]

197. Zhang, Y.; Meng, X.; Chen, L.; Li, D.; Zhao, L.; Zhao, Y.; Li, L.; Shi, H. Age and sex-specific relationships between phthalate exposures and obesity in Chinese children at puberty. *PLoS ONE* **2014**, *9*, e104852. [CrossRef] [PubMed]

198. Hatch, E.; Nelson, J.; Qureshi, M. Association of urinary phthalate metabolite concentrations with body mass index and waist circumference: A crosssectional study of NHANES data 1999–2002. *Environ. Health* **2008**, *15*, 1–15. [CrossRef] [PubMed]

199. Dirtu, A.C.; Geens, T.; Dirinck, E.; Malarvannan, G.; Neels, H.; Van Gaal, L.; Jorens, P.G.; Covaci, A. Phthalate metabolites in obese individuals undergoing weight loss: Urinary levels and estimation of the phthalates daily intake. *Environ. Int.* **2013**, *59*, 344–353. [CrossRef] [PubMed]

200. Braun, J.M.; Sathyanarayana, S.; Hauser, R. Phthalate exposure and children's health. *Curr. Opin. Pediatr.* **2013**, *25*, 247–254. [CrossRef] [PubMed]

201. Hoppin, J.A.; Jaramillo, R.; London, S.J.; Bertelsen, R.J.; Salo, P.M.; Sandler, D.P.; Zeldin, D.C. Phthalate exposure and allergy in the U.S. population: Results from NHANES 2005-2006. *Environ. Health Perspect.* **2013**, *121*, 1129–1134. [CrossRef] [PubMed]

202. Ait Bamai, Y.; Shibata, E.; Saito, I.; Araki, A.; Kanazawa, A.; Morimoto, K.; Nakayama, K.; Tanaka, M.; Takigawa, T.; Yoshimura, T.; et al. Exposure to house dust phthalates in relation to asthma and allergies in both children and adults. *Sci. Total Environ.* **2014**, *485–486*, 153–163. [CrossRef] [PubMed]

203. Wang, I.J.; Karmaus, W.J.; Chen, S.L.; Holloway, J.W.; Ewart, S. Effects ofphthalate exposure on asthma may be mediated through alterations in DNAmethylation. *Clin. Epigenet.* **2015**, *7*, 27–36. [CrossRef]

204. Gascon, M.; Casas, M.; Morales, E.; Valvi, D.; Ballesteros-Gómez, A.; Luque, N.; Rubio, S.; Monfort, N.; Ventura, R.; Martínez, D.; et al. Prenatal exposure to bisphenol A and phthalates and childhood respiratory tract infections and allergy. *J. Allergy Clin. Immunol.* **2015**, *135*, 370–378. [CrossRef] [PubMed]

205. Frederiksen, H.; Sørensen, K.; Mouritsen, A.; Aksglaede, L.; Hagen, C.P.; Petersen, J.H.; Skakkebaek, N.E.; Andersson, A.M.; Juul, A. High urinary phthalate concentration associated with delayed pubarche in girls. *Int. J. Androl.* **2012**, *35*, 216–226. [CrossRef]

206. Upson, K.; Sathyanarayana, S.; De Roos, A.J.; Thompson, M.L.; Scholes, D.; Dills, R.; Holt, V.L. Phthalates and risk of endometriosis. *Environ. Res.* **2013**, *126*, 91–97. [CrossRef]

207. Durmaz, E.; Özmert, E.N.; Erkekoğlu, P.; Giray, B.; Derman, O.; Hıncal, F.; Yurdakök, K. Plasma phthalate levels in pubertal gynecomastia. *Pediatrics* **2010**, *125*, 122–129. [CrossRef] [PubMed]

208. Radke, E.G.; Braun, J.M.; Meeker, J.D.; Cooper, G.S. Phthalate exposure and male reproductive outcomes: A systematic review of human epidemiological evidence. *Environ. Int.* **2018**, *121*, 764–793. [CrossRef] [PubMed]

Review

Bioanalytical and Mass Spectrometric Methods for Aldehyde Profiling in Biological Fluids

Romel P. Dator, Morwena J. Solivio, Peter W. Villalta * and Silvia Balbo *

Masonic Cancer Center, University of Minnesota, 2231 6th Street SE, Minneapolis, MN 55455, USA;
rpdator@umn.edu (R.P.D.); msolivio@umn.edu (M.J.S.)
* Correspondence: villa001@umn.edu (P.W.V.); balbo006@umn.edu (S.B.);
  Tel.: +1-612-626-8165 (P.W.V.); +1-612-624-4240 (S.B.)

Received: 22 March 2019; Accepted: 22 May 2019; Published: 4 June 2019

Abstract: Human exposure to aldehydes is implicated in multiple diseases including diabetes, cardiovascular diseases, neurodegenerative disorders (i.e., Alzheimer's and Parkinson's Diseases), and cancer. Because these compounds are strong electrophiles, they can react with nucleophilic sites in DNA and proteins to form reversible and irreversible modifications. These modifications, if not eliminated or repaired, can lead to alteration in cellular homeostasis, cell death and ultimately contribute to disease pathogenesis. This review provides an overview of the current knowledge of the methods and applications of aldehyde exposure measurements, with a particular focus on bioanalytical and mass spectrometric techniques, including recent advances in mass spectrometry (MS)-based profiling methods for identifying potential biomarkers of aldehyde exposure. We discuss the various derivatization reagents used to capture small polar aldehydes and methods to quantify these compounds in biological matrices. In addition, we present emerging mass spectrometry-based methods, which use high-resolution accurate mass (HR/AM) analysis for characterizing carbonyl compounds and their potential applications in molecular epidemiology studies. With the availability of diverse bioanalytical methods presented here including simple and rapid techniques allowing remote monitoring of aldehydes, real-time imaging of aldehydic load in cells, advances in MS instrumentation, high performance chromatographic separation, and improved bioinformatics tools, the data acquired enable increased sensitivity for identifying specific aldehydes and new biomarkers of aldehyde exposure. Finally, the combination of these techniques with exciting new methods for single cell analysis provides the potential for detection and profiling of aldehydes at a cellular level, opening up the opportunity to minutely dissect their roles and biological consequences in cellular metabolism and diseases pathogenesis.

Keywords: aldehydes; genotoxicity; cancer; diseases; oxidative stress; exposure biomarkers; high-resolution mass spectrometry; data-dependent profiling; derivatization; biological fluids; isotope labeling

---

1. Introduction

Sources of Human Exposure to Aldehydes

Aldehydes are characterized by the presence of a $-HC = O$ reactive site and often exist in combination with other functional groups. They are ubiquitous in the environment, originating from man-made sources, as well as through natural processes (Figure 1). The hydroxyl radical mediated-photochemical oxidation of hydrocarbons generates aldehydes in the atmosphere [1–3]. For instance, formaldehyde is produced from the oxidation of methane and naturally occurring compounds, such as terpenoids and isoprenoids from tree foliage [2]. In industrialized areas, the majority of aldehydes are produced from motor vehicle exhaust (internal diesel engine combustion), which either directly yields aldehydes or generates hydrocarbons, which are eventually converted to aldehydes by photochemical oxidation reactions [1,4–8]. Formaldehyde, acetaldehyde, and acrolein are significant contributors to the overall summed risk of mobile sources of air toxicants according to the United States Environmental Protection Agency (U.S. EPA) [1]. Other sources of aldehydes include agricultural and forest fires, incinerators, and coal-based power plants [9–13]. Additionally, humans are exposed to aldehydes in residential and occupational settings where aldehydes are present in confined spaces [14] due to the release of fumes from indoor furniture, carpets, fabrics, household cleaning agents, cosmetic products, and paints [12,15–18]. Aldehydes are also widely used as fumigants and for biological specimen preservation [1]. Another major source of aldehyde exposure comes from cigarette smoke. Mainstream tobacco smoke (MTS) is composed of significant amounts of acetaldehyde as the major component, followed by acrolein, formaldehyde, and crotonaldehyde [19–26]. Similarly, popular devices such as e-cigarettes, which are advocated as safer alternatives to tobacco, have been found to generate high concentrations of aldehydes [27–37]. Aldehydes are also present in food and beverages (as flavorings), and in alcoholic drinks either as congeners or, in the case of acetaldehyde, as the oxidative by-product of ethanol [38–40]. Biotransformation is another source of aldehyde exposure. This includes metabolism of a sizeable number of environmental agents, such as drugs, tobacco smoke, alcohol, and other forms of xenobiotics [41–43]. Of note, exposure also comes from the metabolism of a number of widely used anticancer drugs such as cyclophosphamide, ifosfamide, and misonidazole as well as other drugs used for the treatment of diseases such as epilepsy and HIV-1 infection [1]. The production of aldehydes is proposed to be an important contributor to the toxicity and undesirable side effects of treatment with these drugs.

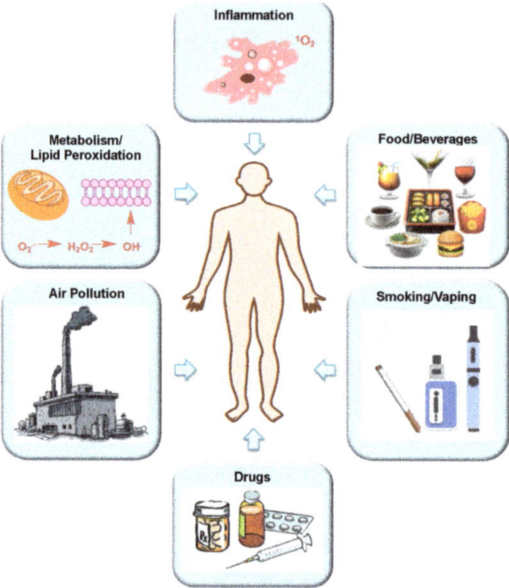

Figure 1. Exogenous and endogenous sources of human exposure to aldehydes.

Finally, normal cellular metabolic pathways such as lipid peroxidation, Alk-B type repair, histone demethylation, carbohydrate or ascorbate autoxidation, carbohydrate metabolism, and amine oxidase-, cytochrome P-450-, and myeloperoxidase-catalyzed metabolic pathways produce aldehydes endogenously [1,44,45]. The metabolism of molecules such as amino acids, vitamins, and steroids, to name a few, also generates aldehydes [46]. Aldehydes are generally formed during conditions of high oxidative stress. Oxidants are generated as a result of normal intracellular metabolism in the mitochondria, peroxisomes, and a number of cytosolic enzyme systems [47]. These metabolic free radicals and oxidants are referred to as reactive oxygen species (ROS). A balance between ROS production and removal by the antioxidant defense systems is essential to maintaining redox homeostasis. A disturbance in the balance favoring pro-oxidative conditions results to oxidative stress. An elevated level of ROS and the resulting oxidative stress leads to biological damage and is implicated in aging and pathologies of various conditions including cancer, cardiovascular, inflammatory, and neurodegenerative diseases [47,48]. The generation of aldehydes is one important consequence of sustained oxidative stress, which can result to the auto-oxidation of lipids (damaging cell membranes) and fatty acids within cells. Lipid peroxidation occurs when a variety of ROS and/or reactive nitrogen species (RNS) oxidize lipids containing carbon-carbon double bonds, especially polyunsaturated fatty acids, resulting in free radical chain reactions and subsequent formation of by-products such as lipid radicals, hydrocarbons, and aldehydes [49]. The correlation between elevated ROS and aldehyde production has been extensively studied and is known to contribute to a multitude of disease pathologies by altering proteomic, genomic, cell signaling, and metabolic processes [50,51]. Indeed, 4-hydroxy-2-nonenal (4-HNE) and malondialdehyde (MDA) are both used as markers of the magnitude of oxidative stress and lipid peroxidation [52]. Dietary consumption of polyunsaturated fatty acids and the subsequent oxidation of these molecules can also result to the formation of aldehydes. The carbohydrate or ascorbate autoxidation pathways generate endogenous glyoxal, which is a major lipid and DNA oxidative degradation product [1]. Likewise, methylglyoxal is produced through the enzymatic reactions of triose phosphate intermediates such as glyceraldehyde-3-phosphate and dihydroxyacetone phosphate during glycolysis or from the metabolism of ketone bodies or threonine [53]. The serum amine oxidase (SAO) and polyamine oxidase (PAO) also generate endogenous aldehydes by catalyzing the deamination of biogenic amines [1]. In summary, the dysregulation of metabolic processes and oxidative stress result in lipid peroxidation, carbohydrate auto-oxidation, protein oxidation, as well as polyamine catabolism, all result in aldehyde formation.

2. Biological Consequences of Aldehyde Exposure on Genome Integrity, Carcinogenesis, and Other Diseases

Low molecular weight aldehydes such as formaldehyde, acetaldehyde, and acrolein are generally toxic compounds. The majority of the most abundant aldehydes are irritants at high doses, and, due to their volatility, induce acute inhalation toxicity. Additionally, aldehydes are believed to play major roles in various debilitating diseases such as cancer and neurodegeneration. Aldehydes are highly reactive, electrophilic compounds, which can exert their detrimental role through interactions with various biomolecules such as phospholipids, peptides, regulatory proteins, enzymes, and DNA forming covalent modifications, affecting their normal functions and leading to mutations and chromosomal aberrations. These mediated effects vary from physiological and homeostatic, to cytotoxic, mutagenic, and carcinogenic [54,55]. Figure 2 shows the structures of common aldehydes implicated in the pathogenesis of multiple human diseases.

Formaldehyde and acetaldehyde, from alcohol consumption, have been classified as Group 1 human carcinogens by the International Agency for Research on Cancer (IARC) [56–59]. Both compounds are believed to exert their carcinogenic effects by reacting with DNA, forming covalent modifications known as DNA adducts [60–66]. These adducts if not repaired or eliminated may translate into mutations and ultimately into dysregulation of normal cellular growth. Aldehyde toxicity is also implicated in aging, and age-related diseases such as cardiovascular and neurological disorders [67–70]. Unlike free radicals with shorter half-lives ranging from nanoseconds to milliseconds, reactive carbonyl compounds (RCCs) including aldehydes are more stable with half-lives ranging from minutes to hours. Because of this relative stability, aldehydes are long-lived and can therefore diffuse from the point of origin and intracellularly and extracellularly attack targets, which are distant from the radical events [71,72].

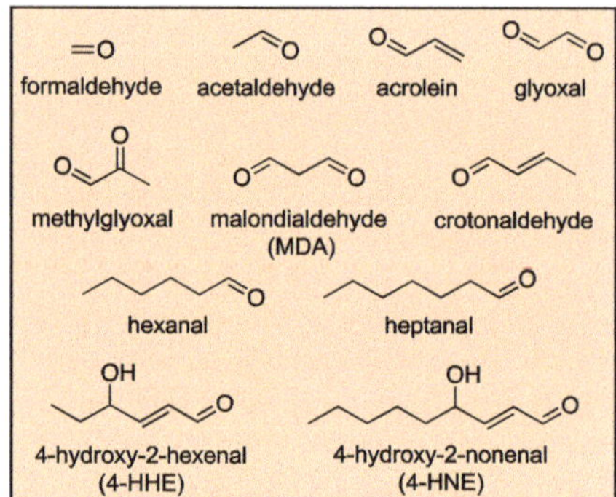

Figure 2. Structures of common aldehydes associated with various human diseases.

Mounting evidence indicates that endogenous aldehydes, such as MDA, 4-HNE, 3-aminopropanal (3-AP), acrolein, formaldehyde, and methylglyoxal, are mediators of neurodegeneration [73] and aldehydes formed during lipid peroxidation (advanced lipoxidation end-products, ALEs) and sugar glycoxidation (advanced glycoxidation end-products, AGEs) accumulate in several oxidative stress and aging disorders [74]. Aldehydes foster oligomerization of proteins and peptides found in neuritic plaques, which is a characteristic of Alzheimer's disease (AD) [75–77]. Physiological concentrations of these aldehydes range from nM to several hundred μM [78,79]. Methylglyoxal concentration in human blood is estimated to be in the 100–120 nM range, while its cellular concentration is about 1–5 μM and 0.1–1 μM for glyoxal [80–82]. MDA concentration in serum is 0.93 ± 0.39 μM [83] and 4-HNE concentration in cells is less than 1 μM [52]. Likewise, the levels of acrolein formed by metabolism are hard to quantify and may reach very high levels in certain microenvironments [84]. Increased levels of these aldehydes in brain and cerebrospinal fluid (CSF) were reported for various neurodegenerative disorders [85]. The levels of 4-HNE are found to increase in the brain regions of deceased AD patients compared to age-matched controls [86], and are elevated in CSF of AD patients compared to healthy controls [87]. Likewise, acrolein is found to be elevated in the amygdala and hippocampus/parahippocampal gyrus in brains of AD patients compared to controls [88]. Protein carbonylation has been associated with the progression of several neurodegenerative disorders including AD, Parkinson's disease (PD), multiple sclerosis (MS) and amyotrophic lateral sclerosis (ALS).

Methylglyoxal is found at significantly higher levels in diabetic patients compared to healthy controls [89], while 4-HNE, is known to form adducts with mitochondrial proteins, (specifically through interactions with cysteine, histidine, and lysine residues), lipids, and DNA resulting to mitochondrial malfunction. The mitochondrial electron transport chain is the most important source of endogenous ROS, converting 1–2% of the total oxygen consumed into superoxide anions [90,91]. An estimate of 1–8% of 4-HNE produced in cells will form adducts with proteins, with 30% of it occurring in the mitochondria, making it consequential in ROS production [51,92,93]. In some cases, ROS overproduction has been associated with mutations in a mitochondrial gene that encodes a component of the electron transport chain [94]. Increasing damage to mitochondrial DNA inevitably results to compromised mitochondrial function and integrity, leading to a vicious cycle of ROS generation and DNA damage [91]. Oxidative damage to mitochondrial DNA in the heart and the brain has been shown to decrease the lifespan in mammals, and mitochondrial dysfunction has been associated with some neurological disorders including AD, PD, Huntington's Diseases (HD), and ALS [48,95].

Finally, endogenous aldehydes may also play a role in the free radical theory of aging at the molecular level, which has gained widespread attention and acceptance. In this context, aging is viewed as a process related to an imbalance favoring pro-oxidant over antioxidant molecules (either by ROS elevation or an age-related downregulation of antioxidant molecules and ROS-mitigating enzymes) and consequently an increase in oxidative stress and the level of aldehydes resulting from it [72].

Despite the fact that these molecules fundamentally underlie early events driving the initiation and propagation of various pathologies, their exact role and diagnostic or prognostic value as clinical biomarkers have been underexploited [96]. The complete cellular "aldehydic load" is considered an important parameter for appraisal of these pathologic statuses [97,98]. Developing methods to detect free aldehydes in biological systems is important in understanding the roles and functions of these molecules in cellular processes and disease pathogenesis. The measurement of free aldehydes has the potential to be used to characterize exposure, but also to identify biomarkers for early disease diagnosis, monitor disease progression and response to therapy, and investigate physiological malfunctions such as high oxidative stress.

3. Metabolism of Aldehydes

As outlined in the previous section, excessive exposure to aldehydes can result in the disruption of a number of cellular functions, which can ultimately contribute to human diseases. The balance between the activation and detoxification of aldehydes will dictate their toxicity, which is dependent on the aldehyde itself and the presence of aldehyde metabolizing enzymes in cells. Several metabolic pathways and metabolizing enzymes are responsible for the metabolism and detoxification of aldehydes. These enzymes include aldehyde-oxidizing enzymes, aldehyde-reducing enzymes, and glutathione (GSH)-dependent aldehyde metabolizing enzymes, as previously reviewed by O'Brien [1]. For instance, 4-HNE is metabolized by glutathione S-transferase (GST) and aldehyde dehydrogenase 2 (ALDH2), and to a minor extent alcohol dehydrogenase (ADH) in rat hepatocytes [92,99–102]. Methylglyoxal is likely metabolized by glyoxalase (GLOX) and reduced by aldo-keto reductase (AKR) 1A2 [1]. The inhibition of ALDH2 activity, with the consequent increase in the level of aldehydes by oxidative stress was also observed in humans and diabetic mice during aging and is associated with cardiac dysfunction [103]. Elimination and in vivo metabolism of alkanals and aromatic aldehydes is via dehydrogenase-catalyzed oxidation. Likewise, the main in vivo elimination and metabolism of alkenals such as acrolein is via glutathione conjugation catalyzed by glutathione transferases [1].

In the case of formaldehyde, its metabolism is known to be mediated by alcohol and aldehyde dehydrogenases, ADH5 and ALDH2, respectively. Depletion of GSH levels in hepatocytes and inhibition of these enzymes result in a marked increase in formaldehyde cytotoxicity [104]. Formaldehyde is a potent DNA and protein cross-linking molecule that organisms produce in vast quantities, through one carbon metabolism (1C-metabolism), and in processes such as enzymatic demethylation of histones and nucleic acids [105]. This is supported by the blood formaldehyde concentration, which ranges from 20–100 μM, and 200–400 μM in a healthy human brain, indicating a substantial source of this molecule [106–109]. A study on mice revealed a two-tier protection mechanism, shielding mice from high levels of endogenous formaldehyde. The first tier involved the enzyme ADH5, which eliminates formaldehyde, while the Fanconi Anemia pathway for cross-link repair reverts DNA damage due to formaldehyde. It was hypothesized that ADH5-dependent formaldehyde oxidation into formate could provide 1C units to enable nucleotide synthesis [110]. Formaldehyde reacts spontaneously with intracellular GSH, present in substantial amounts to form S-hydroxymethylglutathione (HMGSH), which undergoes oxidation by ADH5 and NAD(P)$^+$ to generate S-formylglutathione (FGSH), which is subsequently converted by S-formylglutathione hydrolase (FGH) regenerating GSH and yielding formate. The formate formed in this process is eventually used in biosynthetic reactions [111], thus showing that formaldehyde detoxification produces a 1C unit sustaining essential metabolism [55], including the biosynthesis of purines and thymidine, homeostasis of amino acids glycine, serine, and methionine, epigenetic maintenance, and redox defense [112]. This biochemical route of formaldehyde detoxification can therefore provide the cell with utilizable 1C units [111]. Since this genotoxic molecule is generated in large amounts in the human body, a steady-state balance between formaldehyde generation and removal is established due to detoxification by cellular enzymes including alcohol dehydrogenase 1 (ADH1), which reduces cytosolic formaldehyde to methanol, mitochondrial ALDH2, cytosolic alcohol dehydrogenase 3 (ADH3), also known as glutathione-dependent formaldehyde dehydrogenase, as well the previously mentioned ADH5, all responsible for formaldehyde metabolism [113–116].

Aldehydes are oxidized by the aldehyde dehydrogenase superfamily, of which 16 genes and 3 pseudogenes have been identified in the human genome, including ALDH1A, ALDH2, ALDH1B1, ALDH3A1, and ALDH3A2. ALDH2, for example, is efficient at metabolizing acetaldehyde, a reactive metabolite of ethanol, to acetate and likely plays a major role in reducing the toxicity of aldehydes in humans [117]. Likewise, the aldehyde-reducing enzymes are another superfamily of enzymes responsible for the reduction of aldehydes to alcohol using NADH as a cofactor, and which can be divided into several classes corresponding to the necessary cofactors. The ADH superfamily preferentially uses NADH to reduce aldehydes to alcohols, while using NAD+ to do the reverse reaction but to a lesser extent [1]. This class of enzymes is located in the cytosol and includes ADH1, ADH2, and ADH3. The aldo-keto reductase superfamily uses NADPH solely while others use both NADPH and NADH. This class of enzymes includes AKR1A1, AKR1C, and AKR7A1. The short-chain dehydrogenase/reductase superfamily is another class of aldehyde reducing enzymes responsible for the detoxification of aldehydes in cells. This class of enzymes includes carbonyl reductase (CR) and hydroxypyruvate reductase (GRHPR). CR is considered the main quinone oxidoreductase in human liver and catalyzes the two-electron reductive detoxification of quinones, including PAHs [118]. Another class of aldehyde metabolizing enzymes are GSH-dependent, including ADH5, GSTs, and glyoxalase 1 (GLO1). The class III alcohol dehydrogenase detoxifies formaldehyde via glutathione conjugation. Glutathione conjugation is catalyzed by glutathione transferases and predominantly forms conjugates with alkenals and hydroxyalkenals. Glyoxal and methylglyoxal are metabolized by glutathione conjugation and subsequent isomerization by glyoxalases [1]. The activities of these enzymes in living cells dictate the toxicity of aldehydes. Given these well-established associations of reactive carbonyls in cellular metabolism and contributions in human diseases, methods that will allow the elucidation of their roles and functions in biological systems are needed. This panel of biomarkers could be used to determine exposure, early disease diagnosis, and for monitoring disease progression, as well as therapeutic efficacy.

4. Bioanalytical and Mass Spectrometric Methods for Characterizing Aldehydes

There are a wide variety of analytical and biochemical techniques used to identify and quantify aldehydes. Traditionally, the analysis of aldehydes or carbonyl compounds is performed on matrices such as air, water, and soil for environmental monitoring of air and water quality by US federal agencies such as the US EPA, NIOSH, and ASTM (see Section 4.2 below) [119–123]. Because aldehydes play important roles in cellular processes and are linked to various diseases, these methods were further extended for the identification and characterization of these compounds in biological fluids such as plasma, cerebrospinal fluid (CSF), urine, exhaled breath condensate (EBC), and saliva. One challenging aspect in the measurement of aldehydes in biological matrices is their inherent volatility, polarity, and biochemical instability. Thus, derivatization is commonly used for the analysis of low molecular weight aldehydes in complex matrices to improve chromatographic separation, MS ionization, and MS/MS fragmentation detectability [119,124–127]. A wide range of derivatization reagents, as previously reviewed by Santa [124], and analytical methods are being applied for the analysis of carbonyl compounds in food and beverages, as previously reviewed by Osorio [39]. The different derivatization techniques and analytical methods used to identify and measure these compounds have their strengths and limitations, and, depending on the information one wants to obtain, there are techniques and experimental strategies that are suitable for each specific application. Nonetheless, methods to improve the overall sensitivity and detection of aldehydes in complex biological matrices are still being developed to enable trace level analysis and allow elucidation of their contributions and impact on human health.

4.1. Colorimetric/Fluorimetric/Amperometric Methods

One of the most commonly used methods for the analysis of aldehydes in biological fluids is the assay of thiobarbituric acid reactive substances (TBARS), which are produced under high oxidative stress conditions resulting from lipid peroxidation. Oxidation of lipids generates reactive and unstable lipid hydroperoxides and further decomposition of these hydroperoxides yields MDA, a well-known biomarker of oxidative stress. MDA forms a 1:2 adduct with 2-thiobarbituric acid (2-TBA) and can be measured spectrophotometrically or fluorimetrically [128,129] (Figure 3). Although the specificity of this approach is in question as TBA can react with compounds other than MDA, it is still widely applied to measure lipid peroxidation in various biological samples including animal and human tissues and biofluids, as well as food and drugs [129]. One strategy employed to overcome the limitation of this assay is the prior precipitation of lipoproteins to eliminate interfering soluble 2-TBA-reactive substances. As TBARS are minimized, the assay becomes quite specific for lipid peroxidation [129,130]. In addition, extraction of MDA-reactant adducts is also employed, however, this approach introduces another time-consuming step and adversely affects precision of the assay [130].

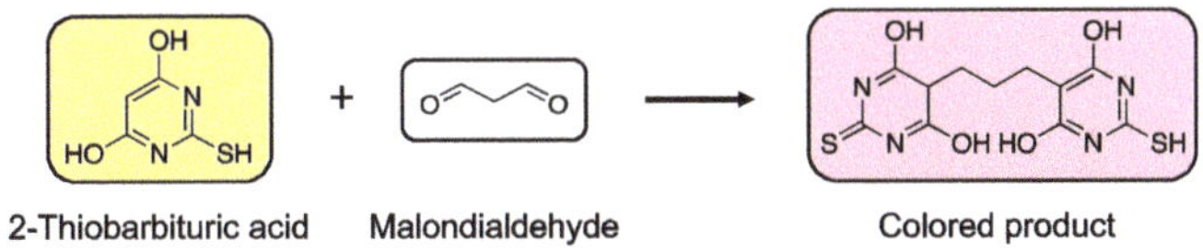

Figure 3. Reaction of 2-thiobarbituric acid (2-TBA) with malondialdehyde (MDA), a biomarker of oxidative stress. 2-TBA reacts with MDA to form a colored product, which is measured spectrophotometrically at 532 nm. The intensity of the colored product reflects the level of lipid peroxidation in the sample.

Another rapid and simple strategy to determine aldehydes in biological fluids, such as saliva, is the development of a microfluidic paper-based analytical device (μPAD) [131]. This device is based on the reaction of aldehydes with 3-methyl-2-benzothiazolinone hydrazine (MBTH) and iron (III) to form a blue formazan complex, which can be evaluated visually (Figure 4) [131]. This approach is simple, rapid, and non-invasive for the analysis of salivary aldehydes, which could be useful in assessing oral cancer risk in population-based studies and point-of-care diagnostics for aldehyde exposure. Methods based on capillary electrophoresis, coupled with amperometric detection (CE-AD) and using electroactive 2-TBA, have been developed and used to analyze two non-electroactive aldehydes, methylglyoxal and glyoxal in urine and water samples. This method demonstrates good specificity for methylglyoxal and glyoxal with the formation of stable pink-chromophore adducts with 2-TBA. Using this approach, the LODs (limit of detection) obtained are 0.2 μg L^{-1} (0.6 nmol L^{-1}) and 1.0 μg L^{-1} (3.2 nmol L^{-1}) for methylglyoxal and glyoxal, respectively [132]. The approaches described above are simple and the instrumentation is easy to use and operate for rapid screening of aldehydes in various matrices. In addition, these analytical techniques can be applied for remote monitoring of aldehydes where more sophisticated bioanalytical tools and mass spectrometry instrumentation are not available. The limitations of these techniques, however, are their low specificity and selectivity for identifying aldehydes, which can be further confounded with increased matrix complexity.

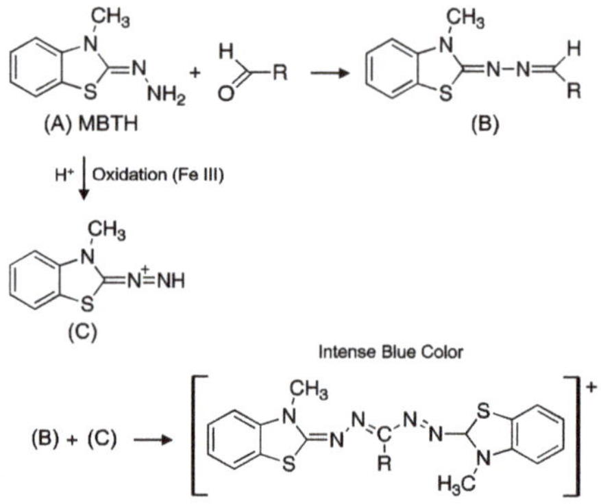

Figure 4. Reaction of MBTH with aldehydes to form an intense blue-colored complex. Figure adapted from Reference [131] (Copyright 2016, Elsevier).

4.2. High-Performance Liquid Chromatography (HPLC) with Ultraviolet (UV)/Fluorescence Detection

Historically, HPLC-UV has been the method of choice for characterizing and quantifying aldehydes in a wide array of matrices and were originally developed for environmental analysis. However, characterization and quantification of aldehydes has gained widespread use in the food and beverage industry, and in the biomedical field, where aldehydes have been shown to play major roles in cellular processes and disease pathogenesis. In addition, the derivatization of carbonyl compounds is typically accomplished using 2,4-dinitrophenylhydrazine (DNPH) to form their corresponding carbonyl-hydrazones. The carbonyl-hydrazones are then analyzed by HPLC with ultraviolet detection. HPLC-UV detection is commonly used to characterize and quantify carbonyl compounds in various matrices because of its simplicity, robustness, and reproducibility. DNPH derivatization and HPLC-UV analysis are used in environmental monitoring of air and water quality and used for screening and monitoring carbonyl compounds in various matrices by the US federal agencies (Table 1) [119,133–137]. The HPLC-UV technique is also being used in the food industry to measure aldehydes in food and beverages [39,138–142] and in biomedical research to measure aldehydes and carbonyls in various matrices such as urine, plasma and serum samples [40,143–152]. DNPH derivatization is also used in conjunction with a reducing agent, 2-picoline borane (2-PB) to stabilize carbonyl-hydrazones and to resolve isomeric compounds produced during the reaction that might interfere with subsequent quantitative analysis by HPLC-UV [153]. DNPH and hydroquinone impregnated into silica cartridges has been used for the determination of acrolein and other carbonyl compounds in cigarette smoke [22]. This approach is useful for characterizing carbonyls in air samples for environmental analysis as well as for the characterization of other α,β-unsaturated aldehydes in tobacco smoke. DNPH derivatization was also used for the analysis and measurement of acetaldehyde in plasma and red blood cells [154], formaldehyde determination in human tissue [151], carbonyl compounds in exhaled breath of e-cigarette users [35], and for the measurement of formaldehyde released from heated hair straightening cosmetic products [18]. Other reagents such as the previously mentioned 2-thiobarbituric acid (2-TBA) and diaminonapthhalene (DAN) are also being used for HPLC-UV analysis of carbonyl compounds from biological matrices and environmental samples [155–157].

To improve sensitivity and allow for simultaneous derivatization and extraction of derivatized carbonyls for HPLC-UV analysis, a wide array of sample preparation techniques have been introduced into the analytical workflows. For instance, a method for the quantification of early lung cancer biomarkers, hexanal and heptanal in urine, has been developed using a bar adsorptive microextraction (BAµE) technique and DNPH derivatization. This approach uses an adsorptive bar impregnated with the derivatization reagent for simultaneous derivatization and extraction of derivatized carbonyls. The LODs obtained for hexanal and heptanal are 0.80 µmol L^{-1} (800 nmol L^{-1}) and 0.40 µmol L^{-1}

(400 nmol L^{-1}), respectively [145]. Similarly, magnetic solid phase extraction coupled with in-situ DNPH derivatization (MSPE-ISD) was developed for the determination of hexanal and heptanal in urine. The extraction, purification, and derivatization of aldehydes are integrated into a single analytical step, simplifying the measurement workflow. The LODs are 1.7 and 2.5 nmol L^{-1} for hexanal and heptanal, respectively. Using this approach, the levels of hexanal and heptanal in urine of lung cancer patients were found to be higher compared to healthy controls [147]. Another method for the analysis of hexanal and heptanal in plasma used DNPH adsorbed on a polymer monolith composed of poly(methacrylic acid-co-ethylene glycol dimethacrylate) for simultaneous derivatization and microextraction, followed by HPLC-UV analysis. The LODs obtained are 2.4 and 3.6 nmol L^{-1} for hexanal and heptanal, respectively [150]. This monolith microextraction technique was further extended and used for the analysis of 5-hydroxymethylfurfural (5-HMF) in beverages such as coffee, honey, beer, soda, and urine [142]. In addition, a method using dispersive liquid–liquid microextraction with 1-dodecanol of DNPH derivatized aldehydes has been developed. Centrifugation of the sample and subsequent solidification of the droplet on an ice bath for easy removal of derivatized compounds for HPLC-UV analysis was performed. The LODs obtained for hexanal and heptanal are 7.90 nmol L^{-1} and 2.34 nmol L^{-1}, respectively. This approach afforded higher sensitivity compared to the conventional liquid-liquid microextraction methods [146]. An alternative approach developed by the same group uses ultrasound-assisted headspace liquid-phase microextraction with in-drop derivatization for the extraction and determination of hexanal and heptanal in blood. This technique uses a polychloroprene PCR tube containing the extraction solvent, methyl cyanide and the derivatization reagent, DNPH. Volatile aldehydes are then headspace extracted and derivatized simultaneously in the droplet and analyzed by HPLC-UV. The LODs for hexanal and heptanal are 0.79 nmol L^{-1} and 0.80 nmol L^{-1}, respectively [148].

Table 1. DNPH derivatization and HPLC-UV analysis of carbonyl compounds for environmental analysis.

Method Number	Matrix	Detection
EPA T0-11	Ambient air	HPLC-UV
EPA 8315A	Liquid, solid, and gas samples	HPLC-UV
ASTM D5197	Ambient air	HPLC-UV
NIOSH 2016 and 2532	Ambient indoor air	HPLC-UV
EPA 554	Drinking water	HPLC-UV

In addition to UV detection, fluorogenic derivatization reagents for the HPLC analysis of aldehydes are widespread in the literature. These tagging reagents are used either as pre-column labeling reagents or in one-pot derivatization of aldehydes. For instance, the labeling reagent 1,3,5,7-tetramethyl-8-aminozide-difluoroboradiaza-s-indacence (BODIPY-aminozide) is used as a pre-column derivatization reagent to monitor aldehydes in human serum by HPLC with fluorescence detection [158]. The BODIPY-based reagent reacts with aldehydes to form stable and highly fluorescent BODIPY hydrazone derivatives, which are easily separated and detected by HPLC with fluorescence detection at 495 nm (maximum excitation wavelength) and 505 nm (maximum emission wavelength). This approach is used to measure trace aliphatic aldehydes in serum samples without pretreatment or enrichment method [158]. Other reagents used for pre-column labeling are 2,2'-furil to label aldehydes [159] and 4-(N,N-dimethylaminosulfonyl)-7-hydrazino-2,1,3-benzoxadiazole to label 4-HNE in human serum [160]. For the one-pot-derivatization of aldehydes, rhodamine B hydrazide (RBH) [161], 2-aminoacridone [162], 9-fluorenylmethoxycarbonyl hydrazine (FMOC-hydrazine) [163], and 2-TBA [164] are used for the determination of malondialdehyde in biological fluids [161] by HPLC with fluorescence detection. For the determination of methylglyoxal, glyoxal, and diacetyl using HPLC-fluorescence, the most commonly used derivatization reagents are 4-methoxy-*o*-phenylenediamine (4-MPD) [165] and 1,2-diamino-4,5-dimethoxybenzene (DDB) [166].

Monitoring of methylglyoxal and glyoxal in diabetic patients has been proposed to help assess the risk of development of diabetic complications. Additionally, an increase in oxidative stress biomarkers has been reported in juvenile swimmers but no prior data has been reported on α-ketoaldehydes in urine associated with swim training. Thus, these methods were applied to compare the levels of these molecules in urine samples from healthy volunteers, diabetic subjects, and juvenile swimmers [165]. For acrolein analysis, luminarin 3 [167] and *m*-aminophenol [168] were used for the derivatization and HPLC-fluorimetric analysis in plasma resulting from the metabolism of drugs such as cyclophosphamide and ifosfamide [167]. HPLC coupled with UV or fluorescence detection are widely used techniques for aldehyde analysis in various environmental and biological matrices. These techniques have been the methods of choice as they offer good sensitivity and robustness. Along with innovative sample pre-treatment incorporated into the assays, low detection limits were obtained for quantifying specific biomarkers associated with various diseases. However, these methods do not provide structural information relating to the analyte of interest and require synthetic standards for analyte identification and confirmation. Finally, co-eluting peaks during HPLC separation can further confound the identification and quantitation of known and unknown carbonyl compounds via UV or fluorescence.

4.3. Aldehyde Visualization in Cells

In addition to HPLC with fluorimetric detection, fluorescent probes were designed and synthesized for real-time visualization of aldehydes in cells such as FP1 and FAP-1 for formaldehyde detection [169,170]. These formaldehyde probes are based on the 2-aza-Cope sigmatropic rearrangement, which yields highly fluorescent signal for the selective and sensitive detection of aldehydes in cells [169,170]. Recently, a novel technique based on real-time imaging of aldehydes in cells using multicolor fluorogenic hydrazone transfer ("DarkZone") was developed (Figure 5). This approach used a cell permeable DarkZone dye (7-(diethylamino)coumarin; DEAC) as a quenched hydrazone, which lights up when the quencher-aldehyde is replaced by the target aldehyde. The fluorescence signals are then detected by flow cytometry or microscopy without the need for washing or cell lysis. This strategy is useful for determining the aldehyde load associated with human diseases [171]. Recently, a novel fluorescent probe to visualize specific and total biogenic carbonyls was developed based on the pattern and fluorescence spectral profile unique to the target carbonyl compound. The probe is based on an *N*-aminoanthranilate methyl ester moiety [96]. These techniques offer real time monitoring of total aldehydes in cells and identification of specific aldehydes based on their unique fluorescence excitation and emission spectra. Overall, real-time imaging of aldehyde production in cells using aldehyde-specific probes allows elucidation of the roles and functions of these compounds in cellular processes and their involvement in disease pathogenesis. These techniques, however, lack the selectivity and specificity for the identification of specific carbonyls in cells as no structural information can be obtained. Finally, these techniques are not applicable to biological matrices such as blood, urine, CSF or saliva.

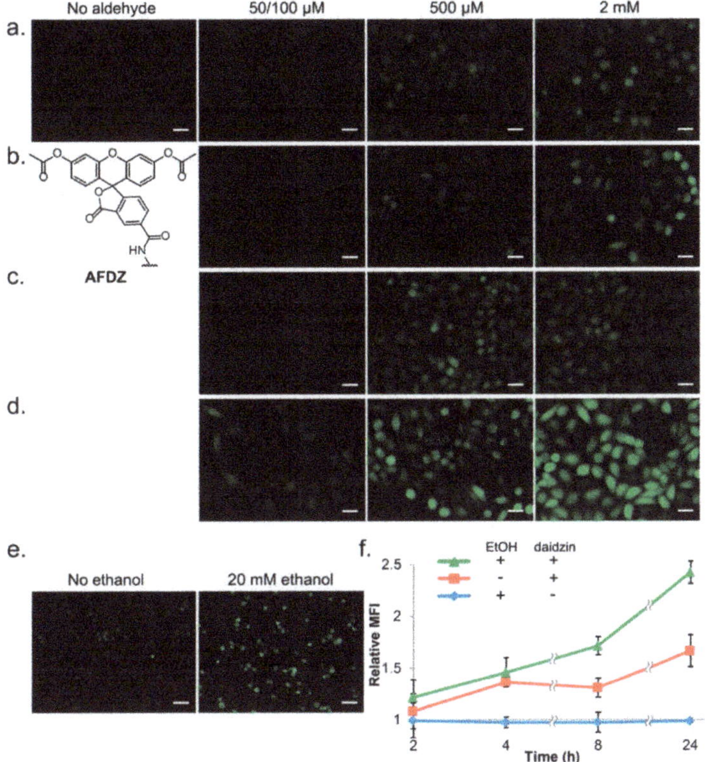

Figure 5. Real-time imaging of total aldehydic load in cells. Cellular aldehyde labeling fluorescence images and flow cytometry data. Hela cells were exposed to varying concentrations of: (**a**) formaldehyde; (**b**) glycolaldehyde; (**c**) acrolein; and (**d**) acetaldehyde along with 20 μM of the dye AFDZ and 10 mM catalyst (2,4-dimethoxyaniline) with images taken after 1 h of incubation. Note that 50 μM was used with acrolein and 100 μM for the other aldehydes tested. (**e**) K562 cells pretreated with 250 μM daidzin and incubated with 40 μM of AFDZ dye, 10 mM catalyst (2,4-dimethoxyaniline), and with/without 20 mM ethanol. (**f**) Flow cytometry data monitoring the production of aldehyde over time in K562 cells with/without ethanol. The fluorescence intensities were compared to that obtained from $t = 0$ without added ethanol and daidzin. Scale bars (20 μM) are shown. Reprinted from [171] (Copyright 2016, American Chemical Society).

4.4. Gas Chromatography (GC)/Gas Chromatography-Mass Spectrometry (GC-MS)

Mass spectrometry is widely used for the characterization and quantification of carbonyl compounds providing more selectivity, specificity, and sensitivity than is possible with UV or fluorescence detection [39,124,172]. There are a wide variety of derivatization reagents and sample preparation methods used to enhance the detection and sensitivity for mass spectrometric analysis of aldehydes (Table 2). For GC-MS analysis, derivatization increases the volatility of aldehydes in biological fluids and is most commonly done with O-2,3,4,5,6-pentafluorobenzyl hydroxylamine hydrochloride (PFBHA) as has been used for the analysis of saliva-available carbonyls in chewing tobacco products [173], to measure methylglyoxal and glyoxal in plasma of diabetic patients [174], formaldehyde in urine [175], and for the determination of MDA and 4-HNE levels in plasma [176]. In addition, PFBHA derivatization is often performed using headspace microextraction with subsequent derivatization on-fiber, on droplet, or for simultaneous extraction, derivatization, and GC-MS of volatile carbonyls. For instance, a quantitative method for the analysis of hexanal, heptanal, and volatile aldehydes in human blood was developed using headspace solid-phase microextraction with on-fiber derivatization with PFBHA and subsequent analysis by GC-MS. This approach afforded LODs of 0.006 nM (0.006 nmol L^{-1}) and 0.005 nM (0.005 nmol L^{-1}) for hexanal and heptanal, respectively [177,178]. Similarly, this approach is implemented for the determination of hexanal, heptanal, octanal, nonanal, and decanal in exhaled breath [179,180] and for the analysis of volatile low molecular weight carbonyls in urine [181]. Likewise, several volatile organic compounds (C3–C9 aldehydes) as promising biomarkers of non-small cell lung cancer (NSCLC) are identified in exhaled breath of patients with lung cancer using on-fiber-derivatization with PFBHA. The LOD and LOQ obtained for all aldehydes are 0.001 nM and 0.003 nM, respectively [182]. On-fiber derivatization using 2,2,2-trifluoroethylhydrazine (TFEH) as derivatization reagent is also used for the analysis of MDA in blood [183].

In addition, PFBHA derivatization on droplet is used for the analysis of hexanal and heptanal in blood [184]. This strategy involves the dissolution of the derivatizing agent in an organic solvent such as decane, and volatile aldehydes are headspace extracted and derivatized in the droplet with subsequent injection for GC-MS analysis. Likewise, a stir bar sorptive extraction (SBSE) for the GC-MS analysis of 4-HNE in urine was developed. This approach used a stir bar impregnated with the derivatization agent, PFBHA. The resulting oximes were further acylated using sulfuric acid and thermally desorbed and analyzed by GC-MS. This approach affords LOD of 22.5 pg mL^{-1} (0.06 nmol L^{-1}) and LOQ of 75 pg mL^{-1} (0.19 nmol L^{-1}) for the target carbonyl, 4-HNE [185]. PFBHA is also used in combination with other derivatization reagents. For example, a novel two-step derivatization approach using PFBHA as the first derivatizing agent followed by *N*-Methyl-*N*-trimethylsilyl-trifluoroacetamide (MSTFA) was developed for the analysis of glyoxal, methylglyoxal, and 3-deoxyglucosone in human plasma by GC-MS [186]. Other derivatization reagents used for GC-MS are 2,3,4,5,6-pentafluorobenzyl bromide (PFB-Br) [187,188] and 2,4,6-trichlorophenylhydrazine (TCPH) [189] for the analysis of MDA in urine; phenylhydrazine (PH) for the analysis of MDA in plasma and rat liver microsomes [190]; pentafluorophenyl hydrazine (PFPH) for the analysis of carbonyls in MTS [23]; 2,3-diaminonaphthalene along with salting-out assisted liquid–liquid extraction (SALLE) and dispersive liquid–liquid microextraction (DLLME) for the analysis of glyoxal and methylglyoxal in urine [191]; and meso-stilbenediamine [192] and 1,2-diaminopropane [193] for the analysis of methylglyoxal serum of diabetic patients and healthy controls by capillary GC-FID.

Methods based on gas chromatography without prior derivatization are also used for the analysis of volatile aldehydes. For example, a GC-MS coupled to a headspace generation autosampler is used for the analysis of endogenous aldehydes in urine as potential biomarkers of oxidative stress [194] and carbonyls such as acetaldehyde, propionaldehyde, acrolein, and crotonaldehyde in MTS [195]. Similarly, acetaldehyde in saliva of subjects after alcohol consumption is determined without prior derivatization using headspace extraction and GC coupled with flame ionization detector (FID) [40]. No prior derivatization is also applied to characterize toxic compounds such as benzene, toluene, butyraldehyde, benzaldehyde, and tolualdehyde in saliva using micro-solid-phase extraction (μSPE) and GC-IMS [196]. Gas chromatography coupled with various detection systems such as FID and mass spectrometry are ideal tools in the direct analyses of volatile carbonyl compounds in complex matrices. These techniques are useful for low molecular weight, volatile aldehydes. However, these methods require derivatization for the analysis of high-molecular weight, less volatile carbonyls.

Table 2. Bioanalytical techniques for characterizing carbonyl compounds.

Analytes	Matrix	Derivatization Reagent	Analytical Method	LOD	LOQ	Reference
Colorimetric/Fluorimetric/Amperometric						
Malondialdehyde	Plasma	2-TBA	Fluorimetric	NR	NR	Yagi 1976 [128]
Malondialdehyde	Plasma, Serum, Tissue	2-TBA	Fluorimetric	NR	NR	Armstrong et al. 1994 [129]
Malondialdehyde	Plasma	2-TBA	Fluorimetric	0.015 µmol L^{-1}	0.025 µmol L^{-1}	Del Rio et al. 2003 [130]
Aldehydes	Saliva	MBTH	Colorimetric	6.1 µM	NR	Ramdzan et al. 2016 [131]
Methylglyoxal and glyoxal	Urine and water	2-TBA	CE-AD	0.2 µg L^{-1} (methylglyoxal) 0.5 µg L^{-1} (glyoxal)	1.0 µg L^{-1} (methylglyoxal) 2.0 µg L^{-1} (glyoxal)	Zhang et al. 2010 [132]
HPLC-UV						
Acrolein, carbonyls	Cigarette smoke	HQ/2,4-DNPH	HPLC-UV	0.015–0.074 µg	0.05–0.25 µg	Uchiyama et al. 2010 [22]
Acetaldehyde	Plasma, red blood cells	2,4-DNPH	HPLC-UV	NR	NR	Di Padova et al. 1986 [154]
Malondialdehyde	Plasma	2-TBA	HPLC-UV	0.05 µM	0.17 µM	Grotto et al. 2007 [156]
Hexanal and heptanal	Urine	2,4-DNPH	HPLC-UV	1.0 µmol L^{-1} (hexanal); 0.7 µmol L^{-1} (heptanal)	3.0 µmol L^{-1} (hexanal); 2.2 µmol L^{-1} (heptanal)	Oenning et al. 2017 [145]
Hexanal and heptanal	Blood	2,4-DNPH	HPLC-UV	7.9 nmol L^{-1} (hexanal); 2.3 nmol L^{-1} (heptanal)	NR	Lili et al. 2010 [146]
Hexanal and heptanal	Urine	2,4-DNPH	HPLC-UV	1.7 nmol L^{-1} (hexanal); 2.5 µmol L^{-1} (heptanal)	5.7 nmol L^{-1} (hexanal); 8.3 µmol L^{-1} (heptanal)	Liu et al. 2015 [147]
Malondialdehyde	Plasma	2-TBA	HPLC-UV	0.02 µmol L^{-1}	NR	Nielsen et al. 1997 [155]
Malondialdehyde	Plasma, Serum	2,3-DAN	HPLC-UV	< 50 pM	NR	Steghens et al. 2001 [157]
5-Hydroxymethylfurfural	Beverages	2,4-DNPH	HPLC-UV	1.0 µg L^{-1}	3.4 µg L^{-1}	Wu et al. 2009 [142]
Hexanal and heptanal	Blood	2,4-DNPH	HPLC-UV	0.8 nmol L^{-1} (hexanal); 0.8 nmol L^{-1} (heptanal)	NR	Xu et al. 2010 [148]
Hexanal and heptanal	Urine and Serum	2,4-DNPH	HPLC-UV	0.8 nmol L^{-1} (hexanal); 0.8 nmol L^{-1} (heptanal)	NR	Xu et al. 2011 [149]
Hexanal and heptanal	Plasma	2,4-DNPH	HPLC-UV	2.4 nmol L^{-1} (hexanal); 3.6 µmol L^{-1} (heptanal)	NR	Zhang et al. 2007 [150]
Formaldehyde	Human Tissue	2,4-DNPH	HPLC-UV	1.5 mg L^{-1}	5.0 mg L^{-1}	Yilmas et al. 2016 [151]
Acetaldehyde	Cell culture media, rat blood and plasma	2,4-DNPH	HPLC-UV	> 3 µM	NR	Guan et al. 2012 [152]
Carbonyls	Air	2,4-DNPH/2-PB	HPLC-UV	NR	NR	Uchiyama et al. 2009 [153]
Carbonyls	Exhaled breath	2,4-DNPH	HPLC-UV	0.001-0.01 µg puff^{-1}	NR	Samburova et al. 2018 [35]
Formaldehyde	Cosmetic products	2,4-DNPH	HPLC-UV	NR	10 mg kg^{-1}	Galli et al. 2015 [18]
HPLC-Fluorescence/Fluorescence						
Glyoxal and methylglyoxal	Urine	DDB	HPLC-Fluorescence	NR	NR	Akira et al. 2004 [166]
Malondialdehyde	Serum, Plasma	2-TBA	HPLC-Fluorescence	NR	0.05 µmol L^{-1}	Seljeskog et al. 2006 [164]
Acrolein	Urine	*m*-aminophenol	HPLC-Fluorescence	NR	NR	Al-Rawithi et al. 1993 [168]
Aliphatic aldehydes	Serum	2,2'-furil	HPLC-Fluorescence	0.19–0.50 nM	NR	Ali et al. 2013 [159]
4-HNE	Serum	DBD-H	HPLC-Fluorescence	0.06 µM	NR	Imazato et al. 2014 [160]
Malondialdehyde	Plasma, Urine	RBH	HPLC-Fluorescence	0.25 nM	0.80 nM	Li et al. 2013 [161]
Malondialdehyde	Plasma	FMOC-hydrazine	HPLC-Fluorescence	4.0 nmol L^{-1}	NR	Mao et al. 2006 [163]

Table 2. *Cont.*

Analytes	Matrix	Derivatization Reagent	Analytical Method	LOD	LOQ	Reference
HPLC-Fluorescence/Fluorescence						
Glyoxal, methylglyoxal, and diacetyl	Urine	4-MPD	HPLC-Fluorescence	1.82–2.31 µg L^{-1}	3.06–3.88 µg L^{-1}	Ojeda et al. 2014 [165]
Acrolein	Plasma	luminarin 3	HPLC-Fluorescence	100 nM	300 nM	Paci et al. 2000 [167]
Aldehydes	Serum	BODIPY-aminozide	HPLC-Fluorescence	0.43–0.69 nM	NR	Xiong et al. 2010 [158]
Malondialdehyde	Urine	2-AA	HPLC-Fluorescence	1.8 nM	5.8 nM	Giera et al. 2011 [162]
Formaldehyde	Cells	FAP-1	Fluorescence	NR	NR	Brewer et al. 2015 [170]
Formaldehyde	Cells	FP1	Fluorescence	NR	NR	Roth et al. 2015 [169]
Total aldehydes	Cells	DarkZone dye/DEAC	Fluorescence	NR	NR	Yuen et al. 2016 [171]
Biogenic aldehydes	Aldehyde standards	methyl-5-methoxy -*N*-aminoanthranilate	Fluorescence	NR	NR	Lazurko et al. 2018 [96]
Gas Chromatography (GC)/Gas Chromatography-Mass Spectrometry (GC-MS)						
Methylglyoxal	Serum	1,2-diaminopropane	GC-FID	40 µg L^{-1}	NR	Khuhawar et al. 2008 [193]
Methylglyoxal	Serum	meso-stilbenediamine	GC-FID	25 µg L^{-1}	NR	Kandhro et al. 2008 [192]
Acetaldehyde	Saliva, blood	no derivatization	GC-FID	NR	NR	Yokohama et al. 2008 [40]
Butyraldehyde, Benzaldehyde, Tolu|aldehyde	Saliva	no derivatization	GC-IMS	0.38–0.49 mg L^{-1}	1.26–1.66 mg L^{-1}	Criado-Garcia et al. 2016 [196]
Pentanal, Hexanal, Heptanal, Octanal, Benzaldehyde	Urine	no derivatization	GC-MS	0.04–0.08 µg L^{-1}	0.12–0.24 µg L^{-1}	Anton et al. 2014 [194]
Acetaldehyde, propionaldehyde, acrolein, crotonaldehyde	MTS	no derivatization	GC-MS	0.014–0.12 µg cig^{-1}	0.045–0.38 µg cig^{-1}	Zhang et al. 2019 [195]
Volatile aldehydes	Urine	PFBHA	GC-MS	0.009–15 µg L^{-1}	0.029–50 µg L^{-1}	Calejo et al. 2016 [181]
Malondialdehyde	Plasma, RLM	Phenylhydrazine (PH)	GC-MS	5 pmol injection^{-1} (LLOD)	NR	Cighetti et al. 1999 [190]
Hexanal and heptanal	Blood	PFBHA	GC-MS	0.006 nM (hexanal); 0.005 nM (heptanal	NR	Deng et al. 2004 [177]
Aldehydes	Blood	PFBHA	GC-MS	0.001–0.006 nM	NR	Deng et al. 2004 [178]
Malondialdehyde	Urine	PFB-Br	GC-MS	0.7 nM	NR	Hanff et al. 2017 [187]
Glyoxal and methylglyoxal	Plasma	PFBOA	GC-MS	NR	NR	Lapolla et al. 2003 [174]
Hexanal and heptanal	Blood	PFBHA	GC-MS	0.12 nM (hexanal); 0.16 nM (heptanal	NR	Li et al. 2005 [184]
Glyoxal and methylglyoxal	Urine	2,3-DAN	GC-MS	0.12 µg L^{-1} (glyoxal); 0.06 µg L^{-1} (methylglyoxal)	0.40 µg L^{-1} (glyoxal); 0.2 µg L^{-1} (methylglyoxal)	Pastor-Belda et al. 2017 [191]
Malondialdehyde	Blood	TFEH	GC-MS	0.4 µg L^{-1}	NR	Shin 2009 [183]
Malondialdehyde	Plasma, Urine	TCPH	GC-MS	0.4 µM (MSD); 0.03 µM (ECD)	NR	Stalikas et al. 2001 [189]
C6-C10 aldehydes	Exhaled breath	PFBHA	GC-MS	0.01–0.03 nM	0.02–0.1 nM	Svensson et al. 2007 [180]
Glyoxal, methylglyoxal, and 3-dG	Plasma	PFBOA; MSTFA	GC-MS	12.8–31.2 µg L^{-1}	12.8–31.2 µg L^{-1}	Wu et al. 2008 [186]
Formaldehyde	Urine	PFBHA	GC-MS	1.08 µg L^{-1}	3.6 µg L^{-1}	Takeuchi et al. 2007 [175]
Volatile aldehydes	Exhaled breath	PFBHA	GC-MS	1.3–56 pmol L^{-1}	4.3–226 pmol L^{-1}	Fuchs et al. 2010 [179]
Aldehydes (C3-C9)	Exhaled breath	PFBHA	GC-MS	1 × 10^{-12} M	3 × 10^{-12} M	Poli et al. 2010 [182]
4-HNE	Urine	PFBHA; sulfuric acid	GC-MS	22.5 ng L^{-1}	75 ng L^{-1}	Stopforth et al. 2006 [185]
Malondialdehyde and 4-HNE	Plasma	PFBHA	GC-MS	NR	NR	Tsikas et al. 2017 [176]
Carbonyls	Chewing Tobacco	PFBHA	GC-MS	NR	100-1000 ppb	Chou et al. 1994 [173]
Carbonyls	MTS	PFPH	GC-MS	NR	NR	Pang et al. 2011 [23]
Malondialdehyde	Plasma	PFB-Br	GC-MS	2 amol	200 nM (LLOQ)	Tsikas et al. 2016 [188]

NR, not reported; MTS, mainstream tobacco smoke; RLM, rat liver microsomes.

4.5. Liquid Chromatography-Mass Spectrometry (LC-MS)

4.5.1. Methods Based on Selected Reaction Monitoring (SRM)

Liquid chromatography–mass spectrometry-based approaches have been used extensively to quantify derivatized carbonyl compounds, and recently for screening of unknown carbonyl compounds. Aldehyde derivatizations using 2,4-DNPH [143,197–202], dansylhydrazine (DnsHz) [203,204], *N*-(1-chloroalkyl)pyridinium [205], *o*-phenyldiamine [206], D-cysteine [207], 9,10-phenanthrenequinone (PQ) [208], 3-nitrophenylhydrazine [209], and 3,4-diaminobenzophenone [210] have been used to provide chromatographic retention and separation, efficient MS ionization, and MS/MS detectability. Typically, LC-MS analysis has been performed using selected reaction monitoring (SRM) with either atmospheric pressure chemical ionization (APCI), atmospheric pressure photoionization (APPI) or electrospray ionization (ESI). For example, D-cysteine has been used to generate alkyl-thiazolidine-carboxylic acid derivatives and analyzed by LC-SRM to quantify aldehydes in beverages with an LOD and LOQ of 0.2–1.9 µg L^{-1} (1.36–8.76 nmol L^{-1}) and 0.7–6.0 µg L^{-1} (4.76–27.6 nmol L^{-1}), respectively [207]. Alternatively, a method for profiling lipophilic reactive carbonyls in biological samples based on dansylhydrazine derivatization and LC-SRM has been developed with monitoring of the characteristic product ion, *m/z* 236.1 corresponding to 5-dimethylaminonaphthalene-1-sulfonyl moiety. This approach detects 400 free reactive carbonyls in plasma samples from mice, of which 34 are confirmed by synthetic standards [204]. Furthermore, charged derivatization reagents, such as 4-(2-(trimethylammonio) ethoxy) benzenaminium halide (4-APC), 4-(2-((4-bromophenethyl)dimethylammonio)ethoxy)benzenaminium dibromide (4-APEBA), N-[2-(aminooxy)ethyl]-N,N-dimethyl-1-dodecylammonium (QDA), and *N,N,N*-triethyl-2-hydrazinyl-2-oxoethanaminium bromide (HIQB), have been used to enhance ionization of the carbonyls for LC-MS analysis. For example, 4-APC, which contains an aniline moiety for reaction with aliphatic aldehydes, and a quaternary ammonium group for improved ionization efficiency and sensitivity, was developed for the analysis and quantitation of aldehydes in biological fluids [211] (Figure 6). Similarly, a second-generation derivatization reagent, 4-APEBA, consisting of a bromophenethyl group for isotopic signature incorporation and additional fragmentation identifiers, has been developed [212]. Another labeling reagent using *N*-(1-chloroalkyl)pyridinium quaternization to provide a charged tag was developed for quantifying aliphatic fatty aldehydes. This approach is used to measure the levels of long-chain non-volatile fatty acids in thyroid carcinoma tissues [205].

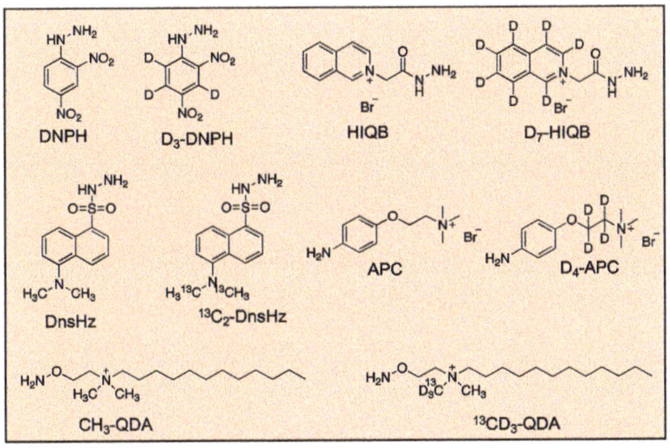

Figure 6. Commonly used differential isotope labeling reagents for profiling and relative quantitation of carbonyl compounds.

Assays with simultaneous derivatization and analysis have been developed. For example, a fully automated in-tube solid phase microextraction/liquid chromatography-post column derivatization with hydroxylamine hydrochloride and mass spectrometry was developed for the analysis of hexanal and heptanal in human urine as potential biomarkers for lung cancer [213]. In addition, this approach has been extended to the analysis of urinary malondialdehyde by DNPH derivatization and LC-SRM [198]. Similarly, an approach based on magnetic solid phase extraction coupled with in-situ derivatization with 2,4-DNPH was developed for the determination of hexanal and heptanal in urine of lung cancer patients [147]. Likewise, an Alternate Isotope-Coded Derivatization (AIDA) was developed to quantify malondialdehyde and 4-HNE in exhaled breath condensate by LC-SRM. This approach affords good quantitation of MDA and 4-HNE and is in good agreement with quantitation of the same samples using external calibration [199].

4.5.2. Screening LC-MS Methods

SRM analysis provides excellent sensitivity and good specificity for quantitative analysis but lacks the ability to screen for unknown aldehydes and requires a knowledge of unique SRM transitions of the known carbonyl compounds to be measured. Thus, data-dependent LC-MS/MS analysis (DDA) with DNPH derivatization is frequently used for untargeted profiling with MS^n spectra used for identification and structural elucidation [135,215,229]. Studies using negative ionization have described the MS and MS/MS behavior of DNPH-derivatized carbonyls [215,216,229]. Studies using positive electrospray ionization have characterized DNPH-derivatized malondialdehyde [198,199,217,230] and 4-HNE [199], and recently we characterized the positive ionization and fragmentation of a wide range of DNPH-derivatized carbonyls to establish consistent fragmentation rules applicable to this class of compounds, allowing for screening of unknown carbonyl compounds and comprehensive detection [218] (Table 3).

Differential Isotope Labeling for Profiling and Relative Quantitation of Aldehydes

To allow simultaneous identification and quantitation of carbonyl compounds in biological fluids and alcoholic beverages, isotopically labeled counterparts are used for differential labeling (Figure 6). 4-APC and its labeled counterpart, D_4-4-APC, have been used for untargeted profiling of aldehydes by differential stable isotope labeling using liquid chromatography-double neutral loss scan-mass spectrometry (SIL-LC-DNLS-MS). Pooled control samples are labeled with isotope labeled compounds, while the individual samples are derivatized with the unlabeled versions. This approach involves scanning of the two characteristic neutral fragments of 87 Da and 91 Da generated upon CID corresponding to the unlabeled 4-APC and labeled D_4-4-APC-derivatized carbonyls, respectively. This strategy enables profiling of 16 and 19 aldehyde-containing compounds in human urine and white wine, respectively. Finally, five aldehydes in human urine and four aldehydes in white wine are confirmed by comparison with synthetic standards [219]. This approach was further extended using an enrichment step by solid phase-extraction using stable isotope labeling–solid phase extraction–liquid chromatography–double precursor ion scan/double neutral loss scan–mass spectrometry analysis (SIL-SPE-LC-DPIS/DNLS-MS) for profiling and relative quantitation of aldehydes in beer. The pair of isotope reagents, 4-APC and D_4-4-APC, are used for differential labeling of the samples and co-eluting m/z pairs separated by 4 Da were detected and identified in the mass spectral data obtained by high resolution LC-QTOF-MS. Using this approach, 25 candidate aldehydes are detected in beer. The 25 candidate aldehydes are then quantified in different beer samples using a targeted MRM approach by monitoring the MRM transitions $[M]^+ \rightarrow [M]^+ - 87$ and $[M+4]^+ \rightarrow [M + 4]^+ - 91$ corresponding to 4-APC and D_4-4-APC, respectively. Fifteen aldehydes are identified and confirmed by comparison with synthetic standards and MS/MS analysis [220]. Likewise, differential labeling for profiling and relative quantitation of fatty aldehydes in biological samples using 2,4-bis-(diethylamino)-6-hydrazino-1,3,5-triazine and its deuterated counterpart has been developed. Using the 2VO dementia rat model system, 43 and 19 fatty aldehydes are significantly altered between the controls and models groups' plasma and brain tissue, respectively [214].

Table 3. LC-MS-based Methods for Characterizing Aldehydes.

Analytes	Matrix	Derivatization Reagent	Ionization Technique	Ionization Mode	Flow rate (μL min^{-1})	MS Technique	LOD	LOQ	Reference
Fatty aldehydes	Plasma, brain tissue	T3	ESI	(+)	500	LC-MS/MS	0.1–1 ng L^{-1}	NR	Tie et al. 2016 [214]
Carbonyls	Air	2,4-DNPH	APCI	(−)	1000	LC-MS^n	10 pg	NR	Kolliker et al. 1998 [215]
Carbonyls	Air	2,4-DNPH	APCI	(−)	1400	LC-MS	20–60 pg	200-600 pg	Grosjean et al. 1999 [135]
Carbonyls	Air in smog chamber	2,4-DNPH	APCI	(−)	560	LC-MS^n	0.5–1 ng	NR	Brombacher et al. 2001 [216]
Carbonyls	Standards	2,4-DNPH	APCI	(−)	550	LC-MS/MS	2.13–30.9 pg	NR	Ochs et al. 2015 [137]
Malondialdehyde	Plasma	2,4-DNPH	ESI	(+)	400	UHPLC-HRMS	32 nM	100 nM	Mendonca et al. 2017 [217]
Carbonyls	Saliva	2,4-DNPH; D$_3$-2,4-DNPH	ESI	(+)	0.3	HR/AM DDA NL MS^3	0.19–3.24 fmol	NR	Dator et al. 2017 [218]
Carbonyls	Engine exhaust, polymers, liquid soaps	2,4-DNPH	APCI	(+) & (−)	200	LC-MS	NR	NR	Olson et al. 1985 [202]
Carbonyls	Automobile exhaust and cigarette smoke	2,4-DNPH	APPI, APCI	(−)	500	LC-MS	2.9–24 nmol L^{-1}	9.7–80 nmol L^{-1}	Van Leeuwen et al. 2004 [136]
Carbonyls	MTS	2,4-DNPH	ESI, APCI, APPI	(−)	500	UHPLC-MS	NR	0.022–0.134 μg mL^{-1}	Miller et al. 2010 [144]
Aldehydes	EBC	2,4-DNPH	APCI	(+) & (−)	800	LC -MS/MS	0.3–1.0 nM	NR	Andreoli et al. 2003 [197]
LMM aldehydes	Urine	2,4-DNPH	ESI	(−)	300	LC-MS/MS	15–65 ng L^{-1}	50–200 ng L^{-1}	Banos et al. 2010 [143]
Malondialdehyde	Urine	2,4-DNPH	ESI	(+)	200	LC-MS/MS	1.6 nmol L^{-1}	6.4 nmol L^{-1}	Chen et al. 2011 [198]
Malondialdehyde and 4-HNE	EBC	2,4-DNPH; D$_3$-2,4-DNPH	ESI	(+)	500	LC-MS/MS	NR	NR	Manini et al. 2010 [199]
Trifluoroacetaldehyde	Human liver microsomes	2,4-DNPH; D$_3$-2,4-DNPH; ^{15}N$_4$-2,4-DNPH	ESI	(−)	200	LC-MS	16 ± 4 μg L^{-1} (SIM) 23 ± 6 μg L^{-1} (NRS) * 59 ± 32 μg L^{-1} (SRM)	NR	Prokai et al. 2012 [201]
Aldehydes and ketones	Drinking water	2,4-DNPH	ESI	(−)	300	LC-MS	25–50 pg	NR	Richardson et al. 2000 [119]
Carbonyls	Air	2,4-DNPH	ESI	(−)	600	LC-MS/MS	0.4–9.4 ng (m^3)$^{-1}$	NR	Chi et al. 2007 [134]
Carbonyls	Water	2,4-DNPH	ESI	(−)	300	LC-MS	0.13–0.76 μg L^{-1}	0.48–2.69 μg L^{-1}	Zwiener et al. 2002 [133]
Aldehydes	Cigarette smoke	2,4-DNPH	ESI	(−)	300	LC-MS	NR	NR	Van der Toorn et al. 2013 [200]
Malondialdehyde	Plasma	3-nitrophenylhydrazine	ESI	(+)	350	LC-MS/MS	0.007 μM (LLOD)	0.02 μM (LLOQ)	Sobsey et al. 2016 [209]
Malondialdehyde	Urine, saliva	3,4-diaminobenzophenone	ESI	(+)	200	LC-MS/MS	0.03–0.1 μg L^{-1}	0.1–0.3 μg L^{-1}	Oh et al. 2017 [210]
Aldehydes	Urine and white wine	4-APC; D$_4$-4-APC	ESI	(+)	200	SIL-LC-DNLS-MS	1.2–10 nmol L^{-1}	NR	Yu et al. 2015 [219]
Aldehydes	Beverages	4-APC; D$_4$-4-APC	ESI	(+)	200	LC-DPIS/DNLS-MS	NR	NR	Zheng et al. 2017 [220]

Table 3. *Cont.*

Analytes	Matrix	Derivatization Reagent	Ionization Technique	Ionization Mode	Flow rate (μL min^{-1})	MS Technique	LOD	LOQ	Reference
Aldehydes	Plasma	4-APC; NaBH$_3$CN	ESI	(+)	150	LC-MS/MS	0.5–2.5 nM	NR	Eggink et al. 2009 [221]
Aliphatic aldehydes	Urine	4-APC; NaBH$_3$CN	ESI	(+)	150	LC-MS	3–33 nM	NR	Eggink et al. 2008 [211]
Aldehydes	Plasma, Urine	4-APEBA; NaBH$_3$CN	ESI	(+)	150	LC-MS/MS	NR	NR	Eggink et al. 2010 [212]
Aldehydes	Beverages	4-HBA	ESI	(+)	500	LC-MS	NR	NR	De Lima et al. 2018 [141]
Aldehydes	Serum	9,10-PQ	ESI	(+)	500	LC-MS/MS	0.004–0.1 nM	0.05–0.25 nM	El-Maghrabey et al. 2016 [208]
Aldehydes	Beverages	D-cysteine	ESI	(+)	200	LC-MS/MS	0.2–1.9 μg L^{-1}	0.7–6.0 μg L^{-1}	Kim et al. 2011 [207]
Aldehydes	Synthesis	DAABD-MHz	ESI	(+)	200	LC-MS/MS	30–60 fmol	NR	Santa et al. 2008 [125]
Malondialdehyde	plasma	dansylhydrazine (DnsHz)	ESI	(+)	300-1500	LC-MS/MS	0.016 mg L^{-1}	0.054 mg L^{-1}	Lord et al. 2009 [203]
Carbonyls	Plasma	DnsHz	ESI	(+)	200	LC-MS/MS	1-20 fmol	2.5-50 fmol	Tomono et al. 2015 [204]
Carbonyls	Urine	DnsHz; ^{13}C$_2$-DnsHz	ESI	(+)	180	LC-MS	NR	NR	Zhao et al. 2017 [222]
Carbonyls	Serum	HIQB; D$_7$-HIQB	ESI	(+)	200	IL-LC-DPIS-MS	0.1–0.21 fmol	NR	Guo et al. 2017 [223]
Aldehydes and ketones	Urine, plasma	HTMOB	ESI	(+)	Infusion	LC-MS/MS	NR	NR	Johnson 2007 [126]
Hexanal and heptanal	Urine	hydroxylamine hydrochloride	ESI	NR	200	UHPLC-MS/MS	15 nM (hexanal); 9 nM (heptanal)	NR	Chen et al. 2017 [213]
Fatty aldehydes	Tissue	N-(1-chloroalkyl)pyridinium	ESI	(+)	300	LC-MS/MS	< 0.3 ng L^{-1}	NR	Cao et al. 2016 [205]
α-dicarbonyls	Plasma	o-phenyldiamine	ESI	(+)	1000	LC-MS/MS	0.5–42.2 nmol L^{-1}	1.5–126.6 nmol L^{-1}	Henning et al. 2014 [206]
Aldehydes and ketones	Yeast extract	p-toluenesulfonylhydrazine	ESI	(+) & (−)	350	SWATH-QqTOF	0.31 μM (ESI + only) 0.36 μM (ESI − only) 0.19 μM (ESI+ or ESI−)	NR	Siegel et al. 2014 [224]
Carbonyls	Tissue	QDA; ^{13}CD$_3$ labeled QDA	ESI	(+)	Infusion	UHR-FT-MS	0.07–0.66 nM	0.2–1.99 nM	Deng et al. 2018 [225]
Carbonyls	Cell extract	QDA; ^{13}CD$_3$ labeled QDA	ESI	(+)	Infusion	FT-ICR-MS	NR	NR	Mattingly et al. 2012 [226]
Carbonyls	Exhaled breath	ATM	ESI	(+)	Infusion	FT-ICR-MS	NR	NR	Fu et al. 2011 [227]
Carbonyls	Exhaled breath	AMAH	ESI	(+)	Infusion	FT-ICR-MS	NR	NR	Knipp et al. 2015 [228]
Aldehydes and ketones	Synthesis	TMPP-AcPFP; TMPP-PrG	ESI	(+)	500	LC-MS/MS	NR	NR	Barry et al. 2003 [127]

NR, not reported; * NRS, narrow range scans; EBC, Exhaled breath condensate; MTS, mainstream tobacco smoke; LMM, low molecular mass.

A high-performance chemical isotope labeling (CIL)-LC-MS method for profiling and quantitative analysis of carbonyl sub-metabolome in human urine using dansylhydrazine (DnsHz) as labeling reagent has been developed [222]. Identification and relative quantitation of carbonyl metabolites was performed using differential tagging with ^{12}C-DnsHz and ^{13}C-DnsHz in urine samples and subsequent analysis using LC-QTOF-MS. In-house software program was developed to process the CIL LC-MS mass spectral and a custom library of DnsHz-labeled standards was constructed (www.mycompoundid.org) for carbonyl metabolites identification. In total, 1737 peak pairs are detected in human urine, of which 33 are confirmed [222]. In addition, a strategy based on isotope labeling and liquid chromatography–double precursor ion scan mass spectrometry (IL-LC-DPIS-MS) was developed for the comprehensive profiling and relative quantitation of carbonyl compounds in human serum using the labeling reagent, HIQB and its corresponding isotope-labeled analog, D_7-HIQB [222]. The characteristic products ions, *m/z* 130.1/137.1 are monitored in the double precursor ion scans during mass spectrometry analysis upon collision-induced dissociation (CID). In total, 156 candidate carbonyl compounds are detected in human serum, of which 12 are further identified by synthetic standards. Using a targeted MRM mode, 44 carbonyls are found to be statistically different in myelogenous leukemia patients compared to healthy controls [223].

Methods Using High-Resolution/Accurate Mass Data Dependent Acquisition (DDA) and Data Independent Acquisition (DIA)

High-resolution mass spectrometry-based methods for metabolomics profiling provide accurate masses of both precursor and MS/MS fragment ions, and thus allow confident identification of detected metabolites in complex biological matrices. Recently, we have developed a high-resolution accurate mass data-dependent MS^3 neutral loss (NL) screening strategy to characterize DNPH-derivatized carbonyls in biological fluids, allowing for the simultaneous detection and quantitation of suspected and unknown/unanticipated carbonyl compounds [218]. Previous analyses of DNPH-derivatized carbonyls were mostly performed in negative ionization mode and at relatively high-flow rates, which limit the sensitivity of detection and quantitation of trace level analytes (Table 3). We found that, in positive mode, these compounds showed a characteristic neutral loss of hydroxyl radical ($^\bullet$OH) upon CID. This NL is not observed in negative mode. The characteristic neutral loss, $^\bullet$OH from DNPH-derivatized carbonyls, is then used as a screening approach during MS acquisition allowing unambiguous identification of RCCs (Figure 7). Furthermore, a relative quantitation strategy by differential isotope labeling using D_0-DNPH and D_3-DNPH is implemented to determine the relative levels of carbonyls after specific exposures. Using this approach, pre-exposure samples are labeled with D_0-DNPH, while post-exposure samples are labeled with D_3-DNPH. The samples are combined in a 1:1 (v/v) ratio and analyzed by our HR-AM NL screening strategy. The MS-based workflow provides an accurate, rapid, and robust method to identify and quantify toxic carbonyls in various biological matrices for exposure risk assessment. This is in contrast to previous work, which used relatively high flow rates (0.2–1.5 mL min^{-1}) and low-resolution MS analysis, limiting their sensitivity and identification confidence at trace analyte levels. We applied this method to characterize the levels of carbonyls after alcohol consumption in humans and showed that acetaldehyde levels are increased after exposure. This strategy is currently being used to characterize the carbonyls associated with e-cigarette use (vaping) as well as tobacco smoking.

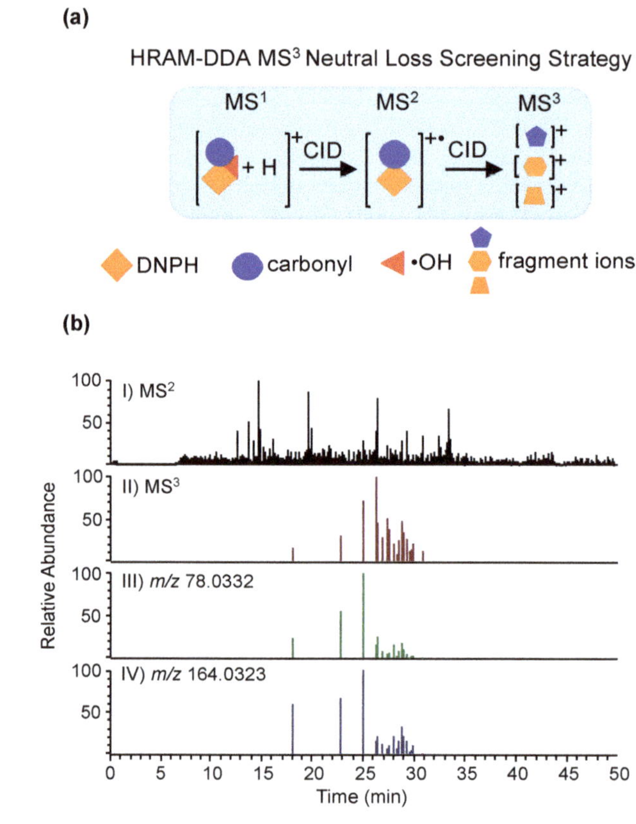

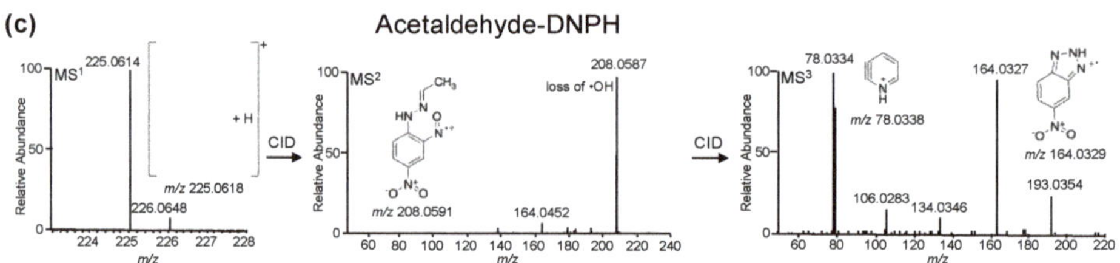

Figure 7. Development of a high-resolution accurate mass data-dependent MS³ neutral loss screening strategy for profiling and quantitative analysis of aldehydes in biological fluids. (**a**) The high-resolution accurate mass of •OH (17.0027 Da) was used to screen for all DNPH-derivatized aldehydes. (**b**) Monitoring of specific fragment ions (*m/z* 78.0332 and *m/z* 164.0323) minimizes possible false positive identification. (**c**) Representative MS, MS², and MS³ spectra of DNPH-derivatized acetaldehyde and proposed structures of major fragment ions. Reprinted with permission from Ref. [218] (Copyright 2017, Springer).

Another strategy based on ultra-high-resolution fourier transform mass spectrometry (UHR FT-MS) method using the tribrid orbitrap fusion was developed for profiling carbonyl metabolites in crude biological extracts. This approach uses a chemoselective tagging reagent, QDA, and its labeled counterpart, $^{13}CD_3$-QDA, for differential isotope labeling of biological samples. Data-dependent TopN MS/MS of the targeted mass difference of 4.0219 Da (QDA and $^{13}CD_3$-QDA metabolite pairs) is performed with direct infusion allowing for long acquisition times, resolved isotopic peaks and high-quality MS and MS/MS data. MS and MS/MS spectral data are processed using a custom software Precalculated Exact Mass Isotopologue Search Engine (PREMISE) for QDA–^{13}CD$_3$-QDA ion pairs and isotopologue identification. The workflow identifies 66 carbonyls in mouse tumor tissues, of which 14 carbonyls are quantified using authentic standards [231]. A similar derivatization and differential labeling approach is applied for the profiling and untargeted metabolomics of carbonyl compounds in cell extracts [226]. Likewise, direct infusion and FT-ICR-MS are used for the analysis of aldehydes and ketones in exhaled breath using 2-(aminooxy)ethyl-*N,N,N*-trimethylammonium iodide (ATM) and 4-(2-aminooxyethyl)-morpholin-4-ium chloride (AMAH) as derivatizing agents [227,228]. ATM is chemically functionalized on a novel microreactor to selectively preconcentrate volatile aldehydes and ketones. This approach demonstrated detection of C1-C12 aldehydes and applicable to any gaseous samples [227]. Similarly, AMAH is used as derivatizing agent coated within a silicon microreactor to capture volatile carbonyls to form AMAH-carbonyl adducts and analyzed by FT-ICR-MS. Subsequent treatment of the derivatized-carbonyl adducts with poly(4-vinylpyridine) yielded volatile carbonyl adducts, which can be analyzed using GC-MS. These complementary approaches using FT-ICR-MS and GC-MS provide a convenient and flexible identification and quantification of isomeric volatile organic compounds in exhaled breath [228]. In addition, an on-line weak-cation exchange liquid chromatography–tandem mass spectrometry using the LC-QTOF-MS^2 has been developed for screening aldehydes in plasma and urine samples. This strategy involves derivatization of aldehydes with 4-APC and subsequent reduction by NaBH$_3$CN. The characteristic MS/MS fragmentation of 4-APC derivatized aldehydes allows confirmation of known aldehydes as well as differentiation of hydroxylated and non-hydroxylated aldehydes [221]. Finally, a novel DIA strategy has been developed for the global analysis of aldehydes and ketones in biological samples. The strategy is based on TSH (*p*-toluenesulfonylhydrazine) derivatization of carbonyl compounds and Sequential Window Acquisition of All Theoretical Fragment-Ion spectra (SWATH) detection. Although the TSH-derivatized carbonyls are efficiently detected in both positive and negative modes, the negative ion mode data acquisition exhibits the signature fragment ion at *m/z* 155.0172, which is monitored using ESI-QqTOF-SWATH allowing chemo-selective identification of carbonyl compounds. Using this strategy, 61 target carbonyls were successfully identified and quantified in biological samples. In addition, SWATH MS data acquisition provides high resolution accurate mass measurements of both the precursor and fragment ions, allowing for confident identification of derivatized compounds [224].

Overall, HPLC coupled with mass spectrometry techniques are powerful tools for profiling and performing quantitative analysis of aldehydes in various biological matrices. The high selectivity and specificity of these methods along with structural information obtained from MS and MS^n mass spectral data are ideal for identifying knowns and unknowns. The more recent LC-MS-based methods presented here offer improved sensitivity, selectivity, and specificity for the detection of aldehydes in complex biological matrices. Although these techniques are highly sensitive, they are also susceptible to matrix interferences requiring rigorous sample clean-up. In addition, these techniques require expensive instrumentation and highly trained users, and are less portable. The development of new and innovative MS-based techniques is continuously evolving towards novel applications, in particular, for trace level analysis ideal for human exposure assessment, allowing for elucidation of their contributions and impact on human health.

5. Future Perspectives

The increased emphasis on the need to improve methods to comprehensively characterize exposures, and the parallel development of enhanced technology is resulting in a number of exciting new analytical techniques and approaches. The introduction of the concept of the exposome, intended as the totality of chemical exposures in an individual's life-time [232], has brought to light new analytical challenges related to the complexity of capturing the totality of various exposures, which are often chemically diverse, present in trace levels, and, in some cases, are resulting from the combination of endogenous and exogenous sources. To address this complexity, tools have been developed to analyze for specific classes of compounds resulting in a number of complementary approaches. Aldehydes are a major component of the exposome, and aldehyde exposure is important in the pathogenesis of several diseases, including certain cancers. Profiling and characterizing these compounds is particularly difficult due to their reactivity and the ubiquitous presence of many of them. The improvement of tools for the investigation of the "aldehydome", the sum of all exogenous and endogenously-formed aldehydes, is needed to elucidate the complex roles these compounds play in physiological and pathological events. With the availability of more advanced MS instrumentation, high performance chromatographic separation, and improved bioinformatics tools, the data acquired allow for increased sensitivity, identification of specific aldehydes, and the establishment of new biomarkers of exposure and effect. Additionally, the combination of these techniques with exciting new methods for single cell detection provides the potential for detection and profiling of aldehydes at a cellular level, opening up the opportunity to minutely dissect their roles and functions in biological systems and in pathogenesis.

Author Contributions: All authors critically reviewed all relevant literature and contributed to writing of the manuscript.

Funding: The work presented in this review carried out in the Balbo Research group was supported by NIOSH-funded MCOHS ERC Pilot Research Training Program (OH008434).

Acknowledgments: Mass spectrometry was carried out in the Analytical Biochemistry Shared Resource of the Masonic Cancer Center, University of Minnesota, funded in part by Cancer Center Support Grant CA-077598 and S10 RR-024618 (Shared Instrumentation Grant).

Conflicts of Interest: The authors declare no conflict of interest.

Abbreviations

2,4-DNPH	2,4-dinitrophenylhydrazine
PFPH	pentafluorophenyl hydrazine
2-PB	2-picoline borane
4-APC	4-(2-(trimethylammonio)ethoxy)benzenaminium halide
4-APEBA	4-(2-((4-bromophenethyl)dimethylammonio)ethoxy)benzenaminium dibromide
HIQB	*N,N,N*-triethyl-2-hydrazinyl-2-oxoethanaminium bromide
QDA	*N*-[2-(aminooxy)ethyl]-*N,N*-dimethyl-1-dodecylammonium
TSH	*p*-toluenesulfonylhydrazine
PFB-Br	2,3,4,5,6-pentafluorobenzyl bromide
HTMOB	4-hydrazino-*N,N,N*-trimethyl-4-oxobutanaminium iodide
DBD-H	4-(*N,N*-dimethylaminosulfonyl)-7-hydrazino-2,1,3-benzoxadiazole
4-MPD	4-methoxy-*o*-phenylenediamine
FMOC-hydrazine	9-fluorenylmethoxycarbonyl hydrazine
PFBHA/PFBOA	*o*-2,3,4,5,6-pentafluorobenzyl)hydroxylamine hydrochloride
ATM	2-(aminooxy)ethyl-*N,N,N*-trimethylammonium iodide
AMAH	4-(2-aminooxyethyl)-morpholin-4-ium chloride
TBARS	thiobarbituric acid reactive substances
2-TBA	2-thiobarbituric acid
TFEH	2,2,2-trifluoroethylhydrazine
FAP-1/FP 1	formaldehyde probe 1
DEAC	diethylaminocoumarin
DAN	diaminonapththalene
RBH	rhodamine B hydrazide
DDB	1,2-diamino-4,5-dimethoxybenzene
MSTFA	*N*-methyl-*N*-trimethylsilyl-trifluoroacetamide
2-AA	2-aminoacridone
3-dG	3-deoxyglucosone
BODIPY aminozide	1,3,5,7-tetramethyl-8-aminozide-difluoroboradiaza-*s*-indacence
5-HMF	5-hydroxymethylfurfural
4-HBA	4-hydrazinobenzoic acid

DnsHz	dansylhydrazine
9,10-PQ	9,10-phenanthrenequinone
T3	2,4-bis-(diethylamino)-6-hydrazino-1,3,5-triazine
TMPP-AcPFP	S-pentafluorophenyl tris(2,4,6-trimethoxyphenyl)phosphonium acetate bromide
TMPP-PrG	4-hydrazino-4-oxobutyl)[tris(2,4,6-trimethoxyphenyl)phosphonium bromide
TCPH	2,4,6-trichlorophenylhydrazine
SALLE	salting-out assisted liquid–liquid extraction
DLLME	dispersive liquid–liquid microextraction
SIL-SPE-LC-DPIS-MS	stable isotope labeling-solid phase extraction-liquid chromatography-double precursor-ion scan mass spectrometry
DNLS-MS	double neutral loss scan mass spectrometry
HR-AM	high resolution accurate mass
MTS	mainstream tobacco smoke
EBC	exhaled breath condensate
ESI	electrospray ionization
APCI	atmospheric pressure chemical ionization
APPI	atmospheric pressure photoionization
UHPLC	ultra-high performance liquid chromatography
SWATH	sequential window acquisition of all theoretical fragment-ion spectra
QqTOF	quadrupole time-of-flight
UHR-FT MS	ultra-high resolution fourier transform mass spectrometry
LC-MS^n	liquid chromatography tandem mass spectrometry
4-HNE	4-hydroxy-2-nonenal
4-HHE	4-hydroxy-2-hexenal
MDA	malondialdehyde
CID	collision-induced dissociation
HCD	high-energy C-trap dissociation
AIDA	alternate isotope-coded derivatization

References

1. O'Brien, P.J.; Siraki, A.G.; Shangari, N. Aldehyde sources, metabolism, molecular toxicity mechanisms, and possible effects on human health. *Crit. Rev. Toxicol.* **2005**, *35*, 609–662. [CrossRef] [PubMed]
2. Atkinson, R. Gas-Phase Tropospheric Chemistry of Organic Compounds: A Review. *Atmos. Environ.* **1990**, *24A*, 1–41. [CrossRef]
3. Riedel, K.; Weller, R.; Schrems, O. Variability of formaldehyde in the Antarctic troposphere. *Phys. Chem. Chem. Phys.* **1999**, *1*, 5523–5527. [CrossRef]
4. Cecinato, A.; Yassaa, N.; Di Palo, V.; Possanzin, M. Observation of volatile and semi-volatile carbonyls in an Algerian urban environment using dinitrophenylhydrazine/silica-HPLC and pentafluorophenylhydrazine/silica-GC-MS. *J. Environ. Monit.* **2002**, *4*, 223–228. [CrossRef] [PubMed]
5. Maldotti, A.; Chiorboli, C.; Bignozzi, C.; Bartocci, C.; Carassiti, V. Photooxidation of 1,3-butadiene containing systems: Rate constant determination for the reaction of acrolein with·OH radicals. *Int. J. Chem. Kinet.* **1980**, *12*, 905–913. [CrossRef]
6. Destaillats, H.; Spaulding, R.S.; Charles, M.J. Ambient air measurement of acrolein and other carbonyls at the Oakland-San Francisco Bay Bridge toll plaza. *Environ. Sci. Technol.* **2002**, *36*, 2227–2235. [CrossRef] [PubMed]
7. Rao, X.; Kobayashi, R.; White-Morris, R.; Spaulding, R.; Frazey, P.; Charles, M.J. GC/ITMS measurement of carbonyls and multifunctional carbonyls in PM2.5 particles emitted from motor vehicles. *J. AOAC Int.* **2001**, *84*, 699–705. [PubMed]
8. Grosjean, E.; Grosjean, D.; Fraser, M.; Cass, G. Air Quality Model Evaluation Data for Organics. 2. C_1–C_{14} Carbonyls in Los Angeles Air. *Environ. Sci. Technol.* **1996**, *30*, 2687–2703. [CrossRef]
9. Dost, F.N. Acute toxicology of components of vegetation smoke. *Rev. Environ. Contam. Toxicol.* **1991**, *119*, 1–46. [PubMed]

10. Materna, B.L.; Jones, J.R.; Sutton, P.M.; Rothman, N.; Harrison, R.J. Occupational exposures in California wildland fire fighting. *Am. Ind. Hyg. Assoc. J.* **1992**, *53*, 69–76. [CrossRef] [PubMed]

11. Dempsey, C. A Comparison of Organic Emissions from Hazardous Waste Incinerators Versus the 1990 Toxics Release Inventory Air Releases. *J. Air Waste Manage. Assoc.* **1993**, *43*, 1374–1379. [CrossRef]

12. Sverdrup, G.; Riggs, K.; Kelley, T.; Barrett, R.; Peltier, R. Toxic Emissions from a Cyclone Burner Boiler with an ESP and with the SNOX Demonstration and from a Pulverized Coal Burner Boiler with an ESP/Wet Flue Gas Desulfurization System. *Gov. Rep. Announc. Indexes* **1994**, *21*, 1–16.

13. Wheeler, R.; Head, F.; McCawley, M. An Industrial Hygiene Characterization of Exposure to Diesel Emissions in an Underground Coal Mine. *Environ. Int.* **1981**, *5*, 485–488. [CrossRef]

14. James, J.T. Carcinogens in spacecraft air. *Radiat. Res.* **1997**, *148*, S11–S16. [CrossRef] [PubMed]

15. Brown, S.K. Chamber assessment of formaldehyde and VOC emissions from wood-based panels. *Indoor Air* **1999**, *9*, 209–215. [CrossRef] [PubMed]

16. Kelley, T.; Sadola, J.; Smith, D. Emission rates of formaldehyde and other carbonyls from consumer and industrial products found in California homes. *Proc. Int. Spec. Conf. Air Waste Manage. Assoc.* **1996**, 521–526.

17. Pickrell, J.A.; Mokler, B.V.; Griffis, L.C.; Hobbs, C.H.; Bathija, A. Formaldehyde release rate coefficients from selected consumer products. *Environ. Sci. Technol.* **1983**, *17*, 753–757. [CrossRef] [PubMed]

18. Galli, C.L.; Bettin, F.; Metra, P.; Fidente, P.; De Dominicis, E.; Marinovich, M. Novel analytical method to measure formaldehyde release from heated hair straightening cosmetic products: Impact on risk assessment. *Regul. Toxicol. Pharmacol.* **2015**, *72*, 562–568. [CrossRef]

19. Rickert, W.S.; Robinson, J.C.; Young, J.C. Estimating the hazards of "less hazardous" cigarettes. I. Tar, nicotine, carbon monoxide, acrolein, hydrogen cyanide, and total aldehyde deliveries of Canadian cigarettes. *J. Toxicol. Environ. Health* **1980**, *6*, 351–365. [CrossRef]

20. Mansfield, C.T.; Hodge, B.T.; Hege, R.B.; Hamlin, W.C. Analysis of formaldehyde in tobacco smoke by high performance liquid chromatography. *J. Chromatogr. Sci.* **1977**, *15*, 301–302. [CrossRef]

21. Smith, C.J.; Hansch, C. The relative toxicity of compounds in mainstream cigarette smoke condensate. *Food Chem. Toxicol.* **2000**, *38*, 637–646. [CrossRef]

22. Uchiyama, S.; Inaba, Y.; Kunugita, N. Determination of acrolein and other carbonyls in cigarette smoke using coupled silica cartridges impregnated with hydroquinone and 2,4-dinitrophenylhydrazine. *J. Chromatogr. A* **2010**, *1217*, 4383–4388. [CrossRef] [PubMed]

23. Pang, X.; Lewis, A.C. Carbonyl compounds in gas and particle phases of mainstream cigarette smoke. *Sci. Total Environ.* **2011**, *409*, 5000–5009. [CrossRef] [PubMed]

24. Borgerding, M.F.; Bodnar, J.A.; Chung, H.L.; Mangan, P.P.; Morrison, C.C.; Risner, C.H.; Rogers, J.C.; Simmons, D.F.; Uhrig, M.S.; Wendelboe, F.N.; et al. Chemical and biological studies of a new cigarette that primarily heats tobacco. Part 1. Chemical composition of mainstream smoke. *Food Chem. Toxicol.* **1998**, *36*, 169–182. [CrossRef]

25. Swauger, J.E.; Steichen, T.J.; Murphy, P.A.; Kinsler, S. An analysis of the mainstream smoke chemistry of samples of the U.S. cigarette market acquired between 1995 and 2000. *Regul. Toxicol. Pharmacol.* **2002**, *35*, 142–156. [CrossRef] [PubMed]

26. Hecht, S.S. Tobacco carcinogens, their biomarkers and tobacco-induced cancer. *Nat. Rev. Cancer* **2003**, *3*, 733–744. [CrossRef]

27. Uchiyama, S.; Ohta, K.; Inaba, Y.; Kunugita, N. Determination of carbonyl compounds generated from the e-cigarette using coupled silica cartridges impregnated with hydroquinone and 2,4-dinitrophenylhydrazine, followed by high-performance liquid chromatography. *Anal. Sci.* **2013**, *29*, 1219–1222. [CrossRef]

28. Lee, M.S.; LeBouf, R.F.; Son, Y.S.; Koutrakis, P.; Christiani, D.C. Nicotine, aerosol particles, carbonyls and volatile organic compounds in tobacco- and menthol-flavored e-cigarettes. *Environ. Health* **2017**, *16*, 42. [CrossRef]

29. Hahn, J.; Monakhova, Y.B.; Hengen, J.; Kohl-Himmelseher, M.; Schüssler, J.; Hahn, H.; Kuballa, T.; Lachenmeier, D.W. Electronic cigarettes: Overview of chemical composition and exposure estimation. *Tob. Induc. Dis.* **2014**, *12*, 23. [CrossRef]

30. Ogunwale, M.A.; Chen, Y.; Theis, W.S.; Nantz, M.H.; Conklin, D.J.; Fu, X.A. A novel method of nicotine quantification in electronic cigarette liquids and aerosols. *Anal. Methods* **2017**, *9*, 4261–4266. [CrossRef]

31. Pankow, J.F.; Kim, K.; McWhirter, K.J.; Luo, W.; Escobedo, J.O.; Strongin, R.M.; Duell, A.K.; Peyton, D.H. Benzene formation in electronic cigarettes. *PLoS One* **2017**, *12*, e0173055. [CrossRef]

32. Cheng, T. Chemical evaluation of electronic cigarettes. *Tob. Control* **2014**, *23* (Suppl. 2), 11–17. [CrossRef]
33. Bekki, K.; Uchiyama, S.; Ohta, K.; Inaba, Y.; Nakagome, H.; Kunugita, N. Carbonyl compounds generated from electronic cigarettes. *Int. J. Environ. Res. Public Health* **2014**, *11*, 11192–11200. [CrossRef]
34. Khlystov, A.; Samburova, V. Flavoring Compounds Dominate Toxic Aldehyde Production during E-Cigarette Vaping. *Environ. Sci. Technol.* **2016**, *50*, 13080–13085. [CrossRef]
35. Samburova, V.; Bhattarai, C.; Strickland, M.; Darrow, L.; Angermann, J.; Son, Y.; Khlystov, A. Aldehydes in exhaled breath during e-cigarette vaping: Pilot study results. *Toxics* **2018**, *6*, 46. [CrossRef]
36. Sleiman, M.; Logue, J.M.; Montesinos, V.N.; Russell, M.L.; Litter, M.I.; Gundel, L.A.; Destaillats, H. Emissions from electronic cigarettes: Key parameters affecting the release of harmful chemicals. *Environ. Sci. Technol.* **2016**, *50*, 9644–9651. [CrossRef]
37. Salamanca, J.C.; Meehan-Atrash, J.; Vreeke, S.; Escobedo, J.O.; Peyton, D.H.; Strongin, R.M. E-cigarettes can emit formaldehyde at high levels under conditions that have been reported to be non-averse to users. *Sci. Rep.* **2018**, *8*, 7559. [CrossRef]
38. Bauer, R.; Cowan, D.A.; Crouch, A. Acrolein in wine: Importance of 3-hydroxypropionaldehyde and derivatives in production and detection. *J. Agric. Food Chem.* **2010**, *58*, 3243–3250. [CrossRef]
39. Osorio, V.M.; Cardeal, Z.L. Analytical methods to assess carbonyl compounds in foods and beverages. *J. Braz. Chem. Soc.* **2013**, *24*, 1711–1718. [CrossRef]
40. Yokoyama, A.; Tsutsumi, E.; Imazeki, H.; Suwa, Y.; Nakamura, C.; Mizukami, T.; Yokoyama, T. Salivary acetaldehyde concentration according to alcoholic beverage consumed and aldehyde dehydrogenase-2 genotype. *Alcohol. Clin. Exp. Res.* **2008**, *32*, 1607–1614. [CrossRef]
41. Li, Y.; Steppi, A.; Zhou, Y.; Mao, F.; Miller, P.C.; He, M.M.; Zhao, T.; Sun, Q.; Zhang, J. Tumoral expression of drug and xenobiotic metabolizing enzymes in breast cancer patients of different ethnicities with implications to personalized medicine. *Sci. Rep.* **2017**, *7*, 4747. [CrossRef]
42. Weng, M.W.; Lee, H.W.; Park, S.H.; Hu, Y.; Wang, H.T.; Chen, L.C.; Rom, W.N.; Huang, W.C.; Lepor, H.; Wu, X.R.; et al. Aldehydes are the predominant forces inducing DNA damage and inhibiting DNA repair in tobacco smoke carcinogenesis. *Proc. Natl. Acad. Sci. USA* **2018**, *115*, E6152–E6161. [CrossRef]
43. Garaycoechea, J.I.; Crossan, G.P.; Langevin, F.; Mulderrig, L.; Louzada, S.; Yang, F.; Guilbaud, G.; Park, N.; Roerink, S.; Nik-Zainal, S.; et al. Alcohol and endogenous aldehydes damage chromosomes and mutate stem cells. *Nature* **2018**, *553*, 171–177. [CrossRef]
44. Trewick, S.C.; Henshaw, T.F.; Hausinger, R.P.; Lindahl, T.; Sedgwick, B. Oxidative demethylation by *Escherichia coli* Alkb directly reverts DNA base damage. *Nature* **2002**, *419*, 174–178. [CrossRef]
45. Kooistra, S.M.; Helin, K. Molecular mechanisms and potential functions of histone demethylases. *Nat. Rev. Mol. Cell Biol.* **2012**, *13*, 297–311. [CrossRef]
46. Vasiliou, V.; Nebert, D.W. Analysis and update of the human aldehyde dehydrogenase (ALDH) gene family. *Hum. Genom.* **2005**, *2*, 138–143.
47. Finkel, T.; Holbrook, N.J. Oxidants, oxidative stress and the biology of ageing. *Nature* **2000**, *408*, 239–247. [CrossRef]
48. Lin, M.T.; Beal, M.F. Mitochondrial dysfunction and oxidative stress in neurodegenerative diseases. *Nature* **2006**, *443*, 787–795. [CrossRef]
49. Pizzimenti, S.; Ciamporcero, E.; Daga, M.; Pettazzoni, P.; Arcaro, A.; Cetrangolo, G.; Minelli, R.; Dianzani, C.; Lepore, A.; Gentile, F.; et al. Interaction of aldehydes derived from lipid peroxidation and membrane proteins. *Front. Physiol.* **2013**, *4*, 242. [CrossRef]
50. Guéraud, F.; Atalay, M.; Bresgen, N.; Cipak, A.; Eckl, P.M.; Huc, L.; Jouanin, I.; Siems, W.; Uchida, K. Chemistry and biochemistry of lipid peroxidation products. *Free Radic. Res.* **2010**, *44*, 1098–1124. [CrossRef]
51. Poli, G.; Schaur, R.J.; Siems, W.G.; Leonarduzzi, G. 4-hydroxynonenal: A membrane lipid oxidation product of medicinal interest. *Med. Res. Rev.* **2008**, *28*, 569–631. [CrossRef]
52. Dianzani, M.U. 4-hydroxynonenal from pathology to physiology. *Mol. Aspects Med.* **2003**, *24*, 263–272. [CrossRef]
53. Thornalley, P.J. Pharmacology of methylglyoxal: Formation, modification of proteins and nucleic acids, and enzymatic detoxification—A role in pathogenesis and antiproliferative chemotherapy. *Gen. Pharmacol.* **1996**, *27*, 565–573. [CrossRef]

54. Singh, S.; Brocker, C.; Koppaka, V.; Chen, Y.; Jackson, B.C.; Matsumoto, A.; Thompson, D.C.; Vasiliou, V. Aldehyde dehydrogenases in cellular responses to oxidative/electrophilic stress. *Free Radic. Biol. Med.* **2013**, *56*, 89–101. [CrossRef]

55. Pontel, L.B.; Rosado, I.V.; Burgos-Barragan, G.; Garaycoechea, J.I.; Yu, R.; Arends, M.J.; Chandrasekaran, G.; Broecker, V.; Wei, W.; Liu, L.; et al. Endogenous formaldehyde is a hematopoietic stem cell genotoxin and metabolic carcinogen. *Mol. Cell* **2015**, *60*, 177–188. [CrossRef]

56. Baan, R.; Straif, K.; Grosse, Y.; Secretan, B.; El Ghissassi, F.; Bouvard, V.; Altieri, A.; Cogliano, V.; WHO International Agency for Research on Cancer Monograph Working Group. Carcinogenicity of alcoholic beverages. *Lancet Oncol.* **2007**, *8*, 292–293. [CrossRef]

57. Secretan, B.; Straif, K.; Baan, R.; Grosse, Y.; El Ghissassi, F.; Bouvard, V.; Benbrahim-Tallaa, L.; Guha, N.; Freeman, C.; Galichet, L.; et al. A review of human carcinogens—Part e: Tobacco, areca nut, alcohol, coal smoke, and salted fish. *Lancet Oncol.* **2009**, *10*, 1033–1034. [CrossRef]

58. Wilbourn, J.; Heseltine, E.; Møller, H. IARC evaluates wood dust and formaldehyde. International agency for research on cancer. *Scand. J. Work Environ. Health* **1995**, *21*, 229–232. [CrossRef]

59. IARC Working Group on the Evaluation of Carcinogenic Risks to Humans. Formaldehyde, 2-butoxyethanol and 1-tert-butoxypropan-2-ol. *IARC Monogr. Eval. Carcinog. Risks Hum.* **2006**, *88*, 1–478.

60. Wang, M.; McIntee, E.J.; Cheng, G.; Shi, Y.; Villalta, P.W.; Hecht, S.S. Identification of DNA adducts of acetaldehyde. *Chem. Res. Toxicol.* **2000**, *13*, 1149–1157. [CrossRef]

61. Wang, Y.; Millonig, G.; Nair, J.; Patsenker, E.; Stickel, F.; Mueller, S.; Bartsch, H.; Seitz, H.K. Ethanol-induced cytochrome P4502E1 causes carcinogenic etheno-DNA lesions in alcoholic liver disease. *Hepatology* **2009**, *50*, 453–461. [CrossRef] [PubMed]

62. Linhart, K.; Bartsch, H.; Seitz, H.K. The role of reactive oxygen species (ROS) and cytochrome P-450 2E1 in the generation of carcinogenic etheno-DNA adducts. *Redox Biol.* **2014**, *3*, 56–62. [CrossRef] [PubMed]

63. Balbo, S.; Brooks, P.J. Implications of acetaldehyde-derived DNA adducts for understanding alcohol-related carcinogenesis. *Adv. Exp. Med. Biol.* **2015**, *815*, 71–88. [PubMed]

64. Brooks, P.J.; Theruvathu, J.A. DNA adducts from acetaldehyde: Implications for alcohol-related carcinogenesis. *Alcohol* **2005**, *35*, 187–193. [CrossRef] [PubMed]

65. Tan, S.L.W.; Chadha, S.; Liu, Y.; Gabasova, E.; Perera, D.; Ahmed, K.; Constantinou, S.; Renaudin, X.; Lee, M.; Aebersold, R.; et al. A class of environmental and endogenous toxins induces BRCA2 haploinsufficiency and genome instability. *Cell* **2017**, *169*, 1105–1118. [CrossRef] [PubMed]

66. Hoffman, E.A.; Frey, B.L.; Smith, L.M.; Auble, D.T. Formaldehyde crosslinking: A tool for the study of chromatin complexes. *J. Biol. Chem.* **2015**, *290*, 26404–26411. [CrossRef] [PubMed]

67. Maynard, S.; Fang, E.F.; Scheibye-Knudsen, M.; Croteau, D.L.; Bohr, V.A. DNA damage, DNA repair, aging, and neurodegeneration. *Cold Spring Harb. Perspect. Med.* **2015**, *5*, a025130. [CrossRef] [PubMed]

68. Yang, M.Y.; Wang, Y.B.; Han, B.; Yang, B.; Qiang, Y.W.; Zhang, Y.; Wang, Z.; Huang, X.; Liu, J.; Chen, Y.D.; et al. Activation of aldehyde dehydrogenase 2 slows down the progression of atherosclerosis via attenuation of ER stress and apoptosis in smooth muscle cells. *Acta Pharmacol. Sin.* **2018**, *39*, 48–58. [CrossRef] [PubMed]

69. Uchida, K. Role of reactive aldehyde in cardiovascular diseases. *Free Radic. Biol. Med.* **2000**, *28*, 1685–1696. [CrossRef]

70. Barrera, G.; Pizzimenti, S.; Daga, M.; Dianzani, C.; Arcaro, A.; Cetrangolo, G.P.; Giordano, G.; Cucci, M.A.; Graf, M.; Gentile, F. Lipid peroxidation-derived aldehydes, 4-hydroxynonenal and malondialdehyde in aging-related disorders. *Antioxidants (Basel)* **2018**, *7*, 102. [CrossRef]

71. Jaganjac, M.; Tirosh, O.; Cohen, G.; Sasson, S.; Zarkovic, N. Reactive aldehydes—Second messengers of free radicals in diabetes mellitus. *Free Radic. Res.* **2013**, *47* (Suppl. 1), 39–48. [CrossRef] [PubMed]

72. Hill, B.G.; Bhatnagar, A. Beyond reactive oxygen species: Aldehydes as arbitrators of alarm and adaptation. *Circ. Res.* **2009**, *105*, 1044–1046. [CrossRef] [PubMed]

73. Matveychuk, D.; Dursun, S.M.; Wood, P.L.; Baker, G.B. Reactive aldehydes and neurodegenerative disorders. *Klin. Psikofarmakol. Bülteni Bull. Clin. Psychopharmacol.* **2011**, *21*, 277–288. [CrossRef]

74. Negre-Salvayre, A.; Coatrieux, C.; Ingueneau, C.; Salvayre, R. Advanced lipid peroxidation end products in oxidative damage to proteins. Potential role in diseases and therapeutic prospects for the inhibitors. *Br. J. Pharmacol.* **2008**, *153*, 6–20. [CrossRef] [PubMed]

75. Burke, W.J.; Li, S.W.; Chung, H.D.; Ruggiero, D.A.; Kristal, B.S.; Johnson, E.M.; Lampe, P.; Kumar, V.B.; Franko, M.; Williams, E.A.; et al. Neurotoxicity of MAO metabolites of catecholamine neurotransmitters: Role in neurodegenerative diseases. *Neurotoxicology* **2004**, *25*, 101–115. [CrossRef]

76. Burke, W.J.; Kumar, V.B.; Pandey, N.; Panneton, W.M.; Gan, Q.; Franko, M.W.; O'Dell, M.; Li, S.W.; Pan, Y.; Chung, H.D.; et al. Aggregation of alpha-synuclein by DOPAL, the monoamine oxidase metabolite of dopamine. *Acta Neuropathol.* **2008**, *115*, 193–203. [CrossRef] [PubMed]

77. Panneton, W.M.; Kumar, V.B.; Gan, Q.; Burke, W.J.; Galvin, J.E. The neurotoxicity of DOPAL: Behavioral and stereological evidence for its role in Parkinson disease pathogenesis. *PLoS ONE* **2010**, *5*, e15251. [CrossRef] [PubMed]

78. Chaplen, F.W.; Fahl, W.E.; Cameron, D.C. Evidence of high levels of methylglyoxal in cultured Chinese hamster ovary cells. *Proc. Natl. Acad. Sci. USA* **1998**, *95*, 5533–5538. [CrossRef] [PubMed]

79. Niki, E. Lipid peroxidation: Physiological levels and dual biological effects. *Free Radic. Biol. Med.* **2009**, *47*, 469–484. [CrossRef]

80. Beisswenger, P.J.; Howell, S.K.; Touchette, A.D.; Lal, S.; Szwergold, B.S. Metformin reduces systemic methylglyoxal levels in type 2 diabetes. *Diabetes* **1999**, *48*, 198–202. [CrossRef]

81. Strzinek, R.A.; Scholes, V.E.; Norton, S.J. The purification and characterization of liver glyoxalase I from normal mice and from mice bearing a lymphosarcoma. *Cancer Res.* **1972**, *32*, 2359–2364. [PubMed]

82. Dobler, D.; Ahmed, N.; Song, L.; Eboigbodin, K.E.; Thornalley, P.J. Increased dicarbonyl metabolism in endothelial cells in hyperglycemia induces anoikis and impairs angiogenesis by RGD and GFOGER motif modification. *Diabetes* **2006**, *55*, 1961–1969. [CrossRef] [PubMed]

83. Bhutia, Y.; Ghosh, A.; Sherpa, M.L.; Pal, R.; Mohanta, P.K. Serum malondialdehyde level: Surrogate stress marker in the Sikkimese diabetics. *J. Nat. Sci. Biol. Med.* **2011**, *2*, 107–112. [PubMed]

84. Moghe, A.; Ghare, S.; Lamoreau, B.; Mohammad, M.; Barve, S.; McClain, C.; Joshi-Barve, S. Molecular mechanisms of acrolein toxicity: Relevance to human disease. *Toxicol. Sci.* **2015**, *143*, 242–255. [CrossRef] [PubMed]

85. Zarkovic, K. 4-hydroxynonenal and neurodegenerative diseases. *Mol. Asp. Med.* **2003**, *24*, 293–303. [CrossRef]

86. Markesbery, W.R.; Lovell, M.A. Four-hydroxynonenal, a product of lipid peroxidation, is increased in the brain in Alzheimer's disease. *Neurobiol. Aging* **1998**, *19*, 33–36. [CrossRef]

87. Lovell, M.A.; Ehmann, W.D.; Mattson, M.P.; Markesbery, W.R. Elevated 4-hydroxynonenal in ventricular fluid in Alzheimer's disease. *Neurobiol. Aging* **1997**, *18*, 457–461. [CrossRef]

88. Lovell, M.A.; Xie, C.; Markesbery, W.R. Acrolein is increased in Alzheimer's disease brain and is toxic to primary hippocampal cultures. *Neurobiol. Aging* **2001**, *22*, 187–194. [CrossRef]

89. Nagaraj, R.H.; Shipanova, I.N.; Faust, F.M. Protein cross-linking by the Maillard reaction. Isolation, characterization, and in vivo detection of a lysine-lysine cross-link derived from methylglyoxal. *J. Biol. Chem.* **1996**, *271*, 19338–19345. [CrossRef] [PubMed]

90. Boveris, A.; Chance, B. The mitochondrial generation of hydrogen peroxide. General properties and effect of hyperbaric oxygen. *Biochem. J.* **1973**, *134*, 707–716. [CrossRef] [PubMed]

91. Solivio, M.J. Investigation of DNA-Protein Cross-Links Generated in Biologically Relevant Oxidant Systems. Ph.D. Thesis, University of Cincinnati, Cincinnati, OH, USA, 2013.

92. Siems, W.; Grune, T. Intracellular metabolism of 4-hydroxynonenal. *Mol. Asp. Med.* **2003**, *24*, 167–175. [CrossRef]

93. Zhao, Y.; Miriyala, S.; Miao, L.; Mitov, M.; Schnell, D.; Dhar, S.K.; Cai, J.; Klein, J.B.; Sultana, R.; Butterfield, D.A.; et al. Redox proteomic identification of HNE-bound mitochondrial proteins in cardiac tissues reveals a systemic effect on energy metabolism after doxorubicin treatment. *Free Radic. Biol. Med.* **2014**, *72*, 55–65. [CrossRef] [PubMed]

94. Nathan, C.; Cunningham-Bussel, A. Beyond oxidative stress: An immunologist's guide to reactive oxygen species. *Nat. Rev. Immunol.* **2013**, *13*, 349–361. [CrossRef] [PubMed]

95. Weissman, L.; de Souza-Pinto, N.C.; Stevnsner, T.; Bohr, V.A. DNA repair, mitochondria, and neurodegeneration. *Neuroscience* **2007**, *145*, 1318–1329. [CrossRef] [PubMed]

96. Lazurko, C.; Radonjic, I.; Suchý, M.; Liu, G.; Rolland-Lagan, A.G.; Shuhendler, A. Fingerprinting biogenic aldehydes through pattern recognition analyses of excitation-emission matrices. *Chembiochem* **2019**, *20*, 543–554. [CrossRef]

97. Gomes, K.M.; Bechara, L.R.; Lima, V.M.; Ribeiro, M.A.; Campos, J.C.; Dourado, P.M.; Kowaltowski, A.J.; Mochly-Rosen, D.; Ferreira, J.C. Aldehydic load and aldehyde dehydrogenase 2 profile during the progression of post-myocardial infarction cardiomyopathy: Benefits of Alda-1. *Int. J. Cardiol.* **2015**, *179*, 129–138. [CrossRef]

98. Zambelli, V.O.; Gross, E.R.; Chen, C.H.; Gutierrez, V.P.; Cury, Y.; Mochly-Rosen, D. Aldehyde dehydrogenase-2 regulates nociception in rodent models of acute inflammatory pain. *Sci. Transl. Med.* **2014**, *6*, 251ra118. [CrossRef]

99. Grune, T.; Siems, W.; Kowalewski, J.; Zollner, H.; Esterbauer, H. Identification of metabolic pathways of the lipid peroxidation product 4-hydroxynonenal by enterocytes of rat small intestine. *Biochem. Int.* **1991**, *25*, 963–971.

100. Grune, T.; Siems, W.G.; Zollner, H.; Esterbauer, H. Metabolism of 4-hydroxynonenal, a cytotoxic lipid peroxidation product, in Ehrlich mouse ascites cells at different proliferation stages. *Cancer Res.* **1994**, *54*, 5231–5235.

101. Siems, W.G.; Zollner, H.; Grune, T.; Esterbauer, H. Metabolic fate of 4-hydroxynonenal in hepatocytes: 1,4-dihydroxynonene is not the main product. *J. Lipid. Res.* **1997**, *38*, 612–622.

102. Hartley, D.P.; Ruth, J.A.; Petersen, D.R. The hepatocellular metabolism of 4-hydroxynonenal by alcohol dehydrogenase, aldehyde dehydrogenase, and glutathione S-transferase. *Arch. Biochem. Biophys.* **1995**, *316*, 197–205. [CrossRef]

103. Wang, J.; Wang, H.; Hao, P.; Xue, L.; Wei, S.; Zhang, Y.; Chen, Y. Inhibition of aldehyde dehydrogenase 2 by oxidative stress is associated with cardiac dysfunction in diabetic rats. *Mol. Med.* **2011**, *17*, 172–179. [CrossRef]

104. Teng, S.; Beard, K.; Pourahmad, J.; Moridani, M.; Easson, E.; Poon, R.; O'Brien, P.J. The formaldehyde metabolic detoxification enzyme systems and molecular cytotoxic mechanism in isolated rat hepatocytes. *Chem. Biol. Interact.* **2001**, *130–132*, 285–296. [CrossRef]

105. Walport, L.J.; Hopkinson, R.J.; Schofield, C.J. Mechanisms of human histone and nucleic acid demethylases. *Curr. Opin. Chem. Biol.* **2012**, *16*, 525–534. [CrossRef]

106. Heck, H.D.; Casanova-Schmitz, M.; Dodd, P.B.; Schachter, E.N.; Witek, T.J.; Tosun, T. Formaldehyde (CH_2O) concentrations in the blood of humans and Fischer-344 rats exposed to CH2O under controlled conditions. *Am. Ind. Hyg. Assoc. J.* **1985**, *46*, 1–3. [CrossRef]

107. Luo, W.; Li, H.; Zhang, Y.; Ang, C.Y. Determination of formaldehyde in blood plasma by high-performance liquid chromatography with fluorescence detection. *J. Chromatogr. B Biomed. Sci. Appl.* **2001**, *753*, 253–257. [CrossRef]

108. Nagy, K.; Pollreisz, F.; Takáts, Z.; Vékey, K. Atmospheric pressure chemical ionization mass spectrometry of aldehydes in biological matrices. *Rapid Commun. Mass Spectrom.* **2004**, *18*, 2473–2478. [CrossRef]

109. Tong, Z.; Han, C.; Luo, W.; Wang, X.; Li, H.; Luo, H.; Zhou, J.; Qi, J.; He, R. Accumulated hippocampal formaldehyde induces age-dependent memory decline. *AGE (Dordr)* **2013**, *35*, 583–596. [CrossRef]

110. Bae, S.; Chon, J.; Field, M.S.; Stover, P.J. Alcohol Dehydrogenase 5 Is a Source of Formate for De Novo Purine Biosynthesis in HepG2 Cells. *J. Nutr.* **2017**, *147*, 499–505. [CrossRef]

111. Burgos-Barragan, G.; Wit, N.; Meiser, J.; Dingler, F.A.; Pietzke, M.; Mulderrig, L.; Pontel, L.B.; Rosado, I.V.; Brewer, T.F.; Cordell, R.L.; et al. Mammals divert endogenous genotoxic formaldehyde into one-carbon metabolism. *Nature* **2017**, *548*, 549–554. [CrossRef]

112. Ducker, G.S.; Rabinowitz, J.D. One-carbon metabolism in health and disease. *Cell Metab.* **2017**, *25*, 27–42. [CrossRef] [PubMed]

113. MacAllister, S.L.; Choi, J.; Dedina, L.; O'Brien, P.J. Metabolic mechanisms of methanol/formaldehyde in isolated rat hepatocytes: Carbonyl-metabolizing enzymes versus oxidative stress. *Chem. Biol. Interact.* **2011**, *191*, 308–314. [CrossRef] [PubMed]

114. Friedenson, B. A common environmental carcinogen unduly affects carriers of cancer mutations: Carriers of genetic mutations in a specific protective response are more susceptible to an environmental carcinogen. *Med. Hypotheses* **2011**, *77*, 791–797. [CrossRef] [PubMed]

115. Lee, S.L.; Wang, M.F.; Lee, A.I.; Yin, S.J. The metabolic role of human ADH3 functioning as ethanol dehydrogenase. *FEBS Lett.* **2003**, *544*, 143–147. [CrossRef]

116. Duester, G.; Farrés, J.; Felder, M.R.; Holmes, R.S.; Höög, J.O.; Parés, X.; Plapp, B.V.; Yin, S.J.; Jörnvall, H. Recommended nomenclature for the vertebrate alcohol dehydrogenase gene family. *Biochem. Pharmacol.* **1999**, *58*, 389–395. [CrossRef]

117. Vasiliou, V.; Pappa, A. Polymorphisms of human aldehyde dehydrogenases. Consequences for drug metabolism and disease. *Pharmacology* **2000**, *61*, 192–198. [CrossRef]

118. Wermuth, B.; Platts, K.L.; Seidel, A.; Oesch, F. Carbonyl reductase provides the enzymatic basis of quinone detoxication in man. *Biochem. Pharmacol.* **1986**, *35*, 1277–1282. [CrossRef]

119. Richardson, S.D.; Caughran, T.V.; Poiger, T.; Guo, Y.B.; Crumley, F.G. Application of DNPH derivatization with LC/MS to the identification of polar carbonyl disinfection by-products in drinking water. *Ozone Sci. Eng.* **2000**, *22*, 653–675. [CrossRef]

120. NIOSH. Aldehydes, screening: Method 2539. In *Manual of Analytical Methods (NMAM)*, 4th ed.; NIOSH: Washington, DC, USA, 1994.

121. ASTM. *Standard Test Method for Determination of Formaldehyde and Other Carbonyl Compounds in Air (Active Sampler Methodology)*; ASTM International: West Conshohocken, PA, USA, 2009.

122. US Environmental Protection Agency. *Compendium Method to-11A: Determination of Formaldehye in Ambient Air Using Adsorbent Cartridge Followed by High Performance Liquid Chromatorgraphy (HPLC)*; EPA: Washington, DC, USA, 1999.

123. US Environmental Protection Agency. *National Air Toxics Trends Station Work Plan Template*; EPA: Washington, DC, USA, 2011.

124. Santa, T. Derivatization reagents in liquid chromatography/electrospray ionization tandem mass spectrometry. *Biomed. Chromatogr.* **2011**, *25*, 1–10. [CrossRef]

125. Santa, T.; Al-Dirbashi, O.Y.; Ichibangase, T.; Rashed, M.S.; Fukushima, T.; Imai, K. Synthesis of 4-[2-(N,N-dimethylamino)ethylaminosulfonyl]-7-N-methylhydrazino-2,1,3-benzoxadiazole (DAABD-MHz) as a derivatization reagent for aldehydes in liquid chromatography/electrospray ionization-tandem mass spectrometry. *Biomed. Chromatogr.* **2008**, *22*, 115–118. [CrossRef]

126. Johnson, D.W. A modified Girard derivatizing reagent for universal profiling and trace analysis of aldehydes and ketones by electrospray ionization tandem mass spectrometry. *Rapid Commun. Mass Spectrom.* **2007**, *21*, 2926–2932. [CrossRef]

127. Barry, S.J.; Carr, R.M.; Lane, S.J.; Leavens, W.J.; Manning, C.O.; Monté, S.; Waterhouse, I. Use of S-pentafluorophenyl tris(2,4,6-trimethoxyphenyl)phosphonium acetate bromide and (4-hydrazino-4-oxobutyl) [tris(2,4,6-trimethoxyphenyl)phosphonium bromide for the derivatization of alcohols, aldehydes and ketones for detection by liquid chromatography/electrospray mass spectrometry. *Rapid Commun. Mass Spectrom.* **2003**, *17*, 484–497.

128. Yagi, K. A simple fluorometric assay for lipoperoxide in blood plasma. *Biochem. Med.* **1976**, *15*, 212–216. [CrossRef]

129. Armstrong, D.; Browne, R. The analysis of free radicals, lipid peroxides, antioxidant enzymes and compounds related to oxidative stress as applied to the clinical chemistry laboratory. *Adv. Exp. Med. Biol.* **1994**, *366*, 43–58.

130. Del Rio, D.; Pellegrini, N.; Colombi, B.; Bianchi, M.; Serafini, M.; Torta, F.; Tegoni, M.; Musci, M.; Brighenti, F. Rapid fluorimetric method to detect total plasma malondialdehyde with mild derivatization conditions. *Clin. Chem.* **2003**, *49*, 690–692. [CrossRef]

131. Ramdzan, A.N.; Almeida, M.I.G.S.; McCullough, M.J.; Kolev, S.D. Development of a microfluidic paper-based analytical device for the determination of salivary aldehydes. *Anal. Chim. Acta* **2016**, *919*, 47–54. [CrossRef]

132. Zhang, J.; Zhang, H.; Li, M.; Zhang, D.; Chu, Q.; Ye, J. A novel capillary electrophoretic method for determining methylglyoxal and glyoxal in urine and water samples. *J. Chromatogr. A* **2010**, *1217*, 5124–5129. [CrossRef]

133. Zweiner, C.; Glauner, T.; Frimmel, F.H. Method optimization for the determination of carbonyl compounds in disinfected water by DNPH derivatization and LC-ESI-MS-MS. *Anal. Bioanal. Chem.* **2002**, *372*, 615–621. [CrossRef]

134. Chi, Y.G.; Feng, Y.L.; Wen, S.; Lu, H.X.; Yu, Z.Q.; Zhang, W.B.; Sheng, G.Y.; Fu, J.M. Determination of carbonyl compounds in the atmosphere by DNPH derivatization and LC-ESI-MS/MS detection. *Talanta* **2007**, *72*, 539–545. [CrossRef]

135. Grosjean, E.; Green, P.G.; Grosjean, D. Liquid chromatography analysis of carbonyl (2,4-dinitrophenyl)hydrazones with detection by diode array ultraviolet spectroscopy and by atmospheric pressure negative chemical ionization mass spectrometry. *Anal. Chem.* **1999**, *71*, 1851–1861. [CrossRef]

136. Van Leeuwen, S.M.; Hendriksen, L.; Karst, U. Determination of aldehydes and ketones using derivatization with 2,4-dinitrophenylhydrazine and liquid chromatography-atmospheric pressure photoionization-mass spectrometry. *J. Chromatogr. A* **2004**, *1058*, 107–112. [CrossRef]

137. Ochs, S.D.M.; Fasciotti, M.; Netto, A.D.P. Analysis of 31 hydrazones of carbonyl compounds by RRLC-UV and RRLC-MS(/MS): A comparison of methods. *J. Spectrosc.* **2015**, *2015*, 1–11. [CrossRef]

138. Wang, Y.; Cui, P. Reactive carbonyl species derived from omega-3 and omega-6 fatty acids. *J. Agric. Food Chem.* **2015**, *63*, 6293–6296. [CrossRef]

139. Zhu, H.; Li, X.; Shoemaker, C.F.; Wang, S.C. Ultrahigh performance liquid chromatography analysis of volatile carbonyl compounds in virgin olive oils. *J. Agric. Food Chem.* **2013**, *61*, 12253–12259. [CrossRef]

140. Faizan, M.; Esatbeyoglu, T.; Bayram, B.; Rimbach, G. A fast and validated method for the determination of malondialdehyde in fish liver using high-performance liquid chromatography with a photodiode array detector. *J. Food Sci.* **2014**, *79*, C484–C488. [CrossRef]

141. De Lima, L.F.; Brandão, P.F.; Donegatti, T.A.; Ramos, R.M.; Gonçalves, L.M.; Cardoso, A.A.; Pereira, E.A.; Rodrigues, J.A. 4-hydrazinobenzoic acid as a derivatizing agent for aldehyde analysis by HPLC-UV and CE-DAD. *Talanta* **2018**, *187*, 113–119. [CrossRef]

142. Wu, J.Y.; Shi, Z.G.; Feng, Y.Q. Determination of 5-hydroxymethylfurfural using derivatization combined with polymer monolith microextraction by high-performance liquid chromatography. *J. Agric. Food Chem.* **2009**, *57*, 3981–3988. [CrossRef]

143. Banos, C.E.; Silva, M. Liquid chromatography-tandem mass spectrometry for the determination of low-molecular mass aldehydes in human urine. *J. Chromatogr. B Anal. Technol. Biomed. Life Sci.* **2010**, *878*, 653–658. [CrossRef]

144. Miller, J.H.; Gardner, W.P.; Gonzalez, R.R. UHPLC separation with MS analysis for eight carbonyl compounds in mainstream tobacco smoke. *J. Chromatogr. Sci.* **2010**, *48*, 12–17. [CrossRef]

145. Oenning, A.L.; Morés, L.; Dias, A.N.; Carasek, E. A new configuration for bar adsorptive microextraction (BAμE) for the quantification of biomarkers (hexanal and heptanal) in human urine by HPLC providing an alternative for early lung cancer diagnosis. *Anal. Chim. Acta* **2017**, *965*, 54–62. [CrossRef]

146. Lili, L.; Xu, H.; Song, D.; Cui, Y.; Hu, S.; Zhang, G. Analysis of volatile aldehyde biomarkers in human blood by derivatization and dispersive liquid-liquid microextraction based on solidification of floating organic droplet method by high performance liquid chromatography. *J. Chromatogr. A* **2010**, *1217*, 2365–2370. [CrossRef]

147. Liu, J.F.; Yuan, B.F.; Feng, Y.Q. Determination of hexanal and heptanal in human urine using magnetic solid phase extraction coupled with in-situ derivatization by high performance liquid chromatography. *Talanta* **2015**, *136*, 54–59. [CrossRef]

148. Xu, H.; Lv, L.; Hu, S.; Song, D. High-performance liquid chromatographic determination of hexanal and heptanal in human blood by ultrasound-assisted headspace liquid-phase microextraction with in-drop derivatization. *J. Chromatogr. A* **2010**, *1217*, 2371–2375. [CrossRef]

149. Xu, H.; Wang, S.; Zhang, G.; Huang, S.; Song, D.; Zhou, Y.; Long, G. A novel solid-phase microextraction method based on polymer monolith frit combining with high-performance liquid chromatography for determination of aldehydes in biological samples. *Anal. Chim. Acta* **2011**, *690*, 86–93. [CrossRef]

150. Zhang, H.J.; Huang, J.F.; Lin, B.; Feng, Y.Q. Polymer monolith microextraction with in situ derivatization and its application to high-performance liquid chromatography determination of hexanal and heptanal in plasma. *J. Chromatogr. A* **2007**, *1160*, 114–119. [CrossRef]

151. Yilmaz, B.; Asci, A.; Kucukoglu, K.; Albayrak, M. Simple high-performance liquid chromatography method for formaldehyde determination in human tissue through derivatization with 2,4-dinitrophenylhydrazine. *J. Sep. Sci.* **2016**, *39*, 2963–2969. [CrossRef]

152. Guan, X.Y.; Rubin, E.; Anni, H. An optimized method for the measurement of acetaldehyde by high-performance liquid chromatography. *Alcohol. Clin. Exp. Res.* **2012**, *36*, 398–405. [CrossRef]

153. Uchiyama, S.; Inaba, Y.; Matsumoto, M.; Suzuki, G. Reductive amination of aldehyde 2,4-dinitorophenylhydrazones using 2-picoline borane and high-performance liquid chromatographic analysis. *Anal. Chem.* **2009**, *81*, 485–489. [CrossRef]

154. Di Padova, C.; Alderman, J.; Lieber, C.S. Improved methods for the measurement of acetaldehyde concentrations in plasma and red blood cells. *Alcohol. Clin. Exp. Res.* **1986**, *10*, 86–89. [CrossRef]

155. Nielsen, F.; Mikkelsen, B.B.; Nielsen, J.B.; Andersen, H.R.; Grandjean, P. Plasma malondialdehyde as biomarker for oxidative stress: Reference interval and effects of life-style factors. *Clin. Chem.* **1997**, *43*, 1209–1214.

156. Grotto, D.; Santa Maria, L.D.; Boeira, S.; Valentini, J.; Charão, M.F.; Moro, A.M.; Nascimento, P.C.; Pomblum, V.J.; Garcia, S.C. Rapid quantification of malondialdehyde in plasma by high performance liquid chromatography-visible detection. *J. Pharm. Biomed. Anal.* **2007**, *43*, 619–624. [CrossRef]

157. Steghens, J.P.; van Kappel, A.L.; Denis, I.; Collombel, C. Diaminonaphtalene, a new highly specific reagent for HPLC-UV measurement of total and free malondialdehyde in human plasma or serum. *Free Radic. Biol. Med.* **2001**, *31*, 242–249. [CrossRef]

158. Xiong, X.J.; Wang, H.; Rao, W.B.; Guo, X.F.; Zhang, H.S. 1,3,5,7-Tetramethyl-8-aminozide-difluoroboradiaza-s-indacene as a new fluorescent labeling reagent for the determination of aliphatic aldehydes in serum with high performance liquid chromatography. *J. Chromatogr. A* **2010**, *1217*, 49–56. [CrossRef]

159. Fathy Bakr Ali, M.; Kishikawa, N.; Ohyama, K.; Abdel-Mageed Mohamed, H.; Mohamed Abdel-Wadood, H.; Mohamed Mohamed, A.; Kuroda, N. Chromatographic determination of aliphatic aldehydes in human serum after pre-column derivatization using 2,2′-furil, a novel fluorogenic reagent. *J. Chromatogr. A* **2013**, *1300*, 199–203. [CrossRef]

160. Imazato, T.; Shiokawa, A.; Kurose, Y.; Katou, Y.; Kishikawa, N.; Ohyama, K.; Ali, M.F.; Ueki, Y.; Maehata, E.; Kuroda, N. Determination of 4-hydroxy-2-nonenal in serum by high-performance liquid chromatography with fluorescence detection after pre-column derivatization using 4-(N,N-dimethylaminosulfonyl)-7-hydrazino-2,1,3-benzoxadiazole. *Biomed. Chromatogr.* **2014**, *28*, 891–894. [CrossRef]

161. Li, P.; Ding, G.; Deng, Y.; Punyapitak, D.; Li, D.; Cao, Y. Determination of malondialdehyde in biological fluids by high-performance liquid chromatography using rhodamine B hydrazide as the derivatization reagent. *Free Radic. Biol. Med.* **2013**, *65*, 224–231. [CrossRef]

162. Giera, M.; Kloos, D.P.; Raaphorst, A.; Mayboroda, O.A.; Deelder, A.M.; Lingeman, H.; Niessen, W.M. Mild and selective labeling of malondialdehyde with 2-aminoacridone: Assessment of urinary malondialdehyde levels. *Analyst* **2011**, *136*, 2763–2769. [CrossRef]

163. Mao, J.; Zhang, H.; Luo, J.; Li, L.; Zhao, R.; Zhang, R.; Liu, G. New method for HPLC separation and fluorescence detection of malonaldehyde in normal human plasma. *J. Chromatogr. B Anal. Technol. Biomed. Life Sci.* **2006**, *832*, 103–108. [CrossRef]

164. Seljeskog, E.; Hervig, T.; Mansoor, M.A. A novel HPLC method for the measurement of thiobarbituric acid reactive substances (TBARS). A comparison with a commercially available kit. *Clin. Biochem.* **2006**, *39*, 947–954. [CrossRef]

165. Ojeda, A.G.; Wrobel, K.; Escobosa, A.R.; Garay-Sevilla, M.E. High-performance liquid chromatography determination of glyoxal, methylglyoxal, and diacetyl in urine using 4-methoxy-o-phenylenediamine as derivatizing reagent. *Anal. Biochem.* **2014**, *449*, 52–58. [CrossRef]

166. Akira, K.; Matsumoto, Y.; Hashimoto, T. Determination of urinary glyoxal and methylglyoxal by high-performance liquid chromatography. *Clin. Chem. Lab. Med.* **2004**, *42*, 147–153. [CrossRef]

167. Paci, A.; Rieutord, A.; Guillaume, D.; Traoré, F.; Ropenga, J.; Husson, H.P.; Brion, F. Quantitative high-performance liquid chromatographic determination of acrolein in plasma after derivatization with Luminarin 3. *J. Chromatogr. B Biomed. Sci. Appl.* **2000**, *739*, 239–246. [CrossRef]

168. Al-Rawithi, S.; el-Yazigi, A.; Nicholls, P.J. Determination of acrolein in urine by liquid chromatography and fluorescence detection of its quinoline derivative. *Pharm. Res.* **1993**, *10*, 1587–1590. [CrossRef]

169. Roth, A.; Li, H.; Anorma, C.; Chan, J. A reaction-based fluorescent probe for imaging of formaldehyde in living cells. *J. Am. Chem. Soc.* **2015**, *137*, 10890–10893. [CrossRef]

170. Brewer, T.F.; Chang, C.J. An aza-cope reactivity-based fluorescent probe for imaging formaldehyde in living cells. *J. Am. Chem. Soc.* **2015**, *137*, 10886–10889. [CrossRef]

171. Yuen, L.H.; Saxena, N.S.; Park, H.S.; Weinberg, K.; Kool, E.T. Dark hydrazone fluorescence labeling agents enable imaging of cellular aldehydic load. *ACS Chem. Biol.* **2016**, *11*, 2312–2319. [CrossRef]

172. Vogel, M.; Buldt, A.; Karst, U. Hydrazine reagents as derivatizing agents in environmental analysis—A critical review. *Fresenius J. Anal. Chem.* **2000**, *366*, 781–791. [CrossRef]

173. Chou, C.-C.; Que Hee, S.S. Saliva-available carbonyl compounds in some chewing tobaccos. *J. Agric. Food Chem.* **1994**, *42*, 2225–2230. [CrossRef]

174. Lapolla, A.; Flamini, R.; Dalla Vedova, A.; Senesi, A.; Reitano, R.; Fedele, D.; Basso, E.; Seraglia, R.; Traldi, P. Glyoxal and methylglyoxal levels in diabetic patients: Quantitative determination by a new GC/MS method. *Clin. Chem. Lab. Med.* **2003**, *41*, 1166–1173. [CrossRef]

175. Takeuchi, A.; Takigawa, T.; Abe, M.; Kawai, T.; Endo, Y.; Yasugi, T.; Endo, G.; Ogino, K. Determination of formaldehyde in urine by headspace gas chromatography. *Bull. Environ. Contam. Toxicol.* **2007**, *79*, 1–4. [CrossRef]

176. Tsikas, D.; Rothmann, S.; Schneider, J.Y.; Gutzki, F.M.; Beckmann, B.; Frölich, J.C. Simultaneous GC-MS/MS measurement of malondialdehyde and 4-hydroxy-2-nonenal in human plasma: Effects of long-term L-arginine administration. *Anal. Biochem.* **2017**, *524*, 31–44. [CrossRef] [PubMed]

177. Deng, C.; Li, N.; Zhang, X. Development of headspace solid-phase microextraction with on-fiber derivatization for determination of hexanal and heptanal in human blood. *J. Chromatogr. B Anal. Technol. Biomed. Life Sci.* **2004**, *813*, 47–52. [CrossRef] [PubMed]

178. Deng, C.; Zhang, X. A simple, rapid and sensitive method for determination of aldehydes in human blood by gas chromatography/mass spectrometry and solid-phase microextraction with on-fiber derivatization. *Rapid Commun. Mass Spectrom.* **2004**, *18*, 1715–1720. [CrossRef] [PubMed]

179. Fuchs, P.; Loeseken, C.; Schubert, J.K.; Miekisch, W. Breath gas aldehydes as biomarkers of lung cancer. *Int. J. Cancer* **2010**, *126*, 2663–2670. [CrossRef] [PubMed]

180. Svensson, S.; Lärstad, M.; Broo, K.; Olin, A.C. Determination of aldehydes in human breath by on-fibre derivatization, solid-phase microextraction and GC-MS. *J. Chromatogr. B Anal. Technol. Biomed. Life Sci.* **2007**, *860*, 86–91. [CrossRef] [PubMed]

181. Calejo, I.; Moreira, N.; Araújo, A.M.; Carvalho, M.; Bastos, M.e.L.; de Pinho, P.G. Optimisation and validation of a HS-SPME-GC-IT/MS method for analysis of carbonyl volatile compounds as biomarkers in human urine: Application in a pilot study to discriminate individuals with smoking habits. *Talanta* **2016**, *148*, 486–493. [CrossRef] [PubMed]

182. Poli, D.; Goldoni, M.; Corradi, M.; Acampa, O.; Carbognani, P.; Internullo, E.; Casalini, A.; Mutti, A. Determination of aldehydes in exhaled breath of patients with lung cancer by means of on-fiber-derivatisation SPME-GC/MS. *J. Chromatogr. B Anal. Technol. Biomed. Life Sci.* **2010**, *878*, 2643–2651. [CrossRef] [PubMed]

183. Shin, H.S. Determination of malondialdehyde in human blood by headspace-solid phase micro-extraction gas chromatography-mass spectrometry after derivatization with 2,2,2-trifluoroethylhydrazine. *J. Chromatogr. B Anal. Technol. Biomed. Life Sci.* **2009**, *877*, 3707–3711. [CrossRef] [PubMed]

184. Li, N.; Deng, C.; Yin, X.; Yao, N.; Shen, X.; Zhang, X. Gas chromatography-mass spectrometric analysis of hexanal and heptanal in human blood by headspace single-drop microextraction with droplet derivatization. *Anal. Biochem.* **2005**, *342*, 318–326. [CrossRef]

185. Stopforth, A.; Burger, B.V.; Crouch, A.M.; Sandra, P. Urinalysis of 4-hydroxynonenal, a marker of oxidative stress, using stir bar sorptive extraction-thermal desorption-gas chromatography/mass spectrometry. *J. Chromatogr. B Anal. Technol. Biomed. Life Sci.* **2006**, *834*, 134–140. [CrossRef] [PubMed]

186. Wu, M.Y.; Chen, B.G.; Chang, C.D.; Huang, M.H.; Wu, T.G.; Chang, D.M.; Lee, Y.J.; Wang, H.C.; Lee, C.I.; Chern, C.L.; et al. A novel derivatization approach for simultaneous determination of glyoxal, methylglyoxal, and 3-deoxyglucosone in plasma by gas chromatography-mass spectrometry. *J. Chromatogr. A* **2008**, *1204*, 81–86. [CrossRef] [PubMed]

187. Hanff, E.; Eisenga, M.F.; Beckmann, B.; Bakker, S.J.; Tsikas, D. Simultaneous pentafluorobenzyl derivatization and GC-ECNICI-MS measurement of nitrite and malondialdehyde in human urine: Close positive correlation between these disparate oxidative stress biomarkers. *J. Chromatogr. B Anal. Technol. Biomed. Life Sci.* **2017**, *1043*, 167–175. [CrossRef] [PubMed]

188. Tsikas, D.; Rothmann, S.; Schneider, J.Y.; Suchy, M.T.; Trettin, A.; Modun, D.; Stuke, N.; Maassen, N.; Frölich, J.C. Development, validation and biomedical applications of stable-isotope dilution GC-MS and GC-MS/MS techniques for circulating malondialdehyde (MDA) after pentafluorobenzyl bromide derivatization: MDA as a biomarker of oxidative stress and its relation to 15(S)-8-iso-prostaglandin F2α and nitric oxide (NO). *J. Chromatogr. B Anal. Technol. Biomed. Life Sci.* **2016**, *1019*, 95–111.

189. Stalikas, C.D.; Konidari, C.N. Analysis of malondialdehyde in biological matrices by capillary gas chromatography with electron-capture detection and mass spectrometry. *Anal. Biochem.* **2001**, *290*, 108–115. [CrossRef] [PubMed]

190. Cighetti, G.; Debiasi, S.; Paroni, R.; Allevi, P. Free and total malondialdehyde assessment in biological matrices by gas chromatography-mass spectrometry: What is needed for an accurate detection. *Anal. Biochem.* **1999**, *266*, 222–229. [CrossRef] [PubMed]

191. Pastor-Belda, M.; Fernández-García, A.J.; Campillo, N.; Pérez-Cárceles, M.D.; Motas, M.; Hernández-Córdoba, M.; Viñas, P. Glyoxal and methylglyoxal as urinary markers of diabetes. Determination using a dispersive liquid-liquid microextraction procedure combined with gas chromatography-mass spectrometry. *J. Chromatogr. A* **2017**, *1509*, 43–49. [CrossRef]

192. Kandhro, A.J.; Mirza, M.A.; Khuhawar, M.Y. Capillary gas chromatographic determination of methylglyoxal from serum of diabetic patients by precolumn derivatization using meso-stilbenediamine as derivatizing reagent. *J. Chromatogr. Sci.* **2008**, *46*, 539–543. [CrossRef]

193. Khuhawar, M.Y.; Zardari, L.A.; Laghari, A.J. Capillary gas chromatographic determination of methylglyoxal from serum of diabetic patients by precolumn derivatization with 1,2-diamonopropane. *J. Chromatogr. B Anal. Technol. Biomed. Life Sci.* **2008**, *873*, 15–19. [CrossRef]

194. Antón, A.P.; Ferreira, A.M.; Pinto, C.G.; Cordero, B.M.; Pavón, J.L. Headspace generation coupled to gas chromatography-mass spectrometry for the automated determination and quantification of endogenous compounds in urine. Aldehydes as possible markers of oxidative stress. *J. Chromatogr. A* **2014**, *1367*, 9–15. [CrossRef]

195. Zhang, X.; Wang, R.; Zhang, L.; Wei, J.; Ruan, Y.; Wang, W.; Ji, H.; Liu, J. Simultaneous determination of four aldehydes in gas phase of mainstream smoke by headspace gas chromatography-mass spectrometry. *Int. J. Anal. Chem.* **2019**, *2019*, 2105839. [CrossRef]

196. Criado-García, L.; Arce, L. Extraction of toxic compounds from saliva by magnetic-stirring-assisted micro-solid-phase extraction step followed by headspace-gas chromatography-ion mobility spectrometry. *Anal. Bioanal. Chem.* **2016**, *408*, 6813–6822. [CrossRef] [PubMed]

197. Andreoli, R.; Manini, P.; Corradi, M.; Mutti, A.; Niessen, W.M. Determination of patterns of biologically relevant aldehydes in exhaled breath condensate of healthy subjects by liquid chromatography/atmospheric chemical ionization tandem mass spectrometry. *Rapid Commun. Mass Spectrom.* **2003**, *17*, 637–645. [CrossRef] [PubMed]

198. Chen, J.L.; Huang, Y.J.; Pan, C.H.; Hu, C.W.; Chao, M.R. Determination of urinary malondialdehyde by isotope dilution LC-MS/MS with automated solid-phase extraction: A cautionary note on derivatization optimization. *Free Radic. Biol. Med.* **2011**, *51*, 1823–1829. [CrossRef] [PubMed]

199. Manini, P.; Andreoli, R.; Sforza, S.; Dall'Asta, C.; Galaverna, G.; Mutti, A.; Niessen, W.M. Evaluation of Alternate Isotope-Coded Derivatization Assay (AIDA) in the LC-MS/MS analysis of aldehydes in exhaled breath condensate. *J. Chromatogr. B Anal. Technol. Biomed. Life Sci.* **2010**, *878*, 2616–2622. [CrossRef] [PubMed]

200. Van der Toorn, M.; Slebos, D.J.; de Bruin, H.G.; Gras, R.; Rezayat, D.; Jorge, L.; Sandra, K.; van Oosterhout, A.J. Critical role of aldehydes in cigarette smoke-induced acute airway inflammation. *Respir. Res.* **2013**, *14*, 45. [CrossRef] [PubMed]

201. Prokai, L.; Szarka, S.; Wang, X.; Prokai-Tatrai, K. Capture of the volatile carbonyl metabolite of flecainide on 2,4-dinitrophenylhydrazine cartridge for quantitation by stable-isotope dilution mass spectrometry coupled with chromatography. *J. Chromatogr. A* **2012**, *1232*, 281–287. [CrossRef] [PubMed]

202. Olson, K.L.; Swarin, S.J. Determination of Aldehydes and Ketones by Derivatization and Liquid-Chromatography Mass-Spectrometry. *J. Chromatogr.* **1985**, *333*, 337–347. [CrossRef]

203. Lord, H.L.; Rosenfeld, J.; Volovich, V.; Kumbhare, D.; Parkinson, B. Determination of malondialdehyde in human plasma by fully automated solid phase analytical derivatization. *J. Chromatogr. B Anal. Technol. Biomed. Life Sci.* **2009**, *877*, 1292–1298. [CrossRef]

204. Tomono, S.; Miyoshi, N.; Ohshima, H. Comprehensive analysis of the lipophilic reactive carbonyls present in biological specimens by LC/ESI-MS/MS. *J. Chromatogr. B Anal. Technol. Biomed. Life Sci.* **2015**, *988*, 149–156. [CrossRef]

205. Cao, Y.; Guan, Q.; Sun, T.; Qi, W.; Guo, Y. Charged tag founded in N-(1-chloroalkyl)pyridinium quaternization for quantification of fatty aldehydes. *Anal. Chim. Acta* **2016**, *937*, 80–86. [CrossRef]

206. Henning, C.; Liehr, K.; Girndt, M.; Ulrich, C.; Glomb, M.A. Extending the spectrum of α-dicarbonyl compounds in vivo. *J. Biol. Chem.* **2014**, *289*, 28676–28688. [CrossRef]
207. Kim, H.J.; Shin, H.S. Simple derivatization of aldehydes with D-cysteine and their determination in beverages by liquid chromatography-tandem mass spectrometry. *Anal. Chim. Acta* **2011**, *702*, 225–232. [CrossRef] [PubMed]
208. El-Maghrabey, M.; Kishikawa, N.; Kuroda, N. 9,10-Phenanthrenequinone as a mass-tagging reagent for ultra-sensitive liquid chromatography-tandem mass spectrometry assay of aliphatic aldehydes in human serum. *J. Chromatogr. A* **2016**, *1462*, 80–89. [CrossRef] [PubMed]
209. Sobsey, C.A.; Han, J.; Lin, K.; Swardfager, W.; Levitt, A.; Borchers, C.H. Development and evaluation of a liquid chromatography-mass spectrometry method for rapid, accurate quantitation of malondialdehyde in human plasma. *J. Chromatogr. B Anal. Technol. Biomed. Life Sci.* **2016**, *1029–1030*, 205–212. [CrossRef] [PubMed]
210. Oh, J.A.; Shin, H.S. Simple and sensitive determination of malondialdehyde in human urine and saliva using UHPLC-MS/MS after derivatization with 3,4-diaminobenzophenone. *J. Sep. Sci.* **2017**, *40*, 3958–3968. [CrossRef] [PubMed]
211. Eggink, M.; Wijtmans, M.; Ekkebus, R.; Lingeman, H.; de Esch, I.J.; Kool, J.; Niessen, W.M.; Irth, H. Development of a selective ESI-MS derivatization reagent: Synthesis and optimization for the analysis of aldehydes in biological mixtures. *Anal. Chem.* **2008**, *80*, 9042–9051. [CrossRef] [PubMed]
212. Eggink, M.; Wijtmans, M.; Kretschmer, A.; Kool, J.; Lingeman, H.; de Esch, I.J.; Niessen, W.M.; Irth, H. Targeted LC-MS derivatization for aldehydes and carboxylic acids with a new derivatization agent 4-APEBA. *Anal. Bioanal. Chem.* **2010**, *397*, 665–675. [CrossRef] [PubMed]
213. Chen, D.; Ding, J.; Wu, M.K.; Zhang, T.Y.; Qi, C.B.; Feng, Y.Q. A liquid chromatography-mass spectrometry method based on post column derivatization for automated analysis of urinary hexanal and heptanal. *J. Chromatogr. A* **2017**, *1493*, 57–63. [CrossRef]
214. Tie, C.; Hu, T.; Jia, Z.X.; Zhang, J.L. Derivatization Strategy for the Comprehensive Characterization of Endogenous Fatty Aldehydes Using HPLC-Multiple Reaction Monitoring. *Anal. Chem.* **2016**, *88*, 7762–7768. [CrossRef]
215. Kolliker, S.; Oehme, M.; Dye, C. Structure elucidation of 2,4-dinitrophenylhydrazone derivatives of carbonyl compounds in ambient air by HPLC/MS and multiple MS/MS using atmospheric chemical ionization in the negative ion mode. *Anal. Chem.* **1998**, *70*, 1979–1985. [CrossRef]
216. Brombacher, S.; Oehme, M.; Beukes, J.A. HPLC combined with multiple mass spectrometry (MSn): An alternative for the structure elucidation of compounds and artefacts found in smog chamber samples. *J. Environ. Monit.* **2001**, *3*, 311–316. [CrossRef] [PubMed]
217. Mendonça, R.; Gning, O.; Di Cesaré, C.; Lachat, L.; Bennett, N.C.; Helfenstein, F.; Glauser, G. Sensitive and selective quantification of free and total malondialdehyde in plasma using UHPLC-HRMS. *J. Lipid Res.* **2017**, *58*, 1924–1931. [CrossRef] [PubMed]
218. Dator, R.; Carrà, A.; Maertens, L.; Guidolin, V.; Villalta, P.W.; Balbo, S. A High Resolution/Accurate Mass (HRAM) Data-Dependent MS^3 Neutral Loss Screening, Classification, and Relative Quantitation Methodology for Carbonyl Compounds in Saliva. *J. Am. Soc. Mass Spectrom.* **2017**, *28*, 608–618. [CrossRef] [PubMed]
219. Yu, L.; Liu, P.; Wang, Y.L.; Yu, Q.W.; Yuan, B.F.; Feng, Y.Q. Profiling of aldehyde-containing compounds by stable isotope labelling-assisted mass spectrometry analysis. *Analyst* **2015**, *140*, 5276–5286. [CrossRef] [PubMed]
220. Zheng, S.J.; Wang, Y.L.; Liu, P.; Zhang, Z.; Yu, L.; Yuan, B.F.; Feng, Y.Q. Stable isotope labeling-solid phase extraction-mass spectrometry analysis for profiling of thiols and aldehydes in beer. *Food Chem.* **2017**, *237*, 399–407. [CrossRef] [PubMed]
221. Eggink, M.; Charret, S.; Wijtmans, M.; Lingeman, H.; Kool, J.; Niessen, W.M.; Irth, H. Development of an on-line weak-cation exchange liquid chromatography-tandem mass spectrometric method for screening aldehyde products in biological matrices. *J. Chromatogr. B Anal. Technol. Biomed. Life Sci.* **2009**, *877*, 3937–3945. [CrossRef] [PubMed]
222. Zhao, S.; Dawe, M.; Guo, K.; Li, L. Development of High-Performance Chemical Isotope Labeling LC-MS for Profiling the Carbonyl Submetabolome. *Anal. Chem.* **2017**, *89*, 6758–6765. [CrossRef]

223. Guo, N.; Peng, C.Y.; Zhu, Q.F.; Yuan, B.F.; Feng, Y.Q. Profiling of carbonyl compounds in serum by stable isotope labeling—Double precursor ion scan—Mass spectrometry analysis. *Anal. Chim. Acta* **2017**, *967*, 42–51. [CrossRef]

224. Siegel, D.; Meinema, A.C.; Permentier, H.; Hopfgartner, G.; Bischoff, R. Integrated quantification and identification of aldehydes and ketones in biological samples. *Anal. Chem.* **2014**, *86*, 5089–5100. [CrossRef]

225. Deng, P.; Higashi, R.M.; Lane, A.N.; Bruntz, R.C.; Sun, R.C.; Raju, M.V.R.; Nantz, M.H.; Qi, Z.; Fan, T.W. Correction: Quantitative profiling of carbonyl metabolites directly in crude biological extracts using chemoselective tagging and nanoESI-FTMS. *Analyst* **2018**, *143*, 999. [CrossRef]

226. Mattingly, S.J.; Xu, T.; Nantz, M.H.; Higashi, R.M.; Fan, T.W. A carbonyl capture approach for profiling oxidized metabolites in cell extracts. *Metabolomics* **2012**, *8*, 989–996. [CrossRef] [PubMed]

227. Fu, X.A.; Li, M.; Biswas, S.; Nantz, M.H.; Higashi, R.M. A novel microreactor approach for analysis of ketones and aldehydes in breath. *Analyst* **2011**, *136*, 4662–4666. [CrossRef] [PubMed]

228. Knipp, R.J.; Li, M.; Fu, X.-A.; Nantz, M.H. A versatile probe for chemoselective capture and analysis of carbonyl compounds in exhaled breath. *Anal. Methods* **2015**, *7*, 6027–6033. [CrossRef]

229. Kolliker, S.; Oehme, M.; Merz, L. Unusual MSn fragmentation patterns of 2,4-dinitrophenylhydrazine and its propanone derivative. *Rapid Commun. Mass Spectrom.* **2001**, *15*, 2117–2126. [CrossRef]

230. Szarka, S.; Prokai-Tatrai, K.; Prokai, L. Application of screening experimental designs to assess chromatographic isotope effect upon isotope-coded derivatization for quantitative liquid chromatography-mass spectrometry. *Anal. Chem.* **2014**, *86*, 7033–7040. [CrossRef] [PubMed]

231. Deng, P.; Higashi, R.M.; Lane, A.N.; Bruntz, R.C.; Sun, R.C.; Ramakrishnam Raju, M.V.; Nantz, M.H.; Qi, Z.; Fan, T.W. Quantitative profiling of carbonyl metabolites directly in crude biological extracts using chemoselective tagging and nanoESI-FTMS. *Analyst* **2017**, *143*, 311–322. [CrossRef] [PubMed]

232. Wild, C.P. Complementing the genome with an "exposome": The outstanding challenge of environmental exposure measurement in molecular epidemiology. *Cancer Epidemiol. Biomark. Prev.* **2005**, *14*, 1847–1850. [CrossRef] [PubMed]

 *toxics*

Review

Protein Adductomics: Analytical Developments and Applications in Human Biomonitoring

George W. Preston and David H. Phillips *

Environmental Research Group, Department of Analytical, Environmental and Forensic Science,
School of Population Health and Environmental Sciences, King's College London, Franklin-Wilkins Building,
150 Stamford Street, London SE1 9NH, UK; george.preston@kcl.ac.uk
* Correspondence: david.phillips@kcl.ac.uk

Received: 22 March 2019; Accepted: 20 May 2019; Published: 25 May 2019

Abstract: Proteins contain many sites that are subject to modification by electrophiles. Detection and characterisation of these modifications can give insights into environmental agents and endogenous processes that may be contributing factors to chronic human diseases. An untargeted approach, utilising mass spectrometry to detect modified amino acids or peptides, has been applied to blood proteins haemoglobin and albumin, focusing in particular on the N-terminal valine residue of haemoglobin and the cysteine-34 residue in albumin. Technical developments to firstly detect simultaneously multiple adducts at these sites and then subsequently to identify them are reviewed here. Recent studies in which the methods have been applied to biomonitoring human exposure to environmental toxicants are described. With advances in sensitivity, high-throughput handling of samples and robust quality control, these methods have considerable potential for identifying causes of human chronic disease and of identifying individuals at risk.

Keywords: haemoglobin; albumin; mass spectrometry; biomarkers; protein adducts

---

1. The Exposome and Adductomics

Many decades of epidemiological observations have indicated that incidences of chronic human diseases are likely to result from a combination of environmental exposures to chemical and physical stressors, and predispositions inherent in human genetics. The wide geographical variation of many such diseases implies that it is environmental factors that play the dominant role, and not inherited predisposition, in disease causation [1], but knowledge of what the environmental factors are is often far from complete. As a consequence, estimations of overall risks associated with these factors are inaccurate and important associations may go undetected. These limitations have recently been framed within the context of the *exposome*, which can be thought of as the environmental counterpart of the genome. Conceptually, the exposome aims to reflect the totality of environmental exposures throughout the human lifespan, and to take into account both external components (e.g., exogenous environmental agents) and internal ones (e.g., endogenous cellular processes that give rise to altered stasis or function) [2–5]. For strategies to improve human health to be effective, it is essential to unravel the causes of chronic human diseases and to assess accurately their risks. The goal of studying the exposome (i.e., of *exposomics*) is disease prevention through the acquisition of a broad scientific perspective that encompasses health, environmental, educational, socioeconomic and political factors [6–9].

Two recent collaborative projects have applied the exposome concept to investigating environmental impacts on human health by assessing environmental exposure at personal and population levels within existing short- and long-term population studies. In the *EXPOsOMICS* project the emphasis has been on the measurement and impact of air and water pollution, studied in a number of adult and child study populations [10]. In the *HELIX* project the focus has been on early-life events, examining exposure to a range of chemicals and physical agents in existing birth cohorts [11]. Both these projects utilised a combination of exposure monitoring, using mobile and static monitors, smartphone and satellite data, and omics techniques to investigate biomarkers associated with exposures. The multi-omic approach has included metabolome, proteome, transcriptome, epigenome and adductome profiles. While many of the results of these interrelated analyses have yet to emerge, it is anticipated that new insights into the importance of environmental factors in the aetiology of human diseases will ensue and that the studies will point the way to improved strategies for monitoring human exposures and their health consequences.

While these projects have focused on human exposures and health outcomes, broader ecological issues may also be addressed by the exposome concept. The adverse outcome pathway (AOP) concept seeks to define the initial molecular events that culminate in adverse (toxicological) endpoints [12]. There is currently much discussion of how to assess the properties of complex mixtures of chemicals, taking into consideration possible positive and negative interactions between their components, in order to refine hazard identification and risk assessment. It has been proposed that considering the relative contributions of components of the exposome in relation to complex mixtures combined with a mechanistic understanding of the induced adverse effects, may improve the integrated risk assessment for both human and environmental health [13].

Electrophiles have long been suspected in the causality of cancer and other chronic diseases. Because they are reactive, they can be measured indirectly through the adducts they form with protein and DNA. Indeed, damage to, or modification of, DNA by reactive intermediates of chemical carcinogens or by ionising and non-ionising radiation is a key early event in the carcinogenic process. The exposome concept encompasses a "top-down" approach to identifying environmental factors that determine susceptibility to disease throughout the entire lifespan. In parallel, a "bottom-up" approach can investigate biomarkers specific for certain environmental exposures, based on knowledge of environmental carcinogens and their pathways of metabolic activation. As part of this approach, protein adductomics constitute the untargeted investigation of modification of proteins by endogenous or exogenous agents.

2. Approaches to Protein Adductomics

The concept of an adductome (that is, a collection of additional products) implicates two types of reactant: Those that add, and those to which are added. In the context of the present discussion, these are nucleophilic protein sites (amino acid residues) and electrophilic toxicants, respectively[1]. Reactants of either type are potentially diverse, meaning that the adductome could be vast. From an analytical standpoint, this potential vastness (i.e., structural diversity) is problematic because of the lack of a common 'handle' or 'signature' by which to purify and identify the adducts. Accordingly, investigators have focused on adducts of either specific nucleophiles or specific electrophiles. If the investigator's aim is to discover biomarkers of exposure, a nucleophile is selected and the electrophiles to which it adds are captured; if the aim is instead to discover targets, an electrophile is selected and the nucleophiles that add to it are captured. Given that the focus of this review is on biomarkers of environmental exposure, we will concentrate on the former approach. The latter approach is also important, however, because it is a route by which novel adducts could be accessed, either directly [14,15] or indirectly [16].

Of the methods that capture electrophiles, the most advanced methods are based on haemoglobin (Hb) and human serum albumin (HSA). There have been a number of important methodological

---

[1] A note regarding language. For the reaction of a nucleophile with an electrophile, the view of the chemist is that the *nucleophile* is the active participant, providing electrons for the chemical bond ('nucleophilic addition', 'nucleophilic attack', and so on). Toxicologists, on the other hand, tend to speak of the *toxicant* as active (toxicant 'binds' to target), and since the toxicant is usually an electrophile the roles would seem to switch. This second interpretation is equally logical because the nucleophilic targets are often endogenous and less mobile (e.g., DNA or protein) and therefore seem to be passive entities.

developments since Rappaport et al. reviewed the subject in 2012 [17]. Another, related review [18] was published during the preparation of the present review.

2.1. Hb as A Target of Electrophiles

Hb is found in the erythrocytes, where it functions as an oxygen carrier. Its high concentration and reactivity (see below) make it a likely target of electrophiles, and its long lifetime in vivo (126 days, the lifetime of an erythrocyte [19]) presumably gives the resulting adducts an opportunity to accumulate. Human Hb A, the major form of Hb in adults, is a tetramer composed of two α-chains and two β-chains. The four chains, each of which binds one molecule of haem, all adopt similar folds in the tetramer. The α- and β-chains have several amino acid residues in common, including the N-terminal valine residues [20]. The α-amino groups of these terminal residues are nucleophilic, and have been observed to react with toxicologically-relevant electrophiles [21]. The N-terminal α-amino groups of the α- and β-chains have similar pK_a values and similar reactivity towards certain electrophiles (e.g., the acetylating agent acetic anhydride), but not necessarily towards all electrophiles [22]. For example, another acetylating agent, methyl acetyl phosphate, has been observed to modify the N-terminus of only the β-chain [23].

The β-chain of Hb possesses a cysteine residue (Cys-β93) for which there is no equivalent in the α-chain [20]. Adducts of Hb Cys-β93 have been the subject of both targeted and, to a lesser extent, untargeted adductomic analyses (see below). A targeted adductomic method (i.e., a method involving simultaneous monitoring of multiple known/hypothesised adducts) was used to monitor Hb adducts of 15 different aromatic amines (e.g., 4-aminobiphenyl) in tobacco smokers' blood [24]. These, it should be pointed out, are not adducts of the amines themselves, but rather of the corresponding arylnitroso compounds [25]. Arylnitroso compounds form via oxidation of the amines' N-hydroxy metabolites, in a reaction for which, in the erythrocyte at least, the oxidant is the oxy form of Hb itself. The Cys-β93 adducts of arylnitroso compounds are N-arylsulfinamides, which hydrolyse under acidic conditions to regenerate their corresponding aromatic amines [26]. On this basis, detection of the aromatic amines liberated by acid hydrolysis of N-arylsulfinamides has been used as an indirect way of detecting the adducts [24].

2.2. The N-alkyl Edman Method

The analytical tractability of Hb N-terminal adducts is due to a general property of N-terminal amino acid residues, namely their ability to be detached from the rest of the protein via Edman degradation. This is a procedure that was originally developed for protein sequencing, but which was modified in the 1980s by Ehrenberg and co-workers for the analysis of Hb N-terminal adducts [27]. Ehrenberg and co-workers' procedure has been referred to as the 'N-alkyl Edman method' because of its ability to detect, for example, N_α-methyl and N_α-ethyl substituents [28,29]. In fact, the observed N_α-substituents have not been limited to simple alkyl groups, but for convenience the modified N-terminal amino acid is referred to as N-alkylvaline. Edman's original procedure involved reacting the α-amino group of a peptide with phenyl isothiocyanate, which rendered an acid-labile product [30]. Treatment of this product with anhydrous acid liberates the terminal amino acid as an anilinothiazolinone, which is then isomerised in aqueous acid to a phenylthiohydantoin (PTH) [31]. Ehrenberg and co-workers found that Hb with N-terminal N-alkylvaline (i.e., a secondary amine) reacted with isothiocyanate reagents in the same way as unmodified Hb, but that the resulting derivatives were labile even under neutral conditions [27]. The final product, a substituted PTH, could therefore be isolated using conditions under which unmodified Hb remained intact.

In subsequent iterations of the N-alkyl Edman method, the isothiocyanate reagent was varied so as to generate analytes appropriate for particular analytical methods. The most recent iteration, the 'FIRE procedure', uses fluorescein isothiocyanate ('FIRE' being a contraction of 'fluorescein isothiocyanate', 'R-group' and 'Edman degradation') [32]. The FIRE procedure was initially developed with targeted analysis in mind, but was later adapted for untargeted analyses ('FIRE screening procedure' [28]).

2.3. The Role of Tandem Mass Spectrometry in Protein Adductomics

Like most other adductomic methodologies, the FIRE screening procedure utilises tandem mass spectrometry (MS/MS) for the detection of adducts. MS/MS, as its name suggests, involves two stages of mass analysis. The first stage is for intact precursor ions (e.g., protonated molecules) and the second stage is for product ions (i.e., fragments of precursor ions). A process of fragmentation takes place in between the two stages. Mass analysis can be performed in either a static mode, whereby ions of specified mass-to-charge ratio (m/z) are isolated, or a dynamic mode, whereby a continuous range of m/z values is scanned. Either stage can be performed in either mode, meaning that a number of different types of experiment are possible. In selected reaction monitoring (SRM), a technique commonly used for targeted analyses, ions of pre-specified m/z are isolated at both stages. Isolation is achieved by defining a narrow window of permissible m/z values and is often done using a quadrupole mass filter. An apparatus commonly used for SRM is the triple quadrupole mass spectrometer, which consists of two quadrupole mass filters, with a collision cell between them, connected in series. The first and second stages of mass analysis take place in the first and second filters, respectively, with fragmentation taking place in the collision cell. Other MS/MS techniques of relevance to this review are precursor ion scanning, data-dependent acquisition (DDA) and data-independent acquisition (DIA). These will be covered in more detail in the sections concerning HSA adductomics.

2.4. Stepped MS/MS Methods

Several adductomic studies have employed stepped methods, which can be thought of as hybrids of SRM and scanning. A stepped method consists of a sequence of SRM experiments that collectively resemble a scan. In considering how the methods work, it is instructive to think of adducts' structures in terms of two distinct parts: A constant part that derives from the nucleophile (common to all precursor ions) and a variable part that derives from the electrophile (variable among precursor ions). It follows, therefore, that a given product ion (or neutral fragment) will be either constant or variable depending on how the precursor ion becomes broken up into fragments. Given that the variable parts of the precursor ions are unlikely to be known *a priori*, the constituent SRM experiments of a stepped method must be necessarily arbitrary. For this reason, it is common to see lists of equally-spaced integer or half-integer m/z values [28]. We have referred to these arbitrary values as sampling points [33]. The idea of an arbitrary SRM experiment might strike the reader as odd, since SRM is traditionally used for targeted analyses, but for untargeted analyses it does not matter where the sampling points fall. The important thing is that, collectively, they are able to capture all relevant adducts. The limitation of stepped methods is their low resolution, which means that they are unable to identify adducts unambiguously purely on the basis of mass. Their value, therefore, tends to be in providing a quantitative description of the distribution of adducts.

2.5. The FIRE Screening Procedure

The FIRE screening procedure [28] is a method for untargeted detection of Hb adducts (Figure 1). It is a stepped method akin to the 'adductome approach to detect DNA damage' developed by Kanaly et al. [34]. In the FIRE screening procedure, different precursor ions (protonated fluorescein thiohydantoins, FTHs) are captured at the first stage of mass analysis via one of 136 different windows. Each window is approximately 0.7 m/z units wide, and the m/z values on which the windows are centred are 1 Da apart. Thus, by cycling through all 136 windows, the method can capture a wide range of precursor ions and can, therefore, detect the corresponding range of mass shifts (between +14 and +149 Da). Once captured, a precursor ion is fragmented, and its products are passed to the second stage of mass analysis. Here, a set of fixed windows permit only constant product ions to pass to the detector (implicates loss of variable neutral fragments), and a variable window permits only variable product ions to pass (implicates loss of constant neutral fragments). If the right combination of constant and variable product ions is detected, then the presence of a corresponding FTH, and therefore Hb adduct, can be inferred. This MS/MS is done 'online' following the chromatographic separation of the FTHs, and the data thus generated are, like those reported by Kanaly et al., visualised as an 'adductome map', usually a plot of *m/z* against retention time [28,34,35].

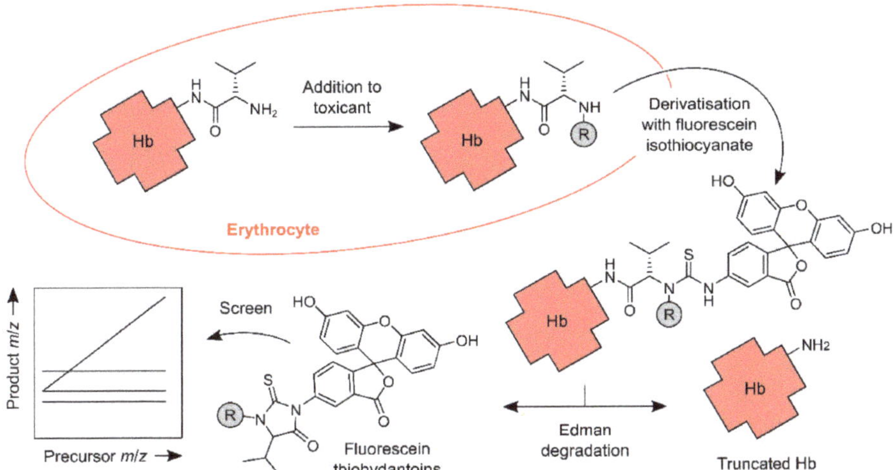

Figure 1. Main steps of the FIRE ('fluorescein isothiocyanate', 'R-group' and 'Edman degradation') screening procedure for Hb *N*-terminal adductomics. The procedure detects 'R' groups, which are generated when an *N*-terminus of Hb reacts with an electrophile in vivo. The *N*-termini are derivatised with fluorescein isothiocyanate, and derivatives with 'R' groups are selectively decomposed to the corresponding fluorescein thiohydantoins. The thiohydantoins are analysed using LC and online 'stepped' triple quadrupole mass spectrometry.

Carlsson et al. used their procedure to screen the blood of smokers and non-smokers and detected 26 features of interest; this study is described below in Section 3.

2.6. HSA as A Target of Electrophiles

HSA is the major protein in human plasma. Its lifetime in vivo, whilst shorter than that of Hb, is presumably still long enough for adducts to accumulate. In vivo, HSA binds fatty acids, scavenges metal ions, and contributes to the oncotic pressure of blood [22]. Extensive use of HSA has been made for the biological monitoring of toxicants, and a detailed account of this can be found in the recent review by Sabbioni and Turesky [36]. For the purposes of the present review, we focus on providing a background to the untargeted HSA adductomics studies.

Thus far, HSA contains a number of nucleophilic sites, including (but not limited to) histidine residues, lysine residues and a single reduced cysteine residue (Cys-34). Notably, histidine residues in HSA, as in Hb, are targets of epoxides [37,38]. Lysine residues in serum albumins are notable targets of aflatoxin B_1 dialdehyde [39,40].

Cys-34 is the only site in HSA for which untargeted adductomic methods have been developed. The motivation to look at this particular site is related to the unique chemistry of thiol groups, and the fact that HSA Cys-34 accounts for the majority of such groups in human plasma [41]. Given that the reacting species is a thiolate anion rather than a thiol group proper [41], adduct formation should be promoted by alkaline conditions and/or basic groups within the local protein environment. The pK_a of the HSA Cys-34 thiol group is controversial, but is generally regarded to be lower than that of a typical thiol group [41]. In the three-dimensional structure of HSA, as determined by X-ray crystallography, the side chain of Cys-34 is partially buried [42]. On this basis, it has been inferred that there might be a limit to the size of the electrophiles that HSA Cys-34 can add to. It has also been recognised, however, that the tertiary structure of HSA is dynamic and that Cys-34 may become less buried upon deprotonation of the thiol group [17,43]. HSA Cys-34 is reactive towards a variety of toxicologically-relevant electrophiles, including sulphur mustard and metabolites of aromatic amines [44,45], and can also undergo oxidative transformations [46,47]. It appears that, in vivo, a substantial proportion of the HSA Cys-34 thiol groups is *S*-thiolated (*S*-[cystein-*S*-yl], *S*-[glutathion-*S*-yl] and so on), and a smaller, but appreciable proportion is found as the corresponding sulfenic, sulfinic or sulfonic acids [41,47].

2.7. HSA Cys-34 Adductomics

To date, methods for HSA Cys-34 adductomics have been based exclusively on peptide analytes (Figure 2). When HSA is digested with trypsin, and no cleavages are missed, Cys-34 and its adducts are found in a 21-amino-acid peptide [48,49]. This peptide, which Rappaport's group has referred to as 'T3' (i.e., the third-heaviest tryptic peptide [49]), has been used as an analyte in a number of studies [49–51]. When a combination of trypsin and chymotrypsin is used, the Cys-34-containing peptide is instead the LQQCPF hexapeptide [43]. The use of Pronase, suggested by Sabbioni and Turesky as a means of generating lower-molecular-weight analytes, has not to our knowledge been implemented for untargeted HSA Cys-34 adductomics [36]. When Noort et al. [44,52] used Pronase to digest HSA adducts of either sulphur mustard or acrylamide, the respective modifications were found in the CPF tripeptide.

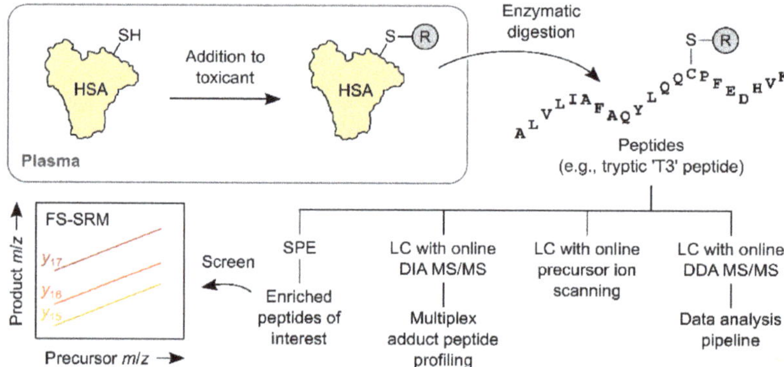

Figure 2. Main steps of published HSA Cys-34 adductomic workflows. The reaction of HSA with an electrophile in the blood plasma installs an 'R' group at the Cys-34 site. The HSA is isolated from plasma or serum and digested—usually with trypsin—to produce a mixture of peptides. Some of the peptides contain 'R' groups and others do not (the introduction of an enrichment step prior to digestion can limit the number of those that do not). Peptides are then separated chromatographically and analysed using MS/MS. One of the MS/MS methods, a stepped triple-quadrupole method termed FS-SRM, is depicted. This method monitors three variable product ions of the tryptic 'T3' peptide (y_{15}, y_{16} and y_{17}).

Some of the first untargeted HSA adductomic analyses were performed by Aldini et al. using the technique of precursor ion scanning [43] (see also the 'chemical modificomics' method proposed by Goto et al. [53]). Precursor ion scanning is an MS/MS technique involving a scan at the first stage of mass analysis and the isolation of a constant product ion at the second stage. The result is a spectrum of the different precursor ions that give rise to a given product. Aldini et al. [43] reacted purified HSA with a mixture of α,β-unsaturated aldehydes (4-hydroxy-2-nonenal, 4-hydroxy-2-hexenal and acrolein), and digested the products with trypsin and chymotrypsin. Analysis of the digestion products, using liquid chromatography (LC) and online precursor ion scanning, revealed peaks corresponding to substituted LQQCPF peptides. These, in turn, corresponded to HSA Cys-34 Michael adducts of the α,β-unsaturated aldehyde reactants.

2.8. Fixed-Step SRM of HSA Adducts

An important development, reported by Li et al. in 2011, was the demonstration of a stepped method called fixed-step SRM (FS-SRM [49]). FS-SRM consists of a sequence of SRM experiments that collectively resemble a linked scan [54]. In developing the method, Li et al. drew on elements of the 'adductome approach to detect DNA damage' described by Kanaly et al. [34,55], and also a method of analysing mercapturic acids described by Wagner et al. [56]. Being a stepped method, FS-SRM is broadly analogous to the FIRE screening procedure (which, in fact, it pre-dates). The analytes in FS-SRM are substituted T3 peptides, and the precursor ions captured in the first stage of mass analysis are triply-protonated peptides. The product ions isolated in the second stage are doubly-charged variable *y*-ions and a singly-charged constant *b*-ion. Together, these precursor and product ions constitute what is effectively a peptide sequence tag [57]. The sampling points used for FS-SRM are 4.5 Da apart and, in the Li and co-workers' study, there were 77 of them. FS-SRM differs from the other stepped methods in that, for FS-SRM, the sample is infused into the mass spectrometer as a mixture of adducts rather than as a series of eluted components. There is still an LC step but it is disconnected from the mass spectrometry, and it serves to capture the entire population of adducts rather than to separate them. The method is therefore freed from a major constraint imposed by LC, namely the need for a full set of SRM experiments to be done within the width of a chromatographic peak.

Our personal experience with protein adductomics has been in the implementation of FS-SRM for epidemiological studies [10,33]. Such studies, which typically involve tens or hundreds of samples, pose challenges that are not necessarily encountered in smaller pilot studies. In implementing the method of Li et al., the main challenge that we faced was the need for higher throughput. This was addressed by evaluating the various stages of sample preparation (HSA purification, adduct enrichment, digestion and peptide clean-up) and optimising these where possible. Notably, we deleted the adduct enrichment step, and we changed the method of sample clean-up from HPLC (serial) to solid-phase extraction (SPE; effectively parallel). A model adduct, prepared by treating HSA with *N*-ethylmaleimide, proved useful for evaluating the performance of the methods.

In parallel with our work on FS-SRM, Grigoryan et al. [50] developed a new analytical workflow based on LC with on-line DDA mass spectrometry. In DDA, the *data* on which the acquisition is *dependent* are precursor ions' *m/z* values, and they are obtained via a high-resolution scan—using, for example, an Orbitrap mass analyser. The *data* are used to direct the isolation of precursor ions, and so only these precursor ions are fragmented. The *acquisition* is the scan via which the resulting product ions are detected. In addition to their analytical method, Grigoryan et al. [50] also developed methods for sample preparation and data analysis (the 'adductomics pipeline'). The method of sample preparation is essentially a streamlined version of the one developed by Li et al. [49]. One major difference with respect to the earlier method, however, was the omission of a reducing agent, which had previously been used to reduce protein disulphide bonds prior to tryptic digestion. The effect of omitting the reducing agent was to preserve *S*-thiolated forms of Cys-34. The method of data analysis begins with the detection of a tag (a combination of constant and variable product ions) in the product-ion scan data. The corresponding precursor ion is then identified, and an ion count chromatogram for this precursor ion is extracted. A particularly innovative part of the pipeline is the method by which the peptide analytes are quantified. Each analyte is quantified relative to a 'housekeeping peptide', which is another tryptic peptide of HSA. In this way, the method is able to control for variation in the quantity of digested HSA. Grigoryan et al. [50] used their pipeline to analyse samples of plasma from smokers and non-smokers, and found a total of 43 putative adducts (see Section 3 below).

2.9. Multiplex Adduct Peptide Profiling

Another promising method for HSA Cys-34 adductomics (and potentially also Hb Cys-β93 adductomics) is 'multiplex adduct peptide profiling' (MAPP [51]). MAPP utilises DIA mass spectrometry, which is perhaps the least prescriptive of all MS/MS techniques. Similar to a stepped

SRM-based method, DIA captures precursor ions via a series of contiguous windows. The windows are, however, rather wider than those used for SRM, and it is therefore likely that a given window will capture multiple precursor ions (in MAPP, for example, the width of each window is 10 m/z units). As in DDA mass spectrometry, the second stage of mass analysis is a scan, and a high-resolution scan is done as an alternative first stage.

The MAPP method, like the 'adductomics pipeline', requires prior knowledge of the peptide analyte's sequence and the site of modification. Series of constant product ions (e.g., b-ions from backbone scission near the N-terminus) are recognised and are linked back to their respective precursor ions via common chromatographic retention times. The substituted peptide's mass shift is then confirmed by the presence of corresponding variable product ions. Although the authors were only able to identify oxidised and S-thiolated forms of HSA Cys-34, their method has the potential to detect toxicologically-relevant adducts (e.g., if the samples could be further enriched for these adducts prior to analysis).

2.10. Hb and HSA Compared

Given that Hb and HSA contain some of the same nucleophilic functional groups, these proteins might be expected to have overlapping reactivity towards electrophiles. The observation that cysteine residues in HSA and Hb can add to comparable amounts of benzene oxide in vivo, for example, is evidence of such overlap [58]. On the other hand, Dingley et al. [59] found that dietary exposure to 2-amino-1-methyl-6-phenylimidazo[4,5-*b*]pyridine (PhIP; see Section 3.3) caused the formation of substantially larger amounts of HSA adducts than Hb adducts. A similar fate has been observed for aflatoxin B_1 in rats: of a given dose of this toxicant, a substantially higher proportion is found bound to serum albumin than to Hb [60,61]. This might also be expected to be the case in humans, and indeed assays for HSA adducts of aflatoxin B_1 dialdehyde have been developed [62]. Possible reasons for differences in the amount or type of adducts include (i) the fact that Hb and HSA are synthesised at different sites in the body (in different cell types), and as a result could be exposed to different electrophiles [36]; (ii) the fact that Hb resides inside the erythrocyte, whereas HSA is secreted [18]; (iii) the influence of neighbouring amino acid side chains and cofactors on the reactivity of the nucleophilic groups (see Sections 2.1 and 2.6); and (iv) the possibility that the erythrocyte membrane could shield Hb from electrophiles, or even sequester electrophiles [63]. It is also worth considering that apparent differences in the extent of adduct formation could reflect differences in chemical and biological stability of the proteins and/or modifications.

2.11. Other Target Proteins

Few proteins other than Hb and HSA have been discussed as candidates for untargeted adductomic analyses, and fewer still have been investigated experimentally. Hb and HSA adducts are probably two of the richest and most accessible sources of potential biomarkers, but this is not to say that other proteins could not provide additional and unique information. Three other proteins of relevance to the present review have been discussed: Collagen, histones and apolipoproteins. Collagen is mentioned by Scheepers in his workshop report [19], presumably because of its abundance in the body and its extremely long lifespan in certain tissues [64]. However, there have been few attempts to use collagen adducts for biological monitoring, probably because of the heterogeneity, physical properties and limited accessibility of collagen [65–67]. Histones, which are also mentioned by Scheepers, represent a more promising source of biomarkers. Work on histone adducts has not been extensive, but some interesting results have been obtained. *N*-Terminal segments of histones are of particular interest because they protrude from nucleosomal core particles, and, on this basis, it is plausible that they could be accessible to electrophiles. Consistent with this idea, SooHoo et al. [68] observed modifications near the N-termini of histones isolated from cultured human lymphoblasts that had been exposed to *anti*-benzo[a]pyrene 7,8-dihydrodiol-9,10-oxide (BPDE). Fabrizi et al. [69] used a model peptide to infer the reactivity of an *N*-terminal segment of histone H2B towards phosgene, and observed the incorporation of carbonyl groups into the peptide.

Apolipoproteins have been investigated as targets of endogenous electrophiles, such as the lipid oxidation product 4-hydroxy-2-nonenal. By definition, endogenous adducts cannot be biomarkers of exposure in the strict sense, but they could potentially be biomarkers of effect. We mention them here because they have been the subject of a recent untargeted adductomics study. This study focused on adducts of histidine and lysine residues in human low density lipoprotein [35]. Unlike the FIRE screening procedure or FS-SRM, the method is not site-specific; rather, it detects modifications to any and all residues of particular amino acid. The analytes are 'free' amino acids, which are prepared from lipoprotein by acid hydrolysis. Consequently, they may represent a mixture of sites, and perhaps a mixture of proteins. The analytical method, like others described elsewhere in this article, involves ultraperformance LC and triple quadrupole mass spectrometry. Apparently it is a stepped method, in which a constant product ion is isolated at the second stage of mass analysis. For adducts of histidine residues, the constant product is the immonium ion of histidine, and for adducts of lysine residues, it is a deaminated immonium ion of lysine. Shibata et al. [35] used their method to analyse low density lipoprotein that had been first purified from human plasma, and then oxidised in vitro. The oxidised lipoprotein was treated with sodium borohydride to reduce imine linkages (as in, for example, a lysine residue adducts of 9-oxononanoic acid), before being hydrolysed and the resulting amino acids analysed. The authors produced adductome maps for lipoprotein with and without oxidation, and by comparing these maps they were able to attribute the formation of the aforementioned 9-oxononanoic acid adduct to the oxidising condition.

2.12. Adduct Enrichment

Enrichment, in the context of untargeted adductomics, entails depletion of the unmodified nucleophile and possibly also other substances that might interfere with the detection of the adducts. In the FIRE screening procedure for Hb adducts, enrichment is facilitated by the detachment of the *N*-alkylvaline residues. This exaggerates the relatively minor difference in structure between Hb and its adducts, thereby allowing the unmodified Hb to be removed readily [28]. For HSA Cys-34 adductomics, methods of enrichment have mainly exploited the reactivity of the Cys-34 thiol group, which is present in the unmodified HSA but not in the adducts. Funk et al. [70] demonstrated the use of a disulfide-functionalised resin for scavenging unmodified HSA, and this method was later used in adductomic workflows [49,51,71]. The main limitation of the thiol scavenging method is that it does not remove *S*-thiolated HSA: If a reducing agent is later added to reduce the other disulfide bonds in HSA (i.e., those of the cystine residues) then the *S*-thiolation is reversed and the Cys-34 thiol would seem to reappear. Funk et al. [70] sought to limit this effect by removing the S-thiolation prior to the scavenging step. In our hands, the thiol scavenging method proved difficult to implement in a high-throughput setting, and so we deleted it from our workflow [33]. Chung et al. [71] used thiol scavenging as the first of two stages of enrichment, the second stage being an antibody-mediated purification of the substituted T3 peptides using a polyclonal antibody raised against the T3 peptide but having cross-reactivity with adducts.

3. Human Biomonitoring

3.1. Methodological Considerations

Human biomonitoring refers to the quantification of xenobiotics or their derivatives (and sometimes their early effects) in human biospecimens [72]. As well as confirming the nature of the exposure, biomonitoring aims to measure the internal dose of the xenobiotic(s). The biomonitoring of protein adducts is usually done as part of the 'bottom up' (targeted) approach (see Section 1). A typical targeted method might involve isotope dilution (i.e., the addition of a known amount of an isotopically-labelled standard) followed by LC-MS/MS. This would require prior characterisation of the adduct and synthesis of a suitable standard.

In principle, data collected via the untargeted approach (e.g., peak areas from LC-MS/MS) could be used in the same way as those collected in targeted studies. However, this would depend on the untargeted method achieving an acceptable accuracy, precision and dynamic range for each relevant adduct. At some stage, a synthetic reference compound would be needed to confirm a particular adduct's identity, and to implicate the corresponding electrophile [18]. For hitherto unknown adducts, possible identities must first be proposed. Methods that have assisted in this endeavour have included database searching, the use of calculator software, and the comparison of measured and predicted physicochemical properties [50,73]. The characterisation of novel adducts—a challenging aspect of the research—has been reviewed in detail by Carlsson et al. [18].

Accuracy, in practice, may suffer as a consequence of the need to capture a range of adducts. It is likely that the use of generic standards (e.g., the S-carbamidomethylated T3 peptide for FS-SRM) affects accuracy, and therefore precludes absolute quantification [33]. Dynamic ranges are dependent on the analytical method, and presumably also on the ability to enrich adducts. As judged from lowest reported adduct concentrations, the detection limits of Grigoryan and co-workers' LC-MS-based method, and of the FIRE screening procedure, are good (<7 and <0.1 adduct molecules per million HSA molecules or Hb chains, respectively [28,50]). The methods should, therefore, be able to detect some xenobiotic adducts, although in practice relatively few such adducts have been observed [50]. For FS-SRM (our implementation), the detection and quantification limits are in the region of one adduct molecule per thousand HSA molecules, and are probably too high to detect xenobiotic adducts [33]. Putative adducts detected by FS-SRM and the other methods may, however, relate to the early effects of exposure.

At the present time, the role of the untargeted methods is to complement the targeted methods, rather than to replace them. Indeed, approaches that combine both methods have been proposed [7]. Some authors advocate a more pragmatic 'fit-for-purpose' approach, which balances methodological rigour with cost. Dennis et al. draw a distinction between regulatory endeavours, which require maximal rigour, and exploratory studies whose aims might be achievable without a fully validated method [7].

While the discipline of untargeted protein adductomics is still a relatively young one, there have been a number of pilot studies that have sought to demonstrate its utility. Additionally, some targeted investigations have looked for adduct formation at the same sites (e.g., Cys-34 of HSA) and these will also be mentioned here.

3.2. Human Biomonitoring of Hb Adducts

In the first adductomic application of the FIRE method (see Section 2.5), Hb samples from smokers and non-smokers were analysed and compared [28]. In all samples seven adducts at the *N*-terminal valine residue were identified; these were the addition of methyl and ethyl groups, and adducts formed by ethylene oxide, acrylonitrile, methyl vinyl ketone, acrylamide and glycidamide; in addition, a further 19 unknown adducts were detected in all samples. Subsequently, one of these unknown adducts has been identified as derived from ethyl methyl ketone [74]. A further four have been attributed to the precursor electrophiles glyoxal, methylglyoxal, acrylic acid and 1-octen-3-one [73]; and recently another adduct, detected in smokers and non-smokers at similar levels, has been identified as *N*-(4-hydroxybenzyl)valine, postulated to have arisen from either 4-quinone methide, which could form the valine adduct via a Michael addition, or 4-hydroxybenzaldehyde, which could form the same adduct via a Schiff base formation followed by reduction [75].

Applying their untargeted Hb adductomic approach to a larger study population, Carlsson et al. [76] analysed blood samples from healthy children about 12 years old ($n = 51$). In this cohort, a total of 24 adducts (12 of them previously identified; see above) were observed and their levels quantified. Relatively large interindividual variations in adduct levels were observed. The frequencies of micronuclei in erythrocytes were also determined. Analysis using a partial least-squares regression model showed that as much as 60% of the micronucleus variation could be explained by the adduct levels. This indicates the ability of such studies to align measurements of internal dose (protein adducts) with endpoints of genotoxicity (micronucleus formation).

3.3. Human Biomonitoring of HSA Adducts

An early study that demonstrated the utility of monitoring HSA for alkylated cysteine involved exposure of human blood to ^{14}C-labelled sulfur mustard (the chemical warfare agent mustard gas) [44]. Isolation and tryptic digestion of albumin produced the 21-amino acid fragment containing a sulfur mustard-cysteine adduct, detected by micro-LC-MS/MS. An alternative method, which employed Pronase for the digestion, yielded a modified tripeptide (Cys-Pro-Phe), which was detected with greater sensitivity than the 21-amino acid fragment. The method was used to analyse samples of blood from nine Iranians exposed to sulfur mustard during the Iran-Iraq war of 1986. In all nine cases, the sulfur mustard-adducted tripeptide was detected.

Application of the FS-SRM method to analyses of archived plasma protein that had been pooled according to subjects' ethnicities and tobacco smoking habits demonstrated differences between pools [49] and suggested that FS-SRM might be able to detect statistically significant differences between groups of individual samples that had not been pooled.

A pilot study of 20 smokers and 20 never-smokers provided evidence of the effect of smoking on levels of putative HSA adducts. Differences between smokers and never-smokers were most apparent in putative adducts with net gains in mass between 105 Da and 114 Da (relative to unmodified HSA) [33].

Further investigations of the effects of tobacco smoking have revealed around 43 adduct features, some of which are positively associated with smoking and but also some that are negatively associated. The former result from genotoxic constituents of tobacco smoke, such as ethylene oxide and acrylonitrile, while the latter, which include Cys-34 oxidation products and disulfides, may reflect alterations in the serum redox state of smokers, resulting in lower adduct levels [50].

Grigoryan et al. used LC and high-resolution mass spectrometry to investigate interactions between the Cys-34 and reactive oxygen species (ROS) [47]. Chronic exposure to ROS is linked to many chronic diseases and, in this study, a number of adducts originating from ROS were detected in human serum: Sulfinic acid, sulfonic acid and a proposed sulfinamide structure (a mono-oxygenated moiety also with the loss of two hydrogen atoms).

Antibody enrichment may pave the way to a more sensitive assay. Using a polyclonal antibody, raised against the T3 peptide, but with cross-reactivity to the peptide containing adducts (see Section 2.12), ten modified T3 peptides were detected in human plasma samples; eight of them were characterised and they included Cys-34 oxidation products, modification involving loss of water or lysine, cysteinylation, and transpeptidation of arginine [71].

In a study of women from the Xuanwei and Fuyuan counties in China, where extensive use of smoky coal for heating and cooking has resulted in very high rates of lung cancer among non-smokers, HSA Cys-34 adducts were compared in 29 females who used smoky coal and 10 controls using other energy sources [77]. Fifty different modified T3 peptides were identified, including oxidation products, mixed disulfides, rearrangements and truncations. Two peptides that were detected at significantly *lower* levels in the smoky coal group were adducts of glutathione and γ-glutamylcysteine. The results are interpreted as evidence that exposure to the indoor combustion products results in depletion of glutathione, an essential antioxidant, as well as its precursor γ-glutamylcysteine [77].

A recent study on the health effects of urban air pollution, the Oxford Street II study [78], involved a randomised crossover design whereby three groups of volunteers (healthy subjects, chronic obstructive pulmonary disease (COPD) sufferers and patients with ischaemic heart disease (IHD)) walked for two hours along a busy street in London where traffic is restricted to diesel buses and taxis. The volunteers also spent two hours walking in a London park on a separate occasion. They were monitored for respiratory and cardiovascular function in both environments and, in addition, two studies have analysed their HSA samples for adducts. In the first report, Liu et al. [79] analysed 50 HSA samples by high-resolution mass spectrometry to determine whether protein modifications differ between COPD or IHD patients and healthy subjects. The untargeted analysis of adducts at the Cys-34 locus of HSA detected 39 adducts with sufficient data, and these adducts were examined for associations with estimated exposures to air pollution and health status. Multivariate linear regression revealed 21 significant associations, mainly with the underlying diseases, but also with air-pollution exposures. Interestingly, most of the associations indicated that adduct levels decreased with the presence of disease or increased pollutant concentrations. Negative associations of COPD and IHD with the Cys-34 disulfide of glutathione and two Cys-34 sulfoxidations were consistent with results from smokers and non-smokers [50] and from non-smoking women exposed to indoor combustion of coal and wood [77].

In the second study, Preston et al. [80] examined a larger number of Oxford Street II samples by the FS-SRM method. Associations between amounts of putative adducts and two types of measure were tested: Pollution (e.g., ambient concentrations of nitrogen dioxide and particulate matter) and health outcome (e.g., measures of lung health and arterial stiffness). There were 11 instances of a response variable being associated with a pollution measurement and eight instances of a response variable being associated with a health outcome measure. However, no two measures of different types were associated with the same adduct amount, suggesting that the internal changes responsible for health outcomes may differ from those that effect changes in adduct amounts.

In a more targeted study, Bellamri et al. [81] investigated the formation in human subjects of HSA adducts at Cys-34 by PhIP, which is formed in cooked meats and may be associated with colorectal, prostate and mammary cancer. Volunteers abstained from eating cooked well-done meat or fish for three weeks, then ate a semi-controlled diet that included cooked beef containing known quantities of PhIP for four weeks. The volunteers then returned to their regular diets, but with the exclusion of cooked well-done meat and fish for a further four weeks. The authors found that an adduct of oxidised PhIP, which was below the limit of detection (LOD) (10 femtograms PhIP/mg HSA) in most subjects before the meat feeding, increased by up to 560-fold at week 4 in subjects who ate meat containing 8.0 to 11.7 µg of PhIP per 150–200 g serving. In contrast, the adduct remained below the LOD in subjects who ingested 1.2 or 3.0 µg PhIP per serving, and PhIP-HSA adduct levels did not correlate with PhIP intake levels across four exposure groups ($p = 0.76$). There were also indications that the PhIP adduct was unstable, having a half-life of fewer than two weeks. Nevertheless, the study demonstrates that the Cys-34 site in HSA is accessible by a relatively large molecule like PhIP, despite concerns about possible steric hindrance (see Section 2.6.).

4. Prospects

A key advantage of monitoring proteins for adducts is the abundance of material that can be obtained from tissue banks; for example, red blood cells are an abundant source of Hb and blood plasma or serum is an abundant source of HSA. The proteins' lifespans in blood mean that there is a substantial "capture period" for monitoring exposure to genotoxicants; and protein adducts, unlike DNA adducts, are not subject to loss through repair processes. Full implementation of the exposome concept requires monitoring individuals or populations at several points in time over the course of their lives [3,4]. This is achievable if biobanks collect material from individuals not just once but multiple times, and such biobanks already exist.

Dried blood spots can also be a suitable source of protein for investigation [82]. If obtained from neonatal blood spots (i.e., Guthrie spots) then the material provides a valuable opportunity for investigating exposures in utero. A single blood spot of about 50 µL is estimated to contain about 9.6 mg of protein, of which about 7.7 mg will be Hb and 1.2 mg HSA [82]. In a proof-of-principle study, Yano et al. [83] identified 26 Cys-34 adducts (oxidation and *S*-thiolation products) in HSA isolated from dried blood spots of 49 newborn babies and were able to distinguish between newborns of smoking and non-smoking mothers on the basis of the levels of a putative cyano modification to Cys-34.

There is also the potential to broaden the scope of adductomics by investigating novel modifiable loci in blood proteins. Cys-34 of HSA and the *N*-terminus of Hb are undoubtedly major targets for electrophiles, but the literature hints at wider reactivity within the blood proteome. Consequently, a hitherto-untapped source of analytes for protein adductomics can be envisaged. Mapping the loci at which adducts can form will be beneficial and, for this purpose, new chemical tools will be required. Identifying adducts detected by the top-down approaches will be a challenge, necessitating chemical synthesis of candidate structures for unequivocal characterisation.

Protein adductomics is a component of the exposome concept that is still relatively novel, but it is one that has already demonstrated the ability to capture electrophiles of both endogenous and exogenous origin; this suggests the potential to contribute meaningfully to the aims of the exposome concept—to describe the totality of all biologically relevant exposures. Rapid advances in mass spectrometry instrumentation, with significant increases in sensitivity and resolution, will drive further advances in protein adductomics methodology. When coupled with other omics approaches, such as proteomics, transcriptomics and metabolomics, all of which have the potential for high-throughput screening of populations, a future can be envisaged in which it will be possible to capture snapshots of human exposure to genotoxicants and the resultant biological consequences at multiple stages throughout life. Building this comprehensive picture should shed significant light on the causes and courses of chronic diseases in humans. Such knowledge will provide new opportunities for early intervention to reduce potentially harmful human exposure, to monitor the effectiveness of intervention strategies and, ultimately, to prevent diseases before they occur.

Author Contributions: Both authors critically reviewed the literature and contributed equally to writing the manuscript.

Funding: The authors' research was funded by the European Community's Seventh Framework Programme (FP7/2007-2013) under grant agreement number 308610 (the EXPOsOMICS project). Additional funding was from Cancer Research UK (Programme Grant CRUK/A14329) and the MRC-PHE Centre for Environment and Health (MRC grant number G0801056/1).

Conflicts of Interest: The authors declare no conflict of interest.

References

1. Rappaport, S.M. Genetic Factors Are Not the Major Causes of Chronic Diseases. *PLoS ONE* **2016**, *11*, e0154387. [CrossRef] [PubMed]
2. Rappaport, S.M.; Smith, M.T. Epidemiology. Environment and disease risks. *Science* **2010**, *330*, 460–461. [CrossRef] [PubMed]
3. Wild, C.P. Complementing the genome with an "exposome": The outstanding challenge of environmental exposure measurement in molecular epidemiology. *Cancer Epidemiol. Biomark. Prev.* **2005**, *14*, 1847–1850. [CrossRef] [PubMed]
4. Wild, C.P. The exposome: From concept to utility. *Int. J. Epidemiol.* **2012**, *41*, 24–32. [CrossRef] [PubMed]
5. Rappaport, S.M.; Barupal, D.K.; Wishart, D.; Vineis, P.; Scalbert, A. The blood exposome and its role in discovering causes of disease. *Environ. Health Perspect.* **2014**, *122*, 769–774. [CrossRef]
6. Rappaport, S.M. Biomarkers intersect with the exposome. *Biomarkers* **2012**, *17*, 483–489. [CrossRef]
7. Dennis, K.K.; Marder, E.; Balshaw, D.M.; Cui, Y.; Lynes, M.A.; Patti, G.J.; Rappaport, S.M.; Shaughnessy, D.T.; Vrijheid, M.; Barr, D.B. Biomonitoring in the Era of the Exposome. *Environ. Health Perspect.* **2017**, *125*, 502–510. [CrossRef]
8. Stewart, B.W.; Bray, F.; Forman, D.; Ohgaki, H.; Straif, K.; Ullrich, A.; Wild, C.P. Cancer prevention as part of precision medicine: 'Plenty to be done'. *Carcinogenesis* **2016**, *37*, 2–9. [CrossRef] [PubMed]
9. Wild, C.P.; Scalbert, A.; Herceg, Z. Measuring the exposome: A powerful basis for evaluating environmental exposures and cancer risk. *Environ. Mol. Mutagen.* **2013**, *54*, 480–499. [CrossRef] [PubMed]
10. Vineis, P.; Chadeau-Hyam, M.; Gmuender, H.; Gulliver, J.; Herceg, Z.; Kleinjans, J.; Kogevinas, M.; Kyrtopoulos, S.; Nieuwenhuijsen, M.; Phillips, D.H.; et al. The exposome in practice: Design of the EXPOsOMICS project. *Int. J. Hyg. Environ. Health* **2017**, *220*, 142–151. [CrossRef]

11. Vrijheid, M.; Slama, R.; Robinson, O.; Chatzi, L.; Coen, M.; van den Hazel, P.; Thomsen, C.; Wright, J.; Athersuch, T.J.; Avellana, N.; et al. The human early-life exposome (HELIX): Project rationale and design. *Environ. Health Perspect.* **2014**, *122*, 535–544. [CrossRef]

12. Ankley, G.T.; Bennett, R.S.; Erickson, R.J.; Hoff, D.J.; Hornung, M.W.; Johnson, R.D.; Mount, D.R.; Nichols, J.W.; Russom, C.L.; Schmieder, P.K.; et al. Adverse outcome pathways: A conceptual framework to support ecotoxicology research and risk assessment. *Environ. Toxicol. Chem.* **2010**, *29*, 730–741. [CrossRef] [PubMed]

13. Escher, B.I.; Hackermüller, J.; Polte, T.; Scholz, S.; Aigner, A.; Altenburger, R.; Böhme, A.; Bopp, S.K.; Brack, W.; Busch, W.; et al. From the exposome to mechanistic understanding of chemical-induced adverse effects. *Environ. Int.* **2017**, *99*, 97–106. [CrossRef] [PubMed]

14. Dennehy, M.K.; Richards, K.A.M.; Wernke, G.R.; Shyr, Y.; Liebler, D.C. Cytosolic and Nuclear Protein Targets of Thiol-Reactive Electrophiles. *Chem. Res. Toxicol.* **2006**, *19*, 20–29. [CrossRef] [PubMed]

15. Weerapana, E.; Simon, G.M.; Cravatt, B.F. Disparate proteome reactivity profiles of carbon electrophiles. *Nat. Chem. Biol.* **2008**, *4*, 405–407. [CrossRef]

16. Medina-Cleghorn, D.; Bateman, L.A.; Ford, B.; Heslin, A.; Fisher, K.J.; Dalvie, E.D.; Nomura, D.K. Mapping Proteome-Wide Targets of Environmental Chemicals Using Reactivity-Based Chemoproteomic Platforms. *Chem. Biol.* **2015**, *22*, 1394–1405. [CrossRef] [PubMed]

17. Rappaport, S.M.; Li, H.; Grigoryan, H.; Funk, W.E.; Williams, E.R. Adductomics: Characterizing exposures to reactive electrophiles. *Toxicol. Lett.* **2012**, *213*, 83–90. [CrossRef]

18. Carlsson, H.; Rappaport, S.M.; Tornqvist, M. Protein Adductomics: Methodologies for Untargeted Screening of Adducts to Serum Albumin and Hemoglobin in Human Blood Samples. *High Throughput* **2019**, *8*, 6. [CrossRef] [PubMed]

19. Scheepers, P.T.J. The use of biomarkers for improved retrospective exposure assessment in epidemiological studies: Summary of an ECETOC workshop. *Biomarkers* **2008**, *13*, 734–748. [CrossRef]

20. Dickerson, R.E.; Geis, I. *Hemoglobin*; Benjamin Cummings: Menlo Park, CA, USA, 1983.

21. Rubino, F.M.; Pitton, M.; Di Fabio, D.; Colombi, A. Toward an "omic" physiopathology of reactive chemicals: Thirty years of mass spectrometric study of the protein adducts with endogenous and xenobiotic compounds. *Mass Spectrom. Rev.* **2009**, *28*, 725–784. [CrossRef] [PubMed]

22. Kaplan, A.; Jack, R.; Opheim, K.E.; Toivola, B.; Lyon, A.W. *Clinical Chemistry*; Williams & Wilkins: Malvern, UK, 1995.

23. Ueno, H.; Pospischil, M.A.; Manning, J.M.; Kluger, R. Site-specific modification of hemoglobin by methyl acetyl phosphate. *Arch. Biochem. Biophys.* **1986**, *244*, 795–800. [CrossRef]

24. Bryant, M.S.; Vineis, P.; Skipper, P.L.; Tannenbaum, S.R. Hemoglobin adducts of aromatic amines: Associations with smoking status and type of tobacco. *Proc. Natl. Acad. Sci. USA* **1988**, *85*, 9788. [CrossRef] [PubMed]

25. Turesky, R.J.; Le Marchand, L. Metabolism and Biomarkers of Heterocyclic Aromatic Amines in Molecular Epidemiology Studies: Lessons Learned from Aromatic Amines. *Chem. Res. Toxicol.* **2011**, *24*, 1169–1214. [CrossRef] [PubMed]

26. Ringe, D.; Turesky, R.J.; Skipper, P.L.; Tannenbaum, S.R. Structure of the single stable hemoglobin adduct formed by 4-aminobiphenyl in vivo. *Chem. Res. Toxicol.* **1988**, *1*, 22–24. [CrossRef] [PubMed]

27. Törnqvist, M.; Mowrer, J.; Jensen, S.; Ehrenberg, L. Monitoring of environmental cancer initiators through hemoglobin adducts by a modified Edman degradation method. *Anal. Biochem.* **1986**, *154*, 255–266. [CrossRef]

28. Carlsson, H.; von Stedingk, H.; Nilsson, U.; Tornqvist, M. LC-MS/MS screening strategy for unknown adducts to N-terminal valine in hemoglobin applied to smokers and nonsmokers. *Chem. Res. Toxicol.* **2014**, *27*, 2062–2070. [CrossRef] [PubMed]

29. Tornqvist, M. Epoxide adducts to N-terminal valine of hemoglobin. *Methods Enzymol.* **1994**, *231*, 650–657. [CrossRef]

30. Edman, P. Method for Determination of the Amino Acid Sequence in Peptides. *Acta Chem. Scand.* **1950**, *4*, 283–293. [CrossRef]

31. Price, N.C.; Stevens, L. *Fundamentals of Enzymology*, 3rd ed.; Oxford University Press: New York, NY, USA, 1999.

32. Von Stedingk, H.; Rydberg, P.; Törnqvist, M. A new modified Edman procedure for analysis of N-terminal valine adducts in hemoglobin by LC–MS/MS. *J. Chromatogr. B* **2010**, *878*, 2483–2490. [CrossRef]

33. Preston, G.W.; Plusquin, M.; Sozeri, O.; van Veldhoven, K.; Bastian, L.; Nawrot, T.S.; Chadeau-Hyam, M.; Phillips, D.H. Refinement of a Methodology for Untargeted Detection of Serum Albumin Adducts in Human Populations. *Chem. Res. Toxicol.* **2017**, *30*, 2120–2129. [CrossRef]

34. Kanaly, R.A.; Hanaoka, T.; Sugimura, H.; Toda, H.; Matsui, S.; Matsuda, T. Development of the adductome approach to detect DNA damage in humans. *Antioxid. Redox Signal.* **2006**, *8*, 993–1001. [CrossRef]

35. Shibata, T.; Shimizu, K.; Hirano, K.; Nakashima, F.; Kikuchi, R.; Matsushita, T.; Uchida, K. Adductome-based identification of biomarkers for lipid peroxidation. *J. Biol. Chem.* **2017**, *292*, 8223–8235. [CrossRef]

36. Sabbioni, G.; Turesky, R.J. Biomonitoring Human Albumin Adducts: The Past, the Present, and the Future. *Chem. Res. Toxicol.* **2017**, *30*, 332–366. [CrossRef] [PubMed]

37. Day, B.W.; Skipper, P.L.; Zaia, J.; Tannenbaum, S.R. Benzo[a]pyrene anti-diol epoxide covalently modifies human serum albumin carboxylate side chains and imidazole side chain of histidine(146). *J. Am. Chem. Soc.* **1991**, *113*, 8505–8509. [CrossRef]

38. Lindh, C.H.; Kristiansson, M.H.; Berg-Andersson, U.A.; Cohen, A.S. Characterization of adducts formed between human serum albumin and the butadiene metabolite epoxybutanediol. *Rapid Commun. Mass Spectrom.* **2005**, *19*, 2488–2496. [CrossRef]

39. Sabbioni, G. Chemical and physical properties of the major serum albumin adduct of aflatoxin B_1 and their implications for the quantification in biological samples. *Chemico-Biol. Interact.* **1990**, *75*, 1–15. [CrossRef]

40. Guengerich, F.P.; Arneson, K.O.; Williams, K.M.; Deng, Z.; Harris, T.M. Reaction of Aflatoxin B_1 Oxidation Products with Lysine. *Chem. Res. Toxicol.* **2002**, *15*, 780–792. [CrossRef]

41. Turell, L.; Radi, R.; Alvarez, B. The thiol pool in human plasma: The central contribution of albumin to redox processes. *Free Radic. Biol. Med.* **2013**, *65*, 244–253. [CrossRef]

42. He, X.M.; Carter, D.C. Atomic structure and chemistry of human serum albumin. *Nature* **1992**, *358*, 209–215. [CrossRef]

43. Aldini, G.; Regazzoni, L.; Orioli, M.; Rimoldi, I.; Facino, R.M.; Carini, M. A tandem MS precursor-ion scan approach to identify variable covalent modification of albumin Cys34: A new tool for studying vascular carbonylation. *J. Mass Spectrom.* **2008**, *43*, 1470–1481. [CrossRef]

44. Noort, D.; Hulst, A.G.; de Jong, L.P.; Benschop, H.P. Alkylation of human serum albumin by sulfur mustard in vitro and in vivo: Mass spectrometric analysis of a cysteine adduct as a sensitive biomarker of exposure. *Chem. Res. Toxicol.* **1999**, *12*, 715–721. [CrossRef] [PubMed]

45. Peng, L.; Dasari, S.; Tabb, D.L.; Turesky, R.J. Mapping Serum Albumin Adducts of the Food-Borne Carcinogen 2-Amino-1-methyl-6-phenylimidazo[4,5-b]pyridine by Data-Dependent Tandem Mass Spectrometry. *Chem. Res. Toxicol.* **2012**, *25*, 2179–2193. [CrossRef] [PubMed]

46. Pathak, K.V.; Bellamri, M.; Wang, Y.; Langouët, S.; Turesky, R.J. 2-Amino-9H-pyrido[2,3-b]indole (AαC) Adducts and Thiol Oxidation of Serum Albumin as Potential Biomarkers of Tobacco Smoke. *J. Biol. Chem.* **2015**, *290*, 16304–16318. [CrossRef]

47. Grigoryan, H.; Li, H.; Iavarone, A.T.; Williams, E.R.; Rappaport, S.M. Cys34 adducts of reactive oxygen species in human serum albumin. *Chem. Res. Toxicol.* **2012**, *25*, 1633–1642. [CrossRef]

48. Dong, Q.; Yan, X.; Kilpatrick, L.E.; Liang, Y.; Mirokhin, Y.A.; Roth, J.S.; Rudnick, P.A.; Stein, S.E. Tandem Mass Spectral Libraries of Peptides in Digests of Individual Proteins: Human Serum Albumin (HSA). *Mol. Cell. Proteom.* **2014**, *13*, 2435. [CrossRef]

49. Li, H.; Grigoryan, H.; Funk, W.E.; Lu, S.S.; Rose, S.; Williams, E.R.; Rappaport, S.M. Profiling Cys34 adducts of human serum albumin by fixed-step selected reaction monitoring. *Mol. Cell. Proteom.* **2011**, *10*, M110.004606. [CrossRef]

50. Grigoryan, H.; Edmands, W.; Lu, S.S.; Yano, Y.; Regazzoni, L.; Iavarone, A.T.; Williams, E.R.; Rappaport, S.M. Adductomics Pipeline for Untargeted Analysis of Modifications to Cys34 of Human Serum Albumin. *Anal. Chem.* **2016**, *88*, 10504–10512. [CrossRef]

51. Porter, C.J.; Bereman, M.S. Data-independent-acquisition mass spectrometry for identification of targeted-peptide site-specific modifications. *Anal. Bioanal. Chem.* **2015**, *407*, 6627–6635. [CrossRef]

52. Noort, D.; Fidder, A.; Hulst, A.G. Modification of human serum albumin by acrylamide at cysteine-34: A basis for a rapid biomonitoring procedure. *Arch. Toxicol.* **2003**, *77*, 543–545. [CrossRef]

53. Goto, T.; Kojima, S.; Shitamichi, S.; Lee, S.H.; Oe, T. Chemical modificomics: A novel strategy for efficient biomarker discovery through chemical modifications on a target peptide. *Anal. Methods* **2012**, *4*, 1945–1952. [CrossRef]

54. Todd, J.F.J. Recommendations for Nomenclature and Symbolism for Mass Spectroscopy. *Pure Appl. Chem.* **1991**, *63*, 1541–1566. [CrossRef]
55. Kanaly, R.A.; Matsui, S.; Hanaoka, T.; Matsuda, T. Application of the adductome approach to assess intertissue DNA damage variations in human lung and esophagus. *Mutat. Res.* **2007**, *625*, 83–93. [CrossRef] [PubMed]
56. Wagner, S.; Scholz, K.; Donegan, M.; Burton, L.; Wingate, J.; Völkel, W. Metabonomics and Biomarker Discovery: LC–MS Metabolic Profiling and Constant Neutral Loss Scanning Combined with Multivariate Data Analysis for Mercapturic Acid Analysis. *Anal. Chem.* **2006**, *78*, 1296–1305. [CrossRef]
57. Mann, M.; Wilm, M. Error-Tolerant Identification of Peptides in Sequence Databases by Peptide Sequence Tags. *Anal. Chem.* **1994**, *66*, 4390–4399. [CrossRef]
58. Yeowell-O'Connell, K.; Rothman, N.; Smith, M.T.; Hayes, R.B.; Li, G.; Waidyanatha, S.; Dosemeci, M.; Zhang, L.; Yin, S.; Titenko-Holland, N.; et al. Hemoglobin and albumin adducts of benzene oxide among workers exposed to high levels of benzene. *Carcinogenesis* **1998**, *19*, 1565–1571. [CrossRef] [PubMed]
59. Dingley, K.H.; Curtis, K.D.; Nowell, S.; Felton, J.S.; Lang, N.P.; Turteltaub, K.W. DNA and protein adduct formation in the colon and blood of humans after exposure to a dietary-relevant dose of 2-amino-1-methyl-6-phenylimidazo[4,5-b]pyridine. *Cancer Epidemiol. Biomark. Prev.* **1999**, *8*, 507–512.
60. Tannenbaum, S.R.; Skipper, P.L. Biological aspects to the evaluation of risk: dosimetry of carcinogens in man. *Fundam. Appl. Toxicol.* **1984**, *4*, S367–S373. [CrossRef]
61. Sabbioni, G.; Skipper, P.L.; Buchi, G.; Tannenbaum, S.R. Isolation and characterization of the major serum albumin adduct formed by aflatoxin B_1 in vivo in rats. *Carcinogenesis* **1987**, *8*, 819–824. [CrossRef]
62. Wild, C.P.; Jiang, Y.Z.; Sabbioni, G.; Chapot, B.; Montesano, R. Evaluation of methods for quantitation of aflatoxin-albumin adducts and their application to human exposure assessment. *Cancer Res.* **1990**, *50*, 245–251.
63. Noort, D.; Fidder, A.; Degenhardt-Langelaan, C.E.; Hulst, A.G. Retrospective detection of sulfur mustard exposure by mass spectrometric analysis of adducts to albumin and hemoglobin: An in vivo study. *J. Anal. Toxicol.* **2008**, *32*, 25–30. [CrossRef]
64. Verzijl, N.; DeGroot, J.; Thorpe, S.R.; Bank, R.A.; Shaw, J.N.; Lyons, T.J.; Bijlsma, J.W.J.; Lafeber, F.P.J.G.; Baynes, J.W.; TeKoppele, J.M. Effect of Collagen Turnover on the Accumulation of Advanced Glycation End Products. *J. Biol. Chem.* **2000**, *275*, 39027–39031. [CrossRef] [PubMed]
65. Jonsson, B.A.G.; Wishnok, J.S.; Skipper, P.L.; Stillwell, W.G.; Tannenbaum, S.R. Lysine Adducts Between Methyltetrahydrophthalic Anhydride and Collagen in Guinea Pig Lung. *Toxicol. Appl. Pharmacol.* **1995**, *135*, 156–162. [CrossRef]
66. Miller, E.J.; Gay, S. Collagen: An overview. *Methods Enzymol.* **1982**, *82*, 3–32. [CrossRef] [PubMed]
67. Miller, E.J.; Kent Rhodes, R. Preparation and characterization of the different types of collagen. *Methods Enzymol.* **1982**, *82*, 33–64. [CrossRef] [PubMed]
68. SooHoo, C.K.; Singh, K.; Skipper, P.L.; Tannenbaum, S.R.; Dasari, R.R. Characterization of benzo[a]pyrene anti-diol epoxide adducts to human histones. *Chem. Res. Toxicol.* **1994**, *7*, 134–138. [CrossRef]
69. Fabrizi, L.; Taylor, G.W.; Cañas, B.; Boobis, A.R.; Edwards, R.J. Adduction of the Chloroform Metabolite Phosgene to Lysine Residues of Human Histone H2B. *Chem. Res. Toxicol.* **2003**, *16*, 266–275. [CrossRef] [PubMed]
70. Funk, W.E.; Li, H.; Iavarone, A.T.; Williams, E.R.; Riby, J.; Rappaport, S.M. Enrichment of cysteinyl adducts of human serum albumin. *Anal. Biochem.* **2010**, *400*, 61–68. [CrossRef] [PubMed]
71. Chung, M.K.; Grigoryan, H.; Iavarone, A.T.; Rappaport, S.M. Antibody enrichment and mass spectrometry of albumin-Cys34 adducts. *Chem. Res. Toxicol.* **2014**, *27*, 400–407. [CrossRef]
72. Angerer, J.; Ewers, U.; Wilhelm, M. Human biomonitoring: State of the art. *Int. J. Hyg. Environ. Health* **2007**, *210*, 201–228. [CrossRef]
73. Carlsson, H.; Tornqvist, M. Strategy for identifying unknown hemoglobin adducts using adductome LC-MS/MS data: Identification of adducts corresponding to acrylic acid, glyoxal, methylglyoxal, and 1-octen-3-one. *Food Chem. Toxicol.* **2016**, *92*, 94–103. [CrossRef]
74. Carlsson, H.; Motwani, H.V.; Osterman Golkar, S.; Tornqvist, M. Characterization of a Hemoglobin Adduct from Ethyl Vinyl Ketone Detected in Human Blood Samples. *Chem. Res. Toxicol.* **2015**, *28*, 2120–2129. [CrossRef] [PubMed]

75. Degner, A.; Carlsson, H.; Karlsson, I.; Eriksson, J.; Pujari, S.S.; Tretyakova, N.Y.; Törnqvist, M. Discovery of Novel N-(4-Hydroxybenzyl)valine Hemoglobin Adducts in Human Blood. *Chem. Res. Toxicol.* **2018**, *31*, 1305–1314. [CrossRef] [PubMed]

76. Carlsson, H.; Aasa, J.; Kotova, N.; Vare, D.; Sousa, P.F.M.; Rydberg, P.; Abramsson-Zetterberg, L.; Tornqvist, M. Adductomic Screening of Hemoglobin Adducts and Monitoring of Micronuclei in School-Age Children. *Chem. Res. Toxicol.* **2017**, *30*, 1157–1167. [CrossRef]

77. Lu, S.S.; Grigoryan, H.; Edmands, W.M.; Hu, W.; Iavarone, A.T.; Hubbard, A.; Rothman, N.; Vermeulen, R.; Lan, Q.; Rappaport, S.M. Profiling the Serum Albumin Cys34 Adductome of Solid Fuel Users in Xuanwei and Fuyuan, China. *Environ. Sci. Technol.* **2017**, *51*, 46–57. [CrossRef] [PubMed]

78. Sinharay, R.; Gong, J.; Barratt, B.; Ohman-Strickland, P.; Ernst, S.; Kelly, F.J.; Zhang, J.J.; Collins, P.; Cullinan, P.; Chung, K.F. Respiratory and cardiovascular responses to walking down a traffic-polluted road compared with walking in a traffic-free area in participants aged 60 years and older with chronic lung or heart disease and age-matched healthy controls: A randomised, crossover study. *Lancet* **2018**, *391*, 339–349. [CrossRef] [PubMed]

79. Liu, S.; Grigoryan, H.; Edmands, W.M.B.; Dagnino, S.; Sinharay, R.; Cullinan, P.; Collins, P.; Chung, K.F.; Barratt, B.; Kelly, F.J.; et al. Cys34 Adductomes Differ between Patients with Chronic Lung or Heart Disease and Healthy Controls in Central London. *Environ. Sci. Technol.* **2018**, *52*, 2307–2313. [CrossRef]

80. Preston, G.W.; Dagnino, S.; Ponzi, E.; Sozeri, O.; van Veldhoven, K.; Barratt, B.; Liu, S.; Grigoryan, H.; Lu, S.S.; Rappaport, S.; et al. Relationships between airborne pollutants, serum albumin adducts and short-term health outcomes in an experimental crossover study. **2019**, submitted.

81. Bellamri, M.; Wang, Y.; Yonemori, K.; White, K.K.; Wilkens, L.R.; Le Marchand, L.; Turesky, R.J. Biomonitoring an albumin adduct of the cooked meat carcinogen 2-amino-1-methyl-6-phenylimidazo[4,5-b]pyridine in humans. *Carcinogenesis* **2018**, *39*, 1455–1462. [CrossRef]

82. Funk, W.E.; Waidyanatha, S.; Chaing, S.H.; Rappaport, S.M. Hemoglobin adducts of benzene oxide in neonatal and adult dried blood spots. *Cancer Epidemiol. Biomark. Prev.* **2008**, *17*, 1896–1901. [CrossRef]

83. Yano, Y.; Grigoryan, H.; Schiffman, C.; Edmands, W.; Petrick, L.; Hall, K.; Whitehead, T.; Metayer, C.; Dudoit, S.; Rappaport, S. Untargeted adductomics of Cys34 modifications to human serum albumin in newborn dried blood spots. *Anal. Bioanal. Chem.* **2019**. [CrossRef]

Article

Internal Doses of Glycidol in Children and Estimation of Associated Cancer Risk

Jenny Aasa [1], Efstathios Vryonidis [1], Lilianne Abramsson-Zetterberg [2] and Margareta Törnqvist [1,*]

[1] Department of Environmental Science and Analytical Chemistry, Stockholm University, 106 91 Stockholm, Sweden; jenny.aasa@aces.su.se (J.A.); efstathios.vryonidis@aces.su.se (E.V.)
[2] National Food Agency, 751 26 Uppsala, Sweden; lilianne.abramsson@slv.se
*   Correspondence: margareta.tornqvist@aces.su.se; Tel.: +46-816-3769

Received: 20 December 2018; Accepted: 29 January 2019; Published: 1 February 2019

Abstract: The general population is exposed to the genotoxic carcinogen glycidol via food containing refined edible oils where glycidol is present in the form of fatty acid esters. In this study, internal (in vivo) doses of glycidol were determined in a cohort of 50 children and in a reference group of 12 adults (non-smokers and smokers). The lifetime in vivo doses and intakes of glycidol were calculated from the levels of the hemoglobin (Hb) adduct N-(2,3-dihydroxypropyl)valine in blood samples from the subjects, demonstrating a fivefold variation between the children. The estimated mean intake (1.4 µg/kg/day) was about two times higher, compared to the estimated intake for children by the European Food Safety Authority. The data from adults indicate that the non-smoking and smoking subjects are exposed to about the same or higher levels compared to the children, respectively. The estimated lifetime cancer risk ($200/10^5$) was calculated by a multiplicative risk model from the lifetime in vivo doses of glycidol in the children, and exceeds what is considered to be an acceptable cancer risk. The results emphasize the importance to further clarify exposure to glycidol and other possible precursors that could give a contribution to the observed adduct levels.

Keywords: glycidol; Hb adduct; N-(2.3-dihydroxypropyl)valine; in vivo; cancer risk; UPLC/MS/MS

1. Introduction

Exposure to genotoxic compounds in interaction with other factors contributes to an increased risk of cancer development [1]. For many cancer types, the onset of the disease are several decades after a specific exposure [2]. Quantification of the cancer risk from exposure to genotoxic compounds is usually based on data from carcinogenicity studies at high doses in rodents, with extrapolation of the obtained cancer risk coefficient to estimated human exposure doses of the studied compound. An improved estimate of the risk would be obtained if the internal (in vivo) dose of the studied compound/metabolite could be used for species extrapolations. This is particularly important for ongoing human exposures, where measured in vivo doses also give an improved estimate of the exposure.

One compound, for which there is an ongoing human exposure, is the genotoxic compound glycidol [3]. This compound has received much attention during the last years, largely due to the detection of glycidyl fatty acid esters in food intended for children, such as as infant formula [4]. Possible exposure sources are food products containing refined edible oils where glycidol is present in the form of glycidyl fatty acid esters [5,6]. The esters are hydrolyzed in the stomach, leading to formation of glycidol (Figure 1) [7]. This is of concern to human health, as glycidol is classified by the International Agency for Research on Cancer (IARC) as probably carcinogenic to humans, Group 2A [8]. Protecting children from exposure to genotoxic compounds is important. Children are considered more vulnerable for exposure to chemical compounds compared with adults, because children are still growing and organ cells are rapidly dividing [9]. This likely increases the rate of fixation of mutations, which can lead to the onset of cancer.

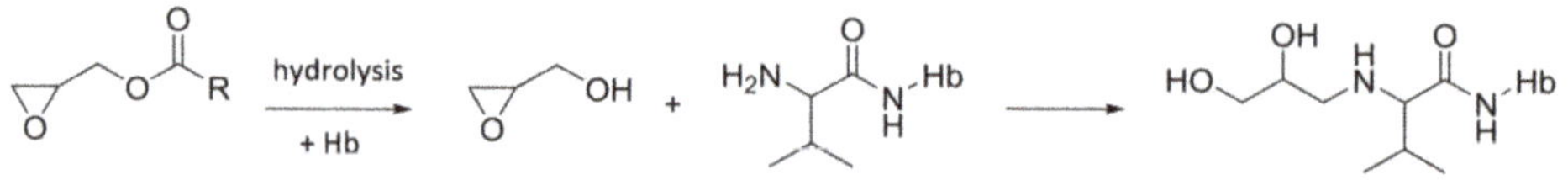

Figure 1. Glycidol is formed in vivo by hydrolysis of glycidyl fatty acid esters [3], where R represents different ester side chains. Glycidol reacts with the N-terminal valine in hemoglobin (Hb) to form *N*-(2,3-dihydroxypropyl)valine (diHOPrVal) as N-terminal.

Here, we will present results from a study of children where the in vivo dose of glycidol has been quantified in blood samples from the individuals. Also, blood samples from 12 adults were analyzed as a reference group. As genotoxic chemicals are usually electrophiles and short-lived in vivo, measurement in vivo of the ultimate genotoxic compound/metabolite is generally difficult per se. Instead, stable adducts to proteins may be used as a biomarker of exposure/internal dose. We have previously developed a method for measurement of adducts to the N-terminal valine in hemoglobin (Hb) using GC/MS/MS [10]. Later, the method was developed for LC/MS/MS, and is referred to as the FIRE procedure [11,12]. If the reaction rate constant for the formation of the specific adduct from the studied electrophile to the N-terminal valine is known, the in vivo dose expressed as the area under the concentration–time curve (AUC) can be calculated for the electrophile from a measured Hb adduct level [13]. This analytical approach has previously been applied in studies of exposed animals and humans for monitoring of internal exposures and quantification of in vivo doses of reactive compounds, for instance present in food [14–16], or for the screening of adducts with an adductomics approach for the investigation of background exposures in humans [17,18]. In the present study, in vivo doses of glycidol were calculated from measured Hb adduct levels in the human blood samples (Figure 1). The in vivo doses were then used for the estimation of the lifetime excess cancer risk, due to glycidol exposure.

2. Materials and Methods

2.1. Chemicals

Glycidol (98%, CAS No. 556-52-5) was purchased from Acros Organics (Geel, Belgium). L-Valine-($^{13}C_5$) (96–98% purity), RS-glyceraldehyde and sodium cyanoborohydride, used for the synthesis of the internal standard *N*-(2,3-dihydroxypropyl)-($^{13}C_5$)valine, were obtained from Cambridge Isotope Laboratories, Inc (Tewksbury, MA, USA) and Sigma Aldrich (St Louis, MO, USA) respectively. Cyanoacetic acid and ammonium hydroxide were purchased from Fluka (Buchs, Switzerland). The analytical standard used for the calibration curve, fluorescein thiohydantoin of *N*-(2,3-dihydroxypropyl)valine (diHOPrVal-FTH), was synthesized previously within the research group [15]. Fluorescein isothiocyanate (FITC) was purchased from Karl Industries (Aurora, OH, USA) and potassium hydrogen carbonate (KHCO$_3$) from Merck (Darmstadt, Germany). All other chemicals (analytical grade) were obtained from Sigma Aldrich (St Louis, MO, USA).

2.2. Study Population

Blood samples from 50 children at the age of about 12 years (35 boys, 15 girls) were obtained from a study in 2014 by the National Food Agency in Sweden regarding food-related exposures in children of school age (reviewed by the Regional Ethical Review Board, Uppsala, Sweden; No. 2013/354 (date of approval: 13 December 2013); following the rules of the Declaration of Helsinki). Venous blood from each individual was sampled at one occasion, and the samples used for they analysis of Hb adducts were centrifuged at 1500× *g* for 10 min to separate red blood cells (RBCs) from the plasma prior to storage of the RBC at −20 °C. In addition to these samples, measurements were also conducted on blood samples from six non-smoking and six smoking adults (males), collected earlier (1997) with ethical approval (from the Regional Ethical Review Board, Stockholm, Sweden; No. 96-312 (date of approval: 14 October 1996). In connection to the blood collection, these blood samples were centrifuged at 1500× *g* for 10 min, to separate RBCs and plasma. The RBCs were washed three times with an equal volume of 0.9% NaCl followed by centrifugation and lysis by the addition of an equal volume of distilled water prior to storage at −20 °C.

2.3. Synthesis of Internal Standard

The internal standard (IS), N-(2,3-dihydroxypropyl)-($^{13}C_5$)valine fluorescein thiohydantoin (FTH), was synthesized in two steps. First, RS-glyceraldehyde (21.6 mg, 240 µmol) and ($^{13}C_5$)valine (13.7 mg, 116 µmol) were mixed in 3.6 mL methanol, followed by the addition of sodium cyanoborohydride (8.9 mg, 142 µmol) for a reduction of the formed Schiff base. The reaction solution was mixed (750 rpm) for 20 hours at 37 °C, followed by evaporation of the solvent. In the second step, the product, N-(2,3-dihydroxypropyl)-($^{13}C_5$)valine, was dissolved in 5 mL acetonitrile (40%, aq.) with 0.125 M $KHCO_3$ and 540 µL FITC (90 mg, 231 µmol, dissolved in dimethylformamide) to generate the final FTH derivative to be used as IS. The derivatization reaction was kept for 20 hours at 37 °C during mixing (700 rpm), before termination by addition of 1 M HCl (250 µL, 250 µmol). The reaction solution was stored in the freezer overnight followed by centrifugation for 10 min (5000 rpm at 4 °C).

The supernatant was filtered and concentrated to ca. 2 mL, and the final product to be used as IS was separated from by-products using semi-preparative HPLC-UV. Eight injections of each 250 µL was applied on a Hichrom C18 column (10 mm × 250 mm, 5 µm). The mobile phase started at isocratic mode for 8 min, followed by gradient mode from 40% A (water) and increasing to 100% B (acetonitrile) in 1 min, which was kept for 4 min before re-equilibrating of the column for 3 min prior to the next injection. The flow rate was 4 mL/min and the total run time 16 min. The UV spectrophotometer (Shimadzu SPD-6A) was set to the wavelength (λ) of 274 nm.

The identity and the purity of the product and the *m/z* was confirmed using LC/UV/MS/MS (API 3200 Q-Trap, AB Sciex, Concord, ON, Canada) running in full scan positive mode (*m/z* 80–800). Five µL was injected onto the column (Discovery HS, C18, 150 × 2.1 mm, 3 µm) with a gradient starting from 90% A (water, 0.1% formic acid) for 0.5 min and increasing to 40% B (acetonitrile, 0.1% formic acid) in 6.5 min. A further increase to 100% B was kept for 2 min followed by re-equilibration of the column. The flow rate was 0.2 mL/min and the total run time was 12 min. All settings for the MS instrument were essentially as in previous studies [11,15]. Finally, the solvent of the verified synthesized product was evaporated until dryness, and the yield of the product was determined gravimetrically to be 7.5 mg (13.2 µmol, 11.4%).

2.4. Procedure for Hemoglobin Adduct Measurement

The blood samples (250 µL) were prepared for analysis according to the FIRE procedure, where the fluorescein isothiocyanate (FITC) reagent is used for the measurement of adducts (*R*) from electrophilic compounds with a modified Edman procedure [11,12]. The hemoglobin (Hb) content was measured in all samples (RBCs diluted with water, 1:1) prior to derivatization and detachment of N-terminals with adducts with FITC during mixing at 37 °C overnight. Internal standard (N-(2,3-dihydroxypropyl)-($^{13}C_5$)valine) was added prior to the work-up procedure that was performed as described in several previous papers [15,17,19]. A final sample volume of 100 µL (40% acetonitrile in water) was used for analysis of Hb adducts with ultra-high performance liquid chromatography (UPLC^{TM}) and high-resolution mass spectrometry (HRMS) (Section 2.6). The intraday variability of the FIRE procedure was investigated by processing three individual blood samples from the children five times in parallel at one occasion.

Calibration samples were prepared by adding known concentrations of diHOPrVal-FTH to bovine blood (Håtunalab AB, Bro, Sweden) followed by the work-up procedure. Two sets of calibration samples were prepared: Set 1) in duplicates at four levels at 0.04–1.6 pmol/sample (for measurement of background levels in human samples) and Set 2) in triplicates at six levels at 11–600 pmol/sample (for k_{val} determination). The preparation and analysis of the calibration samples were as described by Aasa et al. [15], with the exception of the analysis of the samples from Set 1, which were analyzed with HRMS (described in Section 2.6).

2.5. Measurement of Reaction Rate and Calculation of Internal Dose from Adduct Levels

The daily adduct level increment (a) was calculated from the measured steady state level of the adducts (A_{ss}) and the erythrocyte lifetime (t_{er}), according to Equation (1). The t_{er} in humans was assumed to be 126 days [13,20,21].

$$A_{ss} = a\,\frac{t_{er}}{2} \tag{1}$$

The daily adduct increment was then used for calculation of daily AUC, by using the second-order rate constant for the reaction between glycidol and the N-terminal valine, k_{val}. This constant was determined by incubation of glycidol with fresh human blood from four individuals (from Komponentlab, Karolinska University Hospital, Huddinge, Sweden). Triplicate samples of whole blood from each individual at three dose levels, 0 (control), 125 and 250 µM glycidol (Hb: 102–136 g/L), were incubated for one hour at 37 °C during mixing (750 rpm). The incubations were finalized by centrifugation and washing according to previous procedures [15]. The samples (250 µL) were then derivatized with FITC overnight followed by work-up and analysis by LC/MS/MS, as described by Aasa et al. [15] and as the calibration curve Set 2 (Section 2.4). The k_{val} could then be determined and be used for the calculation of the in vivo doses (AUC) from measured adduct levels (A) according to Equation (2), assuming that glycidol at these concentrations is stable during the 1-hour incubation time, as observed earlier [15]. For calculation of the daily AUC the A is replaced with the daily adduct increment (a) calculated in Equation 1.

$$AUC\ (\mu Mh) = \frac{A\ (pmol/g\ Hb)}{k_{val}\ (pmol/g\ Hb/\mu Mh)} \tag{2}$$

2.6. LC/MS/MS System

The analysis of Hb adducts in the in vitro incubations was performed with a triple quadrupole LC/MS/MS instrument, as described by Aasa et al. [15]. Compared to the previous analysis, the m/z transitions 563.1 > 447.1 (from glycidol) and 565.1 > 449.1 (from the internal standard) were also included in the analysis.

The analysis of Hb adducts in blood samples from the studied groups of humans was conducted with a Dionex Ultimate 3000 UHPLC system connected to an Q ExactiveTM HF Hybrid Quadrupole-Orbitrap™ high-resolution mass spectrometer (HRMS) (Thermo Fisher Scientific, MA, USA). The chromatographic separation was performed by injection (20 µL) on an Acquity UPLC™ HSS C18 column, 2.1 × 100 mm, 1.8 µm (Waters, Sollentuna, Sweden). The mobile phase (0.3 mL/min) was running in gradient mode, starting at 80% A (0.1% formic acid in H_2O:ACN; 95:5, v/v) and increasing to 100% B (0.1% formic acid in H_2O:ACN; 5:95, v/v) in 9 min. The final composition was kept for 2.5 min before re-equilibration for 3 min. N-(2,3-Dihydroxypropyl)valine fluorescein thiohydantoin was monitored in parallel reaction monitoring (PRM) mode with the resolution 60 000 and the normalized collision energy (NCE) on 45. The software XCalibur (Thermo Fisher Scientific, MA, USA) was used for processing of the data. The levels of the Hb adduct was quantified by using the accurate masses m/z 563.1463 (diHOPrVal-FTH) and m/z 568.1629 (internal standard, N-(2,3-dihydroxypropyl)-($^{13}C_5$)valine fluorescein thiohydantoin), and the specific fragments associated with diHOPrVal-FTH; m/z 503.0893, 460.0704 and 447.0629 and the internal standard; m/z 505.0955, 462.0772 and 449.0694, with a 3–5 ppm mass tolerance.

2.7. Statistical Analysis

For the measured and calculated parameters (Hb adduct levels, AUC and intakes), the minimum and maximum values along with the mean values and the standard deviations are reported for the studied groups. A *t*-test and the Grubb´s test were used for the analysis of differences between the studied groups and for testing of outliers, respectively.

3. Results

For the present study we used high resolution mass spectrometry (HRMS) operating in parallel reaction monitoring (PMR) mode, which improved the selectivity for detection of the adduct *N*-(2,3-dihydroxypropyl)valine (diHOPrVal). We also synthesized an isotopically substituted internal standard, *N*-(2,3-dihydroxypropyl)-($^{13}C_5$)valine fluorescein thiohydantoin, specific for the quantification of the studied adduct, which is an improvement for the quantification compared to our previous studies [15,18]. A representative example of a chromatogram from analysis of a human blood sample and a *m/z* spectrum of *N*-(2,3-dihydroxypropyl)valine-FTH is shown in Figure 2. The variability of the FIRE method was tested by processing five parallel blood samples from three children, which showed a relative standard deviation of 4.2–7.3% of the replicates (data not shown).

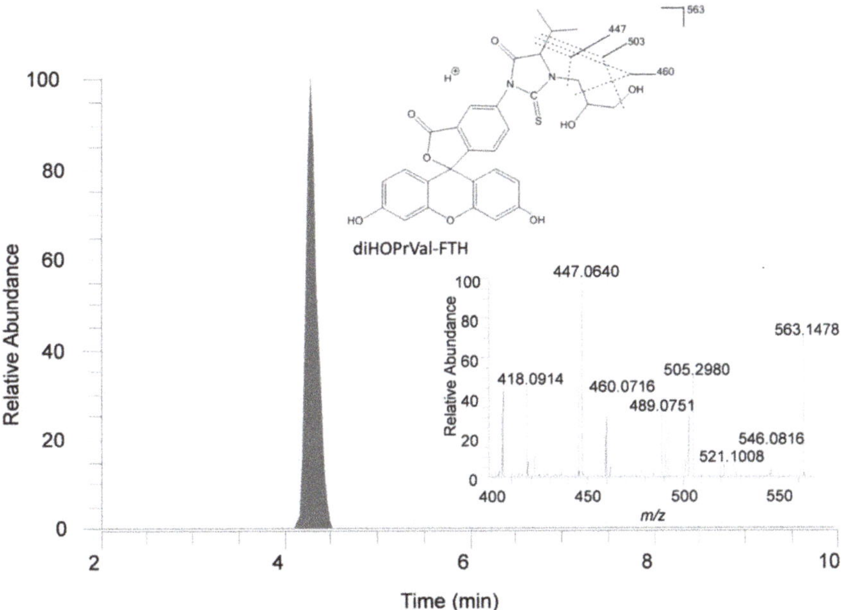

Figure 2. Ion chromatogram from a human sample and exact mass spectrum for the fluorescein thiohydantoin (FTH) derivative (diHOPrVal-FTH) of the N-terminal valine adduct formed by glycidol (6.6 pmol/g Hb at 3 ppm mass tolerance). The *m/z* fragments 503, 460 and 447 were used for quantification of the adduct diHOPrVal-FTH (*m/z* 563).

The Hb adduct was possible to quantify in all studied subjects. The adduct levels observed in the samples from the children (*n* = 50) varied between 4.4 and 20 pmol/g Hb (Figure 3, Table 1A). No statistically significant difference in the levels was observed between the sexes of the children. For the adults, the range of the observed adduct levels was 6.3–31 pmol/g Hb (Figure 3, Table 1A). Although this group consisted of a small number of subjects, there was a statistically significant difference of the adduct levels between the smoking and the non-smoking subjects (adults), with about twice higher mean levels in the smokers (23.4 pmol/g Hb) compared to the non-smokers (10.3 pmol/g Hb) (*p* < 0.01).

The adduct levels of the non-smoking adults were in the range of the adduct levels in the children (Figure 3, Table 1A). Assuming that a chronic exposure (of glycidol or its precursor) is giving rise to the observed adduct levels, the daily adduct increment was calculated using Equation 1 (Table 1A).

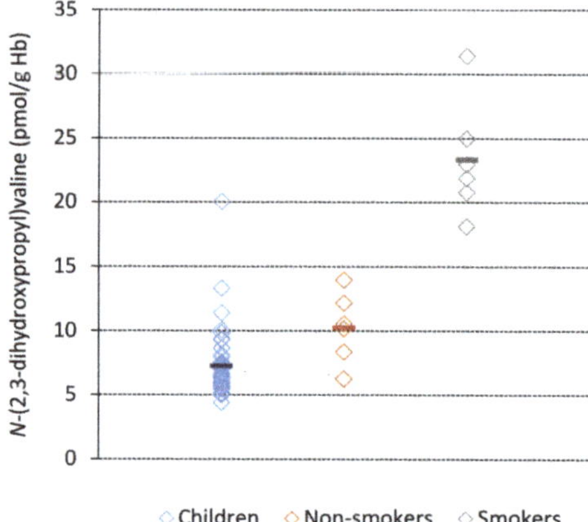

Figure 3. *N*-(2,3-Dihydroxypropyl)valine adduct levels in blood from children (*n* = 50) and, non-smoking (*n* = 6) and smoking (*n* = 6) adults. The mean values are marked as horizontal bars. The higher level observed in the smokers is likely due to presence of glycidol in tobacco smoke (see Section 4).

Table 1. (A) Measured steady state Hb adduct levels (*N*-(2,3-dihydroxypropyl)valine) in blood samples from children and adults and corresponding calculated daily adduct level increments, and (B) the corresponding estimated daily in vivo doses (AUC) and intakes of glycidol.

Group	A			B	
	Hb Adduct (pmol/g Hb)		Daily Hb Adduct Level Increment (pmol/g Hb)	Daily AUC [b] (nMh)	Daily Intake (µg/kg/day)
Subjects (*n*)	Mean [a] $\pm$ SD	Min–Max	Mean [a] $\pm$ SD	Mean $\pm$ SD	Mean $\pm$ SD
All children (50)	7.3 $\pm$ 2.5	4.4–20.1	0.11 $\pm$ 0.04	6.0 $\pm$ 2.1	1.4 $\pm$ 0.5
Boys (35) [c]	7.2 $\pm$ 1.8	4.4–13.3	-	-	-
Girls (15) [c]	7.4 $\pm$ 3.7	5.0–20.1	-	-	-
Non-smokers (6)	10.3 $\pm$ 2.7	6.3–14.0	0.16 $\pm$ 0.04	8.5 $\pm$ 2.3	2.0 $\pm$ 0.5
Smokers (6) [d]	23.4 $\pm$ 4.6	18.1–31.4	0.37 $\pm$ 0.07	19.3 $\pm$ 3.8	4.5 $\pm$ 0.9

[a] Mean levels quantified from four different *m/z* transitions in the MS analysis of the studied Hb adduct. [b] The daily AUC was calculated from the daily adduct level increment (Equation 1) and the second-order reaction rate constant, k_{val} (Equation 2), assumed to be at steady state (from chronic exposure). [c] The calculations of the daily Hb adduct increment, AUC and intake have been reported for boys and girls together as no statistical differences were observed between the sexes. [d] Glycidol in tobacco smoking is likely contributing to the daily intake level (see Section 4).

The daily intake of glycidol (µg/kg per day) was in the next step calculated from the obtained daily Hb adduct level increments. To perform this calculation, the relation between the adduct level (or in vivo dose) and administered dose of glycidol is required. We used the results recently published by Abraham et al., who studied the relation between glycidol-induced diHOPrVal adduct levels and administered dose of glycidol from palm fat in human subjects [22]. The adduct increment in their study was calculated to be 82 pmol/g Hb per mg glycidol/kg body weight (b.w.). This figure was used to estimate the mean daily intakes of glycidol in the different groups of subjects in our study, as presented in Table 1B.

Further, the daily in vivo doses (AUC) of glycidol in the studied human individuals (Table 1B) were calculated from the daily adduct level increments and the k_{val} according to Equation 2. The second-order reaction rate constant (k_{val}) was determined to be 19.2 ± 0.6 pmol/g Hb per µMh from the linear slopes of plots of the Hb adduct levels (*y*-axis) obtained from triplicate incubations of glycidol at two doses (AUC in vitro: *x*-axis) in human fresh whole blood from four individuals (Figure 4). The AUC was used for the calculation of the cancer risk due to glycidol exposure, further discussed in the Discussion part.

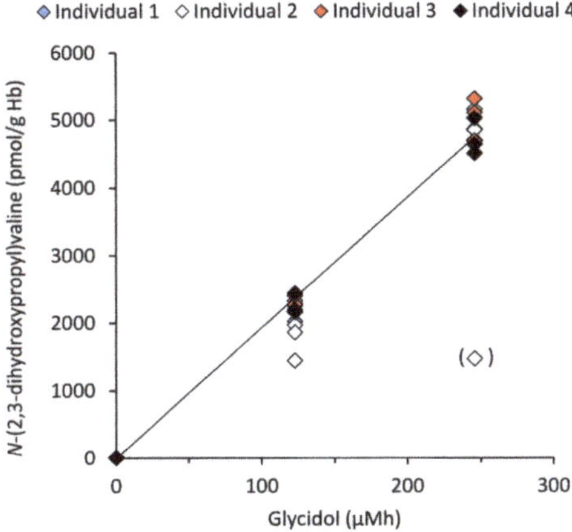

Figure 4. Determination of the second-order reaction rate constant, k_{val}, for the formation of the *N*-(2,3-dihydroxypropyl)valine adduct in Hb. The data points represent the adduct levels from 1-hour incubations of glycidol with fresh human blood from triplicate samples at each dose level from four individuals, where each point corresponds to the mean from two *m/z* transitions. One replicate at the highest dose for individual 2 (data point in parenthesis) was excluded in the analyses as it was judged as an outlier (Grubbs test).

4. Discussion

4.1. N-(2,3-dihydroxypropyl)valine Adduct Levels

In this study, we quantified the levels of the *N*-(2,3-dihydroxypropyl)valine adduct (diHOPrVal) in samples from 50 children of about 12 years of age, showing a fivefold variation in the adduct levels (ca. 4–20 pmol/g Hb). No significant difference was observed between the sexes of the children. The diHOPrVal adduct levels were also quantified in a small number of adults ($n = 12$), where the mean adduct levels where about the double in the smokers compared to the non-smokers. As glycidol is known to be present in tobacco smoke, this was expected [23,24]. The mean Hb adduct level in studied non-smoking adults indicated approximately the same exposure for this group as for the children. A larger sample size from adults should be included for a more reliable comparison.

The diHOPrVal adduct in Hb has earlier been quantified only in small groups of adults in a few published studies, which all show somewhat lower levels compared to our study (Table 2). The observed variation of the mean adduct levels between the different studies may be due to differences in exposure between the studied groups, but also due to the fact that the analyses are performed at different laboratories and with different analytical methods for measurement of the N-terminal adduct in Hb; the *N*-alkyl Edman (GC/MS/MS) and the FIRE procedure (LC/MS/MS), and with no inter-calibration between the laboratories.

Table 2. Steady state levels of the Hb adduct *N*-(2,3-dihydroxypropyl)valine (diHOPrVal) in adult subjects (from different countries) without any known exposure, obtained in published studies.

Studied Group	No. of Subjects	diHOPrVal (pmol/g globin)	Analytical Method	Reference
Non-smokers	11	4.1 ± 0.8 [a]	LC/MS/MS	[22]
Non-smokers	12	3.3 ± 0.8 [a]	LC/MS/MS	[25]
Non-smokers	6	7.1 ± 3.1 [b]	GC/MS/MS	[26]
Non-smokers	3	6.8 ± 3.2 [b]	GC/MS/MS	[27]
Non-smokers	4	2.1 ± 1.1 [b]	GC/MS/MS	[27]
Non-smokers	6	10.3 ± 2.7 [a]	LC/MS/MS	present study
Smokers	6	13.1 ± 12.4 [b]	GC/MS/MS	[27]
Smokers	6	9.5 ± 2.2 [b]	GC/MS/MS	[27]
Smokers	6	23.4 ± 4.6 [a]	LC/MS/MS	present study

[a] FIRE procedure (adduct level expressed as per g Hb, approximately the same as per globin). [b] *N*-alkyl Edman method.

We have assumed that the observed diHOPrVal adduct levels in children and non-smokers originate from the exposure to the genotoxic compound glycidol via food, but this adduct may also theoretically originate from other precursors (Figure 5). One possibility is the food contaminant 3-monochloropropane-1,2-diol (3-MCPD), often occurring in parallel with glycidol in food. 3-MCPD would however give a very small contribution, as 3-MCPD has more than 1000 times lower rate constant for formation of the adduct to the N-terminal in Hb compared to glycidol [28]. The food-related compounds allyl alcohol, found in garlic, or anhydro sugars from carbohydrates, can also theoretically be precursors to the adduct [29,30]. The formation of glycidol from allyl alcohol could be assumed to be possible via a metabolic oxidation. The heating of carbohydrate-rich food (anhydro sugars) could theoretically form glycidol, which was indicated in an animal experiment with feeding with heat-processed feed and measurement of the diHOPrVal adduct [30]. Other theoretically potential precursors are the endogenously produced glyceraldehyde and glycidaldehyde. Both compounds, though, require reduction after formation of a Schiff base to the N-terminal valine in Hb to form the stable diHOPrVal adduct [30]. It is not known whether the reduction of protein adducts from Schiff bases occurs in vivo, but it was observed to occur in blood in vitro [31]. Epichlorohydrin, from occupational exposure, also could form diHOPrVal but it is not a probable exposure source in the presently studied group of humans [27]. Thus, other sources to the measured adduct than glycidol cannot be excluded, which could potentially lead to an overestimation of the in vivo doses (AUC) of glycidol in the studied subjects. It is obvious that low molecular mass adducts in many cases could have several possible precursor electrophiles.

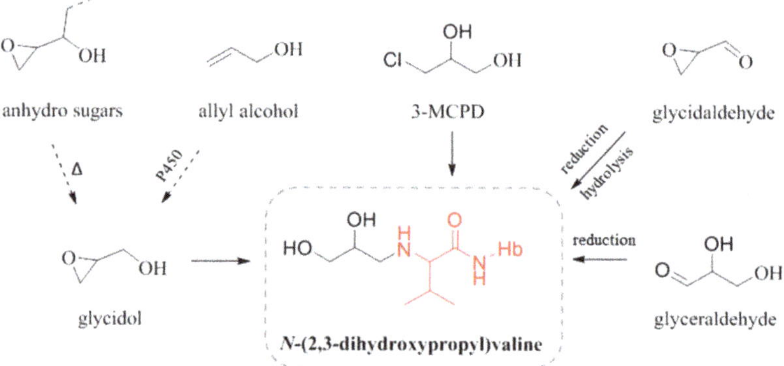

Figure 5. Examples of possible precursors to *N*-(2,3-dihydroxypropyl)valine c.f. [29,30].

In this study, we calculated the AUC of glycidol from the quantified diHOPrVal levels in human subjects, assuming that glycidol is the dominating source of the adduct. The differences in the measured adduct levels and the corresponding intakes and AUC of glycidol (fivefold) between all children (Table 1) could partly be explained by different dietary habits, as glycidyl fatty acid esters are present in different types of food products [3], but also by different genotypes/phenotypes for metabolizing enzymes (e.g., epoxide hydrolase and glutathione transferase, c.f [3]). Furthermore, we could not observe any significant correlations between the diHOPrVal levels in the children and any type of registered food product (from food frequency questionnaires) with potential impact on glycidol exposure (bread, sweets, chips and other fried food) in this limited study (data not shown). Studies of a larger number of subjects and using food frequency questionnaires with more specific questions could possibly enable a sufficient statistical material for such analyses.

4.2. In vivo dose and Intake of Glycidol

Knowledge about the AUC of a particular exposure gives the possibility for a more accurate estimation of the cancer risk. Assuming that the AUCs and the intakes calculated for the studied subjects reflect the exposure to the general European population, the values imply a higher exposure to glycidol than expected, from the mean glycidol intakes estimated by the European Food Safety Authority (EFSA); ca. 0.2 µg/kg b.w./day and 0.6 µg/kg b.w./day for adults and children, respectively [3] and the Swedish National Food Agency (NFA); 0.1 µg/kg b.w./day for adults [32] (Table 3). The relation between adduct level and intake of glycidol in humans was obtained from a recently published human exposure study by Abraham et al., which involved a good number of persons (11) with intake of palm fat oil over 4 weeks corresponding to a mean daily intake of glycidol of 4.3 µg/kg b.w. [22]. The methods for measurement of the diHOPrVal adducts used by Abraham et al. and by us are not inter-calibrated, which might contribute to some uncertainty in this calculation, just like the figure on a mean lifetime of erythrocytes of 126 days (cf., Mitlyng et al. [33] and Abraham et al. [22]).

Table 3. Estimated daily intakes of glycidol and calculated AUC at corresponding estimated lifetime exposures of glycidol, calculated from Hb adduct levels measured in children and adults (non-smokers) in the present study. Daily intakes estimated from dietary patterns by the European Food Safety Authority (EFSA) and the Swedish National Food Agency (NFA) are used for comparison [3,32].

Studied Group	Number of Subjects	Daily Intake (µg/kg b.w./day) Mean [Min–Max]	Estimated Lifetime AUC (µMh) Mean (Approximately)
Present study			
Children	50	1.4 [0.9–3.9]	150
Adults (non-smokers)	6	2.0 [1.2–2.7]	230
Intake estimated by authorities			
Children (EFSA)	n.a.	0.6 [0.4–0.9]	-
Adults (EFSA)	n.a.	0.2 [0.2–0.3]	-
Adults (NFA)	n.a.	0.1	-

[a] n.a.: not available.

The corresponding figures of the AUC per administered dose of glycidol, obtained by different methods, are ca. 35% lower both for rats and monkeys [34,35] compared with the value calculated for humans, which indicates that there are no major differences in the disappearance rate of glycidol between these species and which also supports the reliability of the obtained human data.

4.3. Human Cancer Risk

Different models have been used to assess the human cancer risk of glycidol based on published carcinogenicity data from studies in rodents [36,37]. The European Food and Safety Authority (EFSA) has used the Margin of Exposure (MOE) approach, which is based on estimated intake values of glycidol and the reference point T25 (the dose corresponding to a 25% tumor incidence in the animals used in the carcinogenicity studies) [3]. Using the MOE, EFSA concluded that there is a health concern associated with glycidol exposure. Furthermore, the California Environmental Protection Agency (C. EPA) has calculated a non-significant risk level (NSRL; 1 cancer case in 10^5 individuals over life-time) of 0.54 µg glycidol per day using an additive risk model [38].

Common for these two models are that the estimation of the cancer risk is based on extrapolations from rat to human and the administered doses of glycidol in the carcinogenicity studies. As an improvement of cancer risk estimations, our group at Stockholm University has developed a risk model based on species extrapolation via internal doses (AUC) and background tumor incidence. This model is referred to as the multiplicative risk model and has recently been validated for glycidol, presented in a forthcoming paper [33] and a few other genotoxic carcinogens [39–41]. With this model, described in detail in the referred papers, a cancer risk coefficient (β), which describes the relative tumor risk per in vivo dose, is derived. The relative risk coefficient has been shown to be approximately independent of tumor site, sex, and species for all tested compounds as well as for ionizing radiation. Thus, a common risk coefficient can be derived from the responding sites in the test species in animal cancer tests with a compound. Accordingly, this risk coefficient is assumed to also be valid in humans and can be used for the calculation of the human cancer risk for the studied specific exposure, when in vivo dose and background cancer incidence is known.

For the calculation of the human cancer risk due to glycidol exposure, the in vivo dose over a lifetime (70 years) is compared to the obtained cancer risk coefficient β [33], which is somewhat higher than the risk coefficient obtained by an additive model, as by C.EPA [38]. In the present work, we estimated a mean daily intake of 1.4 µg glycidol per kg bodyweight for the children, which implies about 36 mg/kg during a lifetime (70 years). This is equivalent to the cumulative AUC of ca. 150 µMh (Table 3). Assuming that the background tumor frequency in the general human population is ca. 30% [42], the given lifetime exposure condition lead to an estimate of ca. 200 additional cancer cases in a population of 100,000 (i.e., relative risk increment) at the given exposure condition. In the calculations, we have not considered different contribution to the risk from exposure at different ages, where children in general have higher risk increments per AUC compared with adults. This was observed for subjects exposed to ionizing radiation where the excess relative risk for children is higher compared to adults, as reviewed by Kutanzi et al. [43].

5. Conclusions

This is the first study measuring the Hb adduct *N*-(2,3-dihydroxypropyl)valine and calculation of the corresponding in vivo doses of glycidol in children. The observed variation in the in vivo doses of glycidol within the children´s cohort is likely due to dietary habits and/or different genotypes/phenotypes of metabolic enzymes. The data on diHOPrVal adduct levels in the children as well as in the small group of adults, despite some remaining uncertainties, indicate that calculated intakes of glycidol give contributions that exceed what is considered to be an acceptable cancer risk, using a multiplicative cancer risk model. The obtained data, calculated intakes, and corresponding estimated cancer risks emphasize the importance of further clarifying the background exposure to glycidol from food, as well as possible other sources to the observed diHOPrVal adduct levels in the population, particularly in children.

Author Contributions: Conceptualization, M.T.; Data curation, L.A.-Z. and M.T.; Funding acquisition, M.T.; Investigation, J.A. and E.V.; Methodology, E.V.; Project administration, M.T.; Resources, L.A.-Z. and M.T.; Supervision, J.A. and M.T.; Visualization, J.A. and E.V.; Writing – original draft, J.A.; Writing – review & editing, J.A., E.V., L.A.-Z. and M.T.

Funding: This research was funded by the Research Council Formas grant number 216-2012-1450 and Stockholm University, Stockholm, Sweden.

Acknowledgments: We wish to thank Natalia Kotova for building up the used pilot biobank of blood samples from the children at the Swedish National Food Agency, which was financially supported by the Civil Contingencies Agency.

Conflicts of Interest: The authors declare no conflicts of interests.

References

1. Rappaport, S.M. Genetic Factors Are Not the Major Causes of Chronic Diseases. *PLoS ONE* **2016**, *11*, e0154387. [CrossRef] [PubMed]
2. Nadler, D.L.; Zurbenko, I.G. Estimating Cancer Latency Times Using a Weibull Model. *Adv. Epidemiol.* **2014**, *2014*, 8. [CrossRef]
3. EFSA (European Food Safety Authority). Risks for human health related to the presence of 3- and 2-monochloropropanediol (MCPD), and their fatty acid esters, and glycidyl fatty acid esters in food. *EFSA J.* **2016**, *14*, 159.
4. BfR (Federal Institute for Risk Assessment). Initial Evaluation of the Assessment of Levels of Glycidol Fatty Acid Esters Detected in Refined Vegetable Fats. 2009. Available online: https://mobil.bfr.bund.de/cm/349/initial_evaluation_of_the_assessment_of_levels_of_glycidol_fatty_acid_esters.pdf (accessed on 30 March 2009).
5. Cheng, W.; Liu, G.; Wang, L.; Liu, Z. Glycidyl Fatty Acid Esters in Refined Edible Oils: A Review on Formation, Occurrence, Analysis, and Elimination Methods. *Compr. Rev. Food Sci. Food Saf.* **2017**, *16*, 263–281. [CrossRef]
6. MacMahon, S.; Begley, T.H.; Diachenko, G.W. Occurrence of 3-MCPD and glycidyl esters in edible oils in the United States. *Food Addit Contam. Part A Chem. Anal. Control. Expo. Risk Assess.* **2013**, *30*, 2081–2092. [CrossRef] [PubMed]
7. Appel, K.E.; Abraham, K.; Berger-Preiss, E.; Hansen, T.; Apel, E.; Schuchardt, S.; Vogt, C.; Bakhiya, N.; Creutzenberg, O.; Lampen, A. Relative oral bioavailability of glycidol from glycidyl fatty acid esters in rats. *Arch. Toxicol.* **2013**, *87*, 1649–1659. [CrossRef] [PubMed]
8. IARC (International Agency for Research on Cancer). Glycidol. Some industrial chemicals. In *IARC Monographs on the Evaluation of Carcinogenic Risk of Chemicals to Humans*; International Agency for Research on Cancer: Lyon, France, 2000; Volume 77, pp. 469–486.
9. World Health Organization (WHO). Principles for Evaluating Health Risks in Children Associated with Exposure to Chemicals. In *Environmental Health Criteria 23*; World Health Organization: Geneva, Switzerland, 2006; p. 351. ISBN 978-92-4-157237-8.
10. Törnqvist, M.; Mowrer, J.; Jensen, S.; Ehrenberg, L. Monitoring of environmental cancer initiators through hemoglobin adducts by a modified Edman degradation method. *Anal. Biochem.* **1986**, *154*, 255–266. [CrossRef]
11. von Stedingk, H.; Rydberg, P.; Törnqvist, M. A new modified Edman procedure for analysis of N-terminal valine adducts in hemoglobin by LC-MS/MS. *J. Chromatogr. B Anal. Technol. Biomed. Life. Sci.* **2010**, *878*, 2483–2490. [CrossRef] [PubMed]
12. Rydberg, P. Method for analyzing N-terminal protein adducts using isothiocyanate reagents. European Patent EP1738177 WO/2005/101020, 27 October 2005.
13. Ehrenberg, L.; Moustacchi, E.; Osterman-Golkar, S. Dosimetry of genotoxic agents and dose-response relationships of their effects. *Mutat. Res./Rev. Genetic Toxicol.* **1983**, *123*, 121–182. [CrossRef]
14. Vikström, A.C.; Abramsson-Zetterberg, L.; Naruszewicz, M.; Athanassiadis, I.; Granath, F.N.; Törnqvist, M.A. In vivo doses of acrylamide and glycidamide in humans after intake of acrylamide-rich food. *Toxicol. Sci.* **2011**, *119*, 41–49. [CrossRef] [PubMed]
15. Aasa, J.; Abramsson-Zetterberg, L.; Carlsson, H.; Törnqvist, M. The genotoxic potency of glycidol established from micronucleus frequency and hemoglobin adduct levels in mice. *Food Chem. Toxicol.* **2016**, *100*, 168–174. [CrossRef] [PubMed]
16. Pedersen, M.; von Stedingk, H.; Botsivali, M.; Agramunt, S.; Alexander, J.; Brunborg, G.; Chatzi, L.; Fleming, S.; Fthenou, E.; Granum, B.; et al. Birth weight, head circumference, and prenatal exposure to acrylamide from maternal diet: The European prospective mother-child study (NewGeneris). *Environ. Health Perspect.* **2012**, *120*, 1739–1745. [CrossRef] [PubMed]
17. Carlsson, H.; von Stedingk, H.; Nilsson, U.; Törnqvist, M. LC-MS/MS screening strategy for unknown adducts to N-terminal valine in hemoglobin applied to smokers and nonsmokers. *Chem. Res. Toxicol.* **2014**, *27*, 2062–2070. [CrossRef] [PubMed]

18. Carlsson, H.; Aasa, J.; Kotova, N.; Vare, D.; Sousa, P.F.M.; Rydberg, P.; Abramsson-Zetterberg, L.; Törnqvist, M. Adductomic Screening of Hemoglobin Adducts and Monitoring of Micronuclei in School-Age Children. *Chem. Res. Toxicol.* **2017**. [CrossRef] [PubMed]

19. von Stedingk, H.; Vikström, A.C.; Rydberg, P.; Pedersen, M.; Nielsen, J.K.; Segerbäck, D.; Knudsen, L.E.; Törnqvist, M. Analysis of hemoglobin adducts from acrylamide, glycidamide, and ethylene oxide in paired mother/cord blood samples from Denmark. *Chem. Res. Toxicol.* **2011**, *24*, 1957–1965. [CrossRef] [PubMed]

20. Törnqvist, M.; Fred, C.; Haglund, J.; Helleberg, H.; Paulsson, B.; Rydberg, P. Protein adducts: Quantitative and qualitative aspects of their formation, analysis and applications. *J. Chromatogr. B Anal. Technol. Biomed. Life. Sci.* **2002**, *778*, 279–308. [CrossRef]

21. Shemin, D.; Rittenberg, D. The life span of the human red blood cell. *J. Biol. Chem.* **1946**, *166*, 627–636.

22. Abraham, K.; Hielscher, J.; Kaufholz, T.; Mielke, H.; Lampen, A.; Monien, B. The hemoglobin adduct *N*-(2,3-dihydroxypropyl)-valine as biomarker of dietary exposure to glycidyl esters: A controlled exposure study in humans. *Arch. Toxicol.* **2018**. [CrossRef]

23. Rodgman, A.; Perfetti, T.A. *The Chemical Components of Tobacco and Tobacco Smoke*, 2nd ed.; CRC Press, Taylor & Francis Group: Boca Raton, FL, USA, 2013.

24. Schumacher, J.N.; Green, C.R.; Best, F.W.; Newell, M.P. Smoke composition. An extensive investigation of the water-soluble portion of cigarette smoke. *J. Agric. Food Chem.* **1977**, *25*, 310–320. [CrossRef]

25. Hielscher, J.; Monien, B.H.; Abraham, K.; Jessel, S.; Seidel, A.; Lampen, A. An isotope-dilution UPLC-MS/MS technique for the human biomonitoring of the internal exposure to glycidol via a valine adduct at the N-terminus of hemoglobin. *J. Chromatogr. B Anal. Technol. Biomed. Life. Sci.* **2017**, *1059*, 7–13. [CrossRef]

26. Honda, H.; Onishi, M.; Fujii, K.; Ikeda, N.; Yamaguchi, T.; Fujimori, T.; Nishiyama, N.; Kasamatsu, T. Measurement of glycidol hemoglobin adducts in humans who ingest edible oil containing small amounts of glycidol fatty acid esters. *Food Chem. Toxicol.* **2011**, *49*, 2536–2540. [CrossRef] [PubMed]

27. Hindsø Landin, H.; Grummt, T.; Laurent, C.; Tates, A. Monitoring of occupational exposure to epichlorohydrin by genetic effects and hemoglobin adducts. *Mutat. Res./Fundam. Mol. Mech. Mutagenesis* **1997**, *381*, 217–226. [CrossRef]

28. Aasa, J.; Törnqvist, M.; Abramsson-Zetterberg, L. Measurement of micronuclei and internal dose in mice demonstrates that 3-monochloropropane-1,2-diol (3-MCPD) has no genotoxic potency in vivo. *Food Chem. Toxicol.* **2017**, *109*, 414–420. [CrossRef] [PubMed]

29. Lemar, K.M.; Passa, O.; Aon, M.A.; Cortassa, S.; Muller, C.T.; Plummer, S.; O'Rourke, B.; Lloyd, D. Allyl alcohol and garlic (Allium sativum) extract produce oxidative stress in Candida albicans. *Microbiology* **2005**, *151*, 3257–3265. [CrossRef] [PubMed]

30. Hindsø Landin, H.; Tareke, E.; Rydberg, P.; Olsson, U.; Törnqvist, M. Heating of food and haemoglobin adducts from carcinogens: Possible precursor role of glycidol. *Food Chem. Toxicol.* **2000**, *38*, 963–969. [CrossRef]

31. Degner, A.; Carlsson, H.; Karlsson, I.; Eriksson, J.; Pujari, S.S.; Tretyakova, N.Y.; Törnqvist, M. Discovery of Novel *N*-(4-Hydroxybenzyl)valine Hemoglobin Adducts in Human Blood. *Chem. Res. Toxicol.* **2018**. [CrossRef] [PubMed]

32. NFA (National Food Agency). *2-MCPD, 3-MCPD och Glycidylfettsyraester i Livsmedel på den Svenska Marknaden. Riskhantering, Riskvärdering och Haltdata*; Livsmedelsverkets Rapportserie Nr 35/2017; Livsmedelsverket: Uppsala, Sweden, 2017; p. 42.

33. Mitlyng, B.L.; Singh, J.A.; Furne, J.K.; Ruddy, J.; Levitt, M.D. Use of breath carbon monoxide measurements to assess erythrocyte survival in subjects with chronic diseases. *Am. J. Hematol.* **2006**, *81*, 432–438. [CrossRef]

34. Aasa, J.; Granath, F.; Törnqvist, M. Cancer risk estimation for glycidol based on rodent carcinogenicity studies and in vivo dosimetry. *Food Chem. Toxicol.* **2018**. under review.

35. Wakabayashi, K.; Kurata, Y.; Harada, T.; Tamaki, Y.; Nishiyama, N.; Kasamatsu, T. Species differences in toxicokinetic parameters of glycidol after a single dose of glycidol or glycidol linoleate in rats and monkeys. *J. Toxicol. Sci.* **2012**, *37*, 691–698. [CrossRef]

36. National Toxicology Program. NTP Toxicology and Carcinogenesis Studies of Glycidol (CAS No. 556-52-5) In F344/N Rats and B6C3F1 Mice (Gavage Studies). *Natl. Toxicol. Program. Tech. Rep. Ser.* **1990**, *374*, 1–229.

37. Irwin, R.D.; Eustis, S.L.; Stefanski, S.; Haseman, J.K. Carcinogenicity of glycidol in F344 rats and B6C3F1 mice. *J. Appl. Toxicol.* **1996**, *16*, 201–209. [CrossRef]

38. CEPA (California Environmental Protection Agency), Office of Environmental Health Hazard Assessment (OEHHA). *No Significant Risk Level (NSRL) for the Proposition 65 Carcinogen Glycidol*; California Environmental Protection Agency: Sacramento, CA, USA, 2010; 16p.
39. Granath, F.N.; Vaca, C.E.; Ehrenberg, L.G.; Törnqvist, M.A. Cancer risk estimation of genotoxic chemicals based on target dose and a multiplicative model. *Risk Anal.* **1999**, *19*, 309–320. [CrossRef] [PubMed]
40. Fred, C.; Törnqvist, M.; Granath, F. Evaluation of cancer tests of 1,3-butadiene using internal dose, genotoxic potency, and a multiplicative risk model. *Cancer Res.* **2008**, *68*, 8014–8021. [CrossRef] [PubMed]
41. Törnqvist, M.; Paulsson, B.; Vikström, A.C.; Granath, F. Approach for cancer risk estimation of acrylamide in food on the basis of animal cancer tests and in vivo dosimetry. *J. Agric. Food Chem.* **2008**, *56*, 6004–6012. [CrossRef] [PubMed]
42. Cancerfonden. *Cancerfondsrapporten 2017*; Cancerfonden: Vindspelet Grafiska AB, Sweden, 2017.
43. Kutanzi, K.R.; Lumen, A.; Koturbash, I.; Miousse, I.R. Pediatric Exposures to Ionizing Radiation: Carcinogenic Considerations. *Int. J. Environ. Res. Public. Health.* **2016**, *13*, 1057. [CrossRef] [PubMed]

Review

Biological Evaluation of DNA Biomarkers in a Chemically Defined and Site-Specific Manner

Ke Bian [1], James C. Delaney [2], Xianhao Zhou [1] and Deyu Li [1,*]

[1]   Department of Biomedical and Pharmaceutical Sciences, College of Pharmacy, University of Rhode Island, Kingston, RI 02881, USA; kebian@uri.edu (K.B.); xianhao_zhou@my.uri.edu (X.Z.)
[2]   Visterra, Inc., 275 Second Avenue, Waltham, MA 02451, USA; delaney@mit.edu
*   Correspondence: deyuli@uri.edu; Tel.: +1-(401)-874-9361

Received: 25 May 2019; Accepted: 14 June 2019; Published: 25 June 2019

Abstract: As described elsewhere in this Special Issue on biomarkers, much progress has been made in the detection of modified DNA within organisms at endogenous and exogenous levels of exposure to chemical species, including putative carcinogens and chemotherapeutic agents. Advances in the detection of damaged or unnatural bases have been able to provide correlations to support or refute hypotheses between the level of exposure to oxidative, alkylative, and other stresses, and the resulting DNA damage (lesion formation). However, such stresses can form a plethora of modified nucleobases, and it is therefore difficult to determine the individual contribution of a particular modification to alter a cell's genetic fate, as measured in the form of toxicity by stalled replication past the damage, by subsequent mutation, and by lesion repair. Chemical incorporation of a modification at a specific site within a vector (site-specific mutagenesis) has been a useful tool to deconvolute what types of damage quantified in biologically relevant systems may lead to toxicity and/or mutagenicity, thereby allowing researchers to focus on the most relevant biomarkers that may impact human health. Here, we will review a sampling of the DNA modifications that have been studied by shuttle vector techniques.

Keywords: DNA lesion; DNA damage; shuttle vector technique; replication block; mutagenicity; mutational spectrum; mutational signature; DNA repair; DNA adduct bypass; site-specific mutagenesis

1. Introduction

The human genome is constantly exposed to and damaged by endogenous chemicals, such as reactive oxygen species, lipid peroxidation intermediates, and alkylating agents. These electrophilic reactive chemicals, as well as environmental carcinogens and administered drugs, are known to generate various DNA adducts [1–3]. Some of the adducts block DNA replication or cause mutations and have been used as biomarkers to monitor the level of DNA damage or of disease progression [4–6]. One of the major goals for researchers is to understand the deleterious consequences of those lesions within the cell or animal. Among the different methods for studying the biological effects of the adducts, use of shuttle vectors containing a chemically defined lesion at a specific site has provided information about the biological and toxicological properties of the adduct [4,7]. The shuttle vector-based methods normally involve the steps outlined in Figure 1. *Oligonucleotide synthesis*: An oligonucleotide (oligo) containing a structurally defined lesion at a specific site is made either through a biomimetic route (in situ formation by direct chemical reaction, followed by HPLC purification of site-specifically modified oligo), or purely synthetically using a normal or convertible nucleoside phosphoramidite, etc. *Vector construction*: An ss- or ds-DNA vector containing the modified oligo is built by cutting the parent vector with one or a pair of restriction endonuclease(s), followed by ligation of the 5′-phosphorylated modified oligo. *Cellular processing*: The vector is transfected into different types of cells (e.g., *Escherichia. coli* (*E. coli*) or mammalian), and cellular polymerases are allowed to replicate or transcriptionally bypass the lesion under different repair or bypass conditions. *Data analysis*: DNA is extracted, amplified using PCR, and the biological outcomes are analyzed, which include the ability of the lesion/adduct to block polymerases or cause a mutation when processed by a polymerase during cellular replication. This assessment could be done by plaque or colony counting and picking with Sanger sequencing, ^{32}P-post labeling and thin-layer chromatography (TLC), liquid chromatography-mass spectrometry (LC-MS), next-generation sequencing (NGS), etc. [4,5,7–9]. The shuttle vector-based method was initially introduced by Essigmann [7,9–11], further developed and utilized by Wang [4,5], Moriya [12,13], Livneh [14,15], Greenberg [16,17], Basu [18,19], Lloyd [20,21], Loechler [22,23], Fuchs [24,25], Pagès [26,27], and others. Several informative review articles have been written by these authors on designing and applying the methods.

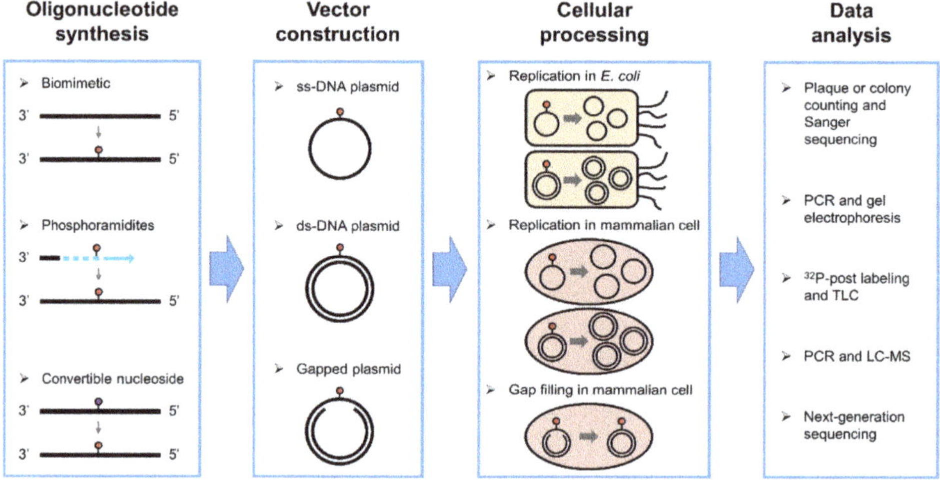

Figure 1. Schematic overview of the shuttle vector-based methods for evaluating DNA biomarkers.

In this work, we will review a variety of DNA biomarkers or probes that have been studied using the shuttle vector techniques and briefly summarize their biological outcomes. In all cases, focus is placed on the effect of the lesion to block replication and to cause mutations. For the details regarding the formation of DNA damage and other properties of the lesions, please refer to the original literature or review articles. We apologize in advance to researchers whose work we could not include in this review. After detailed discussions on individual lesions, we will provide some perspectives on possible future directions.

2. Discussions on Individual Modifications

Below, we will cover modified DNA structures generated from oxidative stress, alkylation, and other processes (Figures 2–5). In the following sections, the biological effects of a certain lesion are briefly summarized. Please see Figures 2–5 for chemical structures and Table 1 for detailed information.

Table 1. Bypass efficiency and mutagenicity of DNA modifications.

Oxidative Lesion	Bypass Efficiency	Mutation	Cell
8-oxo-G	88% [28,29]	G>T 3% (44%, MutY-) [28,29]	*E. coli*
		G>T 8% [30]	Human
Fapy-dG	31–43% (TXN sequences) [31]	G>T 1.2–1.9% (0.7–2.1%, MutM-/MutY-) [31]	*E. coli*
		G>T 10% [30]	Human
NI	7% (57%, SOS) [32]	G>C 8.9%, G>A 19%, G>T 22% (G>C 2.5%, G>A 13%, G>T, 18%, SOS) [32]	*E. coli*
Oa	52% [28], 51% [33], 108% (118% MutY-) [29]	G>T 97% [28], 99% [33], 97% (no change, MutY-) [29]	*E. coli*
Oz	57% [28]	G>T 86% [28]	*E. coli*
Ca	65% [28]	G>T 95% [28]	*E. coli*
Gh	75% [34], 20% (30% MutY-) [29]	G>C 98%, G>T 2% [34], G>T 40%, G>C 57%, G>A 3% (no change, MutY-) [29]	*E. coli*
Sp1	9% [34], 19% (38%, MutY-) [29]	G>C 72%, G>T 27% [34], G>T 78%, G>C 19%, G>A 1% (no change, MutY-) [29]	*E. coli*
Sp2	9% [34], 17% (30%, MutY-) [29]	G>C 57%, G>T 41% [34], G>T 49%, G>C 48%, G>A 3% (no change, MutY-) [29]	*E. coli*
Ur	11% [35], 10% (10% MutY-) [29]	G>T 99% [35], G>T 54%, G>C 35%, G>A 9% (no change, MutY-) [29]	*E. coli*
Iz	60% (71%, SOS) [32]	G>C 88%, G>A 2%, G>T 1.1% (G>C 75%, G>A 3.4%, G>T 5.5%, SOS) [32]	*E. coli*
Cyclo-dG	11% (6% pol V-) [36]	G>A 20% [36]	*E.coli*
S-cdG	4% [37]	G>T 35%, G>A 20% [37]	Human
Cyclo-dA	31% (13% pol V-) [36]	A>T 11% [36]	*E. coli*
S-cdA	6% [37]	A>T 12% [37]	Human
Tg	96% [38]		*E. coli*
5ClC	75% (75% AlkB-) [39]	C>T 5% (same in AlkB-) [39]	*E. coli*
5-OH-C		C>T 0.05%, C>G 0.001% [40]	*E. coli*
5-OH-U		C>T 83% [40]	*E. coli*
Ug		C>T 80% [40]	*E. coli*
THF AP site	6% [28], 5.8% [32], 4% (4% MutY-) [29]	AP>T 50%, AP>C 26%, AP>A 7%, -1 del 13% (no change, MutY-) [29]	*E. coli*
2-deoxyribonolactone	5%, (3% pol II-), (1% pol V-) [41]	T 35%, C 42%, A 12%, G 8%, 5′ T (T 42%, C 38%, A 6%, G 14%, 5′ C) [41]	*E. coli*

Alkyl Modification	Bypass Efficiency	Mutation	Cell
m1G	15% (3%, AlkB-) [9], 20% (2%, AlkB-) [33]	G>T 3% (G>T 57%, G>A 17%, G>C 6%, AlkB-) [9], G>T 4%, G>A 2% (G>T 52%, G>A 20%, G>C 4%, AlkB-) [33]	*E. coli*
m2G	90% (84% AlkB-; 98% DinB-; 96% AlkB- and DinB-) [42]	G>A 3% (2.7%, AlkB-; 3%, DinB-; 3%, AlkB- and DinB-) [42]	*E. coli*
e2G	100% (98% AlkB-), (106% DinB-), (99% AlkB- and DinB-) [42]	G>A 2%, G>C 1% (G>A 2.3%, AlkB-), (G>A 2%, AlkB- and DinB-) [42]	*E. coli*
N^2-CMdG	100% [43]	Not mutagenic [43]	Mouse
R-N^2-CEdG	39% (13% pol V-) [44]	Not mutagenic [44]	*E. coli*
	100% [43]	Not mutagenic (G>A 23%, G>T 15%, pol κ-) [43]	Mouse
S-N^2-CEdG	75% (28% pol V-) [44]	Not mutagenic [44]	*E. coli*
	99% [43]	Not mutagenic (G>A 23%, G>T 15%, pol κ-) [43]	Mouse
FF	101% (100% AlkB-), (28% DinB-), (36% AlkB- and DinB-) [42]	G>C 1%, (G>A 1%, G>T 1%, AlkB-DinB-) [42]	*E. coli*
HF	92% (88% AlkB-), (28% DinB-), (40% AlkB- and DinB-) [42]	G>C 2% [42]	*E. coli*
O^6mG		G>A 99% [45,46]	*E. coli*
O^6-POB-dG	70% [47]	G>A 90%, G>T 2.5% [47]	*E. coli*
O^6-PHB-dG	40% [47]	G>A 95% [47]	*E. coli*
O^6-CM-dG	10% [47]	G>A 10% [47]	*E. coli*
	40% [48]	G>A 6% [48]	Human
O^6-ACM-dG	2% [47]	G>A 30% [47]	*E. coli*

Table 1. *Cont.*

Alkyl Modification	Bypass Efficiency	Mutation	Cell
O^6-HOEt-dG	15% [47]	G>A 40% [47]	*E. coli*
PdG	25% [12]	G>T 6% [12]	Human
α-OH-PdG	17% [12]	G>T 11% [12]	Human
γ-OH-PdG	73% [12]	Not mutagenic [12]	Human
$1,N^2$-eG	4% (2% AlkB-) (1.8% AlkB-DinB-) [8]	G>A 6%, G>T 6%, G>C 2%, −1/2 del 5% (G>A 13%, G>T 13%, G>C 1%, −1/2 del 9%, AlkB-), (same in AlkB-DinB-) [8]	*E. coli*
$2'$-F-N^2,3-eG	21% (26% AlkB-) (14% AlkB-DinB-) [8]	G>A 30% (30% AlkB-), (30% AlkB-DinB-) [8]	*E. coli*
m1A	100% (12%, AlkB-) [9]	A>T 0.06% (0.61%, AlkB-) [9]	*E. coli*
eA	85% (5% AlkB-) [33], 130% (9% AlkB-) [49]	<0.5% (A>T 25%, A>G 5%, A>C 5%, AlkB-) [33], A>C 1%, A>T 1% (A>T 22%, A>C 8%, A>G 7%, AlkB-) [49]	*E. coli*
	17% [50]		Human
EA	135% (14% AlkB-) [49]	A>C 1%, A>G 0.5%, A>T 0.5% (A>C 2%, A>G 1%, A>T 1%, AlkB-) [49]	*E. coli*
N^6-CMdA	98% [36]	Not mutagenic [36]	*E. coli*
	65% (35% pol k-) [48]	Not mutagenic [48]	Human
S-N^6-HB-dA	120% [51]	Not mutagenic [51]	*E. coli*
R,R-N^6,N^6-DHB-dA	100% [51]	<1% [51]	*E. coli*
S,S-N^6,N^6-DHB-dA	60% [51]	A>G 1% [51]	*E. coli*
R,S-$1,N^6$-γ-HMHP-dA	10% [51]	A>T 2% [51]	*E. coli*
O^2-MedT	60% [52], 55% [53]	T>A 1%, T>G 1% [52], T>A 56% [53]	Human
	5% [54]	T>A 10%, T>G 10% [54]	*E. coli*
O^2-EtdT	21% (5% pol V-) [55]	T>C 35%, T>A 15%, T>G 5% (T>C 10%, pol V-) [55]	*E. coli*
	45% [52]	T>A 5%, T>G 3% [52]	Human
O^2-nPrdT	35% [52]	T>A 12%, T>G 5% [52]	Human
O^2-iPrdT	35% [52]	T>A 4%, T>G 1% [52]	Human
O^2-nBudT	30% [52]	T>A 13%, T>G 6% [52]	Human
O^2-iBudT	15% [52]	T>A 4%, T>G 2% [52]	Human
O^2-sBudT	15% [52]	T>A 4%, T>G 2% [52]	Human
O^2-POB-dT	3% [54]	12% T>A, 38% T>G [54]	*E. coli*
	26% [53]	T>A 47% [53]	Human
m3T	6%, (4% AlkB-) [9]	T>A 32%, T>C 6%, T>G 2% (T>A 47%, T>C 9%, T>G 2%, AlkB-) [9]	*E. coli*
N3-EtdT	17% (3% pol V-) [55]	T>C 15%, T>A 21%, T>G 3% (Not mutagenic, pol V-) [55]	*E. coli*
N3-CMdT	55% [36]	T>A 66% [36]	*E. coli*
	40% [48]	T>A 81% [48]	Human
O^4-CMdT	49% [36]	T>C 86% [36]	*E. coli*
	40% [48]	T>C 68% (25% pol ζ-) [48]	Human
O^4-MedT	32% [56]	T>C 58% [56]	Human
O^4-EtdT	76% [55]	T>C 84% (Not mutagenic, pol V-) [55]	*E. coli*
	33% [56]	T>C 82% [56]	Human
O^4-nPrdT	35% [56]	T>C 42% [56]	Human
O^4-iPrdT	30% [56]	T>C 44% [56]	Human
O^4-nBudT	32% [56]	T>C 29% [56]	Human
O^4-iBudT	24% [56]	T>C 42% [56]	Human
O^4-R-sBudT	20% [56]	T>C 25% [56]	Human
O^4-S-sBudT	22% [56]	T>C 25% [56]	Human
m3C	100% (10% AlkB-) [9], 113% (14% AlkB-) [57], 98% (5% AlkB-; 115% DinB-; 7.5% AlkB-DinB-) [42], 100% (15% AlkB-) [39]	C>T 1% (C>T 14%, C>A 14%, C>G 2%, AlkB-) [9], Not mutagenic (C>T 55%, C>A 30%, C>G 1%, AlkB-) [57], Not mutagenic (C>T 41%, C>A 41%, C>G 4%, AlkB-) [42], Not mutagenic (C>T 52%, C>A 30%, AlkB-) [39]	*E. coli*

Table 1. *Cont.*

Alkyl Modification	Bypass Efficiency	Mutation	Cell
e3C	96%, (9% AlkB-) [9]	Not mutagenic (C>T 17%, C>A 11%, C>G 2%, AlkB-) [9]	*E. coli*
N^4-CMdC	83% [36]	Not mutagenic [36]	*E. coli*
	80% [48]	Not mutagenic [48]	Human
5mC	100% (100% AlkB-) [39]	Not mutagenic (same in AlkB-) [39]	*E. coli*
	100% [58]	Not mutagenic [58]	Human
5hmC	100% [59]	Not mutagenic [59]	*E. coli*
	98% [58]	Not mutagenic [58]	Human
5fC	100% [59]	Not mutagenic [59]	*E. coli*
	74% [58]	Not mutagenic [58]	Human
5caC	100% [59]	Not mutagenic [59]	*E. coli*
	72% [58]	Not mutagenic [58]	Human
eC	24% (13% AlkB-) [33]	C>A 24%, C>T 11% (C>A 49%, C>T 31%, AlkB-) [33]	*E. coli*
H-edC	1% [50]	C>G 40% [50]	*E. coli*
	10% [50]	C>A 60%, C>T 32% [50]	Human
5hmU	80% [60]	Not mutagenic [60]	Human
Sp-Me-PTE	110% (Ada-, decreases from 140% to 70%) [61]	TT>GT 50%, TT>GC 15% [61]	*E. coli*
Rp-Me-PTE	30% [61]	Not mutagenic [61]	*E. coli*
Sp-Et-PTE	190% [61]	Not mutagenic [61]	*E. coli*
Rp-Et-PTE	40% [61]	Not mutagenic [61]	*E. coli*
Sp-*n*Pr-PTE	160% [61]	Not mutagenic [61]	*E. coli*
Rp-*n*Pr-PTE	70% [61]	Not mutagenic [61]	*E. coli*
Sp-*n*Bu-PTE	100% [61]	Not mutagenic [61]	*E. coli*
Bulky Lesion	**Bypass Efficiency**	**Mutation**	**Cell**
N^2-MC-dG	38% [62]	G>T 18% [62]	Human
N^2-2,7-DAM-dG	27% [62]	G>T 10% [62]	Human
AL-II-dG	9% [63]	G>T 9% [63]	Mouse
AFB$_1$-N7-dG		G>T 1.5% [64]	*E. coli*
AFB$_1$-FAPY		G>T 14% [65]	*E. coli*
C8-AP-dG	51% [66]	Not mutagenic [66]	Human
C8-AAF-dG	13% [66]	Not mutagenic [66]	Human
C8-AF-dG	97% [66]	Not mutagenic [66]	Human
AL-I-dA	100% (5% Rev3L-) [67]	A>T 50% (Not mutagenic, Rev3L-) [67]	Mouse
AL-II-dA	5% [63]	A>T 22% [63]	Mouse
BPDE-dG	(40% Rev1-); (13% Rev3L-) [68]	G>T 73%, G>A 12%; (G>T 32%, G>A 18%, Rev1-); (G>T 6%, Rev3L-) [68]	Mouse
Crosslinked Lesion	**Bypass Efficiency**	**Mutation**	**Cell**
ICL-RD	43% [69]	5′-G>T 3% [69]	*E. coli*
ICL-R	38% [69]	5′-G>T 3% [69]	*E. coli*
ICL-S	53% [69]	5′-G>T 3% [69]	*E. coli*
AP-dG (dG strand)	38% (43% Pol η-), (13% Pol ι-), (2% Pol κ-), (5% Pol ζ-) [70]	G>A 2-5%, G>T 1–2%, G>C 1% [70]	Human
AP-dG (AP strand)	18% (25% Pol η-), (4% Pol ι-), (1% Pol κ-), (5% Pol ζ-) [70]	AP>T 74%, AP>C 10-20%, AP>G 4–6%, AP>A 1–2% [70]	Human
1,2-GG-*cis*-DDP	11% [71]; 5% (30% SOS) [72]	<0.25% (G>T 1.3%, SOS) [72]	*E. coli*
1,2-AG-*cis*-DDP	22% (32% SOS) [72]	<0.2% (A>T 4.4%, SOS) [72]	*E. coli*
1,3-GTG-*cis*-DDP	13% (14% SOS) [72]	<0.7% [72]	*E. coli*
γ-HOPdG mediated peptide crosslink		G>T 5%, G>C 3% [20]	Human
γ-HOPdA mediated peptide crosslink		Not mutagenic [20]	Human
5fC mediated peptide crosslink		C>T 7%, C>G 1%, C del 2% [73]	Human

Table 1. *Cont.*

Other Nucleotide Analog	Bypass Efficiency	Mutation	Cell
H	5% [74]	T>A 41%, T>C 5%, T>G 4%, −1 del 13% [74]	*E. coli*
F	13% [74]	T>A 9%, T>C 1%, T>G 1% [74]	*E. coli*
L	20% [74]	T>A 5% [74]	*E. coli*
B	12% [74]	T>A 24% [74]	*E. coli*
I	10% [74]	T>A 46%, T>C 1%, T>G 1%, −1 del 6% [74]	*E. coli*
KP1212	128% [57]	C>T 10% [57]	*E. coli*
xG	11% (45% SOS) [75]	G>A 95% [75]	*E. coli*
xA	80% (108% SOS) [75]	<1% [75]	*E. coli*
xT	73% (102% SOS) [75]	T>A 73% [75]	*E. coli*
xC	29% (53% SOS) [75]	C>A 10% [75]	*E. coli*
α-dG	3% [76]	G>A 60%, G>C 6% [76]	*E. coli*
α-dA	20% [76]	Not mutagenic [76]	*E. coli*
α-dT	1% [76]	Not mutagenic [76]	*E. coli*
α-dC	1% [76]	C>A 72% [76]	*E. coli*
dxG	25% [77]	Not mutagenic [77]	*E. coli*
dxA	75% [77]	A>G 10% [77]	*E. coli*
dxT	150% [77]	Not mutagenic [77]	*E. coli*
dxC	125% (CXT), 175%(GXG) [77]	Not mutagenic [77]	*E. coli*
sG	98% [78]	G>A 11% [78]	*E. coli*
sG	98% [79]	G>A 8% [79]	Human
S^6mG	91% [78]	G>A 94% [78]	*E. coli*
S^6mG	95% [79]	G>A 40% [79]	Human
SO₃HG	87% [78]	G>A 77% [78]	*E. coli*
2′-F-G	99% [8]	Not mutagenic [8]	*E. coli*
J	52% [60]	Not mutagenic [60]	Human

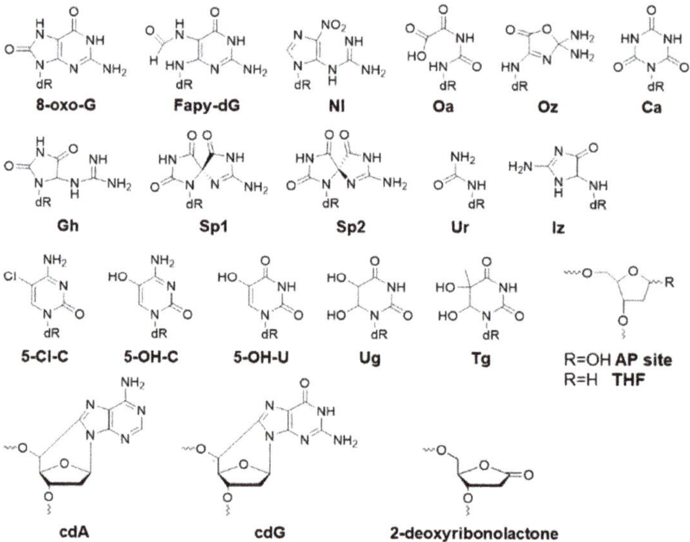

Figure 2. Structures of oxidative lesions.

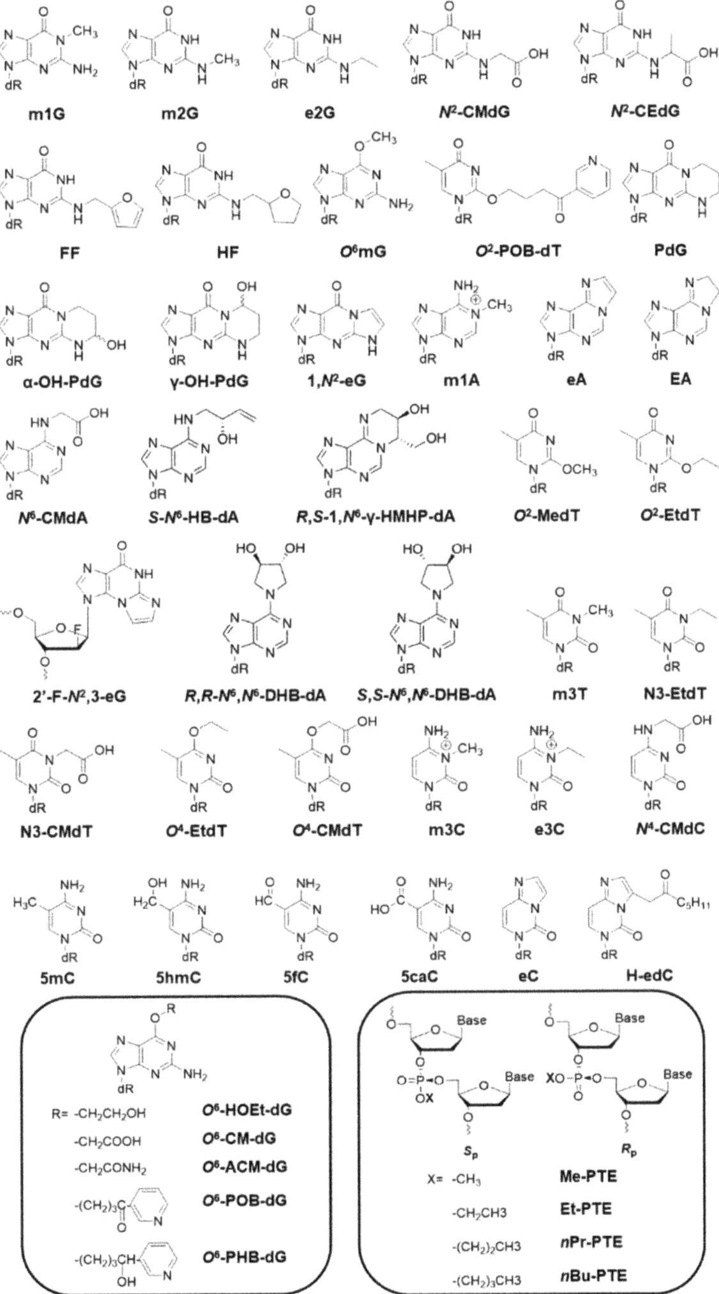

Figure 3. Structures of alkyl modifications.

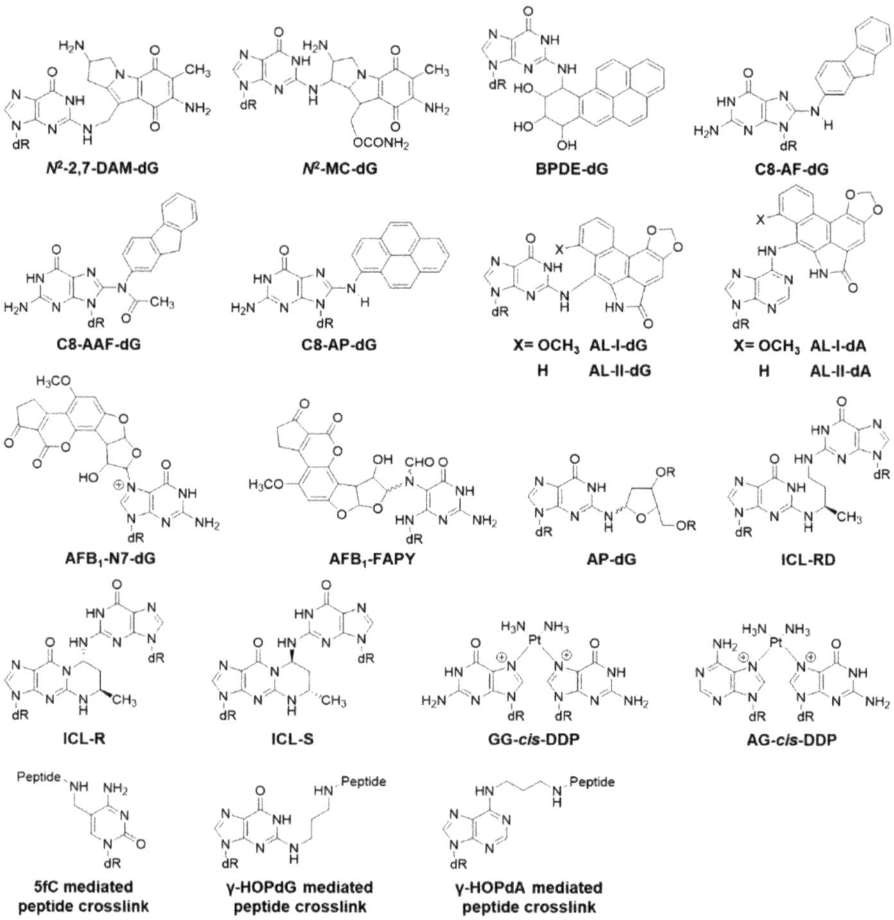

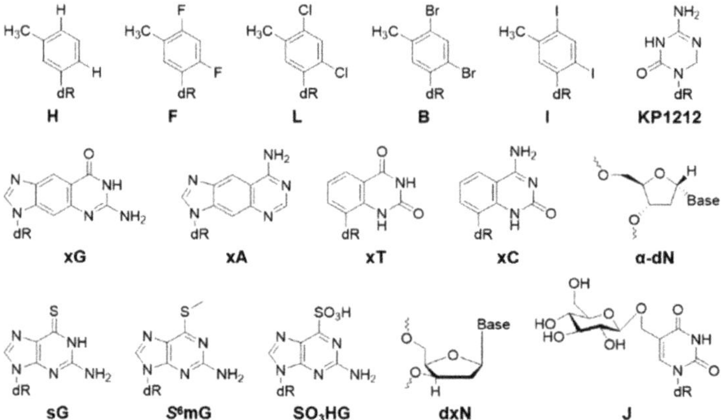

Figure 4. Structures of bulky and crosslinked lesions.

Figure 5. Structures of other nucleotide analogs.

2.1. Oxidative Biomarkers

All the structures of modifications covered in this section are displayed in Figure 2. 8-Oxo-7,8-dihydro-2′-deoxyguanosine (8-oxo-G) is not a strong block to replication, demonstrating greater than 80% bypass efficiency in *E. coli* [28]. Its mutagenic pairing with A during replication in wild type (WT) cells leads to a low amount of G>T mutation (3%) [28]. However, in MutY-cells (MutY: adenine glycosylase in 8-oxo-G:A base excision repair), the G>T mutation increases to 44% [29]. 8-oxo-G causes mainly G>T mutation with a frequency of 8% in human cells [28,29]. Thymidine glycol (Tg) is not a replication block, and it is not mutagenic in *E. coli*; however, tandem lesions of 8-oxoG and Tg are twice as effective as a single 8-oxo-G in blocking DNA replication, and the dual lesion is more mutagenic than 8-oxo-G [38]. Fapy-dG (N-(2-deoxy-α,β-d-erythropentofuranosyl)-N-(2,6-diamino-4-hydroxy-5-formamidopyrimidine)) strongly blocks replication by 60–70% in *E. coli*, but it is not very mutagenic, providing less than 2% G>T mutation [31]. Fapy-dG causes 10% G>T mutation in human cells [30]. 5-Guanidino-4-nitroimidazole (NI) strongly blocks replication (93%) in *E. coli*, giving mainly G>T (22%) and G>A (19%) mutations, and some G>C (9%) mutation as well [32]. Oxaluric acid (Oa) is toxic, blocking replication by 50%, causing nearly 100% G>T mutation in *E. coli* [28,31,34]. Oxazalone (Oz) strongly blocks replication and is very mutagenic, causing 86% G>T mutation [28]. Cyanuric acid lesion (Ca) blocks 35% replication in *E. coli*, and is very mutagenic with 95% G>T mutation [28]. Guanidinohydantoin (Gh) slightly blocks replication (25%), and it is highly mutagenic yielding 97% G>C and 2% G>T mutation [34]. Two stable stereoisomers of spiroiminodihydantoin (Sp1 and Sp2) are strong replication blocks (91%), and are both very mutagenic, causing mainly G>C (72% for Sp1 and 57% for Sp2) and G>T (27% for Sp1 and 41% for Sp2) mutations [34]. Urea lesion (Ur) is a strong replication block (90%) causing 54% G>T, 35% G>C, and 9% G>A mutations [29,35]. Imidazolone adduct (Iz) can be bypassed in *E. coli* with a 40% blockage in replication, essentially causing G>C (88%) mutation, with some G>A (2%) and G>T (1%) mutations [32]. 8,5′-Cyclo-2′-deoxyguanosine (cdG) is a strong replication block (89%) in *E. coli*, and knocking out pol V increases its replication block; it is mutagenic and causes 20% G>A mutation [36]. The 5′ S-diastereomer of cyclo-dG (S-cdG) also strongly blocks DNA replication (96%) in human cells, giving primarily G>T (35%) and G>A (20%) mutations [37]. 8,5′-Cyclo-2′-deoxyadenosine (cdA) is 31% bypassed in *E. coli*, but the bypass efficiency drops to 13% when pol V is removed from the cell [36]. It is mutagenic and causes A>T (11%) mutation [36]. The 5′ S-diastereomer of cyclo-dA (S-cdA) strongly blocks replication in human cells by 94% [37]. Knocking down pol η by siRNA decreases the bypass efficiency and mutagenicity of S-cdA [37]. 5-Chlorocytosine (5-Cl-dC) blocks replication (25%), forming a low level of C>T mutation (5%) in *E. coli* [39]. 5-Hydroxycytosine (5-OH-dC) is not mutagenic in *E. coli* [40]. 5-Hydroxyuracil (5-OH-dU, derived from 5-OH-dC) is very mutagenic providing 83% C>T mutation in *E. coli* [40]. 5,6-Dihydroxy-5,6-dihydrouracil (Ug) is also very mutagenic (80% C>T) in *E. coli* [40].

Tetrahydrofuran (THF) is a stable structural analog to the abasic site (AP site), which is not stable and may lead to further damages to the DNA strand. THF strongly blocks replication (>95%) and causes G>T (50%), G>C (26%), and G>A (7%) mutations; additionally, it causes 13% −1 frame shift mutation [28,29,32,34,80].

2.2. Alkyl Biomarkers

All the structures of modifications covered in this section are displayed in Figure 3. 1-Methyldeoxyguanosine (m1G) is a strong replication block either with or without the repair enzyme AlkB (85% and 97%); it mainly causes ~3% G>T mutation in WT *E. coli*, which increases to more than 50% in AlkB- *E. coli* (AlkB: alkyl DNA adduct direct reversal of damage repair protein) [9,33]. N^2-methylguanine (m2G) weakly blocks replication by 10% in *E. coli*, there is no significant change when knocking out either AlkB or DinB (DinB: DNA polymerase IV), and a small amount of G>A mutation (3%) is seen [42]. N^2-ethylguanine (e2G) does not block replication in *E. coli* and causes a low amount of G>A mutation (2%); eliminating AlkB and DinB does not change the replication bypass and mutagenicity significantly [42]. N^2-carboxymethyl-2′-deoxyguanosine

(N^2-CMdG) and N^2-(1-carboxyethyl)-2′-deoxyguanosine (N^2-CEdG) do not block DNA replication and are not mutagenic in WT mammalian cells; however, each of them causes G>A (23%) and G>T (15%) mutations in mouse embryonic fibroblast (MEF) cells that are deficient in pol κ [43]. N^2-CEdG blocks replication in *E. coli* [44]. The R-N^2-CEdG is a stronger replication block (61%) than S-N^2-CEdG (25%); however, neither of them are mutagenic [44]. N^2-furfurylguanine (N^2-FF-dG) does not block replication in WT *E. coli*; however, it blocks replication about 72% in DinB- cells [42]. It is not very mutagenic with or without DinB [42]. 2-Tetrahydrofuran-2-yl-methylguanine (N^2-HF-dG) is similar in structure to N^2-FF-dG and strongly blocks replication (72%) only when DinB is knocked out, and causes only 2% G>C mutation [42]. O^6-methylguanine (O^6mG) is very mutagenic and leads to almost 100% G>A mutation in Ada/Ogt/UvrB triple knockout *E. coli* (Ada/Ogt: alkyl DNA adduct direct reversal of damage repair protein; UvrB: nucleotide excision repair) [45,46]. N-Nitroso compounds induce DNA lesions: O^6-pyridyloxobutyl-dG (O^6-POB-dG), O^6-pyridylhydroxybutyl-dG (O^6-PHB-dG), O^6-carboxymethyl-dG (O^6-CMdG), which have two structural analogs: O^6-aminocarbonylmethyl-dG (O^6-ACM-dG) and O^6-hydroxyethyl-dG (O^6-HOEt-dG) [47]. O^6-POB-dG slightly blocks DNA replication and induces G>A (90%) transition and G>T (2.5%) transversion in *E. coli* [47]. O^6-PHB-dG is a moderate impediment to DNA replication and causes G>A (95%) mutation exclusively in *E. coli* [47]. O^6-CMdG strongly inhibits replication in *E. coli*, but causes moderate G>A (10%) mutation [47]. O^6-ACM-dG and O^6-HOEt-dG are two analogs of O^6-CM-dG. Both O^6-ACM-dG (2% bypass) and O^6-HOEt-dG (15% bypass) strongly block DNA replication [47]. They also induce G>A mutation with 30% and 40% frequencies, respectively [47]. Major acrolein-dG adducts include 8α and 8β isomers of 3H-8-hydroxy-3-(β-D-2′-deoxyribofuranosyl)-5,6,7,8-tetrahydropyrido[3,2-a]purine-9-one (γ-OH-PdG), 6α and 6β isomers (α-OH-PdG), and 1,N^2-(1,3-propano)-2′-deoxyguanosine (PdG) [12]. The bypass efficiency for γ-OH-PdG is 73% compared to dG control in human cells, and γ-OH-PdG is not very mutagenic (<1%) [12]. α-OH-PdG strongly blocks DNA replication with a bypass efficiency of 17% in human cells and it causes G>T (11%) mutation [13]. PdG strongly blocks replication in human cells and mainly causes 6% G>T mutation [12]. Most of the derivatives of PdG moderately block DNA replication in human cells and cause mainly G>T mutation (2–8%) [81]. 1,N^2-ethenoguanine (1,N^2-eG) is a strong replication blocker (96%) in *E. coli* and causes G>A and G>T mutation by 6% for both, plus a small amount of G>C (2%) mutation; it also causes −1 and −2 frame shift mutations (5%), and knocking out AlkB leads to higher replication block and almost doubles the mutagenicity [8]. 2′-Fluoro-N^2,3-ε-2′-deoxyarabinoguanosine (2′-F-N^2,3-eG), a stable analog of N^2,3-ethenoguanine (N^2,3-eG), blocks replication by 79%, and causes 30% G>A mutation in *E. coli*, with AlkB having no significant influence in its replication bypass and mutagenicity [8].

1-Methyldeoxyadenosine (m1A) strongly blocks replication in AlkB- *E. coli* (88%), but it is not very mutagenic, causing <1% A>T mutation; m1A does not block replication in AlkB+ *E. coli* cells [9]. 1,N^6-ethenoadenine (eA) weakly blocks replication by 4% in WT *E. coli*, but significantly blocks replication (95%) when AlkB is knocked out; likewise, eA is not mutagenic in WT *E. coli*, but shows strong mutagenicity in AlkB- cells (25% A>T mutation) [33,49]. Bypass efficiency of eA in human cells is 17% [50]. 1,N^6-ethanoadenine (EA) does not block replication in WT *E. coli*, but strongly blocks replication by 86% when AlkB is removed; it is not very mutagenic in either WT or AlkB- cells, causing only 2% A>C mutations [49]. N^6-carboxymethyl-2′-deoxyadenosine (N^6-CMdA) minimally blocks replication in *E. coli* and is not mutagenic [36]. S-N^6-HB-dA (HB = 2-hydroxy-3-buten-1-yl) and R,R-N^6,N^6-DHB-dA (DHB = 2,3-dihydroxybutan-1,4-diyl) do not block DNA replication and are not mutagenic in *E. coli* [51]. S,S-N^6,N^6-DHB-dA moderately inhibits replication with a 60% bypass efficiency, and causes minimal 1% A>G mutation [51]. R,S-1,N^6-γ-HMHP-dA (HMHP = 2-hydroxy-3-hydroxymethylpropan-1,3-diyl) strongly inhibits DNA replication but causes only 2% A>T mutation [51].

O^2-Methylthymidine (O^2-Me-dT) can be bypassed by 55% in human cells and mainly causes T>A mutation (56%) [53]. O^2-[4-(3-pyridyl-4-oxobut-1-yl]thymidine (O^2-POB-dT) exhibits genotoxicity showing 26% bypass efficiency and is mutagenic with 47% T>A transversion [53]. Both O^2-Me-dT and O^2-POB-dT strongly block DNA replication in *E. coli* (95% and 97%) [54]. O^2-Me-dT induces 10% T>A and 10% T>G mutations [54]. O^2-POB-dT induces 38% T>G and 12% T>A mutations [54]. O^2-Ethylthymidine (O^2-EtdT) is a strong replication block (79%) in *E. coli*, and knocking out pol IV increases the blocking activity, while knocking out pol V increases the replication block even more [55]. It is very mutagenic and forms T>C (35%), T>A (15%), and T>G (5%) mutations, and mutation frequency drops when pol V is knocked out [55]. The bypass efficiency of O^2-dT alkyl adducts in *E. coli* depends on the size of the alkyl lesion [82]. More than 20% of adducts can be bypassed during replication for ethyl and methyl substitutions, but less than 10% can be bypassed for propyl, and less than 5% for butyl adducts, with the major mutation type being T>C point mutation [82]. O^2-alkyldT lesions strongly inhibit DNA replication (40–85%) in mammalian cells [52]. The blockage effect increases with the size and branching of the alkyl groups [52]. These lesions cause T>A and T>G mutations [52]. 3-Methyldeoxythymidine (m3T) strongly blocks replication in *E. coli* by 94% and is very mutagenic, generating mainly T>A (32%) transversion mutation; eliminating AlkB slightly increases its replication blocking power and mutagenicity [9]. N3-Ethylthymidine (N3-EtdT) strongly blocks replication by 83% in *E. coli*, and knocking out pol V or pol IV increases its blocking activity; it is very mutagenic causing T>A (21%), T>C (15%) and T>G (3%) mutations, and removing pol V eliminates the mutagenicity of this adduct [55]. N3-carboxymethylthymidine (N3-CMdT) strongly blocks replication by 45% in *E. coli*, with the major mutation being T>A (66%); and knocking out pol V slightly increases the mutation rate; however, knocking out pol IV decreases the mutation rate [36]. O^4-carboxymethylthymidine (O^4-CMdT) is a strong replication block (51%) and very mutagenic, causing 86% T>C mutation [36]. N^3-CMdT, O^4-CMdT and O^6-carboxymethyl-dG (O^6-CMdG) moderately block DNA replication in human cells [48]. N^3-CMdT causes T>A (81%) mutation; O^4-CMdT causes T>C (68%) mutation; O^6-CMdG causes G>A (6.4%) mutation; neither N^6-CMdA nor N^4-CMdC block replication or induce mutation [48]. O^4-Ethylthymidine (O^4-EtdT) does not strongly block replication (24%) in WT *E. coli*, but it cannot be efficiently bypassed in pol II/IV/V triple knock out cells [55]. The major mutation of O^4-EtdT is T>C (84%) transition; however, it does not cause mutations in *E. coli* lacking pol V [55]. O^4-Alkylthymidine (O^4-alkyldT) lesions moderately block DNA replication in human cells; pol ι and pol ζ promote the bypass of all O^4-alkyldT lesions except O^4-MedT [56]. The O^4-alkyldT lesions induce only T>C transition mutations in cells [56].

3-Methyldeoxycytidine (m3C) has been demonstrated to strongly block replication (>90%) and generate mainly C to T (50%) and C to A mutations (30%) in the AlkB- *E. coli* cell [9]. However, the lesion is not mutagenic and not blocked by the replicative polymerases in the WT (AlkB+) cell [9]. 3-Ethyldeoxycytidine (e3C) does not block replication in *E. coli*; however, it dramatically blocks replication when knocking out AlkB (91%) [9]. e3C causes 17% C>T, 11% C>A, and 2% C>G mutations in AlkB- *E. coli*, but is not mutagenic in WT cells [9]. The m3C, e3C, and m1A lesions presumably have their methyl or ethyl groups removed by AlkB's direct reversal of DNA alkyl damage mechanism prior to encountering the DNA polymerase [9]. N^4-carboxymethyl-2′-deoxycytidine (N^4-CMdC) weakly blocks replication (17%) and is not mutagenic in *E. coli* [36]. 5-Methylcytosine (5mC) and its derivatives 5-hydroxymethylcytosine (5hmC), 5-formylcytosine (5fC) and 5-carboxylcytosine (5caC) neither block replication nor cause mutation in *E. coli* [39,59]. 5mC also does not block replication in human cells, but there are some blockades of 5hmC (5%), 5fC (25%), and 5caC (28%) towards DNA replication in human cells [58]. 3,N^4-ethenocytosine (eC) is a toxic adduct, which strongly blocks replication (76%) and leads to mutation with a pattern of dominant C>A (24%) and less C>T (11%) mutations in WT *E. coli*; in AlkB- cells, the blockage of replication increases to 87% and mutagenicity rises up to 49% C>A and 31% C>T mutations [33]. Lipid peroxidation-derived product 4-oxo-2(*E*)-nonenal reacts with dG, dA, and dC in DNA to form heptanone (H)-etheno (e) adducts [50]. H-edC shows strong DNA replication blocking in both *E. coli* (99%) and human cells (90%) [50]. It causes mainly C>G (40%) mutation in *E. coli*; however, mostly C>A (60%) and C>T (32%) mutations are seen in human cells [50].

5-Hydroxymethyluracil (5hmU) blocks replication by 20%, but it is not mutagenic in human cells [60]. The S_p alkyl phosphotriester (S_p-alkyl-PTE) lesions display comparable replication bypass efficiency to unmodified DNA in *E. coli*; S_p-Me-PTE is mutagenic causing TT>GT (50%) and TT>GC (15%) mutations [61]. In contrast, R_p-alkyl-PTEs block DNA replication (30–70%) but are not mutagenic [61]. Interestingly, *n*Pr- and *n*Bu-PTEs exhibit higher bypass efficiencies than Me- and Et-PTEs [61].

2.3. Bulky Lesions

All the structures of modifications covered in this section are displayed in Figure 4. *N*-(deoxyguanosin-8-yl)-1-aminopyrene (C8-AP-dG) moderately blocks DNA replication in human cells [66]. *N*-acetyl-2-aminofluorene (C8-AAF-dG) strongly blocks replication [66]. 2-Aminofluorene (C8-AF-dG) slightly blocks replication [66]. All three adducts can be nearly bypassed in error free manner [66]. Aristolochic acids I and II (AA-I, AA-II) are found in all *Aristolochia* species and generate the aristolactam (AL) metabolite for forming DNA adducts with dA and dG. Both AL-II-dA and AL-II-dG strongly block DNA replication in MEF cells [63]. AL-II-dA causes 22% A>T mutation and AL-II-dG causes 9% G>T transversion [63]. Knocking out the rev3L gene dramatically suppresses bypass of AL-I-dA in MEF cells and abolishes A>T transversion [67]. Benzo[*a*]pyrene (BP)-7,8-diol-9,10-epoxide-N^2-deoxyguanosine (BPDE-dG) is an adduct formed by benzo[*a*]pyrene; it predominantly miscodes with G>T (73%) and G>A (12%) mutations in WT MEF cells [68]. Knocking out rev1 gene decreases the bypass efficiency of BPDE-dG to 40% and changes the mutation frequency to 32% G>T and 18% G>A [68]. Knocking out the rev3L gene significantly decreases the bypass efficiency to 13% and decreases the mutation to 6% G>T [68]. Mitomycin C (MC) generates dG-N2-MC and dG-N2-2,7-Diaminomitosene (DAM) adducts, which can be bypassed 38% and 27% in human cells, respectively [62]. The major type of mutation is G>T mutation (18% for dG-N2-MC and 10% for dG-N2-2,7-DAM) [62]. Aflatoxin B_1-N7-dG adduct (AFB$_1$-N7-dG) is weakly mutagenic in *E. coli*, causing 1.5% G>T mutation [64]; and its FAPY adduct causes 14% G>T mutation [65].

2.4. Crosslinked Lesions

All the structures of modifications covered in this section are displayed in Figure 4. N^2-guanine -N^2-guanine interstrand crosslinks (ICLs), 3-(2-deoxyribos-1-yl)-5,6,7,8-(N^2-deoxyguanosyl)-6(either R or S)-methylpyrimido[1,2-R]purine-10(3H)-one is a product induced by acetaldehyde/crotonaldehyde [69]. ICL-S and ICL-R moderately inhibit DNA replication in WT *E. coli*; however, their replication blocking effects increase in uvr- *E. coli* cells [69]. ICL-Rd is a moderate block in WT *E. coli*, but it almost completely blocks replication in uvr- cells [69]. All three lesions are weakly mutagenic in *E. coli* causing exclusively 5′-G>T (3%) transversions; no mutation is observed at the 3′-G site [69]. Similar mutations generated by these lesions are seen in human cells, except ICL-S has a slightly higher mutation frequency (6%) [69]. The crosslinks formed by *cis*-diaminedichloroplatinum (II) (*cis*-DDP, cisplatin) between two guanines or adenine-guanine strongly block DNA replication in *E. coli*, but they are not very mutagenic [72]. 5-Formylcytosine mediated peptide crosslink causes 7% C>T and 1% C>G mutation and 2% C deletion [73]. γ-Hydroxypropanodeoxyguanosine (γ-HOPdG) mediated crosslink between peptide and guanine is mutagenic, causing 5% G>T and 3% G>C mutations; however, the crosslink between peptide and γ-hydroxypropanodeoxyadenine (γ-HOPdA) is not mutagenic [20].

2.5. Other Nucleotide Analogs

All the structures of modifications covered in this section are displayed in Figure 5. A series of unnatural analogs of thymine (T) was developed by the Kool group to probe the biological requirements for DNA polymerases [74]. 3-Toluene-1-β-D-deoxyriboside (H) strongly blocks replication (95%) and is very mutagenic causing T>A (41%), T>C (5%), and T>G (4%) point mutations and −1 frame shift mutation (13%). 2,4-Difluoro-5-toluene-1-β-D-deoxyriboside (F) strongly blocks replication (87%) and is mutagenic causing T>A (9%), T>C (1%), and T>G (1%) mutations. 2,4-Dichloro-5-toluene-1-β-D-deoxyriboside (L) strongly blocks replication (80%) and is slightly mutagenic causing T>A (5%) mutation. 2,4-Dibromo-5-toluene-1-β-D-deoxyriboside

(B) strongly blocks replication (88%) and is mutagenic, causing T>A (24%) mutation. 2,4-Diiodo-5-toluene-1-β-D-deoxyriboside (I) strongly blocks replication (90%) and is very mutagenic causing T>A (46%), T>C (1%), and T>G (1%) point mutations and −1 frame shift mutation (6%) [74]. xG is an 'expanded base' of dG (retaining the hydrogen-bonding face), which strongly blocks replication (89%) and is very mutagenic, causing G>A (95%) mutation [75]. xA (expanded A) weakly blocks replication (20%) and is not mutagenic; xT (expanded T) weakly blocks replication (27%), but is very mutagenic, causing T>A (73%) mutation; xC (expanded dC) strongly blocks replication (71%) and is mutagenic, causing C>A (10%) mutation [75].

The α-anomer of deoxynucleosides (α-dN) can be generated as a result of hydroxyl radical attack on deoxyribose [76]. All α-dNs except α-dA strongly block replication in *E. coli* [76]. α-dC blocks almost 99% replication and causes 72% C>A mutation [76]. α-dG also strongly blocks replication and causes 60% G>A mutation [76]. α-dT blocks almost 99% replication but it is not mutagenic in WT *E. coli* [76]. α-dA is not mutagenic [76]. The anticancer agent 6-thioguanine (sG) and its derivative S^6-methylthioguanine (S^6mG) do not block replication strongly in both *E. coli* and human cells [78]. sG causes 11% G>A mutation and S^6mG causes 94% G>A mutation in *E. coli* [78]. sG is less mutagenic (8%) than S^6mG (40%) in human cells as well [78]. Guanine-S^6-sulfonic acid (SO_3HG) is another derivative of sG [78]. It is not a strong replication block in *E. coli*, but it is very mutagenic, causing 77% G>A mutation [78]. The anti-HIV drug KP1212 is an analog of deoxycytidine [57]. It does not block replication in *E. coli*, but is mutagenic causing 10% C>T mutation [57]. Among the four 2'-deoxyxylonucleosides (xN), only xA and xG exhibit a replication block in *E. coli* [77]. xA is the only mutagenic lesion among the four and causes 10% A>G mutation [77]. Base J strongly blocks replication by 48%, but is not mutagenic in human cells [60].

3. Perspectives

In this review, we survey the biological effects of various DNA lesions or biomarkers studied by the shuttle vector techniques, allowing one to gain insight into how DNA damage or other chemically defined nucleobases are processed by polymerases and repair machinery in a natural cellular environment under physiological conditions. Among the new methods that have been developed or applied in the last decade, MS-based strategies and NGS methods have been demonstrated to be efficient for analyzing the lesion's biological outcomes. LC-MS-based methods are sensitive and accurate for quantifying the degree of lesion bypass and point mutations [4,5]. NGS techniques allow for a large-scale population analysis on many samples at the same time and provide information on a genomic perspective [4,8]. Another possible direction for using vectors as probes to analyze biomarkers is to study the mutational spectrum or mutational signature of a certain chemical or damaging agent [83–86]. LC-MS- and NGS-based analyses not only consider the biological consequences at the lesion site, but also incorporate information from the neighboring bases, such as one or two nucleotides next to the lesion site from both the 5' and 3' direction. An oligonucleotide containing the modified base can be made surrounded by nearest (and next-to nearest) randomized bases and ligated into a shuttle vector. While cellular analysis may pull out a hotspot consensus sequence for poor repair and/or mutagenic replication, this will not answer the primary question of contextual bias in adduct formation. Shuttle vector systems whereby the vector is treated with the chemical to be assessed, followed by quantification of adduct type and amount, and transfection into isogenic cells of varying repair and/or replication backgrounds may tease apart the contribution of local sequence environment to adduct formation, repair, and replication. Such vectors were used over a decade ago [87], and coupled with NGS throughput and bioinformatics, may provide enough reads to make statistically significant claims. Shuttle vectors are currently, to our knowledge, mainly DNA-based; however, one can envision use of RNA-based vectors to study the effect of modified RNA bases on cellular processes such as viral replication, translation, reverse transcription, and possibly even repair. While the role of DNA damage in toxicology focuses mainly on the direct adduction of chemical damage to DNA, pool mutagenesis has often been overlooked, and it would be interesting to leverage shuttle vector techniques to study the incorporation of modified bases from the nucleotide pool in the form of damaged DNA or from DNA-based therapeutics.

Author Contributions: Conceptualization, K.B. and D.L.; writing—original draft preparation, K.B., J.C.D., X.Z. and D.L.; writing—review and editing, K.B., J.C.D., X.Z. and D.L.; supervision, D.L.; funding acquisition, D.L.

Funding: This work was supported by National Institutes of Health under grant numbers R15 CA213042 and R01 ES028865 (to D.L.).

Acknowledgments: The authors want to thank the RI-INBRE program and its director Bongsup Cho for their kind support.

Conflicts of Interest: The authors declare no conflict of interest.

References

1. Hwa Yun, B.; Guo, J.; Bellamri, M.; Turesky, R.J. DNA adducts: Formation, biological effects, and new biospecimens for mass spectrometric measurements in humans. *Mass Spectrom. Rev.* **2018**, 1–28. [CrossRef] [PubMed]
2. Delaney, J.C.; Essigmann, J.M. Biological properties of single chemical-DNA adducts: A twenty year perspective. *Chem. Res. Toxicol.* **2008**, *21*, 232–252. [CrossRef] [PubMed]
3. Miller, E.C. Some current perspectives on chemical carcinogenesis in humans and experimental animals: Presidential Address. *Cancer Res.* **1978**, *38*, 1479–1496.
4. Yu, Y.; Wang, P.; Cui, Y.; Wang, Y. Chemical Analysis of DNA Damage. *Anal. Chem.* **2018**, *90*, 556–576. [CrossRef] [PubMed]
5. You, C.; Wang, Y. Mass Spectrometry-Based Quantitative Strategies for Assessing the Biological Consequences and Repair of DNA Adducts. *Acc. Chem. Res.* **2016**, *49*, 205–213. [CrossRef]
6. Shrivastav, N.; Li, D.; Essigmann, J.M. Chemical biology of mutagenesis and DNA repair: Cellular responses to DNA alkylation. *Carcinogenesis* **2010**, *31*, 59–70. [CrossRef] [PubMed]
7. Delaney, J.C.; Essigmann, J.M. Assays for determining lesion bypass efficiency and mutagenicity of site-specific DNA lesions in vivo. *Meth. Enzymol.* **2006**, *408*, 1–15. [PubMed]
8. Chang, S.; Fedeles, B.I.; Wu, J.; Delaney, J.C.; Li, D.; Zhao, L.; Christov, P.P.; Yau, E.; Singh, V.; Jost, M.; et al. Next-generation sequencing reveals the biological significance of the N^2,3-ethenoguanine lesion in vivo. *Nucleic Acids Res.* **2015**, *43*, 5489–5500. [CrossRef]
9. Delaney, J.C.; Essigmann, J.M. Mutagenesis, genotoxicity, and repair of 1-methyladenine, 3-alkylcytosines, 1-methylguanine, and 3-methylthymine in alkB *Escherichia coli*. *Proc. Natl. Acad. Sci. USA* **2004**, *101*, 14051–14056. [CrossRef]
10. Loechler, E.L.; Green, C.L.; Essigmann, J.M. In vivo mutagenesis by O^6-methylguanine built into a unique site in a viral genome. *Proc. Natl. Acad. Sci. USA* **1984**, *81*, 6271–6275. [CrossRef]
11. Green, C.L.; Loechler, E.L.; Fowler, K.W.; Essigmann, J.M. Construction and characterization of extrachromosomal probes for mutagenesis by carcinogens: Site-specific incorporation of O^6-methylguanine into viral and plasmid genomes. *Proc. Natl. Acad. Sci. USA* **1984**, *81*, 13–17. [CrossRef] [PubMed]
12. Yang, I.-Y.; Johnson, F.; Grollman, A.P.; Moriya, M. Genotoxic mechanism for the major acrolein-derived deoxyguanosine adduct in human cells. *Chem. Res. Toxicol.* **2002**, *15*, 160–164. [CrossRef] [PubMed]
13. Yang, I.-Y.; Chan, G.; Miller, H.; Huang, Y.; Torres, M.C.; Johnson, F.; Moriya, M. Mutagenesis by acrolein-derived propanodeoxyguanosine adducts in human cells. *Biochemistry* **2002**, *41*, 13826–13832. [CrossRef] [PubMed]
14. Livneh, Z.; Cohen, I.S.; Paz-Elizur, T.; Davidovsky, D.; Carmi, D.; Swain, U.; Mirlas-Neisberg, N. High-resolution genomic assays provide insight into the division of labor between TLS and HDR in mammalian replication of damaged DNA. *DNA Repair (Amst.)* **2016**, *44*, 59–67. [CrossRef] [PubMed]
15. Ziv, O.; Diamant, N.; Shachar, S.; Hendel, A.; Livneh, Z. Quantitative measurement of translesion DNA synthesis in mammalian cells. *Methods Mol. Biol.* **2012**, *920*, 529–542. [PubMed]
16. Huang, H.; Greenberg, M.M. Hydrogen bonding contributes to the selectivity of nucleotide incorporation opposite an oxidized abasic lesion. *J. Am. Chem. Soc.* **2008**, *130*, 6080–6081. [CrossRef] [PubMed]
17. Greenberg, M.M. The formamidopyrimidines: Purine lesions formed in competition with 8-oxopurines from oxidative stress. *Acc. Chem. Res.* **2012**, *45*, 588–597. [CrossRef] [PubMed]

18. Bose, A.; Millsap, A.D.; DeLeon, A.; Rizzo, C.J.; Basu, A.K. Translesion Synthesis of the N^2-2′-Deoxyguanosine Adduct of the Dietary Mutagen IQ in Human Cells: Error-Free Replication by DNA Polymerase κ and Mutagenic Bypass by DNA Polymerases η, ζ, and Rev1. *Chem. Res. Toxicol.* **2016**, *29*, 1549–1559. [CrossRef] [PubMed]

19. Basu, A.K. DNA Damage, Mutagenesis and Cancer. *Int. J. Mol. Sci.* **2018**, *19*, 970. [CrossRef]

20. Minko, I.G.; Kozekov, I.D.; Kozekova, A.; Harris, T.M.; Rizzo, C.J.; Lloyd, R.S. Mutagenic potential of DNA-peptide crosslinks mediated by acrolein-derived DNA adducts. *Mutat. Res.* **2008**, *637*, 161–172. [CrossRef]

21. Fernandes, P.H.; Wang, H.; Rizzo, C.J.; Lloyd, R.S. Site-specific mutagenicity of stereochemically defined 1,N^2-deoxyguanosine adducts of trans-4-hydroxynonenal in mammalian cells. *Environ. Mol. Mutagen.* **2003**, *42*, 68–74. [CrossRef] [PubMed]

22. Benasutti, M.; Ezzedine, Z.D.; Loechler, E.L. Construction of an *Escherichia coli* vector containing the major DNA adduct of activated benzo[*a*]pyrene at a defined site. *Chem. Res. Toxicol.* **1988**, *1*, 160–168. [CrossRef] [PubMed]

23. Grueneberg, D.A.; Ojwang, J.O.; Benasutti, M.; Hartman, S.; Loechler, E.L. Construction of a human shuttle vector containing a single nitrogen mustard interstrand, DNA-DNA cross-link at a unique plasmid location. *Cancer Res.* **1991**, *51*, 2268–2272. [PubMed]

24. Fuchs, R.P.; Fujii, S. Translesion DNA synthesis and mutagenesis in prokaryotes. *Cold Spring Harb. Perspect. Biol.* **2013**, *5*, a012682. [CrossRef] [PubMed]

25. Fuchs, R.P.; Fujii, S. Translesion synthesis in *Escherichia coli*: Lessons from the NarI mutation hot spot. *DNA Repair (Amst.)* **2007**, *6*, 1032–1041. [CrossRef] [PubMed]

26. Pagès, V. Single-strand gap repair involves both RecF and RecBCD pathways. *Curr. Genet.* **2016**, *62*, 519–521. [CrossRef] [PubMed]

27. Pagès, V.; Fuchs, R.P. Inserting Site-Specific DNA Lesions into Whole Genomes. *Methods Mol. Biol.* **2018**, *1672*, 107–118. [PubMed]

28. Henderson, P.T.; Delaney, J.C.; Gu, F.; Tannenbaum, S.R.; Essigmann, J.M. Oxidation of 7,8-dihydro-8-oxoguanine affords lesions that are potent sources of replication errors in vivo. *Biochemistry* **2002**, *41*, 914–921. [CrossRef] [PubMed]

29. Delaney, S.; Neeley, W.L.; Delaney, J.C.; Essigmann, J.M. The substrate specificity of MutY for hyperoxidized guanine lesions in vivo. *Biochemistry* **2007**, *46*, 1448–1455. [CrossRef] [PubMed]

30. Pande, P.; Haraguchi, K.; Jiang, Y.-L.; Greenberg, M.M.; Basu, A.K. Unlike catalyzing error-free bypass of 8-oxodGuo, DNA polymerase λ is responsible for a significant part of Fapy·dG-induced G → T mutations in human cells. *Biochemistry* **2015**, *54*, 1859–1862. [CrossRef]

31. Patro, J.N.; Wiederholt, C.J.; Jiang, Y.L.; Delaney, J.C.; Essigmann, J.M.; Greenberg, M.M. Studies on the replication of the ring opened formamidopyrimidine, Fapy·dG in *Escherichia coli*. *Biochemistry* **2007**, *46*, 10202–10212. [CrossRef] [PubMed]

32. Neeley, W.L.; Delaney, J.C.; Henderson, P.T.; Essigmann, J.M. In vivo bypass efficiencies and mutational signatures of the guanine oxidation products 2-aminoimidazolone and 5-guanidino-4-nitroimidazole. *J. Biol. Chem.* **2004**, *279*, 43568–43573. [CrossRef] [PubMed]

33. Delaney, J.C.; Smeester, L.; Wong, C.; Frick, L.E.; Taghizadeh, K.; Wishnok, J.S.; Drennan, C.L.; Samson, L.D.; Essigmann, J.M. AlkB reverses etheno DNA lesions caused by lipid oxidation in vitro and in vivo. *Nat. Struct. Mol. Biol.* **2005**, *12*, 855–860. [CrossRef] [PubMed]

34. Henderson, P.T.; Delaney, J.C.; Muller, J.G.; Neeley, W.L.; Tannenbaum, S.R.; Burrows, C.J.; Essigmann, J.M. The hydantoin lesions formed from oxidation of 7,8-dihydro-8-oxoguanine are potent sources of replication errors in vivo. *Biochemistry* **2003**, *42*, 9257–9262. [CrossRef] [PubMed]

35. Henderson, P.T.; Neeley, W.L.; Delaney, J.C.; Gu, F.; Niles, J.C.; Hah, S.S.; Tannenbaum, S.R.; Essigmann, J.M. Urea lesion formation in DNA as a consequence of 7,8-dihydro-8-oxoguanine oxidation and hydrolysis provides a potent source of point mutations. *Chem. Res. Toxicol.* **2005**, *18*, 12–18. [CrossRef] [PubMed]

36. Yuan, B.; Wang, J.; Cao, H.; Sun, R.; Wang, Y. High-throughput analysis of the mutagenic and cytotoxic properties of DNA lesions by next-generation sequencing. *Nucleic Acids Res.* **2011**, *39*, 5945–5954. [CrossRef] [PubMed]

37. You, C.; Swanson, A.L.; Dai, X.; Yuan, B.; Wang, J.; Wang, Y. Translesion synthesis of 8,5′-cyclopurine-2′-deoxynucleosides by DNA polymerases η, ι, and ζ. *J. Biol. Chem.* **2013**, *288*, 28548–28556. [CrossRef] [PubMed]

38. Yuan, B.; Jiang, Y.; Wang, Y.; Wang, Y. Efficient formation of the tandem thymine glycol/8-oxo-7,8-dihydroguanine lesion in isolated DNA and the mutagenic and cytotoxic properties of the tandem lesions in *Escherichia coli* cells. *Chem. Res. Toxicol.* **2010**, *23*, 11–19. [CrossRef] [PubMed]

39. Fedeles, B.I.; Freudenthal, B.D.; Yau, E.; Singh, V.; Chang, S.; Li, D.; Delaney, J.C.; Wilson, S.H.; Essigmann, J.M. Intrinsic mutagenic properties of 5-chlorocytosine: A mechanistic connection between chronic inflammation and cancer. *Proc. Natl. Acad. Sci. USA* **2015**, *112*, E4571–E4580. [CrossRef]

40. Kreutzer, D.A.; Essigmann, J.M. Oxidized, deaminated cytosines are a source of C –> T transitions in vivo. *Proc. Natl. Acad. Sci. USA* **1998**, *95*, 3578–3582. [CrossRef] [PubMed]

41. Kroeger, K.M.; Jiang, Y.L.; Kow, Y.W.; Goodman, M.F.; Greenberg, M.M. Mutagenic effects of 2-deoxyribonolactone in *Escherichia coli*. An abasic lesion that disobeys the A-rule. *Biochemistry* **2004**, *43*, 6723–6733. [CrossRef] [PubMed]

42. Shrivastav, N.; Fedeles, B.I.; Li, D.; Delaney, J.C.; Frick, L.E.; Foti, J.J.; Walker, G.C.; Essigmann, J.M. A chemical genetics analysis of the roles of bypass polymerase DinB and DNA repair protein AlkB in processing N^2-alkylguanine lesions in vivo. *PLoS ONE* **2014**, *9*, e94716. [CrossRef] [PubMed]

43. Yuan, B.; You, C.; Andersen, N.; Jiang, Y.; Moriya, M.; O'Connor, T.R.; Wang, Y. The roles of DNA polymerases κ and ι in the error-free bypass of N^2-carboxyalkyl-2′-deoxyguanosine lesions in mammalian cells. *J. Biol. Chem.* **2011**, *286*, 17503–17511. [CrossRef] [PubMed]

44. Yuan, B.; Cao, H.; Jiang, Y.; Hong, H.; Wang, Y. Efficient and accurate bypass of N^2-(1-carboxyethyl)-2′-deoxyguanosine by DinB DNA polymerase in vitro and in vivo. *Proc. Natl. Acad. Sci. USA* **2008**, *105*, 8679–8684. [CrossRef] [PubMed]

45. Delaney, J.C.; Essigmann, J.M. Context-dependent mutagenesis by DNA lesions. *Chem. Biol.* **1999**, *6*, 743–753. [CrossRef]

46. Delaney, J.C.; Essigmann, J.M. Effect of sequence context on O^6-methylguanine repair and replication in vivo. *Biochemistry* **2001**, *40*, 14968–14975. [CrossRef] [PubMed]

47. Wang, P.; Leng, J.; Wang, Y. DNA replication studies of *N*-nitroso compound-induced O^6-alkyl-2′-deoxyguanosine lesions in *Escherichia coli*. *J. Biol. Chem.* **2019**, *294*, 3899–3908. [CrossRef] [PubMed]

48. Wu, J.; Wang, P.; Li, L.; Williams, N.L.; Ji, D.; Zahurancik, W.J.; You, C.; Wang, J.; Suo, Z.; Wang, Y. Replication studies of carboxymethylated DNA lesions in human cells. *Nucleic Acids Res.* **2017**, *45*, 7276–7284. [CrossRef] [PubMed]

49. Frick, L.E.; Delaney, J.C.; Wong, C.; Drennan, C.L.; Essigmann, J.M. Alleviation of 1,N^6-ethanoadenine genotoxicity by the *Escherichia coli* adaptive response protein AlkB. *Proc. Natl. Acad. Sci. USA* **2007**, *104*, 755–760. [CrossRef] [PubMed]

50. Pollack, M.; Yang, I.-Y.; Kim, H.-Y.H.; Blair, I.A.; Moriya, M. Translesion DNA Synthesis across the heptanone–etheno-2′-deoxycytidine adduct in cells. *Chem. Res. Toxicol.* **2006**, *19*, 1074–1079. [CrossRef] [PubMed]

51. Chang, S.-C.; Seneviratne, U.I.; Wu, J.; Tretyakova, N.; Essigmann, J.M. 1,3-Butadiene-Induced Adenine DNA Adducts Are Genotoxic but Only Weakly Mutagenic When Replicated in *Escherichia coli* of Various Repair and Replication Backgrounds. *Chem. Res. Toxicol.* **2017**, *30*, 1230–1239. [CrossRef] [PubMed]

52. Wu, J.; Wang, P.; Li, L.; You, C.; Wang, Y. Cytotoxic and mutagenic properties of minor-groove O^2-alkylthymidine lesions in human cells. *J. Biol. Chem.* **2018**, *293*, 8638–8644. [CrossRef] [PubMed]

53. Weerasooriya, S.; Jasti, V.P.; Bose, A.; Spratt, T.E.; Basu, A.K. Roles of translesion synthesis DNA polymerases in the potent mutagenicity of tobacco-specific nitrosamine-derived O^2-alkylthymidines in human cells. *DNA Repair (Amst.)* **2015**, *35*, 63–70. [CrossRef] [PubMed]

54. Jasti, V.P.; Spratt, T.E.; Basu, A.K. Tobacco-specific nitrosamine-derived O^2-alkylthymidines are potent mutagenic lesions in SOS-induced *Escherichia coli*. *Chem. Res. Toxicol.* **2011**, *24*, 1833–1835. [CrossRef] [PubMed]

55. Zhai, Q.; Wang, P.; Wang, Y. Cytotoxic and mutagenic properties of regioisomeric O^2-, N3- and O^4-ethylthymidines in bacterial cells. *Carcinogenesis* **2014**, *35*, 2002–2006. [CrossRef] [PubMed]

56. Wu, J.; Li, L.; Wang, P.; You, C.; Williams, N.L.; Wang, Y. Translesion synthesis of O^4-alkylthymidine lesions in human cells. *Nucleic Acids Res.* **2016**, *44*, 9256–9265.

57. Li, D.; Fedeles, B.I.; Singh, V.; Peng, C.S.; Silvestre, K.J.; Simi, A.K.; Simpson, J.H.; Tokmakoff, A.; Essigmann, J.M. Tautomerism provides a molecular explanation for the mutagenic properties of the anti-HIV nucleoside 5-aza-5,6-dihydro-2'-deoxycytidine. *Proc. Natl. Acad. Sci. USA* **2014**, *111*, E3252–E3259. [CrossRef]

58. Ji, D.; You, C.; Wang, P.; Wang, Y. Effects of tet-induced oxidation products of 5-methylcytosine on DNA replication in mammalian cells. *Chem. Res. Toxicol.* **2014**, *27*, 1304–1309. [CrossRef]

59. Xing, X.-W.; Liu, Y.-L.; Vargas, M.; Wang, Y.; Feng, Y.-Q.; Zhou, X.; Yuan, B.-F. Mutagenic and cytotoxic properties of oxidation products of 5-methylcytosine revealed by next-generation sequencing. *PLoS ONE* **2013**, *8*, e72993. [CrossRef]

60. Ji, D.; Wang, Y. Facile enzymatic synthesis of base J-containing oligodeoxyribonucleotides and an analysis of the impact of base J on DNA replication in cells. *PLoS ONE* **2014**, *9*, e103335. [CrossRef]

61. Wu, J.; Wang, P.; Wang, Y. Cytotoxic and mutagenic properties of alkyl phosphotriester lesions in *Escherichia coli* cells. *Nucleic Acids Res.* **2018**, *46*, 4013–4021. [CrossRef] [PubMed]

62. Bose, A.; Surugihalli, C.; Pande, P.; Champeil, E.; Basu, A.K. Comparative Error-Free and Error-Prone Translesion Synthesis of N^2-2'-Deoxyguanosine Adducts Formed by Mitomycin C and Its Metabolite, 2,7-Diaminomitosene, in Human Cells. *Chem. Res. Toxicol.* **2016**, *29*, 933–939. [CrossRef] [PubMed]

63. Attaluri, S.; Bonala, R.R.; Yang, I.-Y.; Lukin, M.A.; Wen, Y.; Grollman, A.P.; Moriya, M.; Iden, C.R.; Johnson, F. DNA adducts of aristolochic acid II: Total synthesis and site-specific mutagenesis studies in mammalian cells. *Nucleic Acids Res.* **2010**, *38*, 339–352. [CrossRef] [PubMed]

64. Bailey, E.A.; Iyer, R.S.; Stone, M.P.; Harris, T.M.; Essigmann, J.M. Mutational properties of the primary aflatoxin B1-DNA adduct. *Proc. Natl. Acad. Sci. USA* **1996**, *93*, 1535–1539. [CrossRef] [PubMed]

65. Smela, M.E.; Hamm, M.L.; Henderson, P.T.; Harris, C.M.; Harris, T.M.; Essigmann, J.M. The aflatoxin B1 formamidopyrimidine adduct plays a major role in causing the types of mutations observed in human hepatocellular carcinoma. *Proc. Natl. Acad. Sci. USA* **2002**, *99*, 6655–6660. [CrossRef] [PubMed]

66. Watt, D.L.; Utzat, C.D.; Hilario, P.; Basu, A.K. Mutagenicity of the 1-nitropyrene-DNA adduct *N*-(deoxyguanosin-8-yl)-1-aminopyrene in mammalian cells. *Chem. Res. Toxicol.* **2007**, *20*, 1658–1664. [CrossRef] [PubMed]

67. Hashimoto, K.; Bonala, R.; Johnson, F.; Grollman, A.P.; Moriya, M. Y-family DNA polymerase-independent gap-filling translesion synthesis across aristolochic acid-derived adenine adducts in mouse cells. *DNA Repair (Amst.)* **2016**, *46*, 55–60. [CrossRef]

68. Hashimoto, K.; Cho, Y.; Yang, I.-Y.; Akagi, J.; Ohashi, E.; Tateishi, S.; de Wind, N.; Hanaoka, F.; Ohmori, H.; Moriya, M. The Vital Role of Polymerase ζ and REV1 in Mutagenic, but Not Correct, DNA Synthesis across Benzo[*a*]pyrene-dG and Recruitment of Polymerase ζ by REV1 to Replication-stalled Site. *J. Biol. Chem.* **2012**, *287*, 9613–9622. [CrossRef] [PubMed]

69. Liu, X.; Lao, Y.; Yang, I.-Y.; Hecht, S.S.; Moriya, M. Replication-coupled repair of crotonaldehyde/ acetaldehyde-induced guanine-guanine interstrand cross-links and their mutagenicity. *Biochemistry* **2006**, *45*, 12898–12905. [CrossRef]

70. Price, N.E.; Li, L.; Gates, K.S.; Wang, Y. Replication and repair of a reduced 2'-deoxyguanosine-abasic site interstrand cross-link in human cells. *Nucleic Acids Res.* **2017**, *45*, 6486–6493. [CrossRef]

71. Naser, L.J.; Pinto, A.L.; Lippard, S.J.; Essigmann, J.M. Chemical and biological studies of the major DNA adduct of *cis*-diamminedichloroplatinum(II), *cis*-[Pt(NH$_3$)$_2$(d(GpG)], built into a specific site in a viral genome. *Biochemistry* **1988**, *27*, 4357–4367. [CrossRef] [PubMed]

72. Yarema, K.J.; Lippard, S.J.; Essigmann, J.M. Mutagenic and genotoxic effects of DNA adducts formed by the anticancer drug *cis*-diamminedichloroplatinum(II). *Nucleic Acids Res.* **1995**, *23*, 4066–4072. [CrossRef] [PubMed]

73. Ji, S.; Fu, I.; Naldiga, S.; Shao, H.; Basu, A.K.; Broyde, S.; Tretyakova, N.Y. 5-Formylcytosine mediated DNA-protein cross-links block DNA replication and induce mutations in human cells. *Nucleic Acids Res.* **2018**, *46*, 6455–6469. [CrossRef] [PubMed]

74. Kim, T.W.; Delaney, J.C.; Essigmann, J.M.; Kool, E.T. Probing the active site tightness of DNA polymerase in subangstrom increments. *Proc. Natl. Acad. Sci. USA* **2005**, *102*, 15803–15808. [CrossRef] [PubMed]

75. Delaney, J.C.; Gao, J.; Liu, H.; Shrivastav, N.; Essigmann, J.M.; Kool, E.T. Efficient replication bypass of size-expanded DNA base pairs in bacterial cells. *Angew. Chem. Int. Ed. Engl.* **2009**, *48*, 4524–4527. [CrossRef] [PubMed]

76. Amato, N.J.; Zhai, Q.; Navarro, D.C.; Niedernhofer, L.J.; Wang, Y. In vivo detection and replication studies of α-anomeric lesions of 2′-deoxyribonucleosides. *Nucleic Acids Res.* **2015**, *43*, 8314–8324. [CrossRef] [PubMed]

77. Wang, P.; Amato, N.J.; Wang, Y. Cytotoxic and Mutagenic Properties of C3′-Epimeric Lesions of 2′-Deoxyribonucleosides in *Escherichia coli* Cells. *Biochemistry* **2017**, *56*, 3725–3732. [CrossRef]

78. Yuan, B.; Wang, Y. Mutagenic and cytotoxic properties of 6-thioguanine, S^6-methylthioguanine, and guanine-S^6-sulfonic acid. *J. Biol. Chem.* **2008**, *283*, 23665–23670. [CrossRef]

79. Yuan, B.; O'Connor, T.R.; Wang, Y. 6-Thioguanine and S^6-methylthioguanine are mutagenic in human cells. *ACS Chem. Biol.* **2010**, *5*, 1021–1027. [CrossRef]

80. Reuven, N.B.; Tomer, G.; Livneh, Z. The mutagenesis proteins UmuD′ and UmuC prevent lethal frameshifts while increasing base substitution mutations. *Mol. Cell* **1998**, *2*, 191–199. [CrossRef]

81. Stein, S.; Lao, Y.; Yang, I.-Y.; Hecht, S.S.; Moriya, M. Genotoxicity of acetaldehyde- and crotonaldehyde-induced $1,N^2$-propanodeoxyguanosine DNA adducts in human cells. *Mutat. Res.* **2006**, *608*, 1–7. [CrossRef]

82. Zhai, Q.; Wang, P.; Cai, Q.; Wang, Y. Syntheses and characterizations of the in vivo replicative bypass and mutagenic properties of the minor-groove O^2-alkylthymidine lesions. *Nucleic Acids Res.* **2014**, *42*, 10529–10537. [CrossRef]

83. Fedeles, B.I.; Essigmann, J.M. Impact of DNA lesion repair, replication and formation on the mutational spectra of environmental carcinogens: Aflatoxin B1 as a case study. *DNA Repair (Amst.)* **2018**, *71*, 12–22. [CrossRef]

84. Chawanthayatham, S.; Valentine, C.C.; Fedeles, B.I.; Fox, E.J.; Loeb, L.A.; Levine, S.S.; Slocum, S.L.; Wogan, G.N.; Croy, R.G.; Essigmann, J.M. Mutational spectra of aflatoxin B1 in vivo establish biomarkers of exposure for human hepatocellular carcinoma. *Proc. Natl. Acad. Sci. USA* **2017**, *114*, E3101–E3109. [CrossRef]

85. Alexandrov, L.B.; Stratton, M.R. Mutational signatures: The patterns of somatic mutations hidden in cancer genomes. *Curr. Opin. Genet. Dev.* **2014**, *24*, 52–60. [CrossRef]

86. Stratton, M.R. Exploring the genomes of cancer cells: Progress and promise. *Science* **2011**, *331*, 1553–1558. [CrossRef]

87. Kim, M.Y.; Zhou, X.; Delaney, J.C.; Taghizadeh, K.; Dedon, P.C.; Essigmann, J.M.; Wogan, G.N. AlkB influences the chloroacetaldehyde-induced mutation spectra and toxicity in the pSP189 supF shuttle vector. *Chem. Res. Toxicol.* **2007**, *20*, 1075–1083.

Review

Formalin-Fixed Paraffin-Embedded Tissues—An Untapped Biospecimen for Biomonitoring DNA Adducts by Mass Spectrometry

Byeong Hwa Yun, Jingshu Guo and Robert J. Turesky *

Masonic Cancer Center and Department of Medicinal Chemistry, University of Minnesota, 2231 6th St. SE, Minneapolis, MN 55455, USA; bhyun@umn.edu (B.H.Y.); guoj@umn.edu (J.G.)
* Correspondence: rturesky@umn.edu; Tel.: +1-612-626-0141; Fax: +1-612-624-3869

Received: 28 April 2018; Accepted: 25 May 2018; Published: 1 June 2018

Abstract: The measurement of DNA adducts provides important information about human exposure to genotoxic chemicals and can be employed to elucidate mechanisms of DNA damage and repair. DNA adducts can serve as biomarkers for interspecies comparisons of the biologically effective dose of procarcinogens and permit extrapolation of genotoxicity data from animal studies for human risk assessment. One major challenge in DNA adduct biomarker research is the paucity of fresh frozen biopsy samples available for study. However, archived formalin-fixed paraffin-embedded (FFPE) tissues with clinical diagnosis of disease are often available. We have established robust methods to recover DNA free of crosslinks from FFPE tissues under mild conditions which permit quantitative measurements of DNA adducts by liquid chromatography-mass spectrometry. The technology is versatile and can be employed to screen for DNA adducts formed with a wide range of environmental and dietary carcinogens, some of which were retrieved from section-cuts of FFPE blocks stored at ambient temperature for up to nine years. The ability to retrospectively analyze FFPE tissues for DNA adducts for which there is clinical diagnosis of disease opens a previously untapped source of biospecimens for molecular epidemiology studies that seek to assess the causal role of environmental chemicals in cancer etiology.

Keywords: carcinogen; DNA adducts; biomonitoring; formalin-fixed paraffin-embedded tissues; biomarker; mass spectrometry

---

1. Metabolism, Bioactivation, and DNA Adducts as Biomarkers of Exposure and Health Risk

1.1. Xenobiotic Metabolism and Bioactivation of Procarcinogens

Humans are continuously exposed to potentially hazardous chemicals in the environment, diet, medicines, and through occupational exposures. Many of these chemicals undergo biotransformation by phase I and/or phase II enzymes to produce reactive electrophiles that can form adducts with macromolecules [1]. Cytochrome P450s (P450s) are by far the most important Phase I enzymes involved in xenobiotic metabolism [2]. P450s catalyze a variety of reactions, including aliphatic and aromatic hydroxylation, *N*- or *O*-dealkylation, aliphatic desaturation, hetero atom oxidation, and epoxidation reactions [2]. The resulting metabolites can contain functional groups such as –OH, –NH$_2$, and –SH which can undergo conjugation reactions by phase II enzymes, including UDP-glucuronosyltransferases (UGTs), sulfotransferases (SULTs), *N*-acetyltransferases (NATs), glutathione *S*-transferases (GSTs), and methyltransferases [3].

While many Phase I metabolites are detoxification products, some oxidative metabolites are reactive electrophiles, which can induce toxicity or genotoxicity by covalently binding to protein or DNA, or generate free radicals that deplete cellular antioxidants and induce oxidative stress [4,5].

In a similar vein, many phase II enzyme reactions are regarded as detoxification pathways, and the resulting metabolites are efficiently eliminated from the body. However, in some cases, reactive intermediates are generated, and the metabolites can bind to proteins and DNA. The *O*-acetylation or *O*-sulfation of aromatic amines and heterocyclic aromatic hydroxylamines [6], glutathione conjugation of ethylene dibromide [7], *O*-sulfation of hydroxymethyl polycyclic aromatic hydrocarbons [8], and the acyl glucuronidation of carboxylic acid moieties of nonsteroidal anti-inflammatory drugs (NSAIDs) [9] are examples of conjugation reactions leading to reactive intermediates. The metabolic activation of rodent and possible human carcinogens including 2-amino-1-methyl-6-phenylimidazo[4,5-*b*]pyridine (PhIP) [10], aristolochic acid I (AA-I) [11], 5-methylchrysene [12,13] and tamoxifen [14,15], are shown as examples of procarcinogens that require phase I and/or II enzymes to produce penultimate species that bind to DNA (Figure 1).

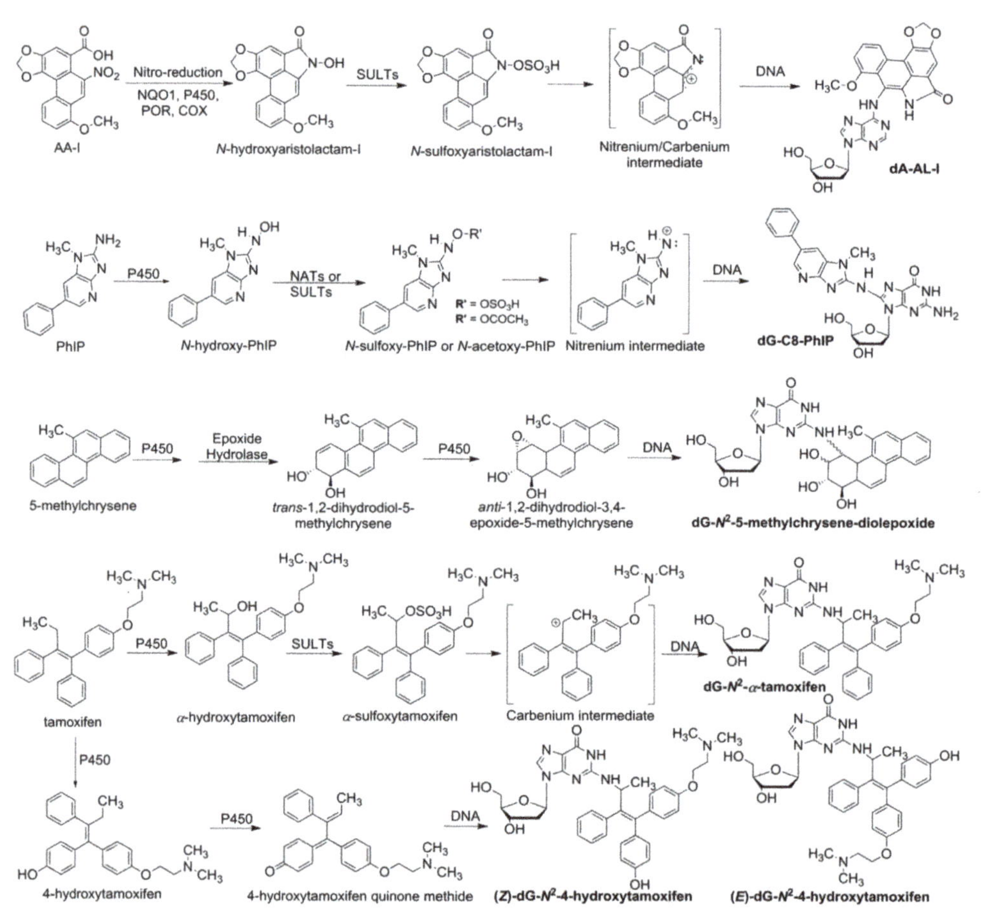

Figure 1. The metabolic activation of aristolochic acid I (AA-I), 2-amino-1-methyl-6-phenylimidazo[4,5-*b*]pyridine (PhIP), 5-methylchrysene, and tamoxifen are shown as prototypes of procarcinogens. Bioactivation is carried out with phase I and/or phase II enzymes, which lead to the formation of DNA adducts. AA-I undergoes nitro-reduction through NAD(P)H:quinone oxidoreductase (NQO1), cytochrome P450s 1A1 and 1A2, NADPH:P450 reductase (POR) or prostaglandin H synthase (COX).

The resulting *N*-hydroxyaristolactam-I is bioactivated by SULTs to form an unstable *N*-sulfoxy ester, which quickly undergoes heterolytic cleavage to produce the reactive nitrenium/carbenium intermediate that forms dA-AL-I and other DNA adducts. PhIP undergoes *N*-hydroxylation by P450s, then it is further bioactivated by NATs or SULTs to form *N*-acetoxy or *N*-sulfoxy esters, which lead to the formation of dG-C8-PhIP through the nitrenium intermediate. 5-Methylchrysene undergoes epoxidation (P450s 1A1 and 1B1) followed by epoxide hydroxylation (epoxide hydrolase) on the bay-region phenyl ring, to form the corresponding *trans*-1,2-dihydrodiol-5-methylchrysene. A subsequent round of monooxygenation leads to the formation of *anti*-1,2-dihydrodiol-3,4-epoxide-5-methylchrysene, which can form a DNA adduct at the N^2-atom of dG (dG-N^2-5-methylchrysene-diolepoxide). Two pathways are involved in the DNA adduct formation of the bioactivated tamoxifen. In the first pathway, oxidation of the allylic ethyl side chain results in the formation of α-hydroxytamoxifen. The subsequent esterification catalyzed by SULTs leads to the reactive carbenium intermediate and the dG-N^2-α-hydroxytamoxifen adduct. The second pathway involves aryl-oxidation of one of the phenyl rings to yield 4-hydroxytamoxifen quinone methide, a reactive electrophile that can form the DNA adducts. Both pathways lead to (*Z*)- or (*E*)-dG-N^2-4-hydroxytamoxifen.

Rodents are often employed as experimental laboratory animals to study metabolism of hazardous chemicals, to screen for DNA adduct formation, and elucidate mechanisms of carcinogenesis [5]. The metabolism of carcinogens and their biological effects in animal models can differ from humans because of species differences in catalytic activities of phase I and II enzymes involved in bioactivation or detoxification [10,16–18]. Thus, animal carcinogen bioassay data may not accurately gauge health risk of some chemicals in humans. However, DNA adducts of carcinogens, which are measures of the biologically effective dose, can serve as biomarkers for the extrapolation of genotoxicity data from animal studies for human risk assessment [19,20].

Epidemiological studies have reported that exposures to different chemicals in the diet and environment, or lifestyle factors, such as tobacco usage and alcohol consumption, are linked to the increased risk of developing certain types of cancers. As examples, polycyclic aromatic hydrocarbons (PAHs) in cigarette smoke are linked to lung cancer [21]; occupational exposures to aromatic amines are linked to bladder cancer [21,22]; usage of traditional Chinese herbal medicines containing AA-I are linked to upper urothelial cancer [23,24]; and consumption of aflatoxin B_1 (AFB$_1$) produced by fungi on agricultural crops, is a risk factor for liver cancer [25,26].

The identification and quantitation of DNA adducts is a first step in elucidating the potential role of a genotoxic chemical in the etiology of cancer [19,20]. The identification DNA adducts in human tissues are likely to represent a combination of recent and longer-term exposures to certain hazardous chemicals. The interpretation of negative findings, or the absence of DNA adducts, must be done with caution, since many adducts can undergo repair [27]. Ideally, the biomonitoring of DNA adducts should be conducted when the multistage process of tumorigenesis began, rather than many years later when the cancer is diagnosed. However, life-style factors such as tobacco smoking, diet, and environmental pollution often represent long-term exposures, and current adduct levels of carcinogens from these exposures are likely to correlate with adduct levels that existed during the time of tumor initiation and progression.

1.2. Methods to Measure DNA Adducts

The measurement of DNA adducts in humans is a challenging analytical task because the levels of DNA adducts generally occur at less than one adduct per 10^7 nucleotides, and the amount of tissue available for measurement is limited. Even for blood, a readily accessible biofluid, the amount of DNA obtained is usually a few up to several tens of micrograms scale. Thus, highly sensitive and specific methods are required to measure DNA adducts in humans. During the past three decades, the major techniques employed to measure DNA adducts have been ^{32}P-postlabeling [28,29], antibody-based immunoassay/immunohistochemistry

(IHC) [30,31], gas chromatography-mass spectrometry (GC-MS) [32], and most recently, liquid chromatography-mass spectrometry (LC-MS) [33–37].

[32]P-postlabeling is a highly sensitive method to detect DNA adducts. The DNA is enzymatically digested to 3′-phospho-2′-deoxyribonucleotides, and ^{32}P-orthophosphate from [γ-^{32}P] ATP is transferred to the 5′-OH position of the 2′-deoxyribonucleotide adduct, by polynucleotide kinase. The adducted 5′-^{32}P-labeled nucleotides are resolved by multi-dimensional thin-layer chromatography with polyethylenimine-modified cellulose plate, or by polyacrylamide electrophoresis, using autoradiography for detection, or by HPLC with radiometric detection [28,29,38,39]. The assay only requires 1–10 μg of DNA, and the sensitivity for some adducts can reach a limit of detection as low as one adduct per 10^{10} nucleotides [29]. Studies in rodents and humans employing ^{32}P-postlabeling methods have shown that many genotoxic chemicals undergo metabolism and covalently adduct to DNA in many organs [29,40,41]. However, there are several limitations of the ^{32}P-postlabeling assay. The technique is labor intensive and its usage requires large amounts of hazardous phosphorous radioactivity. Moreover, the technique is not quantitative [42], and structural information about the identity of the adduct is uncertain, particularly in humans where many overlapping lesions are present [29,40]. Thus, epidemiology studies employing ^{32}P-postlabeling often provide equivocal data about chemical exposures linked to DNA adducts and cancer risk [43–46].

Immunodetection relies on the generation of monoclonal or polyclonal antibodies raised against modified-DNA adducts coupled to carrier proteins, or carcinogen-treated DNA, where usually very high levels of modification, about one modified base to 100 nucleotides, are required for successful generation of a titer [30,47]. The sensitivity of the method depends on the affinity of the antibody, but a detection limit of about one adduct per 10^8 nucleotides for certain DNA adducts can be reached, when detected by fluorescence or chemiluminescence spectroscopy [48,49]. IHC detection of DNA adducts in tissue section-cuts mounted on slides is generally less sensitive than immunoassays performed on isolated DNA; however, IHC allows the visualization of the DNA adduct within specific cell types of a tissue, and is especially suitable for archived human formalin-fixed paraff in-embedded (FFPE) tissues (Section 3) [50]. Cross-reaction of the antibody with DNA adducts of similar structure or cellular components can occur [30,31], which raises concerns about the specificity of the methodology. Immunodetection methods have made significant contributions to the biomonitoring of DNA adducts; however, similarly to the ^{32}P-postlabeling method, immunodetection does not provide structural information to confirm adduct identity, and the method is semi-quantitative.

GC-MS with electron impact ionization, and more recently, negative ion chemical ionization has been employed to measure DNA adducts (primarily used for oxidized DNA bases) where adduct structures can be corroborated from the MS fragmentation spectra [32]. Often, the DNA is hydrolyzed with formic acid or by elevated temperature under neutral pH conditions. Most DNA adducts require chemical derivatization to increase the volatility required for GC analysis. The derivatization process can complicate the analysis and introduce artifact formation, particularly for oxidized DNA base measurements [51]. In contrast, the online coupling of capillary electrophoresis or LC to electrospray ionization (ESI) MS is a breakthrough technology that can measure many DNA adducts which would otherwise undergo thermal decomposition by GC-MS [52].

Currently, LC-ESI-multistage MS (MS^n) is the predominant platform for DNA adduct analyses [33,35,37,53]. The rapidly advancing technologies in LC-MS instrumentation have attained ultra-high sensitivity and selectivity, particularly with ion trap and high resolution accurate mass spectrometry (HRAMS). These platforms include the coupling of nano-flow chromatography and nanoESI source, and versatile and flexible scanning strategies. The detection of DNA adducts at levels as low as one per 10^{11} nucleotides have been reported using a hybrid Orbitrap MS [54]. Both targeted and non-targeted MS scan approaches have been employed to identify many DNA chemical modifications [35–37,55–58].

The DNA is typically digested with a cocktail of nucleases prior to adduct measurements by LC-MS. The digestion products contain adducts formed at the DNA bases of the

2′-deoxyribonucleosides, or in rarer cases, adducts are formed at the phosphate backbones [59,60]. A common feature for many DNA adducts is their tendency to lose the deoxyribose moiety (dR, 116 or 116.0473 Da in HRAMS), when subjected to collision-induced dissociation (CID) [61]. The transition between the adduct precursors ([M + H]$^+$) and their aglycones after losing dR ([M + H − 116]$^+$) is commonly targeted to detect and quantify DNA adducts in MS^n. The constant neutral loss of molecules, such as dR from the 2′-deoxyribonucleosides, serves as the foundation of the "DNA adductomics" approach [37,55,56,62,63]. Figure 2 shows the fragmentation pathways of modified nucleosides, where the major ions are the chemically modified bases after neutral loss of dR, or in less frequent cases the bases are eliminated as the neutral fragment and the carcinogen moieties retain the charge. These types of MS transitions are usually monitored in the targeted and un-targeted approaches by LC-MS [58,63].

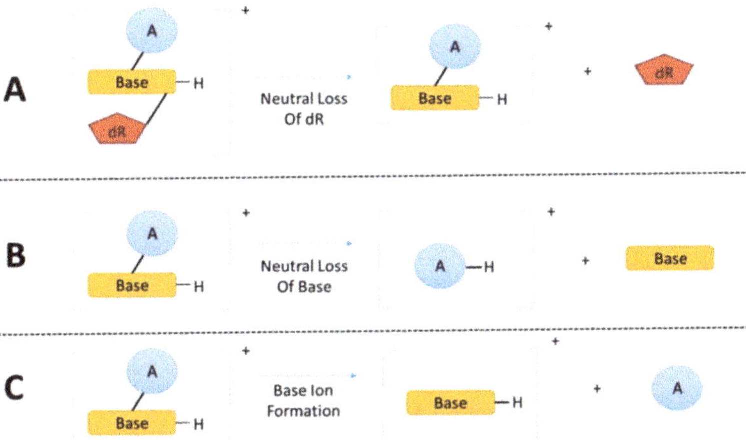

Figure 2. The fragmentation pathways of modified nucleosides analyzed by LC-MS. (**A**) The major fragmentation of the modified nucleosides is the neutral loss of deoxyribose. Other common fragmentations include (**B**) the neutral loss of base and (**C**) the neutral loss of the adduct with the formation of base ions [58].

2. Overview and the History of Formalin Fixation Process

While great strides have been made in the detection of DNA adducts in humans, fresh tissues obtained from biopsies or post-mortem are often not available. The paucity of fresh tissue specimens has hampered the advancement of DNA adduct biomonitoring in human studies. However, archived FFPE tissue specimens with clinical diagnosis of disease are a largely untapped biospecimen and often available for DNA adduct biomarker research.

Formalin, 10% neutral buffered formaldehyde solution, is the most commonly used fixative worldwide [64]. During the process of formalin fixation, formaldehyde undergoes multiple steps of reactions with cellular nucleophilic species to form molecular crosslinks [64,65]. Formaldehyde permeates through the tissue, and the nucleophilic moieties of amino acids and nucleobases attack the formaldehyde yielding unstable intermediates of methylol adducts and Schiff bases [65]. These intermediates are stabilized by forming methylene bridges with a second nucleophilic group, often on another molecule. The methylene bridges formed with DNA and protein are stable crosslinks at room temperature (Figure 3); however, the linkages are reversible by heat treatment and/or under alkaline pH [66,67]. The reversal rate of the crosslink increases exponentially as a function of temperature [66]. The efficacy of reversal of formaldehyde-mediated crosslinks is the most critical feature that impacts the quantitative analysis of RNA, DNA, and protein biomarkers in FFPE tissues.

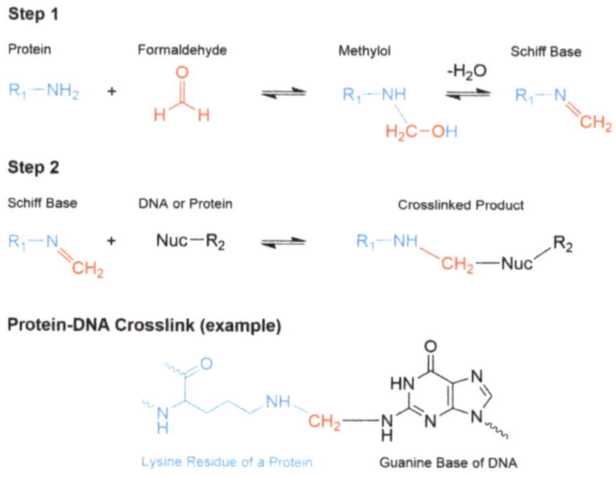

Figure 3. The reactions of formaldehyde mediated crosslinking of DNA and protein. Formaldehyde diffuses through tissue and reacts with a nucleophilic sites of protein and/or DNA base resulting in unstable intermediates of methylol and Schiff base. Then, a second nucleophile from inter- or intramolecular DNA or protein attacks the Schiff base resulting in a crosslinked complex. A specific example of a protein-DNA crosslink is shown. The atoms are color coded: *cyan*, protein; *red*, formaldehyde; and *black*, DNA. Reproduced with permission from [65]. Copyright ASBMB, 2015.

Technical Challenges and Breakthrough Technology in DNA Recovery from FFPE Tissues

FFPE tissues are now widely used in high throughput genomic [64,68–71], proteomics [72–74], and to a lesser extent, metabolomics studies [75,76]. The crucial step in these applications is the quantitative extraction of the molecules of interest. In genomic sequencing studies, the conventional method of DNA isolation from FFPE tissues has often employed elevated temperature (up to 100 °C) and alkaline pH (>9) to achieve a complete reversal of crosslinks. Many automated methods employed in cancer genomics still use elevated temperature to isolate DNA from FFPE tissues. The recovered DNA can serve as a template for PCR amplification. However, these harsh conditions can cause oxidation of nucleobases or depurination of chemically-modified nucleobases, and thus, are not compatible for quantitative measurements of DNA adducts. Moreover, even though formalin-fixation is the most common method of tissue preservation world-wide, the conditions of fixation can vary in different laboratories. A prolonged time of tissue preservation in formalin results in over-fixation of the tissue and leads to inefficient hydrolysis of crosslinks between DNA and protein. Therefore, the yield and quality of the recovered DNA is decreased [77–79]. Thus, the development of robust analytical methods to quantitatively recover DNA adducts from FFPE tissue has been a challenging endeavor.

3. Measurement of DNA Adducts in FFPE Tissues by IHC, ^{32}P-Postlabeling, and LC-MS

3.1. IHC Detection of DNA Adducts

FFPE specimens are often used for immunodetection of DNA adducts, most commonly by IHC methods [30,80]. In contrast to mass spectrometry-based methods, which break down the DNA to the mono 2′-deoxyribonucleoside or DNA base (*vide supra*), IHC methods employ intact DNA. The detection of DNA damage can be carried out on either fixed cells such as lymphocytes, exfoliated oral or bladder cells, or with FFPE tissue section-cuts. The cells or FFPE tissue section-cuts are mounted on glass slides for IHC analysis. Procedures are often used to increase the accessibility of the antibody to the carcinogen DNA adduct to increase the sensitivity of the assay. These procedures can include treatment with proteases to remove histone and other proteins from the DNA, followed by treatment with RNase to eliminate potential cross-reactivity with RNA adducts. Mild acid or base treatment also may be performed to denature the DNA and further increase the accessibility of the antibody to the adduct. It is imperative that the adduct is stable to the denaturing treatment conditions for validation of the IHC technique. The two most commonly used detection systems for visualization of DNA adduct-antibody complexes are immunofluorescence or chromophores, where the secondary antibody is tagged with a chemically conjugated fluorophore, a peroxidase or alkaline phosphatase enzyme. [30,81].

Table 1 summarizes examples of IHC detection of DNA adducts in FFPE tissues. Santella's group detected and quantified DNA adducts of 4-aminobiphenyl (4-ABP), an aromatic amine and a human bladder carcinogen that is formed in tobacco smoke [21,22], and also occurs as a contaminant in some commercial hair dyes [82]. 4-ABP-adducted DNA was detected in uroepithelium of bladder cancer patients [83]. The level of the 4-ABP adduct was correlated with the smoking status and *p53* overexpression, a response to DNA damage. There was linear relationship between the relative degree of DNA adduct staining and the number of cigarettes smoked. The same group also detected DNA adducts of polycyclic aromatic hydrocarbons (PAHs) in archived breast tissues sections using polyclonal antiserum [84,85]. PAHs are incomplete combustion products of organic matter and found in cereal and grain products, some oils, and also found in charred meat and tobacco smoke [42]. PAHs have been linked to human cancers at multiple sites [21]. The most well studied PAH is benzo[*a*]pyrene (B[*a*]P), a human lung carcinogen found as an environmental pollutant, and it also occurs in tobacco smoke, and charred meat [42,86]. The Poirier laboratory developed an antibody raised against DNA modified with *r*7,*t*8-dihydoxy-*t*-9,10-oxy-7,8,9,10-tetrahydrobenzo[*a*]pyrene (BPDE) [87], which later was shown to cross-react with other structurally similar diol-epoxide-PAH-DNA adducts [88]. This PAH-DNA antiserum has been used to screen for DNA adducts in FFPE tissues from human esophagus [81,89], prostate [90], cervix [91], vulva [47], and placenta [80]. A significantly higher level of staining of presumed B[*a*]P adducts was found in benign breast disease in comparison to the cancerous tissues of patients, possibly due to cellular proliferation and dilution of the adduct in cancerous tissue [84,85]. Rundle et al. employed IHC to measure PAH-DNA adducts and examined the associations with alcohol consumption and the influence of GSTM1 genotype on DNA adduct formation in FFPE breast tissues [92]. Subjects harboring the GSTM1-null genotype, which lacks the expression of GTSM1, an enzyme that detoxicates PAH diol-epoxides [93], had increased levels of DNA adducts among current alcohol consumers, but not among nondrinkers. In contrast, in benign tissues from controls, no association was observed between genotype and adduct levels, regardless of drinking status. Poirier also analyzed tamoxifen-DNA adducts in rat hepatocytes by IHC [94]. A steady increase in adduct levels was observed with chronic exposure.

Shirai et al. developed polyclonal antibodies against DNA adducts of 3,2'- dimethyl-4-aminobiphenyl (DMAB), an aromatic amine that induces tumors at multiple sites in rodent models, and PhIP, a probable human carcinogen formed in cooked meat that induces tumors in colorectum and prostate of rodents [95–97]. Dose-related nuclear staining was observed in various acetone-fixed tissues of rodents 24 h after single exposure of DMAB or PhIP. Using the same polyclonal serum, putative DNA adducts of PhIP were detected, by IHC, at high frequency in mammary tissue of women with breast cancer [98] and in prostate tissue of men with prostate cancer [99]. However, these results are at odds with specific mass spectrometry-based methods, where PhIP DNA adducts were detected at considerably lower frequency and at much lower levels of DNA modification in both tissues [100,101]. The discrepancy between the estimates of the PhIP DNA adduct reported by MS and IHC methods suggest the possible cross-reactivity of the polyclonal antibodies with other DNA adducts of similar structure or endogenous cellular components. There is a need to cross-corroborate the identities and levels of DNA adducts measured by IHC and specific MS-based methods. Aoshiba and coworkers raised antibodies against 8-hydroxy-2′-deoxyguanosine (8-OHdG), an oxidative DNA adduct, and 4-hydroxy-2-nonenal (4-HNE), a lipid peroxidation adduct, to evaluate the oxidative stress induced by cigarette smoke in paraffin-embedded pulmonary epithelial cells of mice [102]. There was a dramatic increase in the intensity of signals one hour post cigarette smoke exposure, compared to pre-exposure, which confirmed the causal role of cigarette smoking in oxidative damage to respiratory epithelium.

Table 1. Examples of DNA adducts detected in FFPE tissues.

Detection Methods	DNA Adducts Detected	Tissues	LOD (Per 10^8 Nucleotides)	References
IHC	4-ABP-DNA	Human bladder	NR [a]	[83]
	PAH-DNA	Human breast	NR [a]	[84,85,92]
		Human esophagus	NR [a]	[81,89]
		Human prostate	8	[90]
		Human cervix	20	[91]
		Human vulva	8	[47]
		Human placenta	20	[80]
	DMAB-DNA	Rat multiple tissues	NR [a]	[96]
	PhIP-DNA	Human prostate tissue transplanted to mice	NR [b]	[96]
		Rat multiple tissues	NR [b]	[97]
	8-OHdG	Mouse pulmonary epithelial cells	NR [a]	[102]
	Tamoxifen-DNA	Rat hepatocytes	10	[94]
^{32}P-postlabeling	B[a]P-DNA, 2-AAF-DNA	Rat multiple tissues	NR [c]	[103]
LC-MS^3	dA-AL-I	Mouse liver and kidney, human kidney	0.1	[79,104]
	dG-C8-4-ABP/PhIP, dG-N^2-BPDE, O^6-Me-dG and O^6-POB-dG	Rodent multiple tissues	0.2–0.5	[105]
LC-HR-MS^2	dG-C8-PhIP	Human prostate	0.13	[101]
	dG-C8-4-ABP	Human bladder	0.2	[55]

[a] Adduct levels were reported as relative nuclear stain intensity; [b] Adduct levels were reported as a percentage of positive cells; [c] LOD was reported in the citation, which was one per 10^{10} nucleotides employing 10 µg DNA. NR: Not reported.

3.2. DNA Measurements in FFPE Tissues by ^{32}P-Postlabeling

There is only one report employing ^{32}P-postlabeling to detect DNA adducts in FFPE tissues [103]. In that study, rat tissues were fixed in formalin and embedded in paraffin after dosing with B[a]P or 2-acetylaminofluorene (2-AAF). DNA was extracted from fixed tissues using a modified phenol-chloroform method [106]. The levels of DNA adduct recovered from FFPE tissues were significantly lower than the levels obtained from fresh frozen tissues. The authors concluded that FFPE tissues could be used to screen for DNA adducts but that adduct levels may be underestimated particularly with prolonged time of fixation in formalin.

3.3. Measurement of DNA Adducts in FFPE Tissues of Rodents and Human by LC-MS

The physio-chemical data provided by MS for proof of DNA adduct structure combined with the robust quantitation and high sensitivity makes MS the technique of choice for DNA adduct biomarker measurements. The DNA adducts must be stable towards both the formalin fixation and DNA retrieval processes. Furthermore, the DNA must be fully digestible by nucleases to monodeoxyribonucleosides. Until recently, the recovery of high quality DNA completely devoid of formalin crosslinks was difficult to achieve under mild hydrolysis conditions. However, commercial kits from several vendors now employ mild retrieval conditions at neutral pH to reverse the crosslinks of FFPE DNA. The DNA recovered was shown to be successfully employed as templates for amplification by PCR. We tested commercial kits from several vendors and found the FFPE miniprep kit from Zymo Research (Irvine, CA, USA), with some modifications in manufacturer's protocol, provided high quality DNA that was fully digestible by nucleases [79,101,105].

Our laboratory established a method to isolate DNA from FFPE liver and kidney tissues of C57BL/6J mice, using aristolochic acid I (AA-I) as the model carcinogen [79,104]. AA-I is an upper urinary tract human carcinogen found in Aristolochia plants, some of which have been used in traditional Chinese herbal medicines [79,104]. DNA was isolated from freshly frozen tissue by the phenol-chloroform method, and DNA from FFPE tissue was isolated with the FFPE miniprep kit (Zymo Research). AA-I DNA adducts were measured by ultra-performance liquid chromatography-electrospray ionization-ion trap-multistage MS^n scanning (UPLC-ESI-IT-MS^3). Across all dosing levels, the amounts of AA-I DNA adduct in DNA from FFPE tissues were comparable to those of matching freshly frozen tissues (Figure 4) [104].

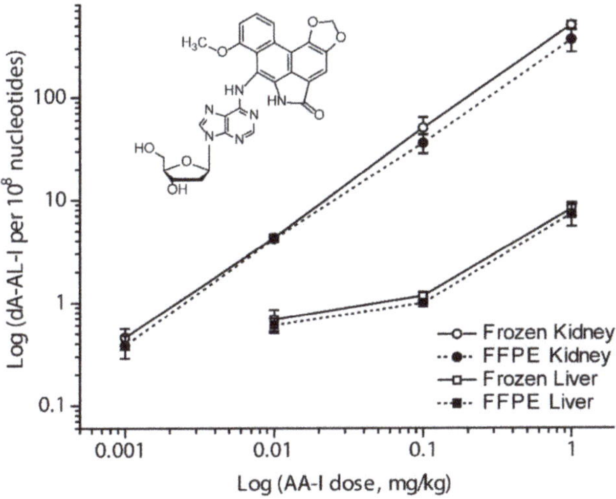

Figure 4. Mean level of dA-AL-I adducts present in mouse kidney and liver following treatment with AA-I (0.001−1 mg/kg body weight). Adduct levels measured in freshly frozen and FFPE mouse kidney (○ and ●) and liver (□ and ■) (mean adduct level, SD, $N = 4$ animals per dose, quadruplicate measurements per animal) were plotted as a function of dose. The overall mean difference in adduct levels between freshly frozen and FFEP kidney and liver tissues across all doses was $21 \pm 14\%$ (mean ± SD). dA-AL-I adduct formation was below the limit of detection in liver of mice dosed with AA-I at 0.001 mg/kg body weight. Mean levels of dA-AL-I adducts were significantly statistically different between freshly frozen and FFPE kidney or liver at the following dose treatments of AA-I: kidney, 1 mg/kg, $p = 0.03$; liver, 0.1 mg/kg, $p = 0.01$; unpaired two-tailed *t*-test. Reproduced with permission from [104]. Copyright ACS, 2013.

Then, we examined the effect of duration of formalin fixation on the recovery of DNA and the level of DNA adducts in rodents treated with AA-I [79]. The yield of DNA retrieved from formalin-fixed tissues decreased as a function of fixation time, and only 30% of DNA was recovered from FFPE tissues after one week of fixation in formalin compared to the freshly frozen tissues. However, the DNA retrieved was completely digested by nucleases and the levels of AA-I DNA adduct were relatively constant between the freshly frozen tissues and FFPE tissues. DNA fragments of 184 and 327 bp extracted from FFPE tissues were readily amplified by PCR, and the quality of sequence data was comparable to that obtained from DNA obtained of fresh frozen tissues [79]. Our findings demonstrate that the DNA can be recovered from FFPE tissue to analyze DNA adducts of AA-I in FFPE tissue, and adducts of AA-I or other carcinogens may be correlated with mutational signatures induced in tumor tissue.

Thereafter, we sought to determine if our method of DNA adduct retrieval from FFPE tissues could be employed to measure DNA adducts of other environmental and dietary genotoxicants. We examined DNA adducts of four important classes of environmental and dietary carcinogens: PAHs (B[*a*]P), aromatic amines (4-ABP), HAAs (PhIP), and *N*-nitroso compounds 4-(methylnitrosamino)-1-(3-pyridyl)-1-butanone (NNK), which is found in tobacco and a lung carcinogen [21,107]. The major DNA adducts of these carcinogens studied were: 10-(2′-deoxyguanosin-N^2-yl)-7,8,9-trihydroxy-7,8,9,10-tetrahydrobenzo[*a*]pyrene (dG-N^2-B[*a*]PPDE), *N*-(2′-deoxyguanosin-8-yl)-4-ABP (dG-C8-4-ABP), *N*-(2′-deoxyguanosin-8-yl)-PhIP (dG-C8-PhIP), O6-[4-oxo-4-(3-pyridy)-butyl]-2′-deoxyguanosine (O^6-POB-dG), O^6-methyl-2′-deoxyguanosine (O^6-methyl-dG) (Figure 5) [105]. All of these adducts and dA-AL-I were measured by UPLC-ESI-IT-MS^3 with the stable isotope dilution method. The levels of DNA adducts in FFPE tissues of rodents preserved in formalin for 24 h were at levels comparable to those levels measured in freshly frozen tissues [105].

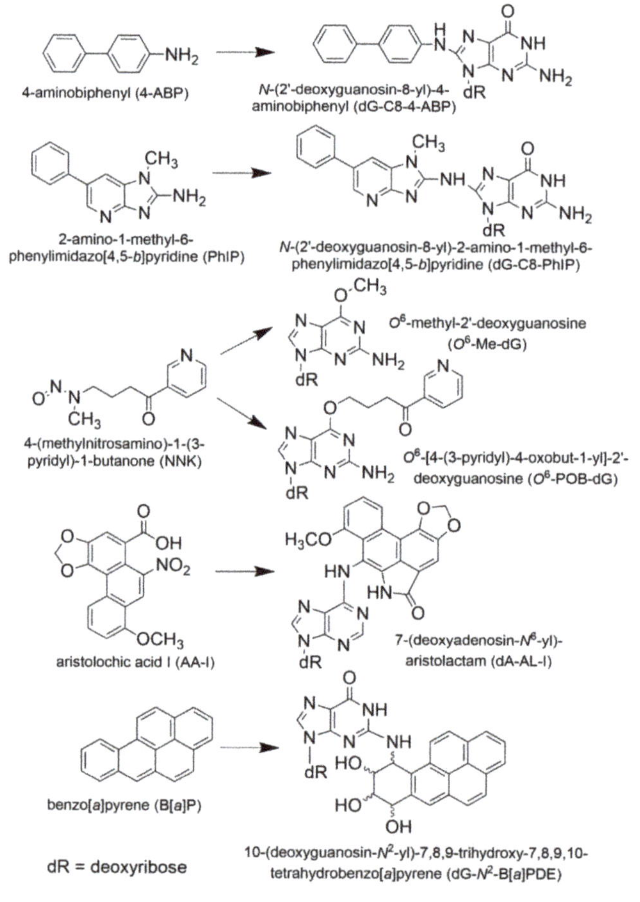

Figure 5. Structures, names, and abbreviations of carcinogens and their adducts used for quantitation of multiclass carcinogenic DNA adducts in freshly frozen and FFPE tissues of rodents. Reproduced with permission from [105]. Copyright ACS, 2016.

The recent improvements in sensitivity of mass spectrometry instrumentation has allowed us to use only 10 to 20 mg of tissue to screen for DNA adducts of environmental and dietary carcinogens in human biopsy samples [100,101,104]. We sought to determine if DNA extraction kits devoted to genomics, such as the FFPE miniprep kit from Zymo Research, could be employed to screen for DNA adducts in human FFPE biospecimens. We applied the method of DNA isolation to assay tissue section-cuts of human FFPE kidney specimens (1.5 cm^2 × 10 μm) from the patients with upper urinary tract carcinoma, who were exposed to AA-I [79,104]. The levels of AA-I DNA adduct measured in FFPE tissues were comparable to those of matching frozen tissues (Figure 6). Some of these FFPE blocks had been stored at room temperature for four to nine years. This was the first report of quantitative measurement of a carcinogen DNA adduct in human FFPE tissue by mass spectrometry. We subsequently showed that DNA adducts of PhIP can be recovered in high yield from human FFPE prostate tissue blocks of prostate cancer patients stored at room temperature for at least 6 months (Figure 7) [101,108]. These findings show that FFPE tissues can be used to retrospectively screen for multiple classes of carcinogen DNA adducts.

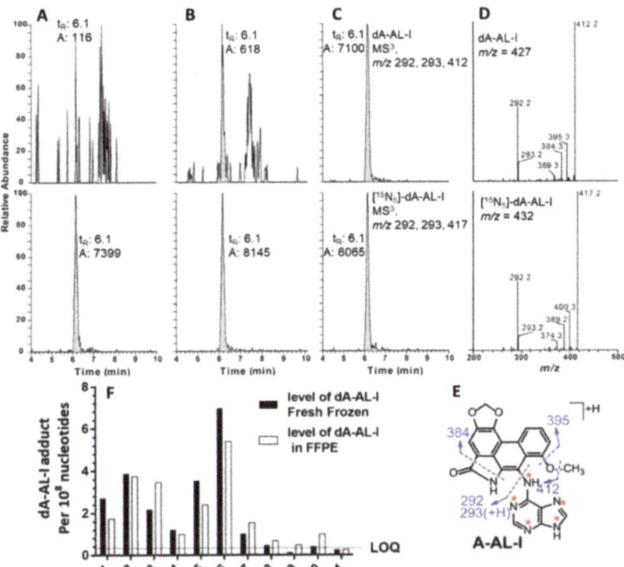

Figure 6. Extracted ion chromatograms of dA-AL-I from human kidney cortex of patients with upper urothelial cancer from Taiwan at levels (**A**) below the LOQ, and positive samples at (**B**) 0.4 adducts, and (**C**) 5.9 adducts per 10^8 bases. The product ion spectra of dA-AL-I obtained from panel C is depicted in (**D**) along with the internal standard [15N5]-dA-AL-I (E, 15N labels of the internal standard of dA-AL-I are depicted with asterisks). Insert (**F**) dA-AL-I adduct levels in matching fresh frozen and FFPE kidney samples, containing both renal cortex and medulla, obtained from 11 individuals residing in endemic regions of Croatia and Serbia who underwent nephroureterectomy for upper urothelial cancer. Reproduced with permission from [104]. Copyright ACS, 2013.

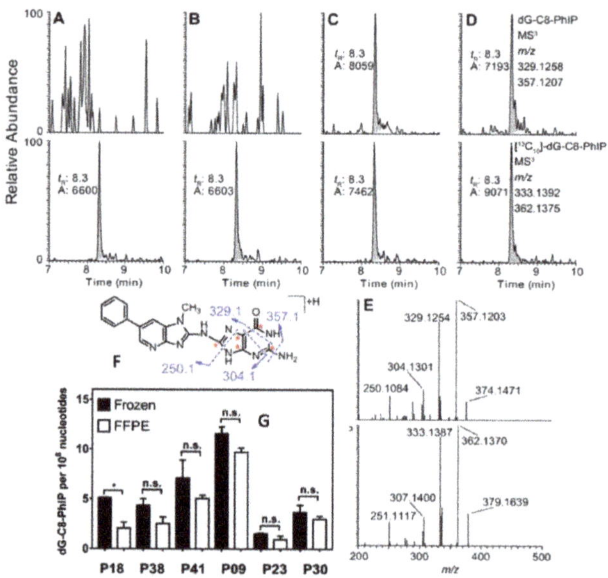

Figure 7. Extracted ion chromatograms of dG-C8-PhIP and ^{13}C-labeled dG-C8-PhIP of DNA from fresh frozen and FFPE human prostate tissues at the MS^3 scan stage. (**A**) Fresh frozen prostate and (**B**) paired FFPE block of a patient who was negative for dG-C8-PhIP; (**C**) fresh frozen prostate and (**D**) paired FFPE block of a patient who was positive dG-C8-PhIP at MS^3 scan stage. The structure and proposed fragmentation mechanism of aglycone of dG-C8-PhIP are depicted fresh frozen prostate and (**D**) paired FFPE block of a patient who was positive for dG-C8-PhIP. (**E**) The product ion spectra at MS^3 of unlabeled and ^{13}C-labeled dG-C8-PhIP are shown. (**F**) Levels of dG-C8-PhIP in paired fresh frozen prostate and FFPE blocks of six patients are shown in (**G**). The levels of adducts are reported as adducts per 10^8 nucleotides. * $p < 0.05$; n.s., statistically not significant. Reproduced with permission from [108]. Copyright ACS, 2017.

3.4. Rapid Throughput Method of DNA Extraction from FFPE Tissue

The method of DNA isolation from FFPE tissues employing the FFPE miniprep kit (Zymo Research) is robust; however, it is a manual and labor-intensive technique and cannot facilely process the large number of samples required for epidemiological studies. We developed a rapid throughput method of DNA isolation from FFPE tissue employing a semi-automated commercial DNA isolation system, Promega Maxwell® 16 MDx system, which is used for genomic research [108]. The system employs silica-magnetic beads technology for DNA isolation and can process 32 samples per hour compared to 4–6 samples per hour by the manual method. The DNA recovered from FFPE tissues using the Promega Maxwell® 16 MDx is fully digestible by nucleases [108]. The levels of dA-AL-I, dG-C8-4-ABP, and dG-C8-PhIP recovered from DNA of FFPE tissues extracted by rapid throughput method are comparable to those levels measured from DNA isolated by phenol-chloroform in matching frozen tissues, and in DNA of FFPE tissues isolated by the commercial manual Zymo kit [108]. With this advancement in DNA isolation technology, we believe that archived FFPE tissues can be used to screen for DNA adducts in large population studies. A scheme and the time of duration of the procedure to isolate DNA from FFPE section cuts or whole tissues, and ensuing chemical analysis by mass spectrometry, are depicted in Figure 8. The recovery of DNA from FFPE tissues and DNA digestion steps require overnight incubation with enzymes to achieve optimal digestion efficiency. The targeted and simultaneous quantification of a selected number of DNA adducts, by UPLC-ESI-IT-MS^3, can be achieved in a 10 to 15 min run time.

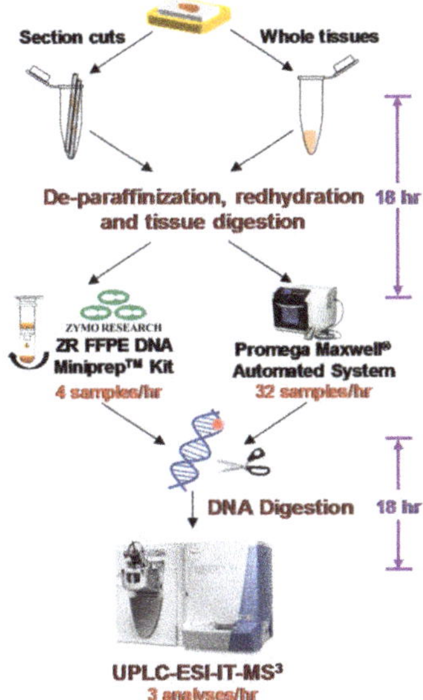

Figure 8. Scheme of FFPE tissue processing for DNA adduct measurements. DNA is extracted from section cuts or excised whole tissues by the FFPE Miniprep kit (Zymo Research) or Promega Maxwell® automated system. After nuclease digestion, the DNA adducts are measured by UPLC-ESI-IT-MS³. The estimated times of the different processes are reported.

3.5. Future Applications of DNA Adduct Measurements in Human Tissues

Although this review has focused on DNA adducts of environmental and dietary carcinogens, the measurements of DNA adducts of chemotherapeutic agents, such as platinum drugs and nitrogen mustards used to treat cancer, also can be measured in fresh frozen and FFPE tissues by mass spectrometry. Drugs that modify the structure of DNA and target cancer cells by interfering with DNA synthesis and cell replication often remain first line of medications used in cancer treatment. Thus, the efficacy of many anticancer drugs is thought to be linked to the levels of specific DNA adducts formed during drug treatment, and the quantitative measurements of the DNA adducts may be used as predictive markers in precision medicine to identify individuals who are most likely to benefit from treatment from those patients who may be less responsive to the therapy [109]. The assessment of DNA adducts of chemotherapeutic drugs and their cellular biological responses has been mostly performed in surrogate specimens, such as white blood cells or in vitro using cell lines rather than in the target cells or tumors, because of the invasiveness of biopsy sampling [109]. However, the exquisite sensitivity of current mass spectrometry instrumentation can allow for measurements that characterize the relationships between level of anticancer drug DNA adducts and pharmacodynamic response in patients using only 10 mg of tissue. As the sensitivity of MS instrumentation continues to improve, the amount of tissue specimen required for analysis will be further reduced, and the application of DNA adduct monitoring of chemotherapeutic drugs in clinical settings can be achieved.

The screening of DNA from FFPE tissues shows great promise to measure DNA adducts of multiple classes of carcinogens and anticancer drugs [37,55,105,109]. While most analyses have focused on one to several adducts, different types of MS scanning approaches are being developed to simultaneous scan for multiple types of DNA adducts in the field of DNA adductomics [63]. Triple quadrupole, quadrupole time-of-flight, ion trap or Orbitrap mass spectrometry instrumentation are being employed in DNA adductomics [55,58,62,63,110]. Our laboratory is developing unbiased non-targeted ion trap and Orbitrap scanning methods to screen for an array of DNA adducts in the human genome in a single assay [55,58]. Some of these adducts are expected to contribute to the tumor mutation burden [111].

A critical need is the development of accompanying informatic tools for data analysis and statistical tools to screen for covalent DNA damage. These scanning technologies and accompanying data analysis tools will provide a wealth of information about the exogenous and endogenous chemicals that damage the genome and may contribute to cancer risk. The implementation of FFPE tissues in DNA adduct biomarker discovery can provide the clues about the origin of human cancers for which an environmental exposure is suspected.

Author Contributions: All authors critically reviewed the literature and contributed to drafting the manuscript.

Acknowledgments: The work cited in this review conducted by the Turesky laboratory has been supported by Grants R01ES019564 and R21ES014438 from the National Institute of Environmental Health Sciences, R01CA122320, R01CA220367, and R33CA186795, and the National Center for Advancing Translational Sciences award number UL1TR000114 from the National Cancer Institute of the National Institutes of Health. Mass spectrometry was carried out in Analytical Biochemistry Shared Resources of the Masonic Cancer Center, University of Minnesota, funded in part by Cancer Center Support Grant CA-077598.

Conflicts of Interest: The authors declare no conflict of interest.

References

1. Stanley, L.A. Drug Metabolism. In *Pharmacognosy: Fundamentals, Applications and Strategies*; Badal, S., Delgoda, R., Eds.; Elsevier: London, UK, 2017; pp. 527–545.
2. Guengerich, F.P. Cytochrome p450 and chemical toxicology. *Chem. Res. Toxicol.* **2008**, *21*, 70–83. [CrossRef] [PubMed]
3. Grillo, M.P. Bioactivation by Phase-II-Enzyme-Catalyzed Conjugation of Xenobiotics. In *Encyclopedia of Drug Metabolism and Interactions*; Lyubimov, A.V., Ed.; Wiley: Hoboken, NJ, USA, 2012; Volume 4.
4. Dekant, W. The role of biotransformation and bioactivation in toxicity. In *Molecular, Clinical and Environmental Toxicology*; Luch, A., Ed.; Birkhäuser: Basel, Switzerland, 2009; Volume 1, pp. 57–86.
5. Miller, E.C. Some current perspectives on chemical carcinogenesis in humans and experimental animals: Presidential address. *Cancer Res.* **1978**, *38*, 1479–1496. [PubMed]
6. Minchin, R.F.; Reeves, P.T.; Teitel, C.H.; McManus, M.E.; Mojarrabi, B.; Ilett, K.F.; Kadlubar, F.F. *N*-and *O*-acetylation of aromatic and heterocyclic amine carcinogens by human monomorphic and polymorphic acetyltransferases expressed in *COS-1* cells. *Biochem. Biophys. Res. Commun.* **1992**, *185*, 839–844. [CrossRef]
7. Thier, R.; Pemble, S.E.; Kramer, H.; Taylor, J.B.; Guengerich, F.P.; Ketterer, B. Human glutathione S-transferase T1-1 enhances mutagenicity of 1,2-dibromoethane, dibromomethane and 1,2,3,4-diepoxybutane in Salmonella typhimurium. *Carcinogenesis* **1996**, *17*, 163–166. [CrossRef] [PubMed]
8. Fu, P.P.; Miller, D.W.; Von Tungeln, L.S.; Bryant, M.S.; Lay, J.O., Jr.; Huang, K.; Jones, L.; Evans, F.E. Formation of C8-modified deoxyguanosine and C8-modified deoxyadenosine as major DNA adducts from 2-nitropyrene metabolism mediated by rat and mouse liver microsomes and cytosols. *Carcinogenesis* **1991**, *12*, 609–616. [CrossRef] [PubMed]
9. Regan, S.L.; Maggs, J.L.; Hammond, T.G.; Lambert, C.; Williams, D.P.; Park, B.K. Acyl glucuronides: The good, the bad and the ugly. *Biopharm. Drug Dispos.* **2010**, *31*, 367–395. [CrossRef] [PubMed]
10. Nauwelaers, G.; Bessette, E.E.; Gu, D.; Tang, Y.; Rageul, J.; Fessard, V.; Yuan, J.M.; Yu, M.C.; Langouet, S.; Turesky, R.J. DNA adduct formation of 4-aminobiphenyl and heterocyclic aromatic amines in human hepatocytes. *Chem. Res. Toxicol.* **2011**, *24*, 913–925. [CrossRef] [PubMed]
11. Sidorenko, V.S.; Attaluri, S.; Zaitseva, I.; Iden, C.R.; Dickman, K.G.; Johnson, F.; Grollman, A.P. Bioactivation of the human carcinogen aristolochic acid. *Carcinogenesis* **2014**, *35*, 1814–1822. [CrossRef] [PubMed]
12. Reardon, D.B.; Prakash, A.S.; Hilton, B.D.; Roman, J.M.; Pataki, J.; Harvey, R.G.; Dipple, A. Characterization of 5-methylchrysene-1,2-dihydrodiol-3,4-epoxide-DNA adducts. *Carcinogenesis* **1987**, *8*, 1317–1322. [CrossRef] [PubMed]
13. Penning, T.M. The aldo-keto reductases (AKRs): Overview. *Chem. Biol. Interact.* **2015**, *234*, 236–246. [CrossRef] [PubMed]

14. Marques, M.M.; Beland, F.A. Identification of tamoxifen-DNA adducts formed by 4-hydroxytamoxifen quinone methide. *Carcinogenesis* **1997**, *18*, 1949–1954. [CrossRef] [PubMed]

15. Osborne, M.R.; Hewer, A.; Hardcastle, I.R.; Carmichael, P.L.; Phillips, D.H. Identification of the major tamoxifen-deoxyguanosine adduct formed in the liver DNA of rats treated with tamoxifen. *Cancer Res.* **1996**, *56*, 66–71. [PubMed]

16. Guengerich, F.P. Comparisons of catalytic selectivity of cytochrome P450 subfamily enzymes from different species. *Chem. Biol. Interact.* **1997**, *106*, 161–182. [CrossRef]

17. Turesky, R.J.; Constable, A.; Fay, L.B.; Guengerich, F.P. Interspecies differences in metabolism of heterocyclic aromatic amines by rat and human P450 1A2. *Cancer Lett.* **1999**, *143*, 109–112. [CrossRef]

18. Edwards, R.J.; Murray, B.P.; Murray, S.; Schulz, T.; Neubert, D.; Gant, T.W.; Thorgeirsson, S.S.; Boobis, A.R.; Davies, D.S. Contribution of CYP1A1 and CYP1A2 to the activation of heterocyclic amines in monkeys and humans. *Carcinogenesis* **1994**, *15*, 829–836. [CrossRef] [PubMed]

19. Himmelstein, M.W.; Boogaard, P.J.; Cadet, J.; Farmer, P.B.; Kim, J.H.; Martin, E.A.; Persaud, R.; Shuker, D.E. Creating context for the use of DNA adduct data in cancer risk assessment: II. Overview of methods of identification and quantitation of DNA damage. *Crit. Rev. Toxicol.* **2009**, *39*, 679–694. [CrossRef] [PubMed]

20. Jarabek, A.M.; Pottenger, L.H.; Andrews, L.S.; Casciano, D.; Embry, M.R.; Kim, J.H.; Preston, R.J.; Reddy, M.V.; Schoeny, R.; Shuker, D.; et al. Creating context for the use of DNA adduct data in cancer risk assessment: I. Data organization. *Crit. Rev. Toxicol.* **2009**, *39*, 659–678. [CrossRef] [PubMed]

21. IARC. Tobacco smoke and involuntary smoking. *IARC Monogr. Eval. Carcinog. Risks Hum.* **2004**, *83*, 1–1438.

22. IARC. *Tobacco smoking IARC Monogr. Eval. Carcinog. Risks Hum.* **1986**, *38*, 35–394.

23. Stiborova, M.; Arlt, V.M.; Schmeiser, H.H. DNA Adducts Formed by Aristolochic Acid Are Unique Biomarkers of Exposure and Explain the Initiation Phase of Upper Urothelial Cancer. *Int. J. Mol. Sci.* **2017**, *18*, 2144. [CrossRef] [PubMed]

24. Rosenquist, T.A.; Grollman, A.P. Mutational signature of aristolochic acid: Clue to the recognition of a global disease. *DNA Repair* **2016**, *44*, 205–211. [CrossRef] [PubMed]

25. Wogan, G.N.; Kensler, T.W.; Groopman, J.D. Present and future directions of translational research on aflatoxin and hepatocellular carcinoma. A review. *Food Addit. Contam. Part A Chem. Anal. Control Expo. Risk Assess.* **2012**, *29*, 249–257. [CrossRef] [PubMed]

26. Kensler, T.W.; Roebuck, B.D.; Wogan, G.N.; Groopman, J.D. Aflatoxin: A 50-year odyssey of mechanistic and translational toxicology. *Toxicol. Sci.* **2011**, *120* (Suppl. 1), S28–S48. [CrossRef] [PubMed]

27. Sancar, A.; Lindsey-Boltz, L.A.; Unsal-Kacmaz, K.; Linn, S. Molecular mechanisms of mammalian DNA repair and the DNA damage checkpoints. *Annu. Rev. Biochem.* **2004**, *73*, 39–85. [CrossRef] [PubMed]

28. Randerath, K.; Reddy, M.V.; Gupta, R.C. ^{32}P-labeling test for DNA damage. *Proc. Natl. Acad. Sci. USA* **1981**, *78*, 6126–6129. [CrossRef] [PubMed]

29. Phillips, D.H. On the origins and development of the (32)P-postlabelling assay for carcinogen-DNA adducts. *Cancer Lett.* **2013**, *334*, 5–9. [CrossRef] [PubMed]

30. Santella, R.M. Immunological methods for detection of carcinogen-DNA damage in humans. *Cancer Epidemiol. Biomark. Prev.* **1999**, *8*, 733–739.

31. Poirier, M.C.; Santella, R.M.; Weston, A. Carcinogen macromolecular adducts and their measurement. *Carcinogenesis* **2000**, *21*, 353–359. [CrossRef] [PubMed]

32. Dizdaroglu, M.; Coskun, E.; Jaruga, P. Measurement of oxidatively induced DNA damage and its repair, by mass spectrometric techniques. *Free Radic. Res.* **2015**, *49*, 525–548. [CrossRef] [PubMed]

33. Singh, R.; Farmer, P.B. Liquid chromatography-electrospray ionization-mass spectrometry: The future of DNA adduct detection. *Carcinogenesis* **2006**, *27*, 178–196. [CrossRef] [PubMed]

34. Klaene, J.J.; Sharma, V.K.; Glick, J.; Vouros, P. The analysis of DNA adducts: The transition from (32)P-postlabeling to mass spectrometry. *Cancer Lett.* **2013**, *334*, 10–19. [CrossRef] [PubMed]

35. Liu, S.; Wang, Y. Mass spectrometry for the assessment of the occurrence and biological consequences of DNA adducts. *Chem. Soc. Rev.* **2015**, *44*, 7829–7854. [CrossRef] [PubMed]

36. Tretyakova, N.; Goggin, M.; Sangaraju, D.; Janis, G. Quantitation of DNA adducts by stable isotope dilution mass spectrometry. *Chem. Res. Toxicol.* **2012**, *25*, 2007–2035. [CrossRef] [PubMed]

37. Guo, J.; Turesky, R.J. Human Biomonitoring of DNA Adducts by Ion Trap Multistage Mass Spectrometry. *Curr. Protoc. Nucleic Acid Chem.* **2016**, *66*, 7.24.21–7.24.25. [PubMed]

38. Shibutani, S.; Kim, S.Y.; Suzuki, N. ^{32}P-Postlabeling DNA damage assays: PAGE, TLC, and HPLC. *Methods Mol. Biol.* **2006**, *314*, 307–321. [PubMed]
39. Pfau, W.; Lecoq, S.; Hughes, N.C.; Seidel, A.; Platt, K.L.; Grover, P.L.; Phillips, D.H. Separation of 32 P-labelled nucleoside 3′,5′-bisphosphate adducts by HPLC. *IARC Sci. Publ.* **1993**, 233–242.
40. Phillips, D.H. DNA adducts as markers of exposure and risk. *Mutat. Res.* **2005**, *577*, 284–292. [CrossRef] [PubMed]
41. Phillips, D.H. Smoking-related DNA and protein adducts in human tissues. *Carcinogenesis* **2002**, *23*, 1979–2004. [CrossRef] [PubMed]
42. Phillips, D.H. Polycyclic aromatic hydrocarbons in the diet. *Mutat. Res.* **1999**, *443*, 139–147. [CrossRef]
43. Agudo, A.; Peluso, M.; Munnia, A.; Lujan-Barroso, L.; Sanchez, M.J.; Molina-Montes, E.; Sanchez-Cantalejo, E.; Navarro, C.; Tormo, M.J.; Chirlaque, M.D.; et al. Aromatic DNA adducts and risk of gastrointestinal cancers: A case-cohort study within the EPIC-Spain. *Cancer Epidemiol. Biomark. Prev.* **2012**, *21*, 685–692. [CrossRef] [PubMed]
44. Gilberson, T.; Peluso, M.E.; Munia, A.; Lujan-Barroso, L.; Sanchez, M.J.; Navarro, C.; Amiano, P.; Barricarte, A.; Quiros, J.R.; Molina-Montes, E.; et al. Aromatic adducts and lung cancer risk in the European Prospective Investigation into Cancer and Nutrition (EPIC) Spanish cohort. *Carcinogenesis* **2014**, *35*, 2047–2054. [CrossRef] [PubMed]
45. Ricceri, F.; Godschalk, R.W.; Peluso, M.; Phillips, D.H.; Agudo, A.; Georgiadis, P.; Loft, S.; Tjonneland, A.; Raaschou-Nielsen, O.; Palli, D.; et al. Bulky DNA adducts in white blood cells: A pooled analysis of 3600 subjects. *Cancer Epidemiol. Biomark. Prev.* **2010**, *19*, 3174–3181. [CrossRef] [PubMed]
46. Ho, V.; Peacock, S.; Massey, T.E.; Godschalk, R.W.; van Schooten, F.J.; Chen, J.; King, W.D. Gene-diet interactions in exposure to heterocyclic aromatic amines and bulky DNA adduct levels in blood leukocytes. *Environ. Mol. Mutagen.* **2015**, *56*, 609–620. [CrossRef] [PubMed]
47. Pratt, M.M.; John, K.; MacLean, A.B.; Afework, S.; Phillips, D.H.; Poirier, M.C. Polycyclic aromatic hydrocarbon (PAH) exposure and DNA adduct semi-quantitation in archived human tissues. *Int. J. Environ. Res. Public Health* **2011**, *8*, 2675–2691. [CrossRef] [PubMed]
48. Mumford, J.L.; Williams, K.; Wilcosky, T.C.; Everson, R.B.; Young, T.L.; Santella, R.M. A sensitive color ELISA for detecting polycyclic aromatic hydrocarbon-DNA adducts in human tissues. *Mutat. Res.* **1996**, *359*, 171–177. [CrossRef]
49. Divi, R.L.; Beland, F.A.; Fu, P.P.; Von Tungeln, L.S.; Schoket, B.; Camara, J.E.; Ghei, M.; Rothman, N.; Sinha, R.; Poirier, M.C. Highly sensitive chemiluminescence immunoassay for benzo[*a*]pyrene-DNA adducts: Validation by comparison with other methods, and use in human biomonitoring. *Carcinogenesis* **2002**, *23*, 2043–2049. [CrossRef] [PubMed]
50. Poirier, M.C. Chemical-induced DNA damage and human cancer risk. *Nat. Rev. Cancer* **2004**, *4*, 630–637. [CrossRef] [PubMed]
51. Ravanat, J.L.; Turesky, R.J.; Gremaud, E.; Trudel, L.J.; Stadler, R.H. Determination of 8-oxoguanine in DNA by gas chromatography-mass spectrometry and HPLC-electrochemical detection: Overestimation of the background level of the oxidized base by the gas chromatography-mass spectrometry assay. *Chem. Res. Toxicol.* **1995**, *8*, 1039–1045. [CrossRef] [PubMed]
52. Turesky, R.J.; Vouros, P. Formation and analysis of heterocyclic aromatic amine-DNA adducts in vitro and in vivo. *J. Chromatogr. B Anal. Technol. Biomed. Life Sci.* **2004**, *802*, 155–166. [CrossRef] [PubMed]
53. Tretyakova, N.; Villalta, P.W.; Kotapati, S. Mass spectrometry of structurally modified DNA. *Chem. Rev.* **2013**, *113*, 2395–2436. [CrossRef] [PubMed]
54. Villalta, P.W.; Hochalter, J.B.; Hecht, S.S. Ultrasensitive High-Resolution Mass Spectrometric Analysis of a DNA Adduct of the Carcinogen Benzo[*a*]pyrene in Human Lung. *Anal. Chem.* **2017**, *89*, 12735–12742. [CrossRef] [PubMed]
55. Guo, J.; Villalta, P.W.; Turesky, R.J. Data-Independent Mass Spectrometry Approach for Screening and Identification of DNA Adducts. *Anal. Chem.* **2017**, *89*, 11728–11736. [CrossRef] [PubMed]
56. Bessette, E.E.; Goodenough, A.K.; Langouet, S.; Yasa, I.; Kozekov, I.D.; Spivack, S.D.; Turesky, R.J. Screening for DNA adducts by data-dependent constant neutral loss-triple stage mass spectrometry with a linear quadrupole ion trap mass spectrometer. *Anal. Chem.* **2009**, *81*, 809–819. [CrossRef] [PubMed]

57. Balbo, S.; Hecht, S.S.; Upadhyaya, P.; Villalta, P.W. Application of a high-resolution mass-spectrometry-based DNA adductomics approach for identification of DNA adducts in complex mixtures. *Anal. Chem.* **2014**, *86*, 1744–1752. [CrossRef] [PubMed]

58. Villalta, P.W.; Balbo, S. The Future of DNA Adductomic Analysis. *Int. J. Mol. Sci.* **2017**, *18*, 1870. [CrossRef]

59. Ma, B.; Zarth, A.T.; Carlson, E.S.; Villalta, P.W.; Upadhyaya, P.; Stepanov, I.; Hecht, S.S. Methyl DNA Phosphate Adduct Formation in Rats Treated Chronically with 4-(Methylnitrosamino)-1-(3-pyridyl)-1-butanone and Enantiomers of Its Metabolite 4-(Methylnitrosamino)-1-(3-pyridyl)-1-butanol. *Chem. Res. Toxicol.* **2018**, *31*, 48–57. [CrossRef] [PubMed]

60. Ma, B.; Zarth, A.T.; Carlson, E.S.; Villalta, P.W.; Stepanov, I.; Hecht, S.S. Pyridylhydroxybutyl and pyridyloxobutyl DNA phosphate adduct formation in rats treated chronically with enantiomers of the tobacco-specific nitrosamine metabolite 4-(methylnitrosamino)-1-(3-pyridyl)-1-butanol. *Mutagenesis* **2017**, *32*, 561–570. [CrossRef] [PubMed]

61. Wolf, S.M.; Vouros, P. Application of capillary liquid chromatography coupled with tandem mass spectrometric methods to the rapid screening of adducts formed by the reaction of *N*-acetoxy-*N*-acetyl-2-aminofluorene with calf thymus DNA. *Chem. Res. Toxicol.* **1994**, *7*, 82–88. [CrossRef] [PubMed]

62. Kanaly, R.A.; Matsui, S.; Hanaoka, T.; Matsuda, T. Application of the adductome approach to assess intertissue DNA damage variations in human lung and esophagus. *Mutat. Res.* **2007**, *625*, 83–93. [CrossRef] [PubMed]

63. Balbo, S.; Turesky, R.J.; Villalta, P.W. DNA adductomics. *Chem. Res. Toxicol.* **2014**, *27*, 356–366. [CrossRef] [PubMed]

64. Fox, C.H.; Johnson, F.B.; Whiting, J.; Roller, P.P. Formaldehyde fixation. *J. Histochem. Cytochem.* **1985**, *33*, 845–853. [CrossRef] [PubMed]

65. Hoffman, E.A.; Frey, B.L.; Smith, L.M.; Auble, D.T. Formaldehyde crosslinking: A tool for the study of chromatin complexes. *J. Biol. Chem.* **2015**, *290*, 26404–26411. [CrossRef] [PubMed]

66. Kennedy-Darling, J.; Smith, L.M. Measuring the formaldehyde Protein-DNA cross-link reversal rate. *Anal. Chem.* **2014**, *86*, 5678–5681. [CrossRef] [PubMed]

67. Boenisch, T. Effect of heat-induced antigen retrieval following inconsistent formalin fixation. *Appl. Immunohistochem. Mol. Morphol.* **2005**, *13*, 283–286. [CrossRef] [PubMed]

68. Graw, S.; Meier, R.; Minn, K.; Bloomer, C.; Godwin, A.K.; Fridley, B.; Vlad, A.; Beyerlein, P.; Chien, J. Robust gene expression and mutation analyses of RNA-sequencing of formalin-fixed diagnostic tumor samples. *Sci. Rep.* **2015**, *5*, 12335. [CrossRef] [PubMed]

69. Ludgate, J.L.; Wright, J.; Stockwell, P.A.; Morison, I.M.; Eccles, M.R.; Chatterjee, A. A streamlined method for analysing genome-wide DNA methylation patterns from low amounts of FFPE DNA. *BMC Med. Genom.* **2017**, *10*, 54. [CrossRef] [PubMed]

70. Robbe, P.; Popitsch, N.; Knight, S.J.L.; Antoniou, P.; Becq, J.; He, M.; Kanapin, A.; Samsonova, A.; Vavoulis, D.V.; Ross, M.T.; et al. Clinical whole-genome sequencing from routine formalin-fixed, paraffin-embedded specimens: Pilot study for the 100,000 Genomes Project. *Genet. Med.* **2018**. [CrossRef] [PubMed]

71. Moran, S.; Vizoso, M.; Martinez-Cardus, A.; Gomez, A.; Matias-Guiu, X.; Chiavenna, S.M.; Fernandez, A.G.; Esteller, M. Validation of DNA methylation profiling in formalin-fixed paraffin-embedded samples using the Infinium HumanMethylation450 Microarray. *Epigenetics* **2014**, *9*, 829–833. [CrossRef] [PubMed]

72. Jiang, X.; Jiang, X.; Feng, S.; Tian, R.; Ye, M.; Zou, H. Development of efficient protein extraction methods for shotgun proteome analysis of formalin-fixed tissues. *J. Proteome Res.* **2007**, *6*, 1038–1047. [CrossRef] [PubMed]

73. Sprung, R.W.; Brock, J.W.C.; Tanksley, J.P.; Li, M.; Washington, M.K.; Slebos, R.J.C.; Liebler, D.C. Equivalence of Protein Inventories Obtained from Formalin-fixed Paraffin-embedded and Frozen Tissue in Multidimensional Liquid Chromatography-Tandem Mass Spectrometry Shotgun Proteomic Analysis. *Mol. Cell. Proteom.* **2009**, *8*, 1988–1998. [CrossRef] [PubMed]

74. Giusti, L.; Lucacchini, A. Proteomic studies of formalin-fixed paraffin-embedded tissues. *Expert Rev. Proteom.* **2013**, *10*, 165–177. [CrossRef] [PubMed]

75. Kelly, A.D.; Breitkopf, S.B.; Yuan, M.; Goldsmith, J.; Spentzos, D.; Asara, J.M. Metabolomic profiling from formalin-fixed, paraffin-embedded tumor tissue using targeted LC/MS/MS: Application in sarcoma. *PLoS ONE* **2011**, *6*, e25357. [CrossRef] [PubMed]

76. Wojakowska, A.; Chekan, M.; Marczak, L.; Polanski, K.; Lange, D.; Pietrowska, M.; Widlak, P. Detection of metabolites discriminating subtypes of thyroid cancer: Molecular profiling of FFPE samples using the GC/MS approach. *Mol. Cell. Endocrinol.* **2015**, *417*, 149–157. [CrossRef] [PubMed]

77. Mortensen, E.; Brown, J. Effects of Fixation on Tissues. In *Prostate Cancer Methods and Protocols*; Russell, P., Jackson, P., Kingsley, E., Eds.; Springer: New York, NY, USA, 2003; Volume 81, pp. 163–179.

78. Xie, R.; Chung, J.Y.; Ylaya, K.; Williams, R.L.; Guerrero, N.; Nakatsuka, N.; Badie, C.; Hewitt, S.M. Factors influencing the degradation of archival formalin-fixed paraffin-embedded tissue sections. *J. Histochem. Cytochem.* **2011**, *59*, 356–365. [CrossRef] [PubMed]

79. Yun, B.H.; Yao, L.; Jelakovic, B.; Nikolic, J.; Dickman, K.G.; Grollman, A.P.; Rosenquist, T.A.; Turesky, R.J. Formalin-fixed paraffin-embedded tissue as a source for quantitation of carcinogen DNA adducts: Aristolochic acid as a prototype carcinogen. *Carcinogenesis* **2014**, *35*, 2055–2061. [CrossRef] [PubMed]

80. Pratt, M.M.; King, L.C.; Adams, L.D.; John, K.; Sirajuddin, P.; Olivero, O.A.; Manchester, D.K.; Sram, R.J.; DeMarini, D.M.; Poirier, M.C. Assessment of multiple types of DNA damage in human placentas from smoking and nonsmoking women in the Czech Republic. *Environ. Mol. Mutagen.* **2011**, *52*, 58–68. [CrossRef] [PubMed]

81. Van Gijssel, H.E.; Divi, R.L.; Olivero, O.A.; Roth, M.J.; Wang, G.Q.; Dawsey, S.M.; Albert, P.S.; Qiao, Y.L.; Taylor, P.R.; Dong, Z.W.; et al. Semiquantitation of polycyclic aromatic hydrocarbon-DNA adducts in human esophagus by immunohistochemistry and the automated cellular imaging system. *Cancer Epidemiol. Biomark. Prev.* **2002**, *11*, 1622–1629.

82. Turesky, R.J.; Freeman, J.P.; Holland, R.D.; Nestorick, D.M.; Miller, D.W.; Ratnasinghe, D.L.; Kadlubar, F.F. Identification of aminobiphenyl derivatives in commercial hair dyes. *Chem. Res. Toxicol.* **2003**, *16*, 1162–1173. [CrossRef] [PubMed]

83. Curigliano, G.; Zhang, Y.J.; Wang, L.Y.; Flamini, G.; Alcini, A.; Ratto, C.; Giustacchini, M.; Alcini, E.; Cittadini, A.; Santella, R.M. Immunohistochemical quantitation of 4-aminobiphenyl-DNA adducts and p53 nuclear overexpression in T1 bladder cancer of smokers and nonsmokers. *Carcinogenesis* **1996**, *17*, 911–916. [CrossRef] [PubMed]

84. Santella, R.M.; Gammon, M.D.; Zhang, Y.J.; Motykiewicz, G.; Young, T.L.; Hayes, S.C.; Terry, M.B.; Schoenberg, J.B.; Brinton, L.A.; Bose, S.; et al. Immunohistochemical analysis of polycyclic aromatic hydrocarbon-DNA adducts in breast tumor tissue. *Cancer Lett.* **2000**, *154*, 143–149. [CrossRef]

85. Motykiewicz, G.; Malusecka, E.; Michalska, J.; Kalinowska, E.; Wloch, J.; Butkiewicz, D.; Mazurek, A.; Lange, D.; Perera, F.P.; Santella, R.M. Immunoperoxidase detection of polycyclic aromatic hydrocarbon-DNA adducts in breast tissue sections. *Cancer Detect. Prev.* **2001**, *25*, 328–335. [PubMed]

86. Hecht, S.S. Progress and challenges in selected areas of tobacco carcinogenesis. *Chem. Res. Toxicol.* **2008**, *21*, 160–171. [CrossRef] [PubMed]

87. Poirier, M.C.; Santella, R.; Weinstein, I.B.; Grunberger, D.; Yuspa, S.H. Quantitation of benzo(*a*)pyrene-deoxyguanosine adducts by radioimmunoassay. *Cancer Res.* **1980**, *40*, 412–416. [PubMed]

88. Weston, A.; Manchester, D.K.; Poirier, M.C.; Choi, J.S.; Trivers, G.E.; Mann, D.L.; Harris, C.C. Derivative fluorescence spectral analysis of polycyclic aromatic hydrocarbon-DNA adducts in human placenta. *Chem. Res. Toxicol.* **1989**, *2*, 104–108. [CrossRef] [PubMed]

89. Van Gijssel, H.E.; Schild, L.J.; Watt, D.L.; Roth, M.J.; Wang, G.Q.; Dawsey, S.M.; Albert, P.S.; Qiao, Y.L.; Taylor, P.R.; Dong, Z.W.; et al. Polycyclic aromatic hydrocarbon-DNA adducts determined by semiquantitative immunohistochemistry in human esophageal biopsies taken in 1985. *Mutat. Res.* **2004**, *547*, 55–62. [CrossRef] [PubMed]

90. John, K.; Ragavan, N.; Pratt, M.M.; Singh, P.B.; Al-Buheissi, S.; Matanhelia, S.S.; Phillips, D.H.; Poirier, M.C.; Martin, F.L. Quantification of phase I/II metabolizing enzyme gene expression and polycyclic aromatic hydrocarbon-DNA adduct levels in human prostate. *Prostate* **2009**, *69*, 505–519. [CrossRef] [PubMed]

91. Pratt, M.M.; Sirajuddin, P.; Poirier, M.C.; Schiffman, M.; Glass, A.G.; Scott, D.R.; Rush, B.B.; Olivero, O.A.; Castle, P.E. Polycyclic aromatic hydrocarbon-DNA adducts in cervix of women infected with carcinogenic human papillomavirus types: An immunohistochemistry study. *Mutat. Res.* **2007**, *624*, 114–123. [CrossRef] [PubMed]

92. Rundle, A.; Tang, D.L.; Mooney, L.; Grumet, S.; Perera, F. The interaction between alcohol consumption and GSTM1 genotype on polycyclic aromatic hydrocarbon-DNA adduct levels in breast tissue. *Cancer Epidemiol. Biomark. Prev.* **2003**, *12*, 911–914.

93. Bartsch, H.; Hietanen, E. The role of individual susceptibility in cancer burden related to environmental exposure. *Environ. Health Perspect.* **1996**, *104* (Suppl. 3), 569–577. [CrossRef] [PubMed]

94. Divi, R.L.; Dragan, Y.P.; Pitot, H.C.; Poirier, M.C. Immunohistochemical localization and semi-quantitation of hepatic tamoxifen-DNA adducts in rats exposed orally to tamoxifen. *Carcinogenesis* **2001**, *22*, 1693–1699. [CrossRef] [PubMed]

95. Felton, J.S.; Jagerstad, I.M.; Knize, M.G.; Skog, K.; Wakabayashi, K. Contents in Foods, Beverages and Tobacco. In *Food Borne Carcinogens: Heterocyclic Amines*; Nagao, M., Sugimura, T., Eds.; John Wiley & Sons Ltd.: Chichester, UK, 2000; pp. 31–71.

96. Shirai, T.; Takahashi, S.; Cui, L.; Yamada, Y.; Tada, M.; Kadlubar, F.F.; Ito, N. Use of polyclonal antibodies against carcinogen-DNA adducts in analysis of carcinogenesis. *Toxicol. Lett.* **1998**, *102–103*, 441–446. [CrossRef]

97. Takahashi, S.; Tamano, S.; Hirose, M.; Kimoto, N.; Ikeda, Y.; Sakakibara, M.; Tada, M.; Kadlubar, F.F.; Ito, N.; Shirai, T. Immunohistochemical demonstration of carcinogen-DNA adducts in tissues of rats given 2-amino-1-methyl-6-phenylimidazo[4,5-*b*]pyridine (PhIP): Detection in paraffin-embedded sections and tissue distribution. *Cancer Res.* **1998**, *58*, 4307–4313. [PubMed]

98. Zhu, J.; Chang, P.; Bondy, M.L.; Sahin, A.A.; Singletary, S.E.; Takahashi, S.; Shirai, T.; Li, D. Detection of 2-amino-1-methyl-6-phenylimidazo[4,5-*b*]-pyridine-DNA adducts in normal breast tissues and risk of breast cancer. *Cancer Epidemiol. Biomark. Prev.* **2003**, *12*, 830–837.

99. Tang, D.; Liu, J.J.; Rundle, A.; Neslund-Dudas, C.; Savera, A.T.; Bock, C.H.; Nock, N.L.; Yang, J.J.; Rybicki, B.A. Grilled meat consumption and PhIP-DNA adducts in prostate carcinogenesis. *Cancer Epidemiol. Biomarkers Prev.* **2007**, *16*, 803–808. [CrossRef] [PubMed]

100. Gu, D.; Turesky, R.J.; Tao, Y.; Langouet, S.A.; Nauwelaers, G.C.; Yuan, J.M.; Yee, D.; Yu, M.C. DNA adducts of 2-amino-1-methyl-6-phenylimidazo[4,5-*b*]pyridine and 4-aminobiphenyl are infrequently detected in human mammary tissue by liquid chromatography/tandem mass spectrometry. *Carcinogenesis* **2012**, *33*, 124–130. [CrossRef] [PubMed]

101. Xiao, S.; Guo, J.; Yun, B.H.; Villalta, P.W.; Krishna, S.; Tejpaul, R.; Murugan, P.; Weight, C.J.; Turesky, R.J. Biomonitoring DNA Adducts of Cooked Meat Carcinogens in Human Prostate by Nano Liquid Chromatography-High Resolution Tandem Mass Spectrometry: Identification of 2-Amino-1-methyl-6-phenylimidazo[4,5-*b*]pyridine DNA Adduct. *Anal. Chem.* **2016**, *88*, 12508–12515. [CrossRef] [PubMed]

102. Aoshiba, K.; Koinuma, M.; Yokohori, N.; Nagai, A. Immunohistochemical evaluation of oxidative stress in murine lungs after cigarette smoke exposure. *Inhal. Toxicol.* **2003**, *15*, 1029–1038. [CrossRef] [PubMed]

103. Hewer, A.; Phillips, D.H. Effect of tissue fixation on recovery of DNA adducts in the ^{32}P-postlabelling assay. *IARC Sci. Publ.* **1993**, 211–214.

104. Yun, B.H.; Rosenquist, T.A.; Nikolic, J.; Dragicevic, D.; Tomic, K.; Jelakovic, B.; Dickman, K.G.; Grollman, A.P.; Turesky, R.J. Human Formalin-Fixed Paraffin-Embedded Tissues: an untapped specimen for biomonitoring of carcinogen DNA adducts by mass spectrometry. *Anal. Chem.* **2013**, *85*, 4251–4258. [CrossRef] [PubMed]

105. Guo, J.; Yun, B.H.; Upadhyaya, P.; Yao, L.; Krishnamachari, S.; Rosenquist, T.A.; Grollman, A.P.; Turesky, R.J. Multiclass carcinogenic DNA adduct quantification in formalin-fixed paraffin-embedded tissues by ultraperformance liquid chromatography-tandem mass spectrometry. *Anal. Chem.* **2016**, *88*, 4780–4787. [CrossRef] [PubMed]

106. Goelz, S.E.; Hamilton, S.R.; Vogelstein, B. Purification of DNA from formaldehyde fixed and paraffin embedded human tissue. *Biochem. Biophys. Res. Commun.* **1985**, *130*, 118–126. [CrossRef]

107. IARC. Smokeless Tobacco and Some Tobacco-Specific *N*-Nitrosamines. *IARC Monogr. Eval. Carcinog. Risks Hum.* **2007**, *89*, 1–152.

108. Yun, B.H.; Xiao, S.; Yao, L.; Krishnamachari, S.; Rosenquist, T.A.; Dickman, K.G.; Grollman, A.P.; Murugan, P.; Weight, C.J.; Turesky, R.J. A rapid throughput method to extract DNA from formalin-fixed paraffin-embedded tissues for biomonitoring carcinogenic DNA adducts. *Chem. Res. Toxicol.* **2017**, *30*, 2130–2139. [CrossRef] [PubMed]

109. Stornetta, A.; Zimmermann, M.; Cimino, G.D.; Henderson, P.T.; Sturla, S.J. DNA adducts from anticancer drugs as candidate predictive markers for precision medicine. *Chem. Res. Toxicol.* **2017**, *30*, 388–409. [CrossRef] [PubMed]

110. Yao, C.; Feng, Y.L. A nontargeted screening method for covalent DNA adducts and DNA modification selectivity using liquid chromatography-tandem mass spectrometry. *Talanta* **2016**, *159*, 93–102. [CrossRef] [PubMed]

111. Stratton, M.R.; Campbell, P.J.; Futreal, P.A. The cancer genome. *Nature* **2009**, *458*, 719–724. [CrossRef] [PubMed]

Review

Recent Studies on DNA Adducts Resulting from Human Exposure to Tobacco Smoke

Bin Ma *,†**, Irina Stepanov and Stephen S. Hecht**

Masonic Cancer Center, University of Minnesota, Minneapolis, MN 55455, USA; stepa011@umn.edu (I.S.); hecht002@umn.edu (S.S.H.)
* Correspondence: bma@umn.edu or bin.ma@gilead.com; Tel.: +1-612-625-4925; Fax: +1-612-624-3869
† Current address: Gilead Sciences, 333 Lakeside Dr, Foster City, CA 94404, USA.

Received: 2 February 2019; Accepted: 13 March 2019; Published: 19 March 2019

Abstract: DNA adducts are believed to play a central role in the induction of cancer in cigarette smokers and are proposed as being potential biomarkers of cancer risk. We have summarized research conducted since 2012 on DNA adduct formation in smokers. A variety of DNA adducts derived from various classes of carcinogens, including aromatic amines, polycyclic aromatic hydrocarbons, tobacco-specific nitrosamines, alkylating agents, aldehydes, volatile carcinogens, as well as oxidative damage have been reported. The results are discussed with particular attention to the analytical methods used in those studies. Mass spectrometry-based methods that have higher selectivity and specificity compared to ^{32}P-postlabeling or immunochemical approaches are preferred. Multiple DNA adducts specific to tobacco constituents have also been characterized for the first time in vitro or detected in vivo since 2012, and descriptions of those adducts are included. We also discuss common issues related to measuring DNA adducts in humans, including the development and validation of analytical methods and prevention of artifact formation.

Keywords: DNA adducts; tobacco smoke; human carcinogen; biomarkers; cancer risk; mass spectrometry

---

1. Introduction

Cigarette smoking causes multiple types of cancers. Despite the advent of advanced cancer genomics and impressive targeted therapies, there were still an estimated 9.6 million cancer deaths worldwide in 2018 with 22% of those deaths being caused by cigarette smoking [1]. In the U.S., where lung cancer is the leading cause of cancer death in both men and women, an estimated 154,000 lung cancer deaths occurred in 2018, with 90% being caused by cigarette smoking [2,3]. Not all smokers develop cancer. While 90% of all lung cancer-related deaths in the U.S. are attributable to cigarette smoking, only 24% of male smokers and 11% of female smokers may die from lung cancer over their lifetime, assuming that there is no competing cause of death [4,5]. A significant research challenge is to identify those smokers who have higher cancer risk, so prevention approaches can be initiated at an early stage before too much damage has been done.

One strategy to address this challenge is to identify and validate smoking-related biomarkers associated with tobacco exposure and cancer risk. We have demonstrated the association of certain urinary tobacco smoke biomarkers with lung cancer risk, independent of smoking intensity and duration. These biomarkers include cotinine and its glucuronide—metabolites of nicotine; 4-(methylnitrosamino)-1-(3-pyridyl)-1-butanol (NNAL) and its glucuronides—metabolites of a tobacco specific lung carcinogen 4-(methylnitrosamino)-1-(3-pyridyl)-1-butanone (NNK); and *r*-1-,*t*-2,3,*c*-4-tetrahydroxy-1,2,3,4-tetrahydrophenanthrene (PheT), a biomarker of polycyclic aromatic hydrocarbon (PAH) exposure and metabolic activation [6,7]. Similarly, levels of a urinary biomarker of exposure to another tobacco-specific carcinogen -*N*′-nitrosonornicotine (NNN)- have been independently and prospectively associated with the risk of esophageal cancer in smokers [8].

In addition, low activity forms of cytochrome P450 2A6, the major nicotine metabolizing enzyme, have been related to decreased lung cancer risk in cigarette smokers [9]. All of these parameters are related to carcinogen uptake by smokers, but do not provide information on the next critical step in the carcinogenic process—DNA adduct formation. DNA adducts associated with tobacco smoke exposure have the potential to significantly contribute to our understanding of the cancer process and could possibly be biomarkers of cancer risk.

DNA adducts are compounds formed when chemicals react with DNA. Figure 1 provides an overview of the central role of DNA adduct formation in tobacco-related cancer. Tobacco smoke contains a highly complex mixture of over 7000 characterized chemical compounds; certain chemicals such as formaldehyde and acetaldehyde are reactive and directly bind to DNA, while some others, including NNK, NNN, and PAH require metabolic activation to reactive intermediates capable of reacting with DNA. The resulting covalent DNA addition products are commonly referred to as "DNA adducts" (Figure 1). Because of the DNA repair enzymes present in the body, the DNA adducts can be removed and the DNA returned to normal. However, if the repair process is overwhelmed or not completely efficient, the DNA adducts can persist and potentially cause miscoding during DNA replication. If the miscoding events occur in critical genes such as *p53* and *RAS*, they can result in the loss of normal growth control mechanisms, and eventually, the development of cancer. Therefore, DNA adduct formation plays a critical role in tobacco smoke-related cancer development. The general protocol to measure DNA adducts includes isolating DNA from biological samples, hydrolyzing the DNA using a cocktail of enzymes, purifying the hydrolyzed samples and enriching the DNA adducts, and then analyzing them by various techniques such as mass spectrometry. The ultimate goal is to quantify these tobacco smoke-related DNA adducts in human biological samples and to incorporate the results into epidemiologic studies to evaluate and validate their role in cancer risk prediction (Figure 1). While this challenging goal has yet to be reached, constantly improving methods for DNA adduct quantitation, as discussed here, provide at least a potential path. We also note that reactive tobacco chemicals can modify other macromolecules such as protein, which can also result in significant biological consequences and can be used in bio-monitoring [10,11].

In 2002, Phillips reviewed previous studies on smoking-related DNA adduct formation in human tissues [12]. A large number of studies mentioned in that review used ^{32}P-postlabeling or immunochemical approaches for the analysis of DNA adducts. In many cases, "bulky DNA adducts" were reported but it was not clear which specific DNA adducts were being measured. Nevertheless, the effects of smoking were still evident by the detection of elevated levels of DNA adducts in many human tissues [12]. In 2012, Phillips updated this topic and reviewed studies on tobacco smoke-related DNA adducts published since 2002 [10]. That review covered the chemical nature and origins of smoking-related DNA adducts and the effect of DNA repair and gene polymorphisms on the levels of DNA adducts in humans. Many studies cited in that review demonstrated convincing evidence of higher levels of certain adducts in smokers than nonsmokers. In the same published issue as Phillips' paper, we also reviewed and discussed structurally characterized DNA adducts in the lungs of smokers and their potential role in lung carcinogenesis by tobacco smoke [13].

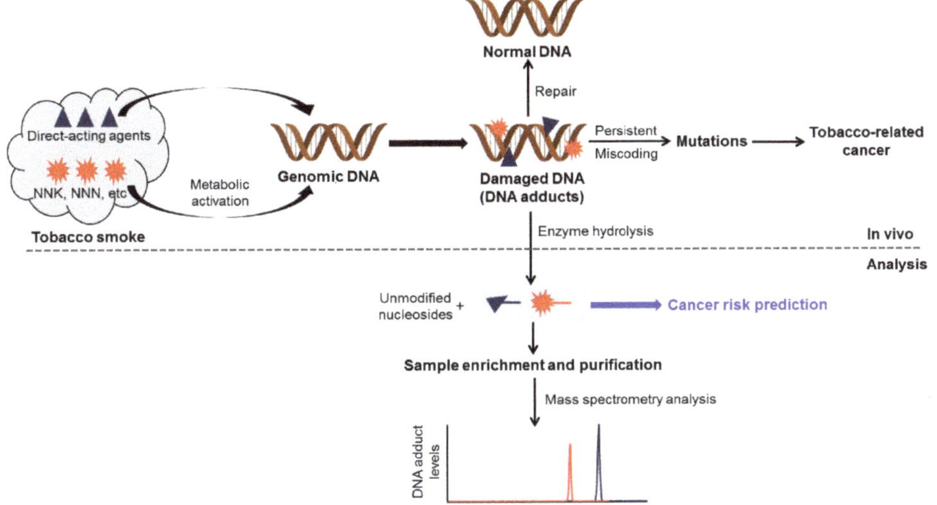

Figure 1. An overview of the central role of DNA adduct formation in tobacco-related cancer. DNA damage also leads to cell apoptosis. A more detailed mechanistic framework was illustrated by Hecht S.S. [13].

This review covers 1) studies on formation of tobacco smoke-related DNA adducts in humans (Figure 2) since Phillips' review in 2012; and 2) newly characterized DNA adducts formed by tobacco-specific chemicals (i.e., NNK and NNN) in in vitro and in vivo models (Figure 3). Future directions of applying DNA adducts as biomarkers of tobacco smoke exposure and associated cancer risk are also discussed.

2. Smoking-Related DNA Adducts in Humans

2.1. Tobacco-specific Nitrosamines (TSNA)

The most studied TSNA are NNK and NNN, which are considered "carcinogenic to humans" by the International Agency for Research on Cancer [14]. NNK causes lung tumors in all species tested, independent of the route of administration [15]. NNN induces tumors of the oral cavity, esophagus, and nasal cavities in rats and respiratory tract tumors in mice, mink, and Syrian golden hamsters [15]. NNK and NNN are present in all tobacco products. Both compounds are specific to tobacco and are not found in any other product, except that NNN can also be formed endogenously from nornicotine, a minor tobacco alkaloid and a nicotine metabolite [16–18].

NNK and NNN are metabolically activated to intermediate **14** (Figure 3), which reacts with DNA to form pyridyloxobutyl (POB) adducts at the four nucleobases and the oxygens of the phosphate backbone (Figure 3) [7,19–22]. Under acid or neutral thermal hydrolysis, most of the POB base adducts release 4-hydroxy-1-(3-pyridyl)-1-butanone (HPB, Figure 2). Multiple in vitro and animal studies demonstrated the relationship of HPB-releasing DNA adducts to NNK and NNN dose as well as biomarkers of NNK exposure [15]. HPB-releasing DNA adducts were also detected by our group and others in human tissues including lung, tracheobronchus, esophagus, and cardia [23–25]. We demonstrated for the first time the presence of HPB-releasing DNA adducts in human lung, with adduct levels being higher in smokers than nonsmokers [25]. Similar results were observed by Schlobe, et al, with HPB levels in the lung being significantly higher in 21 self-reported smokers compared to that in 11 self-reported nonsmokers [24]. However, another study by the same group showed no difference of adduct levels between smokers and nonsmokers [23]. All these studies had a relatively small sample size, probably due to the limited availability of human tissues, which had to be obtained by surgery. Obtaining these human tissues from individuals for smoking exposure evaluation and HPB-releasing DNA adduct analysis is highly impractical in larger studies. Surrogate tissues, which can be less invasively obtained, should be explored in such studies.

Exfoliated oral mucosa cells are relatively simple and noninvasive to be collected, and could be an excellent source of surrogate tissue for evaluating smoking exposure and adduct formation [26,27]. We developed a robust and sensitive liquid chromatography (LC)−electrospray ionization (ESI)−tandem mass spectrometry (MS/MS) method for the analysis of HPB-releasing DNA adducts in human oral cells (Table 1) [28]. Oral cells were collected from 30 smokers and 15 nonsmokers using a mouthwash rinse or buccal brushing with a cytobrush. In the oral cells collected by mouthwash, HPB was detected in 20 out of 28 smoker samples with quantifiable DNA yield, and in 3 out of 15 nonsmoker samples. Higher levels of HPB-releasing DNA adducts were observed in smokers with an average of 12 pmol adducts/mg DNA, compared to nonsmokers with an average of 0.23 pmol adducts/mg DNA. In the buccal brushing samples collected from the same 30 smokers, HPB was detected in 24 out of 27 samples with quantifiable DNA yield, averaging 45 pmol adducts/mg DNA. A correlation of adduct levels ($R = 0.73$, $p < 0.0001$) was also observed between buccal burshings and mouthwash samples from smokers [28]. The LC−ESI−MS/MS method we developed in that study demonstrated the applicability to the analysis of oral cell samples collected by mouthwash or buccal brushing. However, application of this method in studies in which only limited oral cells are available for DNA extraction required further optimization to increase its sensitivity and selectivity.

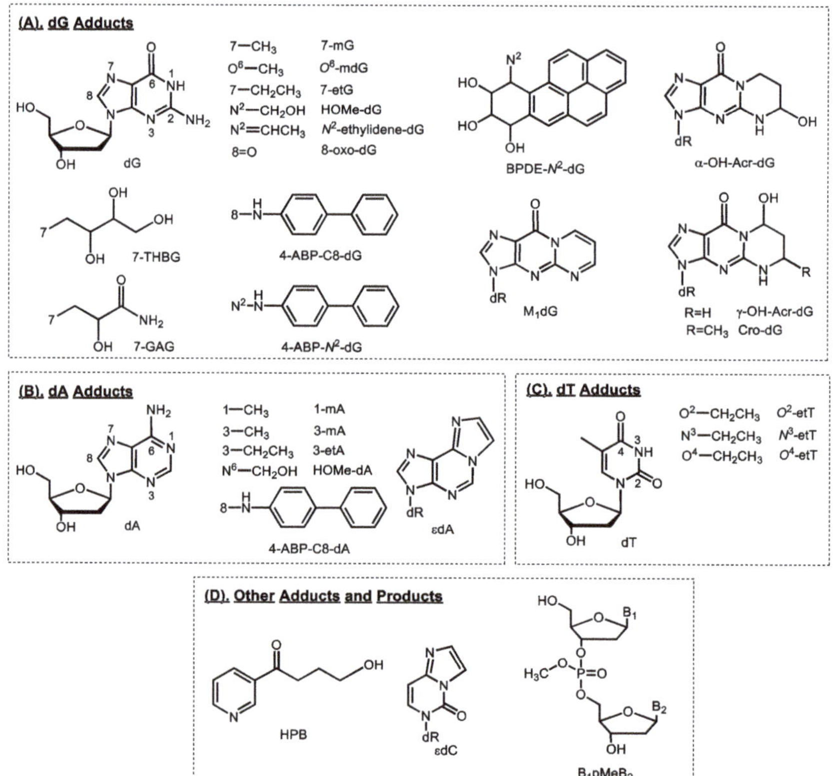

Figure 2. Structures of DNA adducts detected in humans. dR, 2'-deoxyribose; HPB, 4-hydroxy-1-(3-pyridyl)-1-butanone.

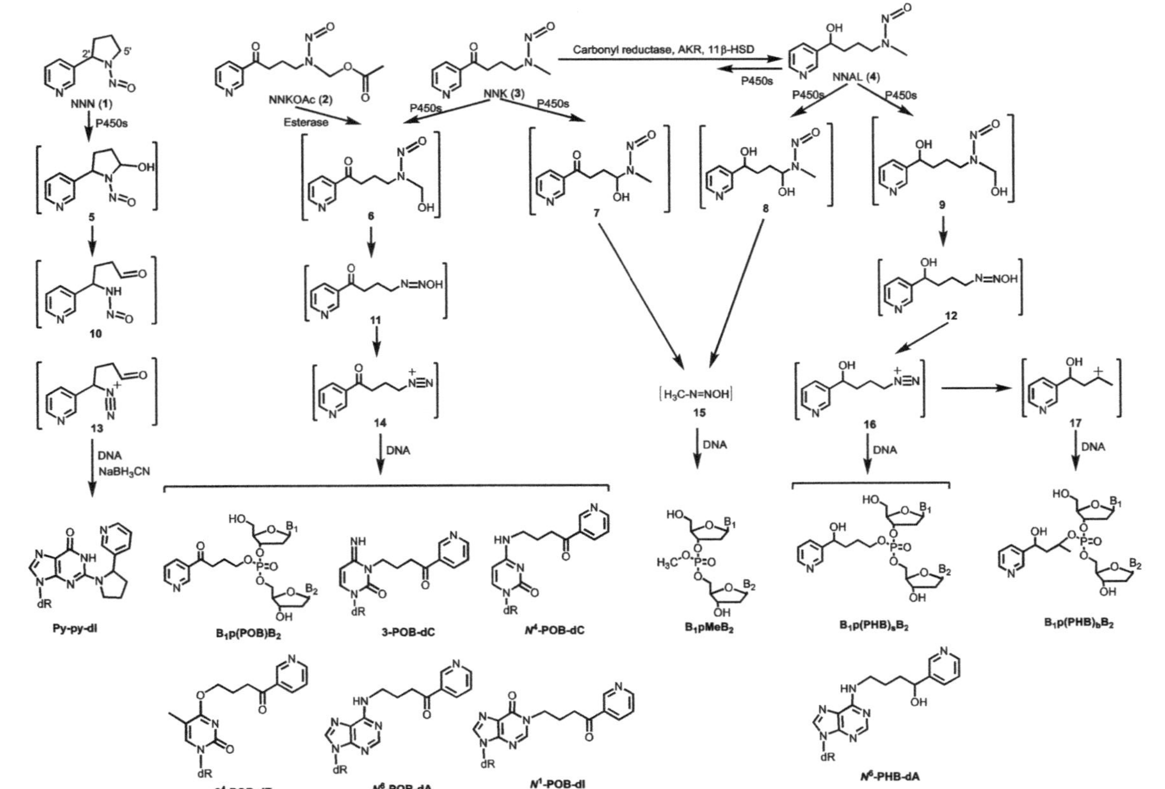

Figure 3. NNK and NNN-derived DNA base and phosphate adducts. NNK and NNN form various DNA adducts, and only newly characterized adducts or adducts detected for the first time in vivo since 2012 are presented. NNN is also metabolized via 2′-hydroxylation to intermediate **14**, which reacts with DNA to form POB base adducts. B_1 and B_2 represent the same or different nucleobases. NNK, 4-(methylnitrosamino)-1-(3-pyridyl)-1-butanone; NNN, *N*′-nitrosonornicotine; POB, pyridyloxobutyl; dR, 2′-deoxyribose.

We then modified the method by optimizing the sample preparation procedure, and more notably by switching from LC−ESI−MS/MS to LC-nanoelectrospray ionization (NSI)−high-resolution tandem mass spectrometry (HRMS/MS) [29]. The LC-NSI-HRMS/MS-based methods for DNA adduct analysis have been very successful in our studies of other DNA adducts [20,30–36]. Using the optimized method, the sensitivity was improved seven-fold compared to the previous LC−ESI−MS/MS method, achieving a limit of quantitation (LOQ) of 0.35 fmol on-column. More importantly, the baseline noise commonly observed in mass spectrometry without high resolution capability was completely eliminated, greatly improving the selectivity of the method. We then applied this method to the analysis of HPB-releasing DNA adducts in oral cells collected by buccal brushing from 65 smokers, including 30 head and neck squamous cell carcinoma (HNSCC) patients and 35 cancer-free controls. HPB-releasing DNA adducts were detected in 29 out of 30 samples of HNSCC patients averaging 8.2 pmol adducts/mg DNA, compared to 20 out of 35 samples of cancer-free smokers averaging 4.5 pmol adducts/mg DNA. The median HPB-releasing DNA adduct level was 6.6 times greater for smokers with HNSCC than for those without HNSCC (p = 0.002) [29]. These results suggest that HPB-releasing DNA adducts may play a critical role in the development of smoking-induced HNSCC, and can potentially be used to identify susceptible smokers.

2.2. Bulky/Aromatic Adducts

The terminology "bulky DNA adducts" comes from early studies using a ^{32}P-postlabeling approach to measure DNA adducts formed by high molecular weight chemical carcinogens, including PAH and probably some other aromatic and nonpolar chemicals [37,38]. PAH are formed primarily by incomplete combustion of tobacco and other organic components during smoking. PAH are also released from burning coal, oil, gasoline, wood, and are present in air, soil and water. Studies on bulky DNA adducts, sometimes referred to as PAH-DNA adducts, were comprehensively covered by Phillips et al. in 2002 [12] and 2012 [10]. The overall trend was that higher adduct levels were observed in smokers compared to nonsmokers. Since 2012, continuous efforts have been made to measure bulky DNA adducts in human samples including lung and leukocytes (Table 1) [39–52]. Smoking is a contributor to adduct formation in these studies. However, since the ^{32}P-postlabeling technique was exclusively used as the detection method, structural information for these adducts was not available, which prevents one from working backwards to evaluate the exposure to responsible chemicals.

One of the PAH adducts is BPDE-N^2-dG (Figure 2), which is formed by BPDE, a metabolite of the most studied carcinogenic PAH—benzo[*a*]pyrene (BaP). In humans, BPDE-N^2-dG was reported to be present in buccal cells [53], sputum [53], lung [53,54], leukocytes [55], lymphocytes [56], tonsil [57], oropharynx [57], gingiva [57], and prostate [58] using immunochemical methods. Another method to measure this adduct is to hydrolyze the DNA in acid and analyze BaP-7,8,9,10-tetraols instead of the intact adducts by LC-fluorescence or gas chromatography/mass spectrometry methods [59,60]. However, the specificity of these methods, although in some cases well designed and validated, has to be considered when interpreting the data. This concern was further highlighted by studies using mass spectrometry-based methods [61–64]. In one study analyzing human lung tissues, BPDE-N^2-dG was not detected in any of the 10 samples with a limit of detection (LOD) of one−three adducts per 10^8 nucleotides [61], and the second study showed detectable adducts in only 1 out 26 human lung DNA samples (LOD, 0.3 adducts per 10^8 nucleotides) [62]. BPDE-N^2-dG was not detected in salivary or oral cell DNA in another two studies [63,64]. Finally, a recent study failed to detect this adduct in human prostate tissue [65].

We recently developed an ultrasensitive LC-NSI-HRMS/MS method for the analysis of BPDE-N^2-dG in human lung DNA, with an LOD of 1 adduct per 10^{11} nucleotides, which is equivalent to 1 adduct per 10 human lung cells (Table 1) [32]. To our knowledge, this is the most sensitive DNA adduct quantitation method yet reported. We applied this method to the analysis of lung samples obtained during surgery for lung cancer from 29 patients. Another unique feature of this study was confirmation of smoking status at the time of lung cancer surgery by the measurement of urinary cotinine and NNAL. BPDE-N^2-dG was detected in 20 out of 29 samples, with smoker and nonsmoker

DNA containing 3.1 and 1.3 adducts per 10^{11} nucleotides, respectively. The adduct levels in our study are in contrast to those obtained by the immunochemical approach, with levels typically ranging from 1–30 adducts per 10^8 nucleotides in lung tissues [53,54]. Compared to mass spectrometry-based methods including our study, significantly higher levels (~1,000-fold) of adducts were observed by immunochemistry-based methods. Although the lung samples analyzed in these studies were from different subjects, the significantly high level observed by immunochemistry-based methods suggests that those methods should be re-evaluated to avoid false-positive results, and the data obtained from those studies should be interpreted with caution. Studies should be conducted using a mass spectrometry-based method and an immunochemistry-based method to analyze the same human lung samples, to compare the levels of BPDE-N^2-dG adducts assessed by the two methods.

2.3. Aromatic and Heterocyclic Aromatic Amines

Aromatic and heterocyclic aromatic amines are structurally related classes of chemicals that are present in tobacco smoke, formed in cooked meats, and occur as contaminants in the atmosphere [38,66]. One such compound, an aromatic amine 4-aminobiphenyl (4-ABP) is a known bladder carcinogen in both animals and humans [66]. The major DNA adduct formed by 4-ABP is 4-ABP-C8-dG, and to a lesser extent, 4-ABP-C8-dA and 4-ABP-N^2-dG (Figure 2) [67].

In a study investigating the mutagenicity of 4-ABP-derived DNA adducts, the three adducts were measured in both normal human urothelial mucosa and bladder tumor tissues by ^{32}P-postlabeling-based methods (Table 1) [68]. All three adducts were detected in the urothelial mucosa samples ($n = 19$) with levels ranging from 2.3–12, 4.6–16, and 2.3–12 adducts per 10^8 nucleotides for 4-ABP-C8-dG, 4-ABP-C8-dA, and 4-ABP-N^2-dG, respectively. The three adducts were also detected in bladder tumor samples ($n = 10$) with levels ranging from 2.3–9.2, 4.6–25, and 2.3–28 adducts per 10^8 nucleotides for 4-ABP-C8-dG, 4-ABP-C8-dA, and 4-ABP-N^2-dG, respectively. No significant difference was observed in either total adduct levels or individual adduct levels between normal human urothelial mucosa and bladder tumor tissue [68].

In contrast to the results obtained by ^{32}P-postlabeling methods, studies using mass spectrometry-based methods showed much lower levels or no detection of 4-ABP adducts in human samples [51,65,67]. Xiao, et al developed and validated an LC-NSI-HRMS/MS method for the analysis of 4-ABP-C8-dG with an LOQ of 0.22 adducts per 10^8 nucleotides using 2.5 µg DNA. The method was applied to the analysis of normal tumor-adjacent prostate tissues from 35 prostate cancer patients. 4-ABP-C8-dG was detected at a level of 2.8 adducts per 10^8 nucleotides in only one out of 35 samples and that one patient was identified as a nonsmoker [65]. The same research group subsequently applied this method to the analysis of normal tumor-adjacent bladder mucosa tissues from bladder cancer patients. 4-ABP-C8-dG was detected in 12 out of 41 samples with levels ranging from 0.14 to 3.4 adducts per 10^8 nucleotides. 4-ABP-C8-dA and 4-ABP-N^2-dG were also detected in the same samples of two subjects who had the highest levels of 4-ABP-C8-dG [67]. In another study, Gu, et al applied an LC-MS/MS^3 method to the analysis of tumor-adjacent normal mammary tissues from 70 breast cancer patients. No 4-ABP-C8-dG was detected in any of the samples with an LOQ of 3 adducts per 10^8 nucleotides [51]. In comparison, an earlier study reported detecting 4-ABP adducts in all the 55 breast tissues by an immunochemical method [69], even though its sensitivity was ~100-fold lower compared to the LC-MS/MS^3 method.

2.4. Methylating Agents

There are several methylating agents present in tobacco products, including the tobacco-specific nitrosamine NNK, *N*-nitrosodimethylamine (NDMA) and methyl chloride. All of them can lead to formation of methyl DNA adducts. NNK and its metabolite NNAL are metabolized to methane diazohydroxide **15** and then methyldiazonium ion (Figure 3), which reacts with DNA forming methyl DNA base adducts including 7-methylguanine (7-mG), O^6-methyldeoxyguanosine (O^6-mdG) (Figure 2) [15] and methyl DNA phosphate adducts [34]. The major mutagenic and toxic methyl base adduct is O^6-mdG, which induces primarily GC→AT transition mutations, with some other minor methyl base adducts also demonstrating mutagenicity and cytotoxicity [70]. In addition to methyl base adducts, some methylating agents also react with the DNA phosphate backbone to form methyl DNA phosphate adducts—B_1pMeB_2 (Figure 2, see details in Section 3.1). B_1pMeB_2 adducts were demonstrated to induce TT→GT and TT→GC mutations [71], inhibit RNA synthesis [72], and affect the binding affinity of DNA to other macromolecules [73].

Methyl DNA base adducts were detected in human lung tissue samples [53,74]. O^6-mdG was measured in the tumor-adjacent normal lung tissue samples from lung cancer patients by an immunochemical approach, and no difference in adduct levels was observed between smokers ($n = 41$) and nonsmokers ($n = 13$) (Table 1) [53]. In another study, 7-mG was detected by an immunochemical approach in lung samples from 14 former and 6 current smokers undergoing surgery for lung cancer [74]. The lung tissues were collected from five different positions of the lung including central bronchus, lung periphery, and three equidistant points along its length. The levels of 7-mG in all the samples averaged 0.75 ± 0.57 adducts per 10^6 dG. No significant difference in adduct levels was observed at different lung positions. Compared to former smokers, the levels of 7-mG were higher ($p = 0.047$) in current smokers at two lung positions including the lung periphery [74].

Multiple methyl DNA base adducts were also detected in human urine samples [75–78]. Wang, et al developed a capillary LC-HRMS/MS method for the simultaneous analysis of 7-mG, 3-methyladenine (3-mA), and 1-methyladenine (1-mA) in human urine samples (Figure 2 and Table 1). They applied the method to the analysis of urine samples from 20 smokers and 14 nonsmokers. The levels of the three adducts were all significantly higher in smokers than nonsmokers, with the difference in 3-mA levels being the most significant (11-fold, $p < 0.0001$) [75]. Higher levels of 3-mA in smokers compared to nonsmokers were also observed in another two studies that used LC-MS/MS-based analytical methods [76,77]. In one study, the level of 3-mA was also correlated with the level of urinary NNAL in smokers ($n = 192$, $r = 0.48$, $p < 0.001$) [77]. The correlation was also observed between adduct (3-mA and 7-mG) levels and urinary NNAL in another study [78]. However, the levels of these methyl adducts detected in urine are far greater than the levels of adducts possibly formed by NNK, suggesting a contribution of methylating agents from other sources such as diet.

We have developed an ultrasensitive LC-NSI-HRMS/MS method for the analysis of methyl DNA phosphate adducts (B_1pMeB_2, Table 1 and Figure 2) in human lung DNA [79]. The adduct levels were measured in both tumor and adjacent normal tissues from 30 lung cancer patients, including 13 current smokers and 17 current nonsmokers, as confirmed by measurements of urinary cotinine and NNAL. Levels of total B_1pMeB_2 in normal lung tissues were higher ($p < 0.05$) in smokers than nonsmokers, with an average of 13 and 8 adducts per 10^9 nucleotides, respectively. More details on DNA phosphate adducts, including B_1pMeB_2, are discussed in Section 3.1.

2.5. Ethylating Agents

Studies have demonstrated the presence of direct-acting ethylating agent(s) in tobacco products and tobacco smoke, though without characterized structure(s). These agents can react with DNA forming ethyl DNA base adducts [80,81]. Compared to methyl DNA base adducts, ethyl DNA base adducts are generally more persistent in vivo with more specificity to smoking [82–84]. Certain ethyl DNA base adducts, for example O^4-ethylthymidine (O^4-etT) (Figure 2), are promutagenic lesions and may contribute to the initiation of hepatocellular carcinomas in animal models [85]. Ethyl DNA base adducts were reported to be present in human leukocytes [84,86], saliva [83,87], and urine samples [76,77,88].

Chen et al. developed a series of LC-NSI-MS/MS methods for the analysis of multiple ethyl DNA base adducts in human biological samples (Table 1) [83,84,86–88]. Two of their studies measured these DNA adducts in human leukocytes. In one study, three ethyl DNA base adducts, O^2-ethylthymidine (O^2-etT, Figure 2), N^3-ethylthymidine (N^3-etT), and O^4-etT, were detected and quantified in leukocyte DNA samples from 20 smokers and 20 nonsmokers [86]. The levels of O^2-etT, N^3-etT and O^4-etT in smokers were 45 ± 52, 41 ± 44, and 48 ± 54 adducts per 10^8 nucleotides, while nonsmokers had significantly lower ($p < 0.001$) adduct levels at 0.19 ± 0.87, 4.1 ± 13, and 1.0 ± 2.9 adducts per 10^8 nucleotides, respectively. Furthermore, the level of O^2-etT was correlated with the smoking index (number of cigarettes per day $\times$ years smoked) ($r = 0.48$, $p < 0.05$) [86]. In the other study, 3-ethyladenine (3-etA), and 7-ethylguanine (7-etG) were detected and quantified in leukocyte DNA samples from 20 smokers and 20 nonsmokers. The levels of 3-etA and 7-etG in smokers were 16 ± 7.8 and 9.7 ± 8.3 adducts per 10^8 nucleotides, significantly higher ($p < 0.0001$) than the levels in nonsmokers at 5.4 ± 2.6 and 0.3 ± 0.8 adducts per 10^8 nucleotides, respectively. Additionally, the levels of both adducts were correlated with the smoking index [84]. Chen et al. also measured ethyl DNA base adducts in human saliva. In one study, O^2-etT, N^3-etT and O^4-etT were detected in saliva samples from 20 smokers at 5.3 ± 6.2, 4.5 ± 5.7, and 4.2 ± 8.0 adducts per 10^8 nucleotides, respectively, while none of the adducts were detected in saliva samples from 13 nonsmokers [83]. In another study, 3-etA and 7-etG were detected and quantified in saliva samples from 15 smokers and 15 nonsmokers. The levels of 3-etA and 7-etG in smokers were 13 ± 7.0 and 14 ± 8.2 adducts per 10^8 nucleotides, significantly higher ($p < 0.0001$) than in nonsmokers with adduct levels at 9.7 ± 5.3 and 3.8 ± 2.8 adducts per 10^8 nucleotides, respectively. In addition, the levels of 7-etG in smokers were correlated with the number of cigarettes per day ($r = 0.76$, $p < 0.0001$) and the smoking index ($r = 0.85$, $p < 0.0001$) [87].

Ethyl DNA base adducts have also been reported to be present in human urine samples (Table 1) [76,77,88]. In addition to leukocytes and saliva, Chen, et al also developed an LC-NSI-MS/MS method for the analysis of 3-etA and 7-etG in urine samples from 21 smokers and 20 nonsmokers. 3-etA and 7-etG were detected in all smokers with levels at 69 ± 29 and 19 ± 14 pg/mL urine, respectively. In nonsmokers, the adducts were detected in 16 out of 20 samples, with levels at 3.5 ± 3.8 and 2.4 ± 3.0 pg/mL urine for 3-etA and 7-etG, respectively. Higher adduct levels were observed in smokers compared to nonsmokers [88]. Similar results were also observed in another two studies that measured 3-etA and 7-etG in urine samples [76,77].

2.6. 1,3-Butadiene

1,3-Butadiene is a colorless gas that is widely used in the polymer industry and is present in cigarette smoke, the urban environment, and automobile exhaust. 1,3-Butadiene is metabolized to epoxides, which react with DNA forming DNA adducts with the most abundant one in animals being 7-(2, 3, 4-trihydroxybut-1-yl) guanine (7-THBG, Figure 2) [89]. Sangaraju et al. developed a capillary LC-HRMS/MS method for the analysis of 7-THBG in human leukocytes (Table 1) [90]. The method was first applied to the analysis of leukocyte samples from 13 smokers and 13 nonsmokers. The levels of 7-THBG were 0.82 ± 0.51 and 0.71 ± 0.53 adducts per 10^8 nucleotides in smokers and nonsmokers, respectively, a non-significant difference ($p = 0.6$). To further evaluate the influence of smoking on adduct formation, the adduct levels were determined in 10 individuals who participated in a smoking cessation study. The leukocyte samples were collected before smoking cessation, and 28 and 84 days afterwards. The adduct level was not significantly changed by cessation. In the same study, the influence of occupational exposure to 1,3-butadiene on adduct levels was also investigated. 7-THBG was measured in leukocyte samples from 10 workers with known exposure levels and 10 matched controls. Compared to the controls (0.31 ± 0.22 adducts per 10^8 nucleotides), the levels of 7-THBG were significantly elevated in the exposed workers (0.97 ± 0.38 adducts per 10^8 nucleotides, $p < 0.001$) [90].

2.7. Acrolein

Human exposure to the α,β-unsaturated acrolein can be from tobacco smoke, automobile exhaust, plastic waste, heated cooking oil and endogenous lipid peroxidation [91]. Acrolein reacts with deoxyguanosine in DNA to form two pairs of regioisomeric $1,N^2$-propanodeoxyguanosine adducts: $(6R/S)$-3-(2'-deoxyribos-1'-yl)-5,6,7,8-tetrahydro-6-hydroxypyrimido[1,2-*a*]purine-10(3*H*)one (α-OH-Acr-dG, Figure 2) and $(8R/S)$-3-(2'-deoxyribos-1'-yl)-5,6,7,8-tetrahydro-8-hydroxypyrimido[1,2-*a*]purine-10(3*H*)one (γ-OH-Acr-dG). Compared to γ-OH-Acr-dG, α-OH-Acr-dG is more mutagenic and predominantly induces G$\rightarrow$T transversions.

We developed an LC-ESI-MS/MS method to analyze Acr-dG adducts in human leukocyte DNA (Table 1) and applied this method to the analysis of leukocyte DNA samples from 25 smokers and 25 nonsmokers [92]. γ-OH-Acr-dG was the predominant isomer in all samples, while α-OH-Acr-dG was detected in only one nonsmoker and two smokers. The total levels of α-OH-Acr-dG and γ-OH-Acr-dG were not different between smokers and nonsmokers, with levels averaging 0.74 ± 0.34 and 0.98 ± 0.55 adducts per 10^8 nucleotides, respectively [92]. In an earlier study, we also measured Acr-dG adducts in human lung DNA from smokers and nonsmokers using an LC-ESI-MS/MS method [93]. However, the smoking status of some subjects in that study was unknown, which made it difficult to interpret the role of smoking in the adduct formation. Recently, we have developed a new ultrasensitive LC-NSI-HRMS/MS method for the analysis of Acr-dG adducts in human lung DNA samples [94]. In addition to improved sensitivity, we also managed to lower the levels of artifactual formation of adducts during sample preparation and minimized the interference of artifactually formed DNA adducts in quantitation. We analyzed the levels of Acr-dG adducts in lung DNA of 19 smokers and 18 nonsmokers who underwent surgery for lung cancer. The smoking status of these subjects was confirmed by urinary total cotinine and NNAL. The levels of α-OH-Acr-dG averaged 0.86 ± 0.27 and 1.0 ± 0.47 adducts per 10^8 nucleotides in smokers and nonsmokers, respectively, and the levels of γ-OH-Acr-dG averaged 2.0 ± 1.4 and 1.5 ± 0.64 adducts per 10^8 nucleotides in smokers and nonsmokers, respectively. There was no significant difference in the levels of α-OH-Acr-dG or γ-OH-Acr-dG between smokers and nonsmokers [94]. While the sample size was relatively small, our results indicate that acrolein is not a major etiological agent for cigarette smoking related DNA damage.

Li et al. developed an LC-ESI-MS/MS-based method for the analysis of Acr-dG adducts in human saliva samples (Table 1). γ-OH-Acr-dG was detected in all samples, and no difference of adduct levels was observed between smokers ($n = 16$) and nonsmokers ($n = 16$). α-OH-Acr-dG was not detected in any of the samples [95]. γ-OH-Acr-dG as the major acrolein-derived DNA adduct was also confirmed in human lung and liver tissues in another LC-ESI-MS/MS-based study [96]. Using an immunochemical approach, Weng et al. measured γ-OH-Acr-dG in multiple human tissues including buccal cells, sputum, and lung tissue samples [53]. The levels of γ-OH-Acr-dG in buccal cells from smokers ($n = 33$) were significantly higher ($p < 0.0001$) compared to nonsmokers ($n = 17$). Similarly, the adduct levels in sputum were higher levels ($p < 0.05$) in smokers ($n = 22$) compared to nonsmokers ($n = 8$). In the same study, γ-OH-Acr-dG was also detected in noncancerous lung tissues obtained from lung cancer patients with higher levels in smokers ($n = 41$) compared to nonsmokers ($n = 13$) [53]. Compared to our study of γ-OH-Acr-dG in human lung tissues [94], the levels obtained by Weng et al. were 10–20 times higher, suggesting possible artifactual formation of γ-OH-Acr-dG, which was not investigated in their study. Similarly, a drastic difference of BPDE-dG levels was also observed between our study and the study by Weng et al. The average level of BPDE-dG in smokers' lung in our study was ~3 adducts per 10^{11} nucleotides [32], while the average level in their study was ~1000 times higher [53]. Due to the potential significant artifactual formation of adducts in that study, its results and conclusions are questionable. Another study employed both ^{32}P-postlabelling and immunochemical methods for the analysis of α-OH-Acr-dG and γ-OH-Acr-dG in normal human urothelial mucosa and bladder tumor tissues. γ-OH-Acr-dG was the predominant isomer in both tissues, with higher levels observed in tumor tissue compared to normal tissue [68].

2.8. Formaldehyde

Formaldehyde is considered a human carcinogen and is widely present in the environment. It is mainly used in the production of industrial resins. It is also produced during cooking and cigarette smoking as well as by endogenous processes. Formaldehyde reacts with DNA and forms several DNA adducts and cross-links, with the most abundant one being N^6-hydroxymethyldeoxyadenosine (HOMe-dA, Figure 2). We were the first group that demonstrated the presence of this specific formaldehyde-DNA adduct in humans [97].

Li et al. developed an LC-ESI-MS/MS-based method for the simultaneous analysis of two formaldehyde-derived DNA adducts, HOMe-dA and N^2-hydroxymethyldeoxyguanosine (HOMe-dG, Figure 2), in human saliva samples (Table 1). HOMe-dA and HOMe-dG were measured after sodium cyanoborohydride (NaBH$_3$CN) reduction as N^6-methyldeoxyadenosine and N^2-methyldeoxyguanosine, respectively. The levels of HOMe-dA were 996 ± 757 and 670 ± 524 adducts per 10^8 nucleotides in smokers ($n = 16$) and nonsmokers ($n = 16$), respectively, while HOMe-dG was detected in a much lower level in the same samples, with 26 ± 21 and 20 ± 11 adducts per 10^8 nucleotides being present in smokers and nonsmokers, respectively. No statistical difference of adduct levels was observed between smokers and nonsmokers [95].

2.9. Acetaldehyde and Crotonaldehyde

Acetaldehyde associated with the consumption of alcoholic beverages is carcinogenic to humans [91]. Acetaldehyde is ubiquitous in the human environment, and is one of the most prevalent carcinogens in cigarette smoke. It also occurs widely in fruits, vegetables, and cooked meat. Crotonaldehyde is also ubiquitously present in the environment. It is present in mobile source emissions, tobacco smoke, and other thermal degradation mixtures. Crotonaldehyde is mutagenic and carcinogenic. Both acetaldehyde and crotonaldehyde react with DNA forming a pair of diastereomeric adducts, (6*S*,8*S*)- and (6*R*,8*R*)-3-(2′-deoxyribos-1′-yl)-5,6,7,8-tetrahydro-8-hydroxy-6-methylpyrimido[1,2-*a*]purine-10(3*H*)one (Cro-dG, Figure 2). In addition to Cro-dG, acetaldehyde also reacts with DNA to generate its major DNA adduct, N^2-ethylidene-deoxyguanosine (N^2-ethylidene-dG). We have previously developed LC-ESI-MS/MS-based methods and detected both Cro-dG and N^2-ethylidene-dG in human DNA [98,99].

Using an immunochemical approach, Weng et al. detected Cro-dG in human buccal cells, sputum, and lung tissue samples (Table 1) [53]. The levels of Cro-dG in buccal cells from smokers ($n = 33$) were significantly higher ($p < 0.0001$) compared to nonsmokers ($n = 17$). Similarly, the adduct levels in sputum were also higher ($p < 0.05$) in smokers ($n = 22$) compared to nonsmokers ($n = 8$). Cro-dG was also detected in noncancerous lung tissues obtained from lung cancer patients with higher levels in smokers ($n = 41$) compared to nonsmokers ($n = 13$) [53]. However, the levels of Cro-dG reached ~1,000 adducts per 10^8 nucleotides in buccal cells and sputum. Such high levels are surprising, and the specificity of the method as well as artifactual formation during sample preparation should be evaluated. In a second study using an LC-ESI-MS/MS method, Cro-dG was detected in saliva samples from 16 smokers at 2.6 ± 2.1 adducts per 10^8 nucleotides, while no Cro-dG was detected in any of the samples from nonsmokers [94]. In a third study, Cro-dG was analyzed in urine samples using an LC-ESI-MS/MS method to investigate exposure to urban air pollution. Higher adduct levels were observed in subjects ($n = 47$) exposed to air pollution with a median value of 21 fmol of Cro-dG per mg creatinine, compared to controls ($n = 35$) with a median value of 8 fmol of Cro-dG per mg creatinine ($p < 0.05$) [100].

The acetaldehyde-derived N^2-ethylidene-dG is generally measured after reduction (e.g., by NaBH$_3$CN) as N^2-ethyl-dG (Table 1) [99]. In the second study mentioned above, N^2-ethylidene-dG (as N^2-ethyl-dG) was also measured in the same saliva samples, with levels of 6.2 ± 3.5 and 0.59 ± 0.89 adducts per 10^8 nucleotides being present in smokers and nonsmokers, respectively. The difference between smokers and nonsmokers was not statistically significant [94]. The formation of N^2-ethylidene-dG was also investigated in alcohol consumption-related studies [101–103].

2.10. Acrylamide

Acrylamide is a probable human carcinogen and is present in tobacco smoke and in carbohydrate-rich foods processed at high temperatures. Acrylamide is metabolized to glycidamide, which reacts with DNA to form a major DNA adduct, 7-(2-carbamoyl-2-hydroxyethyl) guanine (7-GAG, Figure 2). Huang, et al developed an LC-ESI-MS/MS method for the analysis of 7-GAG in human urine (Table 1). The method was applied to the analysis of urine samples from 30 smokers and 33 nonsmokers. The levels of 7-GAG ranged from 0.61–6.22 (mean: 1.4) and 0.36–3.0 (mean: 0.93) µg/g creatinine in smokers and nonsmokers, respectively, a non-significant difference. However, the urinary

N-acetyl-S-(propionamide)-cysteine, a metabolite of acrylamide, showed significantly higher levels ($p < 0.001$) in smokers compared to nonsmokers [104]. The same research group investigated in another study the effect of occupational exposure to acrylamide on the formation of 7-GAG. The adduct was measured in urine samples from eight workers who were exposed to acrylamide and 36 controls. Higher levels ($p < 0.001$) of 7-GAG were observed in exposed workers, with adduct levels ranging from 1.0–13 (mean: 2.5) µg/g creatinine, compared to controls with adduct levels ranging from 0.20–0.93 (mean: 0.36) µg/g creatinine [105].

2.11. Oxidative Damage

Reactive oxygen species (ROS) constitute an important class of DNA damaging agents, and their generation can be elevated due to exposure to tobacco smoke. In addition to damaging DNA directly, ROS also cause DNA damage indirectly by reacting with other macromolecules [106]. For example, ROS can lead to lipid peroxidation of polyunsaturated fatty acids and produce malondialdehyde, which reacts with DNA to form DNA adducts, with the predominant one being 3-(2-deoxy-β-d-erythro-pentafuranosyl)pyrimido[1,2-α]purin-10(3*H*)-one deoxyguanosine (M_1dG, Figure 2) [107]. In addition to M_1dG, some representative oxidative damage-associated DNA adducts include 8-oxo-dG, $1,N^6$-etheno-2-deoxyadenosine (εdA), and $3,N^4$-etheno-2′-deoxycytidine (εdC). A more comprehensive summary of the formation and biological consequences of oxidative stress-induced DNA damage was covered by a recent review [106].

M_1dG is a premutagenic lesion and induces G→T and G→A mutations, which could be an important step in the etiology of oxidative associated-diseases. M_1dG was detected in human leukocytes [33,48,108–111] and nasal epithelium (Table 1) [112,113]. Studies suggest that smoking contributes to the levels of M_1dG in humans. In a study of workers occupationally exposed to silica dust, the levels of M_1dG in nasal epithelium of smokers ($n = 58$) and former smokers ($n = 63$) were 78 ± 9.8 and 81 ± 9.7 adducts per 10^8 nucleotides, respectively, significantly higher ($p < 0.05$) compared to nonsmokers ($n = 132$) with M_1dG levels of 57 ± 6.2 adducts per 10^8 nucleotides [112]. In a study of occupational exposure to asbestos, higher M_1dG levels ($p = 0.005$) were observed in leukocytes of heavy smokers (>40 packs/year, $n = 12$) compared to nonsmokers ($n = 136$), but the difference was not observed among moderate smokers (20.1–40 packs/year, $n = 29$), light smokers (0.1–20 packs/year, $n = 20$), and nonsmokers [108]. In another study investigating the effects of diet on M_1dG levels among industrial estate workers, no difference was observed between smokers ($n = 46$) and nonsmokers ($n = 17$) among those workers. However, in the control group, the adduct levels were significantly higher ($p < 0.05$) in smokers ($n = 64$) compared to nonsmokers ($n = 53$) [110]. In addition to the findings that the levels of M_1dG were elevated due to smoking, some studies also demonstrated the correlation of M_1dG and DNA methylation in smokers [48,111].

We developed an LC-NSI-HRMS/MS method for the analysis of M_1dG in human leukocyte DNA, and measured M_1dG in buffy coat samples (leukocyte-containing fraction) from 25 smokers and 25 nonsmokers (Table 1). The adduct levels in smokers and nonsmokers averaged 2.2 ± 2.4 and 1.9 ± 2.0 adducts per 10^8 nucleotides, respectively, a non-significant difference [33]. Similarly, the effect of smoking on M_1dG formation was not significant in another two studies [109,113].

8-Oxo-dG is the most prominent and widely studied oxidative damage-related DNA adduct. It is a highly mutagenic lesion and mispairs with A during DNA replication leading to a GC→AT conversion. 8-Oxo-dG was detected in various human sample types, including semen [114,115], retina [116], leukocytes [117–120], and urine samples (Table 1) [121]. In a study investigating the effect of tobacco use on the possible etiology of childhood cancer, 8-oxo-dG was measured by an immunochemical method in semen samples from the children's fathers. The levels of 8-oxo-dG were 66 ± 2.9, 54 ± 4.4, and 179 ± 20 ng/mL in smokers ($n = 33$), tobacco chewers ($n = 31$), and subjects who both smoked and chewed tobacco ($n = 41$), respectively, all significantly higher than nonsmokers ($n = 33$) with adduct levels of 34 ± 1.1 ng/mL [115]. The effect of smoking on 8-oxo-dG formation was either not significant or not investigated in other studies [114,116–121].

Table 1. Detection of tobacco smoke-related DNA adducts in human samples.

DNA Adduct	Source	Tissue	Method of Detection	Reference
HPB-releasing DNA adducts	NNK, NNN	Buccal cells	LC-NSI-HRMS/MS, LC-MS/MS	[28,29]
Bulky adducts	PAH, other aromatic and nonpolar chemicals (?)	Leukocytes, lung	^{32}P-postlabelling	[39–52]
BPDE-N^2-dG	Benzo[a]pyrene	Buccal cells, sputum, lung, leukocytes, lymphocytes, tonsil, oropharynx, gingiva, prostate	Immunochemical, LC-NSI-HRMS/MS	[32,53–58]
4-ABP-C8-dG	4-aminobiphenyl (4-ABP)	Bladder, breast	LC-NSI-HRMS/MS, ^{32}P-postlabelling, LC-ESI-MS/MS^3	[51,67,68]
4-ABP-C8-dA	4-ABP	Bladder	^{32}P-postlabelling	[68]
4-ABP-N^2-dG	4-ABP	Bladder	^{32}P-postlabelling	[68]
O^6-methyldeoxyguanosine (O^6-mdG)	NNK, NDMA, other methylating agents	Lung	Immunochemical	[53]
7-Methylguanine (7-mG)	NNK, NDMA, other methylating agents	Lung, urine,	Capillary LC-HRMS/MS, Immunochemical	[74,75,78]
3-Methyladenine (3-mA)	NNK, other methylating agents	Urine	Capillary LC-HRMS/MS, LC-ESI-MS/MS	[75–78]
1-Methyladenine (1-mA)	NNK, other methylating agents	Urine	Capillary LC-HRMS/MS	[75]
Methyl DNA phosphate adducts (B$_1$pMeB$_2$)	NNK, other methylating agents	Lung	LC-NSI-HRMS/MS	[79]
3-Ethyladenine (3-etA)	Unknown ethylating agents	Urine, saliva	LC-ESI-MS/MS, LC-NSI-MS/MS	[76,77,87]
7-Ethylguanine (7-etG)	Unknown ethylating agents	Urine, leukocytes, saliva	LC-ESI-MS/MS, LC-NSI-MS/MS	[76,84,87,88]
O^2-ethylthymidine (O^2-etT)	Unknown ethylating agents	Leukocytes, saliva	LC-NSI-MS/MS	[83,86]
O^4-ethylthymidine (O^4-etT)	Unknown ethylating agents	Leukocytes, saliva	LC-NSI-MS/MS	[83,86]
N^3-ethylthymidine (N^3-etT)	Unknown ethylating agents	Leukocytes, saliva	LC-NSI-MS/MS	[83,84,86,88]
7-(2, 3, 4-trihydroxybut-1-yl) guanine (7-THBG)	1,3-Butadiene	Leukocytes	Capillary LC-HRMS/MS	[90]
1,N^2-propanodeoxyguanosine (Acr-dG)	Acrolein	Buccal cells, sputum, lung, saliva, bladder, liver	Immunochemical, ^{32}P-postlabelling, LC-MS/MS, LC-NSI-HRMS/MS	[53,68,94–96]
N^6-hydroxymethyldeoxyadenosine (HOMe-dA)	Formaldehyde	Saliva	LC-ESI-MS/MS	[95]
N^2-hydroxymethyldeoxyguanosine (HOMe-dG)	Formaldehyde	Saliva	LC-ESI-MS/MS	[95]
N^2-ethylidene-deoxyguanosine (Ethylidene-dG)	Acetaldehyde	Saliva, leukocytes, oral cells	LC-ESI-MS/MS	[95,101–103]
3-(2′-Deoxyribos-1′-yl)-5,6,7,8-tetrahydro-8-hydroxy-6-methylpyrimido[1,2-a]purine-10(3H)one (Cro-dG)	Crotonaldehyde, acetaldehyde	Buccal cells, sputum, lung, urine	Immunochemical, LC-ESI-MS/MS	[53,100]
7-(2-carbamoyl-2-hydroxyethyl) guanine (7-GAG)	Acrylamide	Urine	LC-ESI-MS/MS	[104,105]
3-(2-deoxy-β-d-erythro-pentafuranosyl)pyrimido[1,2-α]purin-10(3H)-one deoxyguanosine (M$_1$dG)	Malondialdehyde	Leukocytes, nasal epithelium	^{32}P-postlabelling, LC-NSI-HRMS/MS	[33,48,108–113]
8-Oxo-dG	ROS	Semen, retina, leukocytes, urine	Immunochemical, LC-NSI-MS/MS, electrochemical	[114–121]
1,N^6-etheno-2′-deoxyadenosine (εdA)	4-hydroxy-2-nonenal	Urine	LC-ESI-MS/MS	[122]
3,N^4-etheno-2′-deoxycytidine (εdC)	4-hydroxy-2-nonenal	Urine	LC-ESI-MS/MS	[122]

εdA and εdC are two representative DNA adducts formed by 4-hydroxy-2-nonenal, a product of lipid peroxidation. Both adducts are promutagenic and could be useful markers to assess oxidative stress-derived DNA damage. Using an LC-ESI-MS/MS method, Bin, et al measured εdA and εdC in urine to investigate the effect of occupational exposure to diesel engine exhaust on adduct formation (Table 1) [122]. The levels of εdA averaged 0.19 nmol/g creatinine in exposed workers (n = 108), significantly higher ($p < 0.001$) than the control group (n = 109) with εdA levels averaging 0.09 nmol/g creatinine. The levels of εdC showed no difference between the two groups. The contribution of smoking to the formation of εdA or εdC was not significant [122].

3. Newly Characterized Tobacco-Specific DNA Adducts

Since 2012, multiple DNA adducts formed by tobacco-specific carcinogens (i.e., NNK and NNN) have been characterized and quantified in in vitro models and in NNK- or NNN-treated animals [19–21,31,34–36,123,124]. In addition to DNA base adducts, which have been extensively studied, we have recently characterized and measured a panel of DNA phosphate adducts formed by NNK and NNN [20,34–36]. The biological significance of these newly identified DNA adducts is still largely unknown and further studies need to be conducted to better understand the formation, removal, and biological implications of these DNA adducts to facilitate their use as biomarkers in exposure evaluation and cancer risk assessment.

3.1. DNA Phosphate Adducts

In addition to DNA base moieties, some alkylating agents also react with the oxygen of the phosphate backbone to form DNA phosphate adducts [125]. Earlier studies have demonstrated that phosphate adducts formed by certain carcinogens persisted in vivo and had longer half-lives than their corresponding base adducts [126,127], suggesting DNA phosphate adducts may serve as better biomarkers of chronic exposure to those carcinogens. For the formation of DNA phosphate adducts by the tobacco-specific carcinogens NNK and NNN, there was only one study using a transalkylation approach to indirectly measure the adduct levels in [^{3}H]NNK-treated mice [128]. However, that study only provided indirect proof of the presence of NNK-derived DNA phosphate adducts, and the chemical structures were not characterized for individual adducts.

One major challenge of direct measurement of DNA phosphate adducts is their structural complexity. After enzyme hydrolysis of DNA samples, DNA phosphate adducts are measured as phosphotriesters (PTE)—B_1p(alkyl)B_2. B_1 and B_2 are the same or different nucleosides, which can be 10 different combinations of the four nucleosides (Table 2) [125]. Due to the tetrahedral phosphate group in B_1p(alkyl)B_2, there can be Rp or Sp diastereomers present. With the same nucleosides, there can be two isomers depending on which oxygen is alkylated. With different nucleosides, because of the connection of sugar moieties, it can be either B_1-5′-alkyl-3′-B_2 or B_1-3′-alkyl-5′-B_2. Therefore, there can be 32 different isomers of B_1p(alkyl)B_2 (Table 2). In order to characterize and measure all of the 32 possible combinations, a specific and powerful analytical method is required. We have developed a series of LC-NSI-HRMS/MS-based methods and characterized four different types of NNK- and NNN-derived DNA phosphate adducts, including pyridyloxobutyl DNA phosphate adducts [B_1p(POB)B_2], pyridylhydroxybutyl DNA phosphate adducts [B_1p(PHB)B_2: B_1p(PHB)$_s$B2 + B_1p(PHB)$_b$B$_2$], and methyl DNA phosphate adducts [B_1pMeB$_2$] [20,34–36,124].

Table 2. Numbers of possible isomers of NNK-derived DNA phosphate adducts.

B_1-B_2	Number of Possible Isomers				Total
	B_1pMeB_2	$B_1p(POB)B_2$	$B_1p(PHB)_sB_2$	$B_1p(PHB)_bB_2$	
A-A	2	2	4	8	16
C-C	2	2	4	8	16
G-G	2	2	4	8	16
T-T	2	2	4	8	16
A-C	4	4	8	16	32
A-G	4	4	8	16	32
A-T	4	4	8	16	32
C-G	4	4	8	16	32
C-T	4	4	8	16	32
G-T	4	4	8	16	32
Total	32	32	64	128	256

* Chiral center.

3.1.1. $B_1p(POB)B_2$

One metabolic pathway of NNK proceeds via α-hydroxylation of its methyl group to produce the unstable intermediate **6** (Figure 3). This intermediate spontaneously yields **11** and then the alkylating agent **14**, which reacts with DNA to form POB DNA base adducts and POB DNA phosphate adducts—$B_1p(POB)B_2$ (Figure 3) [20]. We first characterized and detected $B_1p(POB)B_2$ in calf thymus DNA (CT-DNA) treated with 4-(acetoxymethylnitrosamino)-1-(3-pyridyl)-1-butanone (NNKOAc, **2**), a regiochemically activated form of NNK. The structures of two combinations of $B_1p(POB)B_2$, Cp(POB)C and Tp(POB)T, were confirmed by comparing with the synthetic standards [20,36], and the identities of the other $B_1p(POB)B_2$ adducts were confirmed on the basis of the accurate masses of the precursor ions and the corresponding fragment ions. A total of 30 out of 32 possible $B_1p(POB)B_2$ adducts were detected in NNKOAc-treated CT-DNA. The $B_1p(POB)B_2$ adducts were also detected and quantified in rats treated with NNK acutely (0.1 mmol/kg once daily for 4 days by subcutaneous injection) and chronically (5 ppm in drinking water for 10, 30, 50, and 70 weeks). In the chronically treated rats, some $B_1p(POB)B_2$ adducts are persistent and abundant over 70 weeks, which suggested that those adducts could potentially be detected in smokers and be used as biomarkers to investigate chronic exposure to NNK [20]. The $B_1p(POB)B_2$ adducts were also detected in rats treated with NNAL [35] or NNN [124].

3.1.2. $B_1p(PHB)B_2$

NNK is extensively metabolized to NNAL, which undergoes α-hydroxylation to produce diazonium ion **16**. Diazonium ion **16** reacts with DNA to form PHB base adducts and PHB DNA phosphate adducts, $B_1p(PHB)_sB_2$, where the subscript s represents 'straight chain' (Figure 3). Diazonium ion **16** further rearranges to carbocation **17**, which also reacts with DNA to form another type of PHB DNA phosphate adducts—$B_1p(PHB)_bB_2$, where the subscript b represents 'branched chain' (Figure 3). Since both $B_1p(PHB)_sB_2$ and $B_1p(PHB)_bB_2$ are derived from NNAL, they are collectively called $B_1p(PHB)B_2$ phosphate adducts [36]. For $B_1p(PHB)_sB_2$, because of the two chiral centers, the phosphorus and carbinol carbon, there can be four and eight isomers with the same and different nucleoside combinations, respectively. Therefore, a total of 64 different possible $B_1p(PHB)_sB_2$ adducts can be formed (Table 2). Compared to $B_1p(PHB)_sB_2$, an additional chiral center at the methyl-bearing carbon is present in the structure of $B_1p(PHB)_bB_2$ adducts, so there can be eight and 16 isomers with the same and different nucleoside combinations, respectively. Consequently, a total of 128 different possible $B_1p(PHB)_bB_2$ adducts can be formed (Table 2). Therefore, a total of 192 possible $B_1p(PHB)B_2$ adducts can be formed from the α-hydroxylation of NNAL's methyl group. Using a similar strategy as in $B_1p(POB)B_2$ characterization, a total of 107 out of 192 possible $B_1p(PHB)B_2$ adducts were detected in the NNK-treated rat lung DNA samples. Similar to $B_1p(POB)B_2$ adducts, certain $B_1p(PHB)B_2$ adducts demonstrated persistence for over 70 weeks, suggesting that they could be potential biomarkers of chronic exposure to NNK and NNAL [36]. Both $B_1p(POB)B_2$ and $B_1p(PHB)B_2$ adducts were also detected in NNAL-treated rats [35].

3.1.3. B_1pMeB_2

In addition to the methyl group, α-hydroxylation also occurs on the α-methylene carbon of NNK to produce methane diazohydroxide **15**, which reacts with DNA to form methyl DNA base and phosphate adducts (B_1pMeB_2) [34]. Similar to $B_1p(POB)B_2$ adducts, there are 32 possible isomers of B_1pMeB_2 phosphate adducts (Table 2). We characterized and detected B_1pMeB_2 adducts in lung DNA of rats treated with NNK in their drinking water (5 ppm) for 10, 30, 50, and 70 weeks, and a total of 23 out of 32 possible isomers were detected. Thus far, we have identified 30 $B_1p(POB)B_2$ adducts and 107 $B_1p(PHB)B_2$ adducts, resulting in a total of 160 different structurally unique DNA phosphate adducts in NNK-treated rats. This is by far the most structurally diverse panel of DNA adducts identified from any carcinogen. Consistent with the observations of $B_1p(POB)B_2$ and $B_1p(PHB)B_2$

adducts, certain B_1pMeB_2 phosphate adducts showed persistence in NNK-treated rats. B_1pMeB_2 adducts were also detected in NNAL-treated rats in our study [34]. The relative levels of $B_1p(POB)B_2$, $B_1p(PHB)B_2$ and B_1pMeB_2 phosphate adducts, and their corresponding base adducts in lung DNA of rats treated with NNK or NNAL were summarized in that study.

3.2. DNA Base Adducts

In addition to DNA phosphate adducts, multiple DNA base adducts that are associated with NNK and NNN were either newly characterized in vitro or detected for the first time in vivo. These DNA adducts include 2-(2-(3-pyridyl)-*N*-pyrrolidinyl)-2′-deoxyinosine (py-py-dI) [31], 3-POB-dC, N^4-POB-dC [21], O^4-POB-dT [123], N^6-POB-dA, and N^1-POB-dI [19].

3.2.1. Py-py-dI

Metabolic activation of NNN occurs via one of two pathways: 2′-hydroxylation or 5′-hydroxylation (Figure 3), and 5′-hydroxylation is likely to be the major metabolic pathway in humans. The major adduct formed in vitro via 5′-hydroxylation was py-py-dI, which forms upon the reaction of the N^2 position of dG with diazonium ion **13** followed by $NaBH_3CN$ reduction [129]. Using an LC-ESI-MS/MS method, we have recently detected py-py-dI in rats treated with NNN in the drinking water (7–500 ppm). Py-py-dI was the major DNA adduct resulting from 5′-hydroxylation, with the highest levels being detected in the lung and nasal cavity of NNN-treated rats. Our study also showed that py-py-dI was more abundant than the 2′-hydroxylation of NNN-derived POB DNA base adducts in the in vitro studies in which human liver S9 fraction or human hepatocytes were incubated with NNN (2–500 μM) [31]. The results of this study identified py-py-dI as the major NNN-derived DNA adduct in the human enzyme systems examined. Thus, py-py-dI could be a useful biomarker to investigate the metabolic activation of NNN in humans.

3.2.2. 3-POB-dC and N^4-POB-dC

The intermediate **14** from NNK also reacts with dC in DNA and forms O^2-POB-C, which has been detected in rats treated with NNK or NNN. In addition to the O^2-position, the 3-position and exocyclic N^4-amino group of dC are also potentially reactive sites for alkylation, possibly forming 3-POB-dC and N^4-POB-dC, respectively. We have recently synthesized the chemical standards of 3-POB-dC and N^4-POB-dC, and confirmed the presence of both adducts in NNKOAc-treated CT-DNA. In agreement with our previous studies, O^2-POB-C was the most abundant dC adduct, while considerably lower amounts of 3-POB-dC and N^4-POB-dC were present in NNKOAc-treated CT-DNA [21]. The results of this study provide a more complete picture of dC adduct formation in the reaction of NNKOAc with DNA.

3.2.3. O^4-POB-dT

The major dT adduct formed by NNK and NNN is O^2-POB-dT. Using an LC-NSI-MS/MS method, Leng, et al detected and characterized for the first time a new dT adduct—O^4-POB-dT, in NNKOAc-treated CT-DNA [123]. The identity of O^4-POB-dT was confirmed by the chemical standard synthesized in the study. O^4-POB-dT was also detected in human skin fibroblasts and Chinese hamster ovary cells treated with NNKOAc. O^2-POB-dT was detected in all the systems with higher levels than O^4-POB-dT. Results from this study indicated that both O^4-POB-dT and O^2-POB-dT were subjected to repair by the nucleotide excision repair pathway [123].

3.2.4. dA Adducts

All the previous studies have identified and measured a panel of POB DNA base adducts of dG, dC, and dT, but not dA. To complete the panel of POB DNA base adducts, we determined the possible formation of POB-dA adducts in the in vitro NNKOAc-treated CT-DNA and in vivo

NNK-treated rats [19]. We hypothesized that the initial alkylation of dA would occur at the N^1 position (Figure 3), which results in an unstable cationic intermediate adduct. This unstable adduct can either deprotonate and undergo spontaneous Dimroth rearrangement to give N^6-POB-dA, or deaminate via an addition−elimination mechanism to give N^1-POB-dI. In support of our hypothesis, both N^6-POB-dA and N^1-POB-dI were detected in NNKOAc-treated CT-DNA using our newly developed LC-ESI-MS/MS method. In contrast, only N^6-POB-dA but not N^1-POB-dI was detected in lung and liver DNA from rats treated with NNK for 50 weeks (5 ppm in the drinking water), possibly due to the rapid repair of N^1-POB-dI. Similarly, we also detected N6-PHB-dA in the NNAL-treated rats [19].

4. Perspectives

4.1. Current Limitations

Among all the adducts described in this review, some have been found to be at the same or overlapping levels in smokers and nonsmokers, while others had significantly higher levels in smokers compared to nonsmokers. One should keep in mind when investigating the effects of smoking on DNA adduct formation, the association may be compromised by a number of confounding variables, such as current smoking status at the time of sample collection, exposures to secondhand smoke, air pollution, occupational exposure, dietary habits, and disease status. For example, self-reported smoking status collected through questionnaires may not be accurate. In addition, in studies which used surgically obtained tissues from cancer patients, smoking status may not be up-to-date if subjects stopped smoking for a few weeks or more before surgery. This would have an unknown effect on DNA adduct levels since their persistence in those human tissues is generally unknown. To overcome this limitation, urine samples can be collected from the same subjects for the measurement of tobacco biomarkers (e.g., NNAL and cotinine) to confirm their current smoking status.

Another potential confounding variable is the presence in the environment of the same chemicals that form DNA adducts. For example, PAH are tobacco smoke carcinogens and DNA adduct formation by PAH is often used to investigate smoking-related cancers. However, PAH are also present in the environment where combustion occurs and in the diet due to high temperature cooking. Therefore, factors such as air pollution and diet need to be taken into account when interpreting the association between PAH-derived DNA adduct formation and tobacco smoke exposure. The only carcinogens that are specific to tobacco smoke are the tobacco-specific nitrosamines such as NNK and NNN [7,15]. DNA adducts formed by NNK and NNN therefore have the potential to be ideal biomarkers to investigate tobacco-related cancer risk.

4.2. Analytical Methodologies

The chemical analysis of DNA adducts has been extensively reviewed [107,130,131]. To utilize DNA adducts as biomarkers of tobacco smoke exposure and cancer risk assessment, reliable analytical methods have to be developed and fully validated for linearity, accuracy, precision, specificity, and ruggedness. High-throughput methods are preferred and will allow the analysis of DNA adducts to be performed in studies with large sample sizes, which are typical in epidemiology. ^{32}P-postlabeling and immunochemical approaches were extensively used for DNA adduct analysis in the past and are still quite prevalent nowadays. However, the uncertainty in labeling efficiency and the lack of physiochemical structural confirmation of the ^{32}P-postlabeling technique are major disadvantages potentially undermining the adequacy of data produced by this method. A critical drawback of the immunochemical approach is that the specificity of the antibodies for some DNA adducts is uncertain, as they may cross-react with other DNA lesions or endogenous components, resulting in errors in characterization and quantitation. The disadvantages of ^{32}P-postlabeling and immunochemical approaches are evident when the results are compared to those obtained by mass spectrometry-based methods. For example, the levels of BPDE-N^2-dG and 4-ABP-C8-dG, as discussed in this review, and the levels of DNA adducts formed by 2-amino-1-methyl-6-phenylimidazo[4,5-*b*]pyridine [65], were all generally 10 to 100-fold higher in the same type of tissues when results of ^{32}P-postlabeling or immunochemical methods were compared to those obtained by using mass spectrometry-based methods. This difference suggests that one should be cautious when interpreting data obtained from ^{32}P-postlabeling or immunochemical methods. We recommend that DNA adduct identity and quantity be confirmed by well-validated mass spectrometry-based methods.

During the past few decades, mass spectrometry coupled with the stable isotope dilution methods has evolved to become the gold standard for identification, characterization, and quantitation of DNA adducts. Further development and validation of mass spectrometry-based methods for high-throughput, sensitive and accurate quantitation of DNA adducts in humans are required in future studies. The LC-NSI-HRMS/MS-based technique can achieve detection limits in the low amol (10^{-18} mol), or even high zmol (10^{-21} mol) range, allowing the highest levels of specificity and sensitivity. To achieve such detection limits in human biological samples, the samples should be thoroughly purified to avoid co-eluting interferences and ion suppression during the mass spectrometry analysis. In addition to the analysis of targeted DNA adducts, mass spectrometry-based DNA adductomics can be used to comprehensively screen for DNA modifications, including both known and unknown/unidentified DNA adducts, and expedite the discovery of novel tobacco smoke-related DNA adducts [132].

4.3. Artifactual Formation

Particularly for DNA adducts formed endogenously, in addition to the efforts on improving the detection sensitivity and sample purification process, precautions should be taken to eliminate artifactual formation during all the steps of sample analysis, including sample collection and storage, DNA isolation and enzymatic hydrolysis, sample purification and enrichment, and mass spectrometry analysis. For instance, artifactual formation is one likely contributor to the relatively high levels of γ-OH-Acr-dG and Cro-dG measured in one study [53]. Another example is 8-oxo-dG. When the DNA is extracted from the cell, the in vivo protecting enzymes and antioxidants are no longer present to neutralize oxygen radicals and repair DNA damage, and 8-oxo-dG can thus be formed by the oxidation of free dG as well as dG in DNA [133].

We suggest the following strategy to systematically evaluate and reduce the possible artifactual formation of endogenous DNA adducts: 1) Sample collection. The samples should be handled by well-trained personnel to reduce the period between collection and storage. The samples should be placed on ice if they cannot be processed immediately; 2) Sample storage. The storage conditions should be evaluated to determine that no artifactual formation or degradation of DNA adducts occurs. A comparison of DNA adduct levels can be performed between the DNA isolated immediately after sample collection and the DNA isolated from the same sample after a certain period of storage; 3) DNA isolation. Commercially available calf thymus DNA (CT-DNA) contains most of the endogenous DNA adducts at certain levels, and can be used to investigate possible artifactual formation. To perform such an experiment, CT-DNA is dissolved in solution and isolated using the proposed protocol [33,116]. The adduct levels in the isolated CT-DNA are then compared to the levels in the same CT-DNA without isolation to determine whether the adduct levels are impacted by the DNA isolation process. This approach can also be used to investigate whether certain chemicals (e.g., antioxidants) can be used to lower or prevent artifactual formation; 4) DNA enzymatic hydrolysis and sample purification. Stable isotope-labeled nucleotides can be added to samples prior to starting DNA hydrolysis, and the possible artifactual formation can be monitored by checking whether correspondingly labeled DNA adducts are present [116]; 5) Mass spectrometry analysis. Depending on the sample purification approach, the free nucleotides may still be in the final samples and form corresponding adducts in the ion source of the mass spectrometer (e.g., dG forms 8-oxo-dG in the ESI ion source) [116]. If the free nucleotide happens to co-elute with the adduct, additional adducts may be formed from the free nucleotide, resulting in the overestimation of the adduct levels. To avoid this, the free nucleotide should be separated from the adduct, either at the sample purification step or on the LC column during the mass spectrometry analysis.

4.4. Other Considerations

Stability of DNA adducts over time. The adduct levels are supposed to reflect the balance between DNA adduct formation and repair. If this is the case, it would be expected that the DNA adduct level would remain relatively constant over time in a similar exposure environment. In the case of tobacco smoke-related DNA adducts, the level of an ideal DNA adduct as a biomarker should stay constant if the smoker maintains his/her habit. A longitudinal study can be performed where the same biological samples are collected at different time intervals (days, weeks, or months apart) and the level of this DNA adduct is measured to determine its stability over time.

Surrogate tissues. To conduct a longitudinal study, surgically obtaining human tissues such as lung becomes highly impractical. Instead, surrogate tissues such as bronchoalveolar lavage (BAL), peripheral white blood cells or oral cells can be used. For example, studies have demonstrated a correlation between changes in the oral cavity and lung in smokers [26,134], so oral cells would potentially be a source of readily obtained biological samples.

Computational approach. In addition to experimental techniques, computational approaches have been used to study conformational preference of the adducts in DNA, their potential to cause mismatch and mutations, and their interactions with DNA polymerases and repair enzymes [135–137]. For example, a multiscale computational approach demonstrated that both O^6-POB-G and O^6-PHB-G pair with C and T, with the latter being more mutagenic because of the difference in the bulky moiety hydrogen-bonding pattern [135]. Such studies provide important insights into the biological significance of smoking-related DNA adducts, which is critical for understanding their role in tobacco-induced cancers.

Author Contributions: All authors critically reviewed the literature and contributed to drafting the manuscript.

Funding: The work cited in this review conducted by the Hecht laboratory has been supported by Grants CA-81301 and CA-138338 from the National Cancer Institute; The work cited in this review conducted by the Stepanov laboratory has been supported by Grants CA-179246 and CA-180880 from the National Cancer Institute. Mass spectrometry was carried out in Analytical Biochemistry Shared Resources of the Masonic Cancer Center, University of Minnesota, funded in part by Cancer Center Support Grant CA-077598.

Conflicts of Interest: The authors declare no conflict of interest.

References

1. World Health Organization. Cancer. Key Facts. Available online: https://www.who.int/news-room/fact-sheets/detail/cancer (accessed on 30 January 2018).
2. Centers for Disease Control and Prevention. Tobacco-Related Mortality. Available online: https://www.cdc.gov/tobacco/data_statistics/fact_sheets/health_effects/tobacco_related_mortality/index.htm (accessed on 30 January 2018).
3. Siegel, R.L.; Miller, K.D.; Jemal, A. Cancer statistics, 2018. *CA Cancer J. Clin.* **2018**, *68*, 7–30. [CrossRef]
4. Thun, M.J.; Henley, S.J.; Calle, E.E. Tobacco use and cancer: An epidemiologic perspective for geneticists. *Oncogene* **2002**, *21*, 7307–7325. [CrossRef]
5. World Health Organization. Tobacco. Available online: https://www.who.int/news-room/fact-sheets/detail/tobacco (accessed on 30 January 2018).
6. Yuan, J.M.; Butler, L.M.; Stepanov, I.; Hecht, S.S. Urinary tobacco smoke-constituent biomarkers for assessing risk of lung cancer. *Cancer Res.* **2014**, *74*, 401–411. [CrossRef]
7. Hecht, S.S.; Stepanov, I.; Carmella, S.G. Exposure and metabolic activation biomarkers of carcinogenic tobacco-specific nitrosamines. *Acc. Chem. Res.* **2016**, *49*, 106–114. [CrossRef] [PubMed]
8. Yuan, J.M.; Knezevich, A.D.; Wang, R.; Gao, Y.T.; Hecht, S.S.; Stepanov, I. Urinary levels of the tobacco-specific carcinogen N'-nitrosonornicotine and its glucuronide are strongly associated with esophageal cancer risk in smokers. *Carcinogenesis* **2011**, *32*, 1366–1371. [CrossRef]
9. Yuan, J.M.; Nelson, H.H.; Carmella, S.G.; Wang, R.; Kuriger-Laber, J.; Jin, A.; Adams-Haduch, J.; Hecht, S.S.; Koh, W.P.; Murphy, S.E. CYP2A6 genetic polymorphisms and biomarkers of tobacco smoke constituents in relation to risk of lung cancer in the Singapore Chinese Health Study. *Carcinogenesis* **2017**, *38*, 411–418. [CrossRef]

10. Phillips, D.H.; Venitt, S. DNA and protein adducts in human tissues resulting from exposure to tobacco smoke. *Int. J. Cancer* **2012**, *131*, 2733–2753. [CrossRef] [PubMed]

11. Sabbioni, G.; Turesky, R.J. Biomonitoring human albumin adducts: The past, the present, and the future. *Chem. Res. Toxicol.* **2017**, *30*, 332–366. [CrossRef]

12. Phillips, D.H. Smoking-related DNA and protein adducts in human tissues. *Carcinogenesis* **2002**, *23*, 1979–2004. [CrossRef] [PubMed]

13. Hecht, S.S. Lung carcinogenesis by tobacco smoke. *Int. J. Cancer* **2012**, *131*, 2724–2732. [CrossRef]

14. International Agency for Research on Cancer. Smokeless tobacco and some tobacco-specific N-nitrosamines. In *IARC Monographs on the Evaluation of Carcinogenic Risks to Humans*; International Agency for Research on Cancer: Lyon, France, 2007; Volume 89, pp. 1–592.

15. Hecht, S.S. Biochemistry, biology, and carcinogenicity of tobacco-specific N-nitrosamines. *Chem. Res. Toxicol.* **1998**, *11*, 559–603. [CrossRef]

16. Bustamante, G.; Ma, B.; Yakovlev, G.; Yershova, K.; Le, C.; Jensen, J.; Hatsukami, D.K.; Stepanov, I. Presence of the carcinogen N'-nitrosonornicotine in saliva of e-cigarette users. *Chem. Res. Toxicol.* **2018**, *31*, 731–738. [CrossRef] [PubMed]

17. Hoffmann, D.; Adams, J.D. Carcinogenic tobacco-specific N-nitrosamines in snuff and in the saliva of snuff dippers. *Cancer Res.* **1981**, *41*, 4305–4308.

18. Porubin, D.; Hecht, S.S.; Li, Z.Z.; Gonta, M.; Stepanov, I. Endogenous formation of N'-nitrosonornicotine in F344 rats in the presence of some antioxidants and grape seed extract. *J. Agric. Food Chem.* **2007**, *55*, 7199–7204. [CrossRef] [PubMed]

19. Carlson, E.S.; Upadhyaya, P.; Villalta, P.W.; Ma, B.; Hecht, S.S. Analysis and identification of 2′-deoxyadenosine-derived adducts in lung and liver DNA of F-344 rats treated with the tobacco-specific carcinogen 4-(methylnitrosamino)-1-(3-pyridyi)-1-butanone and enantiomers of its metabolite 4-(methylnitrosamino)-1-(3-pyridyl)-1-butanol. *Chem. Res. Toxicol.* **2018**, *31*, 358–370.

20. Ma, B.; Villalta, P.W.; Zarth, A.T.; Kotandeniya, D.; Upadhyaya, P.; Stepanov, I.; Hecht, S.S. Comprehensive high-resolution mass spectrometric analysis of DNA phosphate adducts formed by the tobacco-specific lung carcinogen 4-(methylnitrosamino)-1-(3-pyridyl)-1-butanone. *Chem. Res. Toxicol.* **2015**, *28*, 2151–2159. [CrossRef]

21. Michel, A.K.; Zarth, A.T.; Upadhyaya, P.; Hecht, S.S. Identification of 4-(3-pyridyl)-4-oxobutyl-2′-deoxycytidine adducts formed in the reaction of DNA with 4-(acetoxymethylnitrosamino)1-(3-pyridyl)-1-butanone: A chemically activated form of tobacco-specific carcinogens. *ACS Omega* **2017**, *2*, 1180–1190. [CrossRef]

22. Lao, Y.; Yu, N.; Kassie, F.; Villalta, P.W.; Hecht, S.S. Analysis of pyridyloxobutyl DNA adducts in F344 rats chronically treated with (*R*)- and (*S*)-N'-nitrosonornicotine. *Chem. Res. Toxicol.* **2007**, *20*, 246–256. [CrossRef]

23. Schlobe, D.; Holzle, D.; Hatz, D.; von Meyer, L.; Tricker, A.R.; Richter, E. 4-Hydroxy-1-(3-pyridyl)-1-butanone-releasing DNA adducts in lung, lower esophagus and cardia of sudden death victims. *Toxicology* **2008**, *245*, 154–161. [CrossRef]

24. Hoelzle, D.; Schlobe, D.; Tricker, A.R.; Richter, E. Mass spectrometric analysis of 4-hydroxy-1-(3-pyridyl)-1-butanone-releasing DNA adducts in human lung. *Toxicology* **2007**, *232*, 277–285. [CrossRef]

25. Foiles, P.G.; Akerkar, S.A.; Carmella, S.G.; Kagan, M.; Stoner, G.D.; Resau, J.H.; Hecht, S.S. Mass-spectrometric analysis of tobacco-specific nitrosamine DNA adducts in smokers and nonsmokers. *Chem. Res. Toxicol.* **1991**, *4*, 364–368. [CrossRef]

26. Hecht, S.S. Oral cell DNA adducts as potential biomarkers for lung cancer susceptibility in cigarette smokers. *Chem. Res. Toxicol.* **2017**, *30*, 367–375. [CrossRef]

27. Tan, D.J.; Goerlitz, D.S.; Dumitrescu, R.G.; Han, D.F.; Seillier-Moiseiwitsch, F.; Spernak, S.M.; Orden, R.A.; Chen, J.G.; Goldman, R.; Shields, P.G. Associations between cigarette smoking and mitochondrial DNA abnormalities in buccal cells. *Carcinogenesis* **2008**, *29*, 1170–1177. [CrossRef]

28. Stepanov, I.; Muzic, J.; Le, C.T.; Sebero, E.; Villalta, P.; Ma, B.; Jensen, J.; Hatsukami, D.; Hecht, S.S. Analysis of 4-hydroxy-1-(3-pyridyl)-1-butanone (HPB)-releasing DNA adducts in human exfoliated oral mucosa cells by liquid chromatography-electrospray ionization-tandem mass spectrometry. *Chem. Res. Toxicol.* **2013**, *26*, 37–45. [CrossRef]

29. Ma, B.; Ruszczak, C.; Jain, V.P.; Khariwala, S.S.; Lindgren, B.; Hatsukami, D.K.; Stepanoy, I. Optimized liquid chromatography nanoelectrospray-high-resolution tandem mass spectrometry method for the analysis of 4-hydroxy-1-(3-pyridyl)-1-butanone-releasing DNA adducts in human oral cells. *Chem. Res. Toxicol.* **2016**, *29*, 1849–1856. [CrossRef]

30. Yang, J.; Villalta, P.W.; Upadhyaya, P.; Hecht, S.S. Analysis of O^6-[4-(3-pyridyl)-4-oxobut-1-yl]-2'-deoxyguanosine and other DNA adducts in rats treated with enantiomeric or racemic N'-Nitrosonornicotine. *Chem. Res. Toxicol.* **2016**, *29*, 87–95. [CrossRef]

31. Zarth, A.T.; Upadhyaya, P.; Yang, J.; Hecht, S.S. DNA adduct formation from metabolic 5'-hydroxylation of the tobacco-specific carcinogen N'-nitrosonornicotine in human enzyme systems and in rats. *Chem. Res. Toxicol.* **2016**, *29*, 380–389. [CrossRef]

32. Villalta, P.W.; Hochalter, J.B.; Hecht, S.S. Ultrasensitive high-resolution mass spectrometric analysis of a DNA adduct of the carcinogen benzo[*a*]pyrene in human lung. *Anal. Chem.* **2017**, *89*, 12735–12742. [CrossRef]

33. Ma, B.; Villalta, P.W.; Balbo, S.; Stepanov, I. Analysis of a malondialdehyde-deoxyguanosine adduct in human leukocyte DNA by liquid chromatography nanoelectrospray-high-resolution tandem mass spectrometry. *Chem. Res. Toxicol.* **2014**, *27*, 1829–1836. [CrossRef]

34. Ma, B.; Zarth, A.T.; Carlson, E.S.; Villalta, P.W.; Upadhyaya, P.; Stepanov, I.; Hecht, S.S. Methyl DNA phosphate adduct formation in rats treated chronically with 4-(methylnitrosamino)-1-(3-pyridyl)-1-butanone and enantiomers of its metabolite 4-(methylnitrosamino)-1-(3-pyridyi)-1-butanol. *Chem. Res. Toxicol.* **2018**, *31*, 48–57. [CrossRef]

35. Ma, B.; Zarth, A.T.; Carlson, E.S.; Villalta, P.W.; Stepanov, I.; Hecht, S.S. Pyridylhydroxybutyl and pyridyloxobutyl DNA phosphate adduct formation in rats treated chronically with enantiomers of the tobacco-specific nitrosamine metabolite 4-(methylnitrosamino)-1-(3-pyridyl)-1-butanol. *Mutagenesis* **2017**, *32*, 561–570.

36. Ma, B.; Zarth, A.T.; Carlson, E.S.; Villalta, P.W.; Upadhyaya, P.; Stepanov, I.; Hecht, S.S. Identification of more than 100 structurally unique DNA-phosphate adducts formed during rat lung carcinogenesis by the tobacco-specific nitrosamine 4-(methylnitrosamino)-1-(3-pyridyl)-1-butanone. *Carcinogenesis* **2018**, *39*, 232–241. [CrossRef]

37. Munnia, A.; Giese, R.W.; Polvani, S.; Galli, A.; Cellai, F.; Peluso, M.E.M. Bulky DNA adducts, tobacco smoking, genetic susceptibility, and lung cancer risk. *Adv. Clin. Chem.* **2017**, *81*, 231–277.

38. Poirier, M.C. Linking DNA adduct formation and human cancer risk in chemical carcinogenesis. *Environ. Mol. Mutagen.* **2016**, *57*, 499–507. [CrossRef]

39. Rynning, I.; Arlt, V.M.; Vrbova, K.; Neca, J.; Rossner, P., Jr.; Klema, J.; Ulvestad, B.; Petersen, E.; Skare, O.; Haugen, A.; et al. Bulky DNA adducts, microRNA profiles, and lipid biomarkers in Norwegian tunnel finishing workers occupationally exposed to diesel exhaust. *Occup. Environ. Med.* **2019**, *76*, 10–16. [CrossRef]

40. Gilberson, T.; Peluso, M.E.M.; Munia, A.; Lujan-Barroso, L.; Sanchez, M.J.; Navarro, C.; Amiano, P.; Barricarte, A.; Quiros, J.R.; Molina-Montes, E.; et al. Aromatic adducts and lung cancer risk in the European Prospective Investigation into Cancer and Nutrition (EPIC) Spanish cohort. *Carcinogenesis* **2014**, *35*, 2047–2054.

41. Pavanello, S.; Carta, A.; Mastrangelo, G.; Campisi, M.; Arici, C.; Porru, S. Relationship between telomere length, genetic traits and environmental/occupational exposures in bladder cancer risk by structural equation modelling. *Int. J. Environ. Res. Public Health* **2017**, *15*, 5. [CrossRef]

42. Mastrangelo, G.; Carta, A.; Arici, C.; Pavanello, S.; Porru, S. An etiologic prediction model incorporating biomarkers to predict the bladder cancer risk associated with occupational exposure to aromatic amines: A pilot study. *J. Occup. Med. Toxicol.* **2017**, *12*, 23. [CrossRef]

43. Porru, S.; Pavanello, S.; Carta, A.; Arici, C.; Simeone, C.; Izzotti, A.; Mastrangelo, G. Complex relationships between occupation, environment, DNA adducts, genetic polymorphisms and bladder cancer in a case-control study using a structural equation modeling. *PLoS ONE* **2014**, *9*, e94566.

44. Agudo, A.; Peluso, M.; Munnia, A.; Lujan-Barroso, L.; Barricarte, A.; Amiano, P.; Navarro, C.; Sanchez, M.J.; Quiros, J.R.; Ardanaz, E.; et al. Aromatic DNA adducts and breast cancer risk: A case-cohort study within the EPIC-Spain. *Carcinogenesis* **2017**, *38*, 691–698.

45. Molina, E.; Perez-Morales, R.; Rubio, J.; Petrosyan, P.; Cadena, L.H.; Arlt, V.M.; Phillips, D.H.; Gonsebatt, M.E. The *GSTM1null* (deletion) and *MGMT84* rs12917 (Phe/Phe) haplotype are associated with bulky DNA adduct levels in human leukocytes. *Mutat. Res.* **2013**, *758*, 62–68. [CrossRef] [PubMed]

46. Pedersen, M.; Schoket, B.; Godschalk, R.W.; Wright, J.; von Stedingk, H.; Tornqvist, M.; Sunyer, J.; Nielsen, J.K.; Merlo, D.F.; Mendez, M.A.; et al. Bulky DNA adducts in cord blood, maternal fruit-and-vegetable consumption, and birth weight in a European Mother-Child Study (NewGeneris). *Environ. Health Perspect.* **2013**, *121*, 1200–1206. [CrossRef] [PubMed]

47. Nilsson, R.; Antic, R.; Berni, A.; Dallner, G.; Dettbarn, G.; Gromadzinska, J.; Joksic, G.; Lundin, C.; Palitti, F.; Prochazka, G.; et al. Exposure to polycyclic aromatic hydrocarbons in women from Poland, Serbia and Italy—Relation between PAH metabolite excretion, DNA damage, diet and genotype (the EU DIEPHY project). *Biomarkers* **2013**, *18*, 165–173. [CrossRef]

48. Peluso, M.; Bollati, V.; Munnia, A.; Srivatanakul, P.; Jedpiyawongse, A.; Sangrajrang, S.; Piro, S.; Ceppi, M.; Bertazzi, P.A.; Boffetta, P.; et al. DNA methylation differences in exposed workers and nearby residents of the Ma Ta Phut industrial estate, Rayong, Thailand. *Int. J. Epidemiol.* **2012**, *41*, 1753–1760. [CrossRef]

49. Pedersen, M.; Halldorsson, T.I.; Autrup, H.; Brouwer, A.; Besselink, H.; Loft, S.; Knudsen, L.E. Maternal diet and dioxin-like activity, bulky DNA adducts and micronuclei in mother-newborns. *Mutat. Res.* **2012**, *734*, 12–19. [CrossRef]

50. Agudo, A.; Peluso, M.; Munnia, A.; Lujan-Barroso, L.; Sanchez, M.J.; Molina-Montes, E.; Sanchez-Cantalejo, E.; Navarro, C.; Tormo, M.J.; Chirlaque, M.D.; et al. Aromatic DNA adducts and risk of gastrointestinal cancers: A case-cohort study within the EPIC-Spain. *Cancer Epidemiol. Prev. Biomark.* **2012**, *21*, 685–692. [CrossRef]

51. Gu, D.; Turesky, R.J.; Tao, Y.Q.; Langouet, S.A.; Nauwelaers, G.C.; Yuan, J.M.; Yee, D.; Yu, M.M.C. DNA adducts of 2-amino-1-methyl-6-phenylimidazo[4,5-b]pyridine and 4-aminobiphenyl are infrequently detected in human mammary tissue by liquid chromatography/tandem mass spectrometry. *Carcinogenesis* **2012**, *33*, 124–130. [CrossRef]

52. Lee, M.S.; Asomaning, K.; Su, L.; Wain, J.C.; Mark, E.J.; Christiani, D.C. MTHFR polymorphisms, folate intake and carcinogen DNA adducts in the lung. *Int. J. Cancer* **2012**, *131*, 1203–1209. [CrossRef]

53. Weng, M.W.; Lee, H.W.; Park, S.H.; Hu, Y.; Wang, H.T.; Chen, L.C.; Rom, W.N.; Huang, W.C.; Lepor, H.; Wu, X.R.; et al. Aldehydes are the predominant forces inducing DNA damage and inhibiting DNA repair in tobacco smoke carcinogenesis. *Proc. Natl. Acad. Sci. USA* **2018**, *115*, E6152–E6161. [CrossRef]

54. Jin, Y.T.; Xu, P.W.; Liu, X.N.; Zhang, C.Y.; Tan, C.; Chen, C.M.; Sun, X.Y.; Xu, Y.C. Cigarette smoking, BPDE-DNA adducts, and aberrant promoter methylations of tumor suppressor genes (TSGs) in NSCLC from Chinese population. *Cancer Investig.* **2016**, *34*, 173–180. [CrossRef]

55. Niehoff, N.; White, A.J.; McCullough, L.E.; Steck, S.E.; Beyea, J.; Mordukhovich, I.; Shen, J.; Neugut, A.I.; Conway, K.; Santella, R.M.; et al. Polycyclic aromatic hydrocarbons and postmenopausal breast cancer: An evaluation of effect measure modification by body mass index and weight change. *Environ. Res.* **2017**, *152*, 17–25. [CrossRef] [PubMed]

56. Naidoo, R.N.; Makwela, M.H.; Chuturgoon, A.; Tiloke, C.; Ramkaran, P.; Phulukdaree, A. Petrol exposure and DNA integrity of peripheral lymphocytes. *Int. Arch. Occup. Environ Health* **2016**, *89*, 785–792. [CrossRef] [PubMed]

57. Chuang, C.Y.; Tung, J.N.; Su, M.C.; Wu, B.C.; Hsin, C.H.; Chen, Y.J.; Yeh, K.T.; Lee, H.; Cheng, Y.W. BPDE-like DNA adduct level in oral tissue may act as a risk biomarker of oral cancer. *Arch. Oral Biol.* **2013**, *58*, 102–109. [CrossRef] [PubMed]

58. Rundle, A.; Richards, C.; Neslund-Dudas, C.; Tang, D.L.; Rybicki, B.A. Neighborhood socioeconomic status modifies the association between individual smoking status and PAH-DNA adduct levels in prostate tissue. *Environ. Mol. Mutagen.* **2012**, *53*, 384–391. [CrossRef] [PubMed]

59. Pavanello, S.; Favretto, D.; Brugnone, F.; Mastrangelo, G.; Dal Pra, G.; Clonfero, E. HPLC/fluorescence determination of anti-BPDE-DNA adducts in mononuclear white blood cells from PAH-exposed humans. *Carcinogenesis* **1999**, *20*, 431–435. [CrossRef]

60. Ewa, B.; Danuta, M.S. Polycyclic aromatic hydrocarbons and PAH-related DNA adducts. *J. Appl. Genet.* **2017**, *58*, 321–330. [CrossRef]

61. Monien, B.H.; Schumacher, F.; Herrmann, K.; Glatt, H.; Turesky, R.J.; Chesne, C. Simultaneous detection of multiple DNA adducts in human lung samples by isotope-dilution UPLC-MS/MS. *Anal. Chem.* **2015**, *87*, 641–648. [CrossRef]

62. Beland, F.A.; Churchwell, M.I.; Von Tungeln, L.S.; Chen, S.J.; Fu, P.P.; Culp, S.J.; Schoket, B.; Gyorffy, E.; Minarovits, J.; Poirier, M.C.; et al. High-performance liquid chromatography electrospray ionization tandem mass spectrometry for the detection and quantitation of benzo[*a*]pyrene-DNA adducts. *Chem. Res. Toxicol.* **2005**, *18*, 1306–1315. [CrossRef]

63. Bessette, E.E.; Spivack, S.D.; Goodenough, A.K.; Wang, T.; Pinto, S.; Kadlubar, F.F.; Turesky, R.J. Identification of carcinogen DNA adducts in human saliva by linear quadrupole ion trap/multistage tandem mass spectrometry. *Chem. Res. Toxicol.* **2010**, *23*, 1234–1244. [CrossRef]

64. Bessette, E.E.; Goodenough, A.K.; Langouet, S.; Yasa, I.; Kozekov, I.D.; Spivack, S.D.; Turesky, R.J. Screening for DNA adducts by data-dependent constant neutral loss-triple stage mass spectrometry with a linear quadrupole ion trap mass spectrometer. *Anal. Chem.* **2009**, *81*, 809–819. [CrossRef]

65. Xiao, S.; Guo, J.; Yun, B.H.; Villalta, P.W.; Krishna, S.; Tejpaul, R.; Murugan, P.; Weight, C.J.; Turesky, R.J. Biomonitoring DNA adducts of cooked meat carcinogens in human prostate by nano liquid chromatography-high resolution tandem mass spectrometry: Identification of 2-amino-1-methyl-6-phenylimidazo[4,5-b]pyridine DNA adduct. *Anal. Chem.* **2016**, *88*, 12508–12515. [CrossRef] [PubMed]

66. International Agency for Research on Cancer. Some aromatic amines, organic dyes, and related exposures. In *IARC Monographs on the Evaluation of Carcinogenic Risks to Humans*; International Agency for Research on Cancer: Lyon, France, 2010; Volume 99, pp. 1–658.

67. Guo, J.; Villalta, P.W.; Weight, C.J.; Bonala, R.; Johnson, F.; Rosenquist, T.A.; Turesky, R.J. Targeted and untargeted detection of DNA adducts of aromatic amine carcinogens in human bladder by ultra-performance liquid chromatography-high-resolution mass spectrometry. *Chem. Res. Toxicol.* **2018**, *31*, 1382–1397. [CrossRef]

68. Lee, H.W.; Wang, H.T.; Weng, M.W.; Hu, Y.; Chen, W.S.; Chou, D.; Liu, Y.; Donin, N.; Huang, W.C.; Lepor, H.; et al. Acrolein- and 4-aminobiphenyl-DNA adducts in human bladder mucosa and tumor tissue and their mutagenicity in human urothelial cells. *Oncotarget* **2014**, *5*, 3526–3540. [CrossRef] [PubMed]

69. Faraglia, B.; Chen, S.Y.; Gammon, M.D.; Zhang, Y.J.; Teitelbaum, S.L.; Neugut, A.I.; Ahsan, H.; Garbowski, G.C.; Hibshoosh, H.; Lin, D.X.; et al. Evaluation of 4-aminobiphenyl-DNA adducts in human breast cancer: The influence of tobacco smoke. *Carcinogenesis* **2003**, *24*, 719–725. [CrossRef]

70. Peterson, L.A. Context matters: Contribution of specific DNA adducts to the genotoxic properties of the tobacco-specific nitrosamine NNK. *Chem. Res. Toxicol.* **2017**, *30*, 420–433. [CrossRef]

71. Wu, J.; Wang, P.; Wang, Y. Cytotoxic and mutagenic properties of alkyl phosphotriester lesions in Escherichia coli cells. *Nucleic Acids Res.* **2018**, *46*, 4013–4021. [CrossRef]

72. Marushige, K.; Marushige, Y. Template properties of DNA alkylated with *N*-methyl-*N*-nitrosourea and *N*-ethyl-*N*-nitrosourea. *Chem. Biol. Interact.* **1983**, *46*, 179–188. [CrossRef]

73. He, C.; Hus, J.C.; Sun, L.J.; Zhou, P.; Norman, D.P.; Dotsch, V.; Wei, H.; Gross, J.D.; Lane, W.S.; Wagner, G.; et al. A methylation-dependent electrostatic switch controls DNA repair and transcriptional activation by *E. coli* ada. *Mol. Cell* **2005**, *20*, 117–129. [CrossRef]

74. Crosbie, P.A.J.; Harrison, K.; Shah, R.; Watson, A.J.; Agius, R.; Barber, P.V.; Margison, G.P.; Povey, A.C. Topographical study of O^6-alkylguanine DNA alkyltransferase repair activity and N^7-methylguanine levels in resected lung tissue. *Chem.-Biol. Interact.* **2013**, *204*, 98–104. [CrossRef]

75. Wang, Y.; Narayanapillai, S.; Hu, Q.; Fujioka, N.; Xing, C.G. Contribution of tobacco use and 4-(methylnitrosamino)-1-(3-pyridyl)-1-butanone to three methyl DNA adducts in urine. *Chem. Res. Toxicol.* **2018**, *31*, 836–838. [CrossRef]

76. Hu, K.; Zhao, G.; Liu, J.W.; Jia, L.Z.; Xie, F.W.; Zhang, S.S.; Liu, H.M.; Liu, M.Y. Simultaneous quantification of three alkylated-purine adducts in human urine using sulfonic acid poly(glycidyl methacrylate-divinylbenzene)-based microspheres as sorbent combined with LC-MS/MS. *J. Chromatogr. B* **2018**, *1081*, 19–28. [CrossRef] [PubMed]

77. Tian, Y.; Hou, H.; Zhang, X.; Wang, A.; Liu, Y.; Hu, Q. New validated LC-MS/MS method for the determination of three alkylated adenines in human urine and its application to the monitoring of alkylating agents in cigarette smoke. *Anal. Bioanal. Chem.* **2014**, *406*, 5293–5302. [CrossRef]

78. Hu, C.W.; Hsu, Y.W.; Chen, J.L.; Tam, L.M.; Chao, M.R. Direct analysis of tobacco-specific nitrosamine NNK and its metabolite NNAL in human urine by LC-MS/MS: Evidence of linkage to methylated DNA lesions. *Arch. Toxicol.* **2014**, *88*, 291–299. [CrossRef] [PubMed]

79. Ma, B.; Villalta, P.W.; Hochalter, J.B.; Stepanov, I.; Hecht, S.S. Methyl DNA phosphate adduct formation in lung tumor tissue and adjacent normal tissue of lung cancer patients. *Carcinogenesis* **2019**. [CrossRef]

80. Singh, R.; Kaur, B.; Farmer, P.B. Detection of DNA damage derived from a direct acting ethylating agent present in cigarette smoke by use of liquid chromatography-tandem mass spectrometry. *Chem. Res. Toxicol.* **2005**, *18*, 249–256. [CrossRef]

81. Hu, C.W.; Cooke, M.S.; Chang, Y.J.; Chao, M.R. Direct-acting DNA ethylating agents associated with tobacco use primarily originate from the tobacco itself, not combustion. *J. Hazard. Mater.* **2018**, *358*, 397–404. [CrossRef] [PubMed]

82. Shooter, K.V.; Slade, T.A. The stability of methyl and ethyl phosphotriesters in DNA in vivo. *Chem. Biol. Interact.* **1977**, *19*, 353–361. [CrossRef]

83. Chen, H.J.C.; Lee, C.R. Detection and simultaneous quantification of three smoking-related ethylthymidine adducts in human salivary DNA by liquid chromatography tandem mass spectrometry. *Toxicol. Lett.* **2014**, *224*, 101–107. [CrossRef] [PubMed]

84. Chen, H.J.C.; Liu, Y.F. Simultaneous quantitative analysis of N^3-ethyladenine and N^7-ethylguanine in human leukocyte deoxyribonucleic acid by stable isotope dilution capillary liquid chromatography-nanospray ionization tandem mass spectrometry. *J. Chromatogr. A* **2013**, *1271*, 86–94. [CrossRef]

85. Swenberg, J.A.; Dyroff, M.C.; Bedell, M.A.; Popp, J.A.; Huh, N.; Kirstein, U.; Rajewsky, M.F. O^4-Ethyldeoxythymidine, but not O^6-ethyldeoxyguanosine, accumulates in hepatocyte DNA of rats exposed continuously to diethylnitrosamine. *Proc. Natl. Acad. Sci. USA* **1984**, *81*, 1692–1695. [CrossRef]

86. Chen, H.J.C.; Wang, Y.C.; Lin, W.P. Analysis of ethylated thymidine adducts in human leukocyte DNA by stable isotope dilution nanoflow liquid chromatography-nanospray ionization tandem mass spectrometry. *Anal. Chem.* **2012**, *84*, 2521–2527. [CrossRef] [PubMed]

87. Chen, H.J.C.; Lin, C.R. Noninvasive measurement of smoking-associated N^3-ethyladenine and N^7-ethylguanine in human salivary DNA by stable isotope dilution nanoflow liquid chromatography-nanospray ionization tandem mass spectrometry. *Toxicol. Lett.* **2014**, *225*, 27–33. [CrossRef] [PubMed]

88. Chen, H.J.C.; Lin, C.R. Simultaneous quantification of ethylpurine adducts in human urine by stable isotope dilution nanoflow liquid chromatography nanospray ionization tandem mass spectrometry. *J. Chromatogr. A* **2013**, *1322*, 69–73. [CrossRef] [PubMed]

89. Tretyakova, N.Y.; Chiang, S.Y.; Walker, V.E.; Swenberg, J.A. Quantitative analysis of 1,3-butadiene-induced DNA adducts in vivo and in vitro using liquid chromatography electrospray ionization tandem mass spectrometry. *J. Mass Spectrom.* **1998**, *33*, 363–376. [CrossRef]

90. Sangaraju, D.; Villalta, P.; Goggin, M.; Agunsoye, M.O.; Campbell, C.; Tretyakova, N. Capillary HPLC-accurate mass ms/ms quantitation of N^7-(2,3,4-trihydroxybut-1-yl)-guanine adducts of 1,3-butadiene in human leukocyte DNA. *Chem. Res. Toxicol.* **2013**, *26*, 1486–1497. [CrossRef]

91. International Agency for Research on Cancer. Personal habits and indoor combustions. A review of human carcinogens. In *IARC Monographs on the Evaluation of Carcinogenic Risks to Humans*; International Agency for Research on Cancer: Lyon, France, 2012; Volume 100, pp. 1–538.

92. Zhang, S.Y.; Balbo, S.; Wang, M.Y.; Hecht, S.S. Analysis of acrolein-derived 1,N^2-propanodeoxyguanosine adducts in human leukocyte DNA from smokers and nonsmokers. *Chem. Res. Toxicol.* **2011**, *24*, 119–124. [CrossRef]

93. Zhang, S.Y.; Villalta, P.W.; Wang, M.Y.; Hecht, S.S. Detection and quantitation of acrolein-derived 1,N^2-propanodeoxyguanosine adducts in human lung by liquid chromatography-electrospray ionization-tandem mass spectrometry. *Chem. Res. Toxicol.* **2007**, *20*, 565–571. [CrossRef]

94. Yang, J.; Balbo, S.; Villalta, P.W.; Hecht, S.S. Analysis of acrolein-derived 1, N^2-propanodeoxyguanosine adducts in human lung DNA from smokers and non-smokers. *Chem. Res. Toxicol.* **2019**, *32*, 318–325. [CrossRef]

95. Li, X.Y.; Liu, L.J.; Wang, H.J.; Chen, J.; Zhu, B.B.; Chen, H.; Hou, H.W.; Hu, Q.Y. Simultaneous analysis of six aldehyde-DNA adducts in salivary DNA of nonsmokers and smokers using stable isotope dilution liquid chromatography electrospray ionization-tandem mass spectrometry. *J. Chromatogr. B* **2017**, *1060*, 451–459. [CrossRef]

96. Chung, F.L.; Wu, M.Y.; Basudan, A.; Dyba, M.; Nath, R.G. Regioselective formation of acrolein-derived cyclic 1,N^2-propanodeoxyguanosine adducts mediated by amino acids, proteins, and cell lysates. *Chem. Res. Toxicol.* **2012**, *25*, 1921–1928. [CrossRef]

97. Wang, M.Y.; Cheng, G.; Balbo, S.; Carmella, S.G.; Villalta, P.W.; Hecht, S.S. Clear differences in levels of a formaldehyde-DNA adduct in leukocytes of smokers and nonsmokers. *Cancer Res.* **2009**, *69*, 7170–7174. [CrossRef]

98. Zhang, S.; Villalta, P.W.; Wang, M.Y.; Hecht, S.S. Analysis of crotonaldehyde- and acetaldehyde-derived 1,N2-propanodeoxyguanosine adducts in DNA from human tissues using liquid chromatography-electrospray ionization-tandem mass spectrometry. *Chem. Res. Toxicol.* **2006**, *19*, 1386–1392. [CrossRef]

99. Wang, M.; Yu, N.; Chen, L.; Villalta, P.W.; Hochalter, J.B.; Hecht, S.S. Identification of an acetaldehyde adduct in human liver DNA and quantitation as N2-ethyldeoxyguanosine. *Chem. Res. Toxicol.* **2006**, *19*, 319–324. [CrossRef] [PubMed]

100. Garcia, C.C.M.; Freitas, F.P.; Sanchez, A.B.; Di Mascio, P.; Medeiros, M.H.G. Elevated alpha-methyl-γ-hydroxy-1,N^2-propano-2′-deoxyguanosine levels in urinary samples from individuals exposed to urban air pollution. *Chem. Res. Toxicol.* **2013**, *26*, 1602–1604. [CrossRef] [PubMed]

101. Balbo, S.; Meng, L.; Bliss, R.L.; Jensen, J.A.; Hatsukami, D.K.; Hecht, S.S. Time course of DNA adduct formation in peripheral blood granulocytes and lymphocytes after drinking alcohol. *Mutagenesis* **2012**, *27*, 485–490. [CrossRef] [PubMed]

102. Balbo, S.; Meng, L.; Bliss, R.L.; Jensen, J.A.; Hatsukami, D.K.; Hecht, S.S. Kinetics of DNA adduct formation in the oral cavity after drinking alcohol. *Cancer Epidemiol. Prev. Biomark.* **2012**, *21*, 601–608. [CrossRef]

103. Singh, R.; Gromadzinska, J.; Mistry, Y.; Cordell, R.; Juren, T.; Segerback, D.; Farmer, P.B. Detection of acetaldehyde derived N^2-ethyl-2′-deoxyguanosine in human leukocyte DNA following alcohol consumption. *Mutat. Res.* **2012**, *737*, 8–11. [CrossRef] [PubMed]

104. Huang, C.C.J.; Wu, C.F.; Shih, W.C.; Luo, Y.S.; Chen, M.F.; Li, C.M.; Liou, S.H.; Chung, W.S.; Chiang, S.Y.; Wu, K.Y. Potential association of urinary N^7-(2-carbamoyl-2-hydroxyethyl) guanine with dietary acrylamide intake of smokers and nonsmokers. *Chem. Res. Toxicol.* **2015**, *28*, 43–50. [CrossRef] [PubMed]

105. Huang, Y.F.; Huang, C.C.J.; Lu, C.A.; Chen, M.L.; Liou, S.H.; Chiang, S.Y.; Wu, K.Y. Feasibility of using urinary N^7-(2-carbamoyl-2-hydroxyethyl) Guanine as a biomarker for acrylamide exposed workers. *J. Expo. Sci. Environ. Epidemiol.* **2018**, *28*, 589–598. [CrossRef] [PubMed]

106. Yu, Y.; Cui, Y.X.; Niedernhofer, L.J.; Wang, Y.S. Occurrence, biological consequences, and human health relevance of oxidative stress-induced DNA damage. *Chem. Res. Toxicol.* **2016**, *29*, 2008–2039. [CrossRef]

107. Marnett, L.J. Lipid peroxidation-DNA damage by malondialdehyde. *Mutat. Res.* **1999**, *424*, 83–95. [CrossRef]

108. Bonassi, S.; Cellai, F.; Munnia, A.; Ugolini, D.; Cristaudo, A.; Neri, M.; Milic, M.; Bonotti, A.; Giese, R.W.; Peluso, M.E.M. 3-(2-deoxy-β-D-erythro-pentafuranosyl)pyrimido[1,2-alpha]purin-10 (3H)-one deoxyguanosine adducts of workers exposed to asbestos fibers. *Toxicol. Lett.* **2017**, *270*, 1–7. [CrossRef]

109. Saieva, C.; Peluso, M.; Palli, D.; Cellai, F.; Ceroti, M.; Selvi, V.; Bendinelli, B.; Assedi, M.; Munnia, A.; Masala, G. Dietary and lifestyle determinants of malondialdehyde DNA adducts in a representative sample of the Florence City population. *Mutagenesis* **2016**, *31*, 475–480. [CrossRef]

110. Peluso, M.; Munnia, A.; Piro, S.; Jedpiyawongse, A.; Sangrajrang, S.; Giese, R.W.; Ceppi, M.; Boffetta, P.; Srivatanakul, P. Fruit and vegetable and fried food consumption and 3-(2-deoxy-β-D-erythro-pentafuranosyl)pyrimido[1,2-alpha] purin-10(3H)-one deoxyguanosine adduct formation. *Free Radic. Res.* **2012**, *46*, 85–92. [CrossRef]

111. Peluso, M.E.M.; Munnia, A.; Bollati, V.; Srivatanakul, P.; Jedpiyawongse, A.; Sangrajrang, S.; Ceppi, M.; Giese, R.W.; Boffetta, P.; Baccarelli, A.A. Aberrant methylation of hypermethylated-in-cancer-1 and exocyclic DNA adducts in tobacco smokers. *Toxicol. Sci.* **2014**, *137*, 47–54. [CrossRef]

112. Peluso, M.E.M.; Munnia, A.; Giese, R.W.; Chellini, E.; Ceppi, M.; Capacci, F. Oxidatively damaged DNA in the nasal epithelium of workers occupationally exposed to silica dust in Tuscany region, Italy. *Mutagenesis* **2015**, *30*, 519–525. [CrossRef]

113. Bono, R.; Munnia, A.; Romanazzi, V.; Bellisario, V.; Cellai, F.; Peluso, M.E.M. Formaldehyde-induced toxicity in the nasal epithelia of workers of a plastic laminate plant. *Toxicol. Res.* **2016**, *5*, 752–760. [CrossRef]

114. Micillo, A.; Vassallo, M.R.C.; Cordeschi, G.; D'Andrea, S.; Necozione, S.; Francavilla, F.; Francavilla, S.; Barbonetti, A. Semen leukocytes and oxidative-dependent DNA damage of spermatozoa in male partners of subfertile couples with no symptoms of genital tract infection. *Andrology* **2016**, *4*, 808–815. [CrossRef]

115. Kumar, S.B.; Chawla, B.; Bisht, S.; Yadav, R.K.; Dada, R. Tobacco use increases oxidative DNA damage in sperm—Possible etiology of childhood cancer. *Asian Pac. J. Cancer Prev.* **2015**, *16*, 6967–6972. [CrossRef]

116. Ma, B.; Jing, M.; Villalta, P.W.; Kapphahn, R.J.; Montezuma, S.R.; Ferrington, D.A.; Stepanov, I. Simultaneous determination of 8-oxo-2′-deoxyguanosine and 8-oxo-2′-deoxyadenosine in human retinal DNA by liquid chromatography nanoelectrospray-tandem mass spectrometry. *Sci. Rep.* **2016**, *6*, 22375. [CrossRef]

117. De Carvalho, A.M.; Carioca, A.A.F.; Fisberg, R.M.; Qi, L.; Marchioni, D.M. Joint association of fruit, vegetable, and heterocyclic amine intake with DNA damage levels in a general population. *Nutrition* **2016**, *32*, 260–264. [CrossRef] [PubMed]

118. Kendzia, B.; Pesch, B.; Marczynski, B.; Lotz, A.; Welge, P.; Rihs, H.P.; Bruning, T.; Raulf-Heimsoth, M. Pre- and postshift levels of inflammatory biomarkers and DNA damage in non-bitumen-exposed construction workers-subpopulation of the German Human Bitumen Study. *J. Toxicol. Environ. Health A* **2012**, *75*, 533–543. [CrossRef] [PubMed]

119. Kafferlein, H.U.; Marczynski, B.; Simon, P.; Angerer, J.; Rihs, H.P.; Wilhelm, M.; Straif, K.; Pesch, B.; Bruning, T. Internal exposure to carcinogenic polycyclic aromatic hydrocarbons and DNA damage: A null result in brief. *Arch. Toxicol.* **2012**, *86*, 1317–1321. [CrossRef] [PubMed]

120. Kumar, A.; Pant, M.C.; Singh, H.S.; Khandelwal, S. Assessment of the redox profile and oxidative DNA damage (8-OHdG) in squamous cell carcinoma of head and neck. *J. Cancer Res. Ther.* **2012**, *8*, 254–259. [PubMed]

121. Deng, Q.F.; Dai, X.Y.; Guo, H.; Huang, S.L.; Kuang, D.; Feng, J.; Wang, T.; Zhang, W.Z.; Huang, K.; Hu, D.; et al. Polycyclic aromatic hydrocarbons-associated microRNAs and their interactions with the environment: Influences on oxidative DNA damage and lipid peroxidation in coke oven workers. *Environ. Sci. Technol.* **2014**, *48*, 4120–4128. [CrossRef]

122. Bin, P.; Shen, M.L.; Li, H.B.; Sun, X.; Niu, Y.; Meng, T.; Yu, T.; Zhang, X.; Dai, Y.F.; Gao, W.M.; et al. Increased levels of urinary biomarkers of lipid peroxidation products among workers occupationally exposed to diesel engine exhaust. *Free Radic. Res.* **2016**, *50*, 820–830. [CrossRef]

123. Leng, J.; Wang, Y. Liquid chromatography-tandem mass spectrometry for the quantification of tobacco-specific nitrosamine-induced DNA adducts in mammalian cells. *Anal. Chem.* **2017**, *89*, 9124–9130. [CrossRef]

124. Li, Y.; Ma, B.; Cao, Q.; Balbo, S.; Zhao, L.; Upadhyaya, P.; Hecht, S.S. Mass spectrometric quantitation of pyridyloxobutyl DNA phosphate adducts in rats chronically treated with N'-nitrosonornicotine. *Chem. Res. Toxicol.* **2019**. [CrossRef]

125. Jones, G.D.D.; Le Pla, R.C.; Farmer, P.B. Phosphotriester adducts (PTEs): DNA's overlooked lesion. *Mutagenesis* **2010**, *25*, 3–16. [CrossRef]

126. Den Engelse, L.; De Graaf, A.; De Brij, R.J.; Menkveld, G.J. O^2- and O^4-ethylthymine and the ethylphosphotriester dTp(Et)dT are highly persistent DNA modifications in slowly dividing tissues of the ethylnitrosourea-treated rat. *Carcinogenesis* **1987**, *8*, 751–757. [CrossRef]

127. Singer, B. In vivo formation and persistence of modified nucleosides resulting from alkylating agents. *Environ. Health Perspect.* **1985**, *62*, 41–48. [CrossRef]

128. Haglund, J.; Henderson, A.P.; Golding, B.T.; Tornqvist, M. Evidence for phosphate adducts in DNA from mice treated with 4-(N-methyl-N-nitrosamino)-1-(3-pyridyl)-1-butanone (NNK). *Chem. Res. Toxicol.* **2002**, *15*, 773–779. [CrossRef]

129. Upadhyaya, P.; McIntee, E.J.; Villalta, P.W.; Hecht, S.S. Identification of adducts formed in the reaction of 5′-acetoxy-N'-nitrosonornicotine with deoxyguanosine and DNA. *Chem. Res. Toxicol.* **2006**, *19*, 426–435. [CrossRef]

130. Yu, Y.; Wang, P.; Cui, Y.; Wang, Y. Chemical analysis of DNA damage. *Anal. Chem.* **2018**, *90*, 556–576. [CrossRef]

131. Tretyakova, N.; Villalta, P.W.; Kotapati, S. Mass spectrometry of structurally modified DNA. *Chem. Rev.* **2013**, *113*, 2395–2436. [CrossRef]

132. Villalta, P.W.; Balbo, S. The future of DNA adductomic analysis. *Int. J. Mol. Sci.* **2017**, *18*, 1870. [CrossRef]

133. Hofer, T.; Moller, L. Reduction of oxidation during the preparation of DNA and analysis of 8-hydroxy-2′-deoxyguanosine. *Chem. Res. Toxicol.* **1998**, *11*, 882–887. [CrossRef]
134. Bhutani, M.; Pathak, A.K.; Fan, Y.H.; Liu, D.D.; Lee, J.J.; Tang, H.L.; Kurie, J.M.; Morice, R.C.; Kim, E.S.; Hong, W.K.; et al. Oral epithelium as a surrogate tissue for assessing smoking-induced molecular alterations in the lungs. *Cancer Prev. Res.* **2008**, *1*, 39–44. [CrossRef]
135. Wilson, K.A.; Garden, J.L.; Wetmore, N.T.; Wetmore, S.D. Computational insights into the mutagenicity of two tobacco-derived carcinogenic DNA lesions. *Nucleic Acids Res.* **2018**, *46*, 11858–11868. [CrossRef]
136. Manderville, R.A.; Wetmore, S.D. Understanding the mutagenicity of O-linked and C-linked guanine DNA adducts: A combined experimental and computational approach. *Chem. Res. Toxicol.* **2017**, *30*, 177–188. [CrossRef]
137. Millen, A.L.; Sharma, P.; Wetmore, S.D. C8-linked bulky guanosine DNA adducts: Experimental and computational insights into adduct conformational preferences and resulting mutagenicity. *Future Med. Chem.* **2012**, *4*, 1981–2007. [CrossRef]

Perspective

Aristolochic Acids: Newly Identified Exposure Pathways of this Class of Environmental and Food-Borne Contaminants and its Potential Link to Chronic Kidney Diseases

Chi-Kong Chan [1], Yushuo Liu [2], Nikola M. Pavlović [3,4,5,*] and Wan Chan [1,2,6,*]

[1] Department of Chemistry, The Hong Kong University of Science and Technology, Clear Water Bay, Kowloon, Hong Kong; ckchanak@connect.ust.hk

[2] Environmental Science Programs, The Hong Kong University of Science and Technology, Clear Water Bay, Kowloon, Hong Kong; yliuee@connect.ust.hk

[3] Clinic of Nephrology, Clinical Centre Niš, Niš 18000, Serbia

[4] Medical Faculty, University of Niš, Niš 18000, Serbia

[5] Serbian Medical Society, Branch Niš, Niš 18000, Serbia

[6] Division of Environment & Sustainability, The Hong Kong University of Science and Technology, Clear Water Bay, Kowloon, Hong Kong

* Correspondence: nikpavster@gmail.com (N.M.P.); chanwan@ust.hk (W.C.); Tel.: +381-64-352-8583 (N.M.P.); +852-2358-7370 (W.C.); Fax: +381-18-510-363 (N.M.P.); +852-2358-1594 (W.C.)

Received: 4 January 2019; Accepted: 14 March 2019; Published: 19 March 2019

Abstract: Aristolochic acids (AAs) are nitrophenanthrene carboxylic acids naturally produced by *Aristolochia* plants. These plants were widely used to prepare herbal remedies until AAs were observed to be highly nephrotoxic and carcinogenic to humans. Although the use of AA-containing *Aristolochia* plants in herbal medicine is prohibited in countries worldwide, emerging evidence nevertheless has indicated that AAs are the causative agents of Balkan endemic nephropathy (BEN), an environmentally derived disease threatening numerous residents of rural farming villages along the Danube River in countries of the Balkan Peninsula. This perspective updates recent findings on the identification of AAs in food as a result of the root uptake of free AAs released from the decayed seeds of *Aristolochia clematitis* L., in combination with their presence and fate in the environment. The potential link between AAs and the high prevalence of chronic kidney diseases in China is also discussed.

Keywords: aristolochic acids; food contamination; environmental pollution; root uptake; aristolochic acid nephropathy; Balkan endemic nephropathy; chronic kidney disease

---

1. Introduction

Aristolochic acids (AAs, Figure 1) are a class of nitrophenanthrene carboxylic acids naturally produced in *Aristolochia* species (*spp.*) (notably *Aristolochia clematitis* L., *Aristolochia contorta* Bunge, *Aristolochia fangchi* Y. C. Wu ex L. D. Chow & S. M. Hwang, *Aristolochia debilis* Siebold & Zucc. and *Aristolochia manschuriensis* Kom.), *Bragantia spp.* or *Asarum spp.* plants that have been widely used as herbal medicines [1]. As early as 1964, Jackson et al. observed the nephrotoxicity of AAs when high doses were administered to humans [2]. Their human carcinogenicity was not confirmed until the diagnosis of tumors and, later, the identification of DNA-AA adducts (Figure 1) in animal models exposed to AAs and in AA-intoxicated patients [3–5]. The term "Chinese herbs nephropathy" (CHN) was used to describe this unique type of rapidly progressive nephropathy causing end-stage renal failure and urothelial cancer [6]. AAs and AA-containing herbs are currently classified by the

International Agency for Research on Cancer as Group I carcinogens (carcinogenic to humans), and the sale of AA-containing herbal medicines is banned by countries worldwide [1,7].

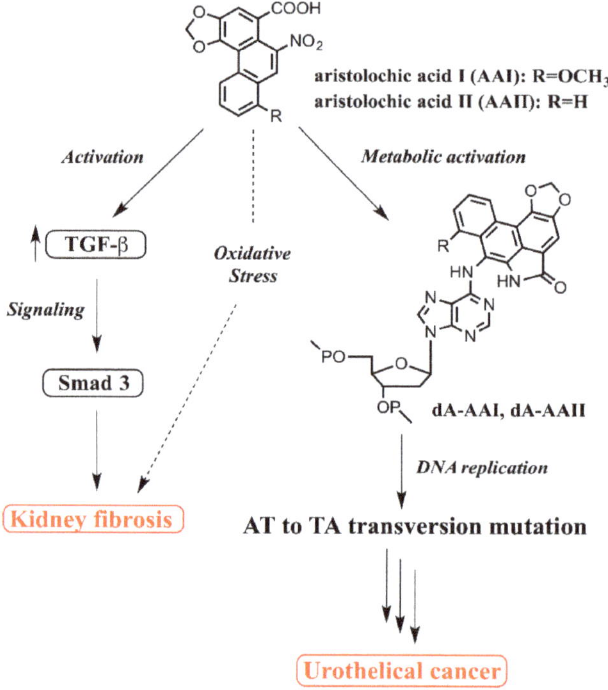

Figure 1. Postulated mechanisms for the nephrotoxicity and carcinogenicity of aristolochic acids.

The origin of CHN was the grim discovery of hundreds of Belgian women who developed severe renal diseases in 1992 after taking diet pills containing extracts of Chinese herbs [8]. Follow-up investigations revealed the presence of AAs in the diet pills and showed that *Stephania tetrandra* S. Moore (Han Fangji) had been inadvertently replaced by *A. fangchi* due to their common Chinese names (both Fangji) [9]. The claim was further supported by the detection of DNA-AA adducts in kidney tissues from the CHN patients that proved their previous exposure to AAs [10]. The term CHN has thus been revised to aristolochic acid nephropathy (AAN) for a more accurate description of the etiological agent of the disease.

After the disclosure of cases in Belgium, similar cases linked to the intake of Chinese herbs were reported in different countries and regions, including Spain, Japan, France, the United Kingdom, Taiwan, the United States, Germany, China, Korea, Australia, Bangladesh, and Hong Kong [11]. In addition, Chinese herbs containing AAs were used as remedies for other purposes such as eczema and pain relief [12]. Although the use and retail sale of AA-containing herbs have been banned worldwide since the discovery of its toxicity, these cases, especially the recent ones [11], revealed that the general population can still obtain AA-containing remedies through online platforms and pharmacies without a prescription.

For decades, residents in farming villages along the Danube River of the Balkan Peninsula have suffered from a mysterious end-stage renal disease, Balkan endemic nephropathy (BEN). A high prevalence of this chronic kidney disease (CKD) was observed among farmers in Bosnia and Herzegovina, Bulgaria, Croatia, Romania, and Serbia (Figure 2) [13,14]. The potential causative roles of polycyclic aromatic hydrocarbons (PAHs), heavy metals, and aflatoxins to the development of the disease were investigated. However, no conclusions could be drawn based on these studies. Since BEN shares common clinical features with AAN, it was later proposed that chronic dietary poisoning by AAs through contaminated food is responsible for the development of BEN. In 1969, Ivić speculated that the seeds of *A. clematitis* (commonly known as birthwort), a widespread AA-containing weed, intermingled with wheat grains during the machine harvesting process and contaminated the food [15]. However, the sources of human exposure pathways to AAs remained unknown for a long time.

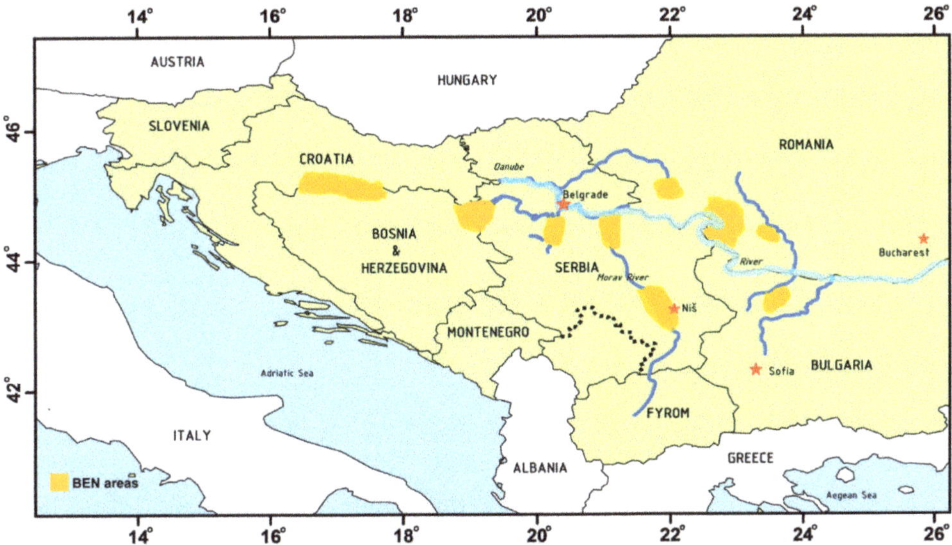

Figure 2. Map showing the geographic distribution of endemic nephropathy in rural farming villages located near tributaries of the Danube River in countries of the Balkan Peninsula.

Environmental exposure to AAs has been proposed to be the main cause of BEN. Only recently, through a collaborative effort led by research groups from Hong Kong and Serbia, scientists have further supported and strengthened the hypothesis that BEN is an environmentally derived disease, by analyzing wheat and corn grains cultivated in endemic villages in Serbia [16]. Both AA-I and AA-II were detected in the food grains and cultivation soil samples collected from the endemic villages, providing the first direct piece of evidence for this novel exposure pathway in the etiology of BEN [16]. It has been estimated that more than 25,000 individuals in the Balkan area are afflicted with BEN, and more than 100,000 individuals living in endemic regions could be at risk [17]. AA-associated renal diseases affect not only residents of the Balkan area but also residents of countries and regions where the practice of traditional Chinese medicine (TCM) is deeply rooted, such as China, Taiwan, and South Korea. It was projected that more than one hundred million Chinese are suffering from CKD, and some of the pathology and disease-causing agents have yet to be elucidated [17]. In this perspective, we claim that AA intake via AA-contaminated food is likely to be one of the primary causes of BEN. Moreover, we propose the novel concept that the consumption of AA-containing TCM could have been and still is the major pathway of human exposure to AAs and is responsible for CKD in China.

2. The Occurrence of Aristolochic Acids in the Environment

AA-containing species of *Aristolochia*, *Bragantia*, and *Asarum* usually grow in valley ditches, on roadsides and on hillsides as bushes at 200–1,500 meters above sea level. They can be widely found as weeds in tropical and subtropical areas, including continental Southeast Asia, Malaysia, China, tropical Africa, and South America. For instance, in China, AA-containing plants are known to grow in the south near the Yangtze River to the Yellow River Basin in Shandong and Henan provinces [18], and they are cultivated to prepare TCM in Guangdong and Guangxi provinces. Despite not yet being reported in the literature, it is possible that the nonmedicinal parts, e.g. leaf and root, of these plants may decompose in the soil and release free AAs into the environment.

Evidence of *Aristolochia spp.* in the Balkan Peninsula is also widely reported in the literature, and these plants are thought to be related to BEN [19]. Although environmental exposure to AAs has long been claimed to be the trigger for the mechanisms of BEN, specifically, in the hypothesis that the fruit of *A. clematitis* comingles with wheat grains during the harvesting process, no scientific evidence has been provided to support this claim [15]. Questions have been raised as to this proposal because of the different ripening times of wheat and *A. clematitis* in the Balkans (Figure 3); while wheat is ripe in early summer (end of June-beginning of July), the fruit of *A. clematitis,* which is rich in AAs, matures in late July [20,21]. Moreover, the significantly different sizes and weights between an *A. clematitis* fruit and wheat would result in most of the coharvested *A. clematitis* seeds being discarded during the threshing and milling processes [21]. Recently, free AAs released from the decay of *A. clematitis* were identified in the soil of a cultivated field in an endemic area of Serbia [16,22]. This identification raised the possibility that the agricultural soil in endemic areas throughout the Balkan region is extensively contaminated with AAs from the decay of *A. clematitis* plants, which are widespread in the endemic areas.

Figure 3. Photos showing (**A**) *Aristolochia clematitis* growing in a wheat field in the village Kutleš in Serbia (photo taken in May 2015), (**B**) enlarged view of *Aristolochia clematitis* growing together with wheat in cultivated field, and (**C**) decaying seed of *Aristolochia clematitis* in the wheat field.

The possibility that water systems in the Balkan regions, including drinking and irrigation water, might be contaminated by AAs that originated from leaching soil is consistent with that finding. However, until now, there was no scientific evidence showing that irrigation water from a river or another water body could contain significant levels of AAs or other environmentally toxic agents released from soil runoff or other sources. The Pliocene lignite hypothesis, an attempt to explain the rise of BEN, postulates that chronic exposure to PAHs or other toxic organic compounds occurs because of leaching from the lignite low-rank coals underlying the endemic settlement into the deep well water [23]. It is not known whether AAs even exist in coal, originating either from the lignite structures and/or soil leaching, to leach into the water. Thus, the analysis of AAs in water bodies, such as wells and streams, is extremely important to public health.

An additional exposure pathway of AAs could result from environmental transport mediated by butterflies. Several species of *Papilionidae* protect themselves from predators by sequestering AA in their body and thus provide their own defense by becoming unpalatable. The larvae of *Papilionidae* in the *Zerynthiini* and *Troidini* tribes feed on *Aristolochia* containing AAs, which are then ingested and stored during the pupal and adult stage of the butterfly. *Zerynthia polyxena*, a butterfly that feeds on *A. clematitis*, accumulates AA-Ia and AA-C in its body tissues. *Battus polydamas cebriones*, belonging to the *Troidini*, which primarily feeds on *A. trilobata*, was found to release AAs in a sample of "Chiniy-trèf", which is a type of medicine produced by maceration of the larvae of *Battus polydamas cebriones* [24].

3. Uptake and Bioaccumulation in Food Crops

Studies have shown that plants absorb and accumulate different environmental pollutants from the soil, such as heavy metals, antibiotics, and pesticides [25–27]. As Chan et al. have demonstrated, the existence of AAs in the cultivated fields in Serbia (Figure 4) [16] raises the concern of whether AAs translocated from the soil into food crops grown in contaminated fields could accumulate in the plants.

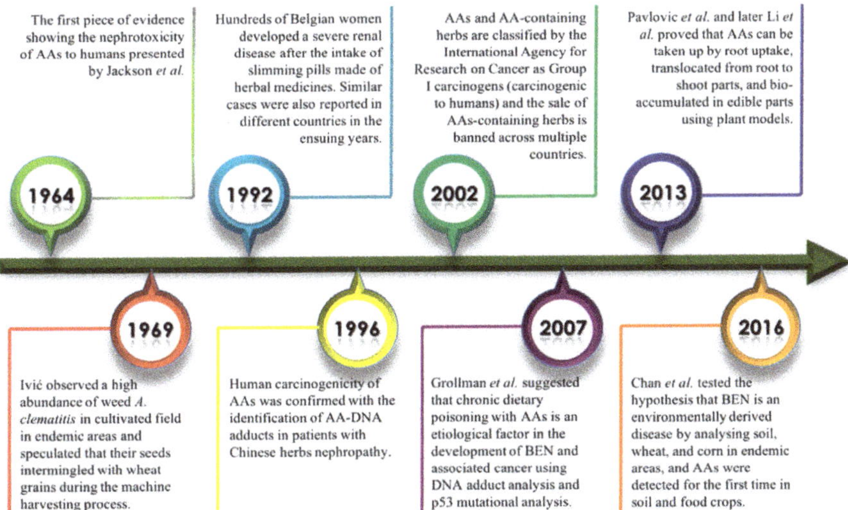

Figure 4. A brief timeline of the discovery of the toxicity of AAs and their occurrence in food and the environment.

The fate and transport of AAs are difficult to predict, given their complex structures. The pKa values of AA-I and AA-II are 3.3 and 3.2, respectively [12], and as soil pH typically is between 5.5 to 8.0, the anionic forms of AAs dominate. The octanol-water partition coefficient (K_{ow}) further describes the fate and transport of AAs. Previous studies recorded the log K_{ow} values of AA-I and AA-II to be 1.65 and 1.23 at natural pH and found that the maximum translocation of chemicals from the soil medium occurs in chemicals with K_{ow} ~1.78 [12]. These characteristics consequently create the possibility that AAs, AA-I in particular, could be taken up by the root, translocated from the root to parts of the shoot and bioaccumulated in the food grains. This hypothesis was further supported by the observation that significantly higher levels of AA-I than of AA-II were observed in wheat grain, tomato fruits, and spring onions.

This hypothesis was recently confirmed by Pavlovic et al. and Li et al., where the former used maize (*Zea mays*) and cucumber (*Cucumis sativus*) as plant models to demonstrate the highly efficient root uptake of AAs from a contaminated nutrient solution in a laboratory setting. The data confirmed that AAs can be absorbed from a nutrient solution by the roots of maize and cucumber [19]. The unequivocal presence of AAs in the roots provided the first line of evidence that root uptake may be one of the principal exposure pathways accounting for the etiology of BEN. To establish a direct relationship between the root uptake of AAs in food crops and the human exposure pathway to AAs in natural settings, Li et al. cultivated lettuce, tomatoes, and spring onions in contaminated soils to mimic realistic environmental exposure to AAs. The results provided the first line of evidence that AAs were transferred from contaminated media to edible parts of common food crops and that they were highly resistant to microbial activity and plant cell metabolism and were therefore able to persist in food crops [28].

Based on these findings, some researchers speculated that if the soil in some endemic areas contained residual AAs released from *A. clematitis*, it would also contaminate the food crops produced in the endemic villages. Therefore, Chan et al. collected soil, wheat, and corn grain samples from the well-known endemic village Kutleš in Serbia. By using high-performance liquid chromatography with fluorescence detection (HPLC-FLD), a method of high sensitivity and selectivity, AAs were detected in soil, corn and wheat samples for the first time [16]. The soil samples collected showed measurable concentrations of AA-I (86.59 ± 27.15 ng/g) and AA-II (25.73 ± 11.46 ng/g), and higher levels of AA-I (91.31 ± 89.02 ng/g) and AA-II (41.01 ± 12.45 ng/g) were detected in wheat grain. A similar observation was also recorded in Romania, where AA-I was detected in soil and soil organic matter samples retrieved from both endemic and non-endemic areas in the country [29], and in another large-scale study in Serbia [22]. These results further support the role of chronic environmental exposure to AAs in the etiology of BEN.

The prospect of finding AAs could be extended to other food crops, such as herbs and vegetables, that are grown in highly endemic regions and countries. It is not yet fully understood whether and to what extent the local and international food chains have been contaminated with AAs, although the prolonged consumption of food harvested from contaminated fields is presumed to be a leading cause of chronic intoxication, with AAs linked to the etiology of BEN and attendant upper tract urothelial cancers (UTUC). Thus, the extent to which nephrotoxic and carcinogenic AAs contaminate the environment remains a significant question to be addressed, suggesting the need to conduct worldwide surveillance of AAs in drinking water and food commodities.

4. Risks to Human Health

The European Agency for the Evaluation of Medicinal Products warns European Union member states "to take steps to ensure that the public is protected from exposure to AAs". However, several publications reported that *A. clematitis* is a common weed in the cultivated fields in endemic areas, in addition to the original observation by Ivić [15,16,19,30]. These findings suggested that AAs could be released into the environment from the decomposition of the *A. clematitis* plant. In the village Kutleš (43°8′22.93″ N, 21°51′39.44″ E, elevation 206–214 m) in Serbia, widespread growth of *A. clematitis* in wheat and corn fields was observed, and the local soil and wheat samples showed a significant level of AAs (Figure 3) [16]. This finding further supported the idea that AAs can be taken up from the environment to contaminate the food crops and that AAs could enter human bodies and lead to BEN via the daily ingestion of AA-tainted food, thus causing public health problems.

A follow-up question could be whether the processing of wheat and maize flour can eliminate the AAs deposited in their respective grains. The typical method to produce flour is by milling, where grains are ground into flour. The other general procedures prior to milling include storing, cleaning, conditioning, and grinding. Cleaning merely removes coarse impurities and fine materials, which could include some fruits and parts of *A. clematitis*, but pollutants that have accumulated inside the grains are unlikely to be removed in the process. For instance, pesticides in wheat flour are often identified and quantified by academic scientists, and we cannot ignore the possibility that AAs contaminate local and international food chains through flour [31]. Similar to the research conducted to identify pesticides in flour samples, research to detect AAs in wheat and maize should be conducted to uncover and confirm the human exposure pathway to AAs and identify the links to BEN. Our group has recently developed methods to quantitate AAs in both wheat and maize flour, and we have collected locally produced samples from supermarkets in Serbia and Bulgaria. The preliminary results have shown that approximately one-fifth of the samples possess part-per-trillion levels of AAs (unpublished results). The positive identification of AAs in flour samples may directly threaten public health locally and globally. As the milling process cannot remove all of the AAs in grains, it is important to determine whether AAs could be eliminated by baking bread and cooking pasta. Methods to lower the quantity of AAs in flour and its products have been reported earlier, and show 30% of AA-I and 20% of AA-II can be removed from food after normal cooking methods such as baking and boiling. Notably, if cysteine is added to the food before boiling, 90% of AAs can be eliminated from AA-tainted food samples boiled in the water [32].

The carcinogenicity and nephrotoxicity of AAs have long been proven to threaten human lives, with numerous CHN, AAN and BEN sufferers identified globally. In addition to BEN, there are still some unknown chronic kidney diseases (CKD*u*), including Mesoamerican nephropathy (MeN), Sri Lankan CKD*u*, Indian CKD*u*, Egyptian CKD*u*, and Tunisian CKD*u*, occurring worldwide, and their etiological mechanisms have not been confirmed [33]. Due to the widespread distribution of *Aristolochia* and *Asarum* species worldwide and their widespread use as herbal medicines, it is possible that these CKD*u* could be linked to AAs at different extents. However, this claim has not been supported by any studies and is awaiting a more in-depth investigation.

5. Possible Metabolites in the Environment

The identification of AA metabolites in the environment, whether in soil or water systems, has not previously been reported. Interestingly, such metabolites have recently been found, for the first time, in the larva, pupa, and imago of Battus feeding on *Aristolochia* [34]. Some sequestered AAs in the Battus body (AA-Ia, AA-IIIa, AA-IVa, and AA-IVb) are converted to AA *O*-glucosides (AA-IaG, AA-IIIaG, AA-IVaG, and AA-IVbG) through insect-mediated *O*-glucosylation, in which the glucosyl group links to those acids, and is known to be the main detoxification pathway for the insects [34,35]. Notably, the larval integument contains a high concentration of AAs, while the adult contains a lower level than the larva, which indicates that molting is another method of detoxification. The integuments decompose and release AAs into the environment after metamorphosis [34]. Ultimately, the major sink of AAs occurs in the natural environment, such as water bodies and soil systems.

Existing research [28] has raised the possibility that AAs may degrade in the soil, but their metabolites and the mechanism of their formation are still unknown, suggesting that the identification of AA metabolites merits additional research. It is hypothesized that experiments such as spiking AAs into a soil model could simulate the degradation of naturally existing AAs in the environment. After the soil incubation process, liquid chromatography coupled with mass spectrometry can be employed to analyze the extracts of the soil samples and identify AA metabolites.

6. Overview of Potential Biomarkers for AA intoxication

Cancer and nephropathy caused by AAs require the development of potential biomarkers to facilitate clinical diagnosis. The biomarkers known to date to assist in diagnosing AA nephropathy are classified into two categories, namely, biomarkers of exposure and biomarkers of effect.

Regarding biomarkers of exposure, carcinogenesis and mutagenicity were found to correlate with the formation of DNA-AA adducts [36]. Once AAs enter rodent or human bodies, they are first activated by enzymatic nitro-reduction to form *N*-hydroxyaristolactams and then become reactive cyclic *N*-acylnitrenium ions (Figure 1) [11]. The enzymes responsible for the nitro-reduction step are NAD(P)H:quinone oxidoreductase, hepatic microsomal cytochrome P450 (CYP) 1A1/2 and kidney microsomal NADPH:CYP reductase [37,38]. The delocalized positive charge on the *N*-acylnitrenium ion then attacks the exocyclic amino groups on dA, dG and dC to give stable and persistent DNA-AA adducts [37,38]. 7-(Deoxyadenosine-N^6-yl)-aristolactam I (dA-AAI) is the most abundant DNA adduct formed and exhibits extended persistence in kidney (Figure 1) [39]. Other specific DNA-AA adducts, such as 7-(deoxyadenosine-N^6-yl)-aristolactam II (dA-AA-II), 7-(deoxyguanosine-N^2-yl)-aristolactam I (dG-AAI), and 7-(deoxyguanosine-N^2-yl)-aristolactam II (dG-AA-II), were found in renal tissues of patients diagnosed with AAN and BEN [40,41]. A number of studies also identified the presence of these adducts in other organs, including the liver and stomach [40,42,43]. These adducts accumulate in cells and hinder DNA replication, eventually causing cell dysfunction and death. Chen et al. observed 151 cases of upper urinary tract epithelial cancer in Taiwan, where AA-laced herbal remedies are used extensively, and found that AAs reacted with DNA to form DNA-AA adducts, which have unique mutation characteristics in tumor suppressor gene *TP53* [44]. The oncogenes *FGFR3* and *HRAS* also seem to contain mutations induced by AAs, as the frequency of A:T crossover at codons 373 and 61 of *FGFR3* and *HRAS* in Taiwanese patients was five times higher than that in patients with urothelial cancer worldwide [45]. In addition, analysis of the data of UUC patients showed that the unique A:T to T:A transversion mutation formed mutation hotspots (53.1% of the total). *FGFR3* and *TP53* mutations define different approaches in the early diagnosis of urothelial cell carcinoma [46].

Another category of biomarkers is the biomarkers of effect. Li et al. discovered that oxidative stress participates in the development of AAN due to the significant decrease in the level of the antioxidant glutathione (GSH) [47]. For instance, they observed an elevated level of methylglyoxal (MGO) in kidneys in their mouse models, and the highly cytotoxic MGO could further modify proteins to form an advanced glycation end product, N^ε-carboxymethyllysine (CML) [47]. The generation of MGO and CML in combination with reduced intrarenal antioxidant capacity can further lead to aging problems and diabetic complications [48]. In the treatment of AAN, we may consider controlling the levels of MGO and CML. A possible alternative mechanism of AA-induced toxicity is that AA-I depletes thiols in cells, whose imbalance impacts reactive oxygen species generation, DNA damage, and mitochondrial dysfunction [49]. AA-I is hydrogenated by reaction with cysteine or glutathione in cells, followed by loss of the nitro group. Because this reaction may occur under conditions similar to those in the human body (pH 7.0 and 37 °C), it is reasonable that AA-I may damage kidney cells by reacting away cysteine or GSH [50]. Cysteine is essential for many peptides and proteins, such as the antioxidant GSH and iron-sulfur cluster alloys [51,52]; GSH helps to maintain the normal function of the immune system and detoxifies some drugs (paracetamol), toxins (free radicals, iodoacetic acid) and heavy metals (lead and mercury) [53]. Therefore, AA-I may reduce thiol levels in the kidneys, leading to nephropathy caused by functional aggregation and oxidation.

7. Postulated Mechanism for AA-associated Kidney Fibrosis

One fundamental clinical feature of AAN is progressive kidney tubulointerstitial fibrosis, which is common in CKD and ultimately leads to end-stage renal disease [11]. AAs have been shown to cause lesions, leading mainly to renal tubular epithelial cell (RTEC) and tubulointerstitial injuries, in both in vitro and in vivo studies [54–56]. Under normal circumstances, these injured epithelial cells demonstrate a strong ability for repair through cell necrosis and apoptosis [54]; the surrounding intact cells then proliferate actively to maintain tubule integrity and renal function [54,57]. However, upon examination of the renal biopsy specimens from AAN patients, Yang et al. discovered that the acute epithelial cell injury caused by AA intoxication was not repaired by normal cell regeneration. Yang et al. then investigated expression of the epidermal growth factor (EGF), which plays an important role in cell repair, modulating cell proliferation and growth by binding to its receptor, in the specimens [54]. The results showed that EGF expression was suppressed in AAN patients and could account for the absence of RTEC regeneration.

Most of the AA-intoxicated rodent models pointed to a significant role of the overexpression of transforming growth factor-β (TGF-β), an essential cytokine in fibrogenesis and stimulator of myofibroblast activation, in the development of interstitial fibrosis, although the underlying molecular events remained poorly understood in past decades [54,58–61]. More recently, TGF-β was proven to induce fibrosis by Smad 3 stimulation under various conditions, such as bleomycin-induced pulmonary fibrosis, obstructive nephropathy, and liver and colon fibrosis [62]. Zhou et al. found AA-induced progressive renal failure and tubulointerstitial fibrosis in Smad3 wild type mice, but not in Smad3 knockout mice, indicating that Smad3 is a key factor in the development of AA-induced nephropathy (Figure 1) and that the specific deletion of TGF-β/Smad3 signaling may be a potential therapeutic target for chronic AAN [62]. Furthermore, Dai et al. reported that the loss of a negative regulator of TGF-β/Smad signaling, Smad 7, could account for the progressive renal injury observed in a mouse model of chronic AAN [63]. This proposal was validated by dosing a Smad7 knockout model with AAs, which resulted in enhanced renal injury progression. The restoration and ultrasound-mediated gene transfer of Smad7 in the knockout mice and injured kidney of wild-type mice, respectively, were later demonstrated to inhibit the further development of chronic renal nephropathy [63]. Tying together these findings with previous studies on other renal disease models highlighting the capability of Smad7 overexpression to attenuate kidney inflammation and fibrosis [64,65], Smad7 could play a protective role in the pathogenesis of chronic AAN.

Recent studies also have shown that BMP-7 inhibited epithelial to mesenchymal transition (EMT) through reducing the production of TGF-β and myofibroblast activation in HK-2 cells initiated by AAs [66]. EMT is associated with injury repair, tissue regeneration, and organ fibrosis. Its main biological function is to produce fibroblasts to repair tissue damage caused by trauma and inflammation. Additionally, the promotion of the AA-induced EMT in HK-2 cells was observed after the addition of gremlin, an antagonist of BMP-7 [67]. Even though several pathways of AA-induced TGF-β have been discovered, there is a gap in understanding the mechanisms of AA-induced renal fibrosis, such as the role of oxidative stress in kidney fibrosis (Figure 1).

8. Challenges and Future Perspectives

The study of AAs in the pathophysiology of BEN and other related CKD*u* presents an array of challenges. There are official and sound medical records of BEN and UTUC in the Balkan regions but no data or records about AA-contaminated crops, soil, or water, presenting difficulties in analyzing and defining their geographical distribution. If there were proper records and statistics on the geographical distribution of BEN patients and AA contamination in addition to the medical records, we could target those areas with high BEN incidences and launch large-scale projects to discover the root cause of the diseases. This strategy could also be extended to countries that are suffering from CKD*u*.

AAN does not endanger only a single country or region, since cases have been reported in many countries worldwide in the past decades [11]. Due to the financial instability of developing countries and lack of evidence regarding the presence and geographical distribution of AAs in developed countries, the situation of AA contamination may not be easily improved. The identification of AAs in the environment and food samples can be challenging since the concentrations of AAs present in the samples are usually at trace levels within complicated sample matrices, and thus require highly sensitive instruments and skilled technicians. Therefore, acquiring instruments and employing experts to perform a comprehensive investigation of all potential sources of AAs in the environment could be costly for local governments, who may be reluctant to prioritize and confront an existing AA contamination problem. In addition, agricultural activities are often the backbone of the economic system in developing countries, and it may be infeasible to terminate the cultivation of food crops in the contaminated areas and implement soil remediation efforts. Additionally, there is no available standard remediation method for AA-contaminated soil clean-up, in contrast to the soil remediation methods for other types of pollution, as recommended by the USA Environmental Protection Agency.

With significant evidence demonstrating that AAs are strong human carcinogens and potent nephrotoxins [1], considered now to be a worldwide problem threatening more than one billion people, regulatory agencies should be alerted to the potential existence of AAs and should classify them as new contaminants in both soil and food crops. This recognition is especially important for the governments of Asian countries with a long history of practicing Chinese herbal medicine, such as China. More than

100 million people in mainland China are estimated to be suffering from CKD [17], but the pathology and disease-causing agents are not fully understood.

Owing to the widespread use of *Aristolochia spp.* in the preparation of traditional herbal remedies, Grollman has speculated on an unrecognized linkage between AAs and global diseases [17]. He and his team conducted a molecular epidemiologic approach in Taiwan and revealed that nearly one-third of the population had been administered AA-containing remedies. A dose-dependent relationship between the consumption of AA-containing herbal remedies and the risk of developing renal cancer was also observed [17]. Such an epidemiologic study should also be carried out in other countries with frequent use of AA-containing herbs, such as Korea and China. In Korea, the prescription of AA-containing herbal medicine to patients by clinics was known to be a cause of local AAN cases, even after the prohibition of AA-containing ingredients in herbal medicine by the Korea Food and Drug Administration and the number of AAN cases was reported to be underestimated [68]. Given that some *Aristolochia spp.* are still allowed to be used according to the 2015 Chinese Pharmacopoeia [17] and the list of already marketed TCMs with AA-containing herbs was recently disclosed by the China Food and Drug Administration, part of the Chinese population could have been exposed to AAs through TCM intake.

Could AAs, therefore, be responsible for the prevalence of CKD in China or of other CKD*u*? To answer this question, at the first stage, large-scale TCM and herbal medicine analyses are needed to differentiate AA-containing and non-AA-containing TCM and quantitate the concentration of AAs in these remedies. Furthermore, to uncover the true cause of the diseases, the clinical features of CKD*u* could be compared to those of AAN and BEN, and the analysis of a robust and sensitive biomarker, DNA-AA adducts, should be performed in patients with CKD, given their high persistence in target organs and unique mutation patterns. Other biomarkers, such as covalent blood protein adducts, should also be developed to obtain simple and routine clinical tests and screening for human exposure to AAs.

Since *Aristolochia* plants grow all over the earth in different geographic environments, worldwide surveillance for the existence of AAs in cultivation fields and in food and drug products is urgently needed. To safeguard public health, methods for remediating AA-contaminated farmland should be developed and implemented in a timely and appropriate manner in contaminated areas. The residents in the endemic areas should be immediately informed of the existence of AAs in their cultivated fields, food, and drug ingredients in order to stop the dietary intake of AAs through AA-contaminated food and to help prevent them from acquiring end-stage kidney disease and carcinoma of the upper urinary tract. Last but not least, we should investigate the molecular events involved in the carcinogenesis and fibrogenesis of AAN and CKD more vigorously to discover new therapeutic agents for appropriate disease prevention and treatment.

Funding: This research was funded by the TUYF Charitable Trust (TUYF18SC01).

Acknowledgments: W. Chan expresses his sincere gratitude to the Hong Kong University of Science and Technology for a Startup Funding (Grant R9310). C.-K. Chan was supported by a Hong Kong PhD Fellowship offered by the University Grants Committee of Hong Kong.

Conflicts of Interest: The authors declare no competing financial or nonfinancial interest.

References

1. IARC Working Group on the Evaluation of Carcinogenic Risks to Humans; World Health Organization; International Agency for Research on Cancer. *Some Traditional Herbal Medicines, Some Mycotoxins, Naphthalene and Styrene*; World Health Organization: Lyon, France, 2002; pp. 1–556.
2. Jackson, L.; Kofman, S.; Weiss, A.; Brodovsky, H. Aristolochic acid (NSC-50413): Phase I clinical study. *Cancer Chemother. Rep.* **1964**, *42*, 65.
3. Mengs, U.; Lang, W.; Poch, J.A. The carcinogenic action of aristolochic acid in rats. *Arch. Toxicol.* **1982**, *51*, 107–119. [CrossRef]
4. Mengs, U. Acute toxicity of aristolochic acid in rodents. *Arch. Toxicol.* **1987**, *59*, 328–331. [CrossRef]

5. Schmeiser, H.H.; Bieler, C.A.; Wiessler, M.; van Ypersele de Strihou, C.; Cosyns, J.P. Detection of DNA adducts formed by aristolochic acid in renal tissue from patients with Chinese herbs nephropathy. *Cancer Res.* **1996**, *56*, 2025–2028. [PubMed]

6. Cosyns, J.P.; Jadoul, M.; Squifflet, J.P.; De Plaen, J.F.; Ferluga, D.; Van Ypersele De Strihou, C. Chinese herbs nephropathy: A clue to balkan endemic nephropathy? *Kidney Int.* **1994**, *45*, 1680–1688. [CrossRef] [PubMed]

7. Kessler, D.A. Cancer and herbs. *N. Engl. J. Med.* **2000**, *342*, 1742–1743. [CrossRef]

8. Vanherweghem, J.-L.; Depierreux, M.F.; Tielemans, C.; Abramowicz, D.; Dratwa, M.; Jadoul, M.; Richard, C.; Vandervelde, D.; Verbeelen, D.; Vanhaelen-Fastre, R.; et al. Rapidly progressive interstitial renal fibrosis in young women: Association with slimming regimen including Chinese herbs. *Lancet* **1993**, *341*, 387–391. [CrossRef]

9. Vanhaelen, M.; Vanhaelen-Fastre, R.; But, P.; Vanherweghem, J.L. Identification of aristolochic acid in Chinese herbs. *Lancet* **1994**, *343*, 174. [CrossRef]

10. Nortier, J.L.; Martinez, M.C.; Schmeiser, H.H.; Arlt, V.M.; Bieler, C.A.; Petein, M.; Depierreux, M.F.; De Pauw, L.; Abramowicz, D.; Vereerstraeten, P.; et al. Urothelial carcinoma associated with the use of a Chinese herb (*Aristolochia fangchi*). *N. Engl. J. Med.* **2000**, *342*, 1686–1692. [CrossRef]

11. Jadot, I.; Declèves, A.-E.; Nortier, J.; Caron, N. An integrated view of aristolochic acid nephropathy: Update of the literature. *Int. J. Mol. Sci.* **2017**, *18*, 297. [CrossRef] [PubMed]

12. Tangtong, C. *Environmental Processes Controlling the Fate and Transport of Aristolochic Acid in Agricultural Soil and Copper in Contaminated Lake Sediment*; Michigan State University: East Lansing, MI, USA, 2014.

13. Tatu, C.A.; Orem, W.H.; Finkelman, R.B.; Feder, G.L. The etiology of balkan endemic nephropathy: Still more questions than answers. *Environ. Health Perspect.* **1998**, *106*, 689–700. [CrossRef] [PubMed]

14. Grollman, A.P.; Shibutani, S.; Moriya, M.; Miller, F.; Wu, L.; Moll, U.; Suzuki, N.; Fernandes, A.; Rosenquist, T.; Medverec, Z.; et al. Aristolochic acid and the etiology of endemic (balkan) nephropathy. *Proc. Natl. Acad. Sci. USA* **2007**, *104*, 12129–12134. [CrossRef] [PubMed]

15. Ivić, M. Etiology of endemic nephropathy. *Lijec. Vjesn.* **1969**, *91*, 1273–1281.

16. Chan, W.; Pavlović, N.M.; Li, W.; Chan, C.K.; Liu, J.; Deng, K.; Wang, Y.; Milosavljević, B.; Kostić, E.N. Quantitation of aristolochic acids in corn, wheat grain, and soil samples collected in serbia: Identifying a novel exposure pathway in the etiology of balkan endemic nephropathy. *J. Agric. Food Chem.* **2016**, *64*, 5928–5934. [CrossRef] [PubMed]

17. Grollman, A.P. Aristolochic acid nephropathy: Harbinger of a global iatrogenic disease. *Environ. Mol. Mutagen.* **2013**, *54*, 1–7. [CrossRef]

18. Ma, J.S. The geographical distribution and the system of *Aristolochiaceae*. *Acta Phytotaxon. Sin.* **1990**, *28*, 345–355.

19. Pavlović, N.M.; Maksimović, V.; Maksimović, J.D.; Orem, W.H.; Tatu, C.A.; Lerch, H.E.; Bunnell, J.E.; Kostić, E.N.; Szilagyi, D.N.; Paunescu, V. Possible health impacts of naturally occurring uptake of aristolochic acids by maize and cucumber roots: Links to the etiology of endemic (balkan) nephropathy. *Environ. Geochem. Health* **2013**, *35*, 215–226. [CrossRef]

20. Mantle, P.G.; Herman, D.; Tatu, C. Is aristolochic acid really the cause of the Balkan endemic nephropathy? *J. Controversies Biomed. Res.* **2016**, *2*, 9–20. [CrossRef]

21. Chan, C.-K.; Liu, Y.; Pavlović, N.M.; Chan, W. Etiology of Balkan endemic nephropathy: An update on aristolochic acids exposure mechanisms. *Chem. Res. Toxicol.* **2018**, *31*, 1109–1110. [CrossRef]

22. Li, W.; Chan, C.-K.; Liu, Y.; Yao, J.; Mitić, B.; Kostić, E.N.; Milosavljević, B.S.; Davinić, I.; Orem, W.H.; Tatu, C.A.; et al. Aristolochic acids as persistent soil pollutants: Determination of risk for human exposure and nephropathy from plant uptake. *J Agric Food Chem.* **2018**, *66*, 11468–11476. [CrossRef]

23. Pavlović, N.M. Balkan endemic nephropathy-current status and future perspectives. *Clin. Kidney J.* **2013**, *6*, 257–265. [CrossRef] [PubMed]

24. Cachet, X.; Langrand, J.; Bottai, C.; Dufat, H.; Locatelli-Jouans, C.; Nossin, E.; Boucaud-Maitre, D. Detection of aristolochic acids I and II in "chiniy-Trèf", a traditional medicinal preparation containing caterpillars feeding on *Aristolochia trilobata* L. in Martinique, French West Indies. *Toxicon* **2016**, *114*, 28–30. [CrossRef]

25. Peralta-Videa, J.R.; Lopez, M.L.; Narayan, M.; Saupe, G.; Gardea-Torresdey, J. The biochemistry of environmental heavy metal uptake by plants: Implications for the food chain. *Int. J. Biochem. Cell Biol.* **2009**, *41*, 1665–1677. [CrossRef]

26. Behrendt, H.; Brüggemann, R. Modelling the fate of organic chemicals in the soil plant environment: Model study of root uptake of pesticides. *Chemosphere* **1993**, *27*, 2325–2332. [CrossRef]
27. Wang, Y.; Chan, K.K.J.; Chan, W. Plant uptake and metabolism of nitrofuran antibiotics in spring onion grown in nitrofuran-contaminated soil. *J. Agric. Food Chem.* **2017**, *65*, 4255–4261. [CrossRef] [PubMed]
28. Li, W.; Hu, Q.; Chan, W. Uptake and accumulation of nephrotoxic and carcinogenic aristolochic acids in food crops grown in *Aristolochia clematitis*-contaminated soil and water. *J. Agric. Food Chem.* **2016**, *64*, 107–112. [CrossRef]
29. Gruia, A.T.; Oprean, C.; Ivan, A.; Cristea, M.; Draghia, L.; Damiescu, R.; Pavlovic, N.M.; Paunescu, V.; Tatu, C.A. Balkan endemic nephropathy and aristolochic acid I: An investigation into the role of soil and soil organic matter contamination, as a potential natural exposure pathway. *Environ. Geochem. Health* **2018**, *40*, 1437–1448. [CrossRef]
30. De Broe, M.E. Chinese herbs nephropathy and balkan endemic nephropathy: Toward a single entity, aristolochic acid nephropathy. *Kidney Int.* **2012**, *81*, 513–515. [CrossRef]
31. González-Curbelo, M.Á.; Hernández-Borges, J.; Borges-Miquel, T.M.; Rodríguez-Delgado, M.Á. Determination of organophosphorus pesticides and metabolites in cereal-based baby foods and wheat flour by means of ultrasound-assisted extraction and hollow-fiber liquid-phase microextraction prior to gas chromatography with nitrogen phosphorus detection. *J. Chromatogr. A* **2013**, *1313*, 166–174. [CrossRef]
32. Li, W.; Chan, C.-K.; Wong, Y.L.; Chan, K.J.; Chan, H.W.; Chan, W. Cooking methods employing natural anti-oxidant food additives effectively reduced concentration of nephrotoxic and carcinogenic aristolochic acids in contaminated food grains. *Food Chem.* **2018**, *264*, 270–276. [CrossRef]
33. Gifford, F.J.; Gifford, R.M.; Eddleston, M.; Dhaun, N. Endemic nephropathy around the world. *Kidney Int. Rep.* **2017**, *2*, 282–292. [CrossRef] [PubMed]
34. Priestap, H.A.; Velandia, A.E.; Johnson, J.V.; Barbieri, M.A. Secondary metabolite uptake by the *Aristolochia*-feeding papilionoid butterfly *Battus polydamas*. *Biochem. Syst. Ecol.* **2012**, *40*, 126–137. [CrossRef]
35. Ahmad, S.A.; Hopkins, T.L. β-Glucosylation of plant phenolics by phenol β-glucosyltransferase in larval tissues of the tobacco hornworm, *Manduca sexta* (L.) *Insect Biochem. Mol. Biol.* **1993**, *23*, 581–589. [CrossRef]
36. Bieler, C.A.; Stiborova, M.; Wiessler, M.; Cosyns, J.-P.; van Ypersele de Strihou, C.; Schmeiser, H.H. ^{32}P-post-labelling analysis of DNA adducts formed by aristolochic acid in tissues from patients with Chinese herbs nephropathy. *Carcinogenesis* **1997**, *18*, 1063–1067. [CrossRef]
37. Rosenquist, T.A.; Grollman, A.P. Mutational signature of aristolochic acid: Clue to the recognition of a global disease. *DNA Repair* **2016**, *44*, 205–211. [CrossRef]
38. Chen, M.; Gong, L.; Qi, X.; Xing, G.; Luan, Y.; Wu, Y.; Xiao, Y.; Yao, J.; Li, Y.; Xue, X.; et al. Inhibition of renal NQO1 activity by dicoumarol suppresses nitroreduction of aristolochic acid I and attenuates its nephrotoxicity. *Toxicol. Sci.* **2011**, *122*, 288–296. [CrossRef]
39. Schmeiser, H.H.; Nortier, J.L.; Singh, R.; Gamboa da Costa, G.; Sennesael, J.; Cassuto-Viguier, E.; Ambrosetti, D.; Rorive, S.; Pozdzik, A.; Phillips, D.H.; et al. Exceptionally long-term persistence of DNA adducts formed by carcinogenic aristolochic acid I in renal tissue from patients with aristolochic acid nephropathy. *Int. J. Cancer* **2014**, *135*, 562–567. [CrossRef]
40. Schmeiser, H.H.; Stiborova, M.; Arlt, V.M. Chemical and molecular basis of the carcinogenicity of *Aristolochia* plants. *Curr. Opin. Drug Discov. Dev.* **2009**, *12*, 141–148.
41. Stiborova, M.; Arlt, V.M.; Schmeiser, H.H. DNA adducts formed by aristolochic acid are unique biomarkers of exposure and explain the initiation phase of upper urothelial cancer. *Int. J. Mol. Sci.* **2017**, *18*, 2144. [CrossRef]
42. Fernando, R.C.; Schmeiser, H.H.; Scherf, H.R.; Wiessler, M. Formation and persistence of specific purine DNA adducts by ^{32}P-postlabelling in target and non-target organs of rats treated with aristolochic acid I. *IARC Sci. Publ.* **1993**, *124*, 167–171.
43. Liu, Y.; Chan, C.-K.; Jin, L.; Wong, S.-K.; Chan, W. Quantitation of DNA adducts in target and nontarget organs of aristolochic acid I-exposed rats: Correlating DNA adduct levels with organotropic activities. *Chem. Res. Toxicol.* **2019**. [CrossRef] [PubMed]
44. Chen, C.H.; Dickman, K.G.; Moriya, M.; Zavadil, J.; Sidorenko, V.S.; Edwards, K.L.; Gnatenko, D.V.; Wu, L.; Turesky, R.J.; Wu, X.R.; et al. Aristolochic acid-associated urothelial cancer in Taiwan. *Proc. Natl. Acad. Sci. USA* **2012**, *109*, 8241–8246. [CrossRef]

45. Bakkar, A.A.; Wallerand, H.; Radvanyi, F.; Lahaye, J.B.; Pissard, S.; Lecerf, L.; Kouyoumdjian, J.C.; Abbou, C.C.; Pairon, J.C.; Jaurand, M.C.; et al. FGFR3 and TP53 gene mutations define two distinct pathways in urothelial cell carcinoma of the bladder. *Cancer Res.* **2003**, *63*, 108–8112.
46. McCormick, F. Ras-related proteins in signal transduction and growth control. *Mol. Reprod. Dev.* **1995**, *42*, 500–506. [CrossRef]
47. Li, Y.C.; Tsai, S.H.; Chen, S.M.; Chang, Y.M.; Huang, T.C.; Huang, Y.P.; Chang, C.T.; Lee, J.A. Aristolochic acid-induced accumulation of methylglyoxal and N^ε-(carboxymethyl)lysine: An important and novel pathway in the pathogenic mechanism for aristolochic acid nephropathy. *Biochem. Biophys. Res. Commun.* **2012**, *423*, 832–837. [CrossRef] [PubMed]
48. Thornalley, P.J. Glyoxalase I-structure, function and a critical role in the enzymatic defence against glycation. *Biochem. Soc. Trans.* **2003**, *31*, 1343–1348. [CrossRef] [PubMed]
49. Liu, J.; Shen, H.M.; Ong, C.N. Role of intracellular thiol depletion, mitochondrial dysfunction and reactive oxygen species in *Salvia Miltiorrhiza*-induced apoptosis in human hepatoma HepG$_2$ cells. *Life Sci.* **2001**, *69*, 1833–1850. [CrossRef]
50. Priestap, H.A.; Barbieri, M.A. Conversion of aristolochic acid I into aristolic acid by reaction with cysteine and glutathione: Biological implications. *J. Nat. Prod.* **2013**, *76*, 965–968. [CrossRef]
51. Lill, R.; Mühlenhoff, U. Iron–sulfur-protein biogenesis in eukaryotes. *Trends. Biochem. Sci.* **2005**, *30*, 133–141. [CrossRef]
52. Baker, D.H.; Czarnecki-Maulden, G.L. Pharmacologic role of cysteine in ameliorating or exacerbating mineral toxicities. *J. Nutr.* **1987**, *117*, 1003–1010. [CrossRef]
53. Pompella, A.; Visvikis, A.; Paolicchi, A.; De Tata, V.; Casini, A.F. The changing faces of glutathione, a cellular protagonist. *Biochem. Pharmacol.* **2003**, *66*, 1499–1503. [CrossRef]
54. Yang, L.; Li, X.; Wang, H. Possible mechanisms explaining the tendency towards interstitial fibrosis in aristolochic acid-induced acute tubular necrosis. *Nephrol. Dial. Transplant.* **2007**, *22*, 445–456. [CrossRef] [PubMed]
55. Lebeau, C.; Arlt, V.M.; Schmeiser, H.H.; Boom, A.; Verroust, P.J.; Devuyst, O.; Beauwens, R. Aristolochic acid impedes endocytosis and induces DNA adducts in proximal tubule cells. *Kidney Int.* **2001**, *6*, 1332–1342. [CrossRef] [PubMed]
56. Lebeau, C.; Debelle, F.D.; Arlt, V.M. Early proximal tubule injury in experimental aristolochic acid nephropathy: Functional and histological studies. *Nephrol. Dial. Transplant.* **2005**, *20*, 2321–2332. [CrossRef]
57. Toback, F.G. Regeneration after acute tubular necrosis. *Kidney Int.* **1996**, *41*, 226. [CrossRef]
58. Pozdzik, A.A.; Salmon, I.J.; Debelle, F.D.; Decaestecker, C.; van den Branden, C.; Verbeelen, D.; Deschodt-Lanckman, M.M.; Vanherweghem, J.-L.; Nortier, J.L. Aristolochic acid induces proximal tubule apoptosis and epithelial to mesenchymal transformation. *Kidney Int.* **2008**, *73*, 595–607. [CrossRef]
59. Pozdzik, A.A.; Salmon, I.J.; Husson, C.P.; Decaestecker, C.; Rogier, E.; Bourgeade, M.-F.; Deschodt-Lanckman, M.M.; Vanherweghem, J.-L.; Nortier, J.L. Patterns of interstitial inflammation during the evolution of renal injury in experimental aristolochic acid nephropathy. *Nephrol. Dial. Transplant.* **2008**, *23*, 2480–2491. [CrossRef]
60. Li, J.; Zhang, Z.; Wang, D.; Wang, Y.; Li, Y.; Wu, G. TGF-β1/Smads signaling stimulates renal interstitial fibrosis in experimental AAN. *J. Recept. Signal Transduct. Res.* **2009**, *29*, 280–285. [CrossRef]
61. Wang, Y.; Zhang, Z.; Shen, H.; Lu, Y.; Li, H.; Ren, X.; Wu, G. TGF-β1/Smad7 Signaling Stimulates Renal Tubulointerstitial Fibrosis Induced by AAI. *J. Recept. Signal Transduct.* **2008**, *28*, 413–428. [CrossRef]
62. Zhou, L.; Fu, P.; Huang, X.R.; Liu, F.; Chung, A.C.K.; Lai, K.N.; Lan, H.Y. Mechanism of chronic aristolochic acid nephropathy: Role of Smad3. *Am. J. Physiol. Ren. Physiol.* **2010**, *298*, F1006–F1017. [CrossRef]
63. Dai, X.-Y.; Zhou, L.; Huang, X.-R.; Fu, P.; Lan, H.-Y. Smad7 protects against chronic aristolochic acid nephropathy in mice. *Oncotarget* **2015**, *6*, 11930–11944. [CrossRef] [PubMed]
64. Chung, A.C.K.; Dong, Y.; Yang, W.; Zhong, X.; Li, R.; Lan, H.Y. Smad7 suppresses renal fibrosis via altering expression of TGF-β/Smad3-regulated microRNAs. *Mol. Ther.* **2013**, *21*, 388–398. [CrossRef] [PubMed]
65. Hou, C.C.; Wang, W.; Huang, X.R.; Fu, P.; Chen, T.H.; Sheikh-Hamad, D.; Lan, H.Y. Ultrasound-microbubble-mediated gene transfer of inducible Smad7 blocks transforming growth factor-beta signaling and fibrosis in rat remnant kidney. *Am. J. Pathol.* **2005**, *166*, 761–771. [CrossRef]
66. Wang, Z.; Zhao, J.; Zhang, J.; Wei, J.; Zhang, J.; Huang, Y. Protective effect of BMP-7 against aristolochic acid-induced renal tubular epithelial cell injury. *Toxicol. Lett.* **2010**, *198*, 348–357. [CrossRef]

67. Li, Y.; Wang, Z.; Wang, S.; Zhao, J.; Zhang, J.; Huang, Y. Gremlin-mediated decrease in bone morphogenetic protein signaling promotes aristolochic acid-induced epithelial-to-mesenchymal transition (EMT) in HK-2 cells. *Toxicology* **2012**, *297*, 68–75. [CrossRef] [PubMed]
68. Ban, T.H.; Min, J.-W.; Seo, C.; Kim, D.R.; Lee, Y.H.; Chung, B.H.; Jeong, K.-H.; Lee, J.W.; Kim, B.S.; Lee, S.-H.; et al. Update of aristolochic acid nephropathy in Korea. *Korean J. Intern. Med.* **2018**, *33*, 961–969. [CrossRef]

Review

Radiocarbon Tracers in Toxicology and Medicine: Recent Advances in Technology and Science

Michael A. Malfatti [1], Bruce A. Buchholz [2], Heather A. Enright [1], Benjamin J. Stewart [1], Ted J. Ognibene [2], A. Daniel McCartt [2], Gabriela G. Loots [1], Maike Zimmermann [3,4], Tiffany M. Scharadin [3,4], George D. Cimino [4], Brian A. Jonas [3], Chong-Xian Pan [3], Graham Bench [2], Paul T. Henderson [3,4,*] and Kenneth W. Turteltaub [1,*]

[1] Biosciences and Biotechnology Division, Lawrence Livermore National Laboratory, Livermore, CA 94550, USA; malfatti1@llnl.gov (M.A.M.); enright3@llnl.gov (H.A.E.); stewart66@llnl.gov (B.J.S.); loots1@llnl.gov (G.G.L.)

[2] Center for Accelerator Mass Spectrometry, Lawrence Livermore National Laboratory, Livermore, CA 94550, USA; buchholz2@llnl.gov (B.A.B.); ognibene1@llnl.gov (T.J.O.); mccartt1@llnl.gov (A.D.M.); bench1@llnl.gov (G.B.)

[3] Department of Internal Medicine, Division of Hematology and Oncology and UC Davis Comprehensive Cancer Center, University of California Davis Medical School, Sacramento, CA 95817, USA; mzimmermann@ucdavis.edu (M.Z.); tscharadin@ucdavis.edu (T.M.S.); bajonas@ucdavis.edu (B.A.J.); cxpan@ucdavis.edu (C.-X.P.)

[4] Accelerated Medical Diagnostics Incorporated, Berkeley, CA 94708, USA; george_cimino@comcast.net

[*] Correspondence: phenderson@ucdavis.edu (P.T.H.); turteltaub2@llnl.gov (K.W.T.)

Received: 4 March 2019; Accepted: 6 May 2019; Published: 9 May 2019

Abstract: This review summarizes recent developments in radiocarbon tracer technology and applications. Technologies covered include accelerator mass spectrometry (AMS), including conversion of samples to graphite, and rapid combustion to carbon dioxide to enable direct liquid sample analysis, coupling to HPLC for real-time AMS analysis, and combined molecular mass spectrometry and AMS for analyte identification and quantitation. Laser-based alternatives, such as cavity ring down spectrometry, are emerging to enable lower cost, higher throughput measurements of biological samples. Applications covered include radiocarbon dating, use of environmental atomic bomb pulse radiocarbon content for cell and protein age determination and turnover studies, and carbon source identification. Low dose toxicology applications reviewed include studies of naphthalene-DNA adduct formation, benzo[a]pyrene pharmacokinetics in humans, and triclocarban exposure and risk assessment. Cancer-related studies covered include the use of radiocarbon-labeled cells for better defining mechanisms of metastasis and the use of drug-DNA adducts as predictive biomarkers of response to chemotherapy.

Keywords: accelerator mass spectrometry; cavity ring down spectrophotometry; radiocarbon; naphthalene; benzo[a]pyrene; cell turnover; triclocarban; metastasis; DNA adducts; biomarkers

---

1. Introduction

Radioisotopes play an important role in advancing our knowledge in the biomedical sciences. The applications are broad, ranging from positron-emission-tomography to the use of scintillation radiometry for determining protein turnover rates [1–4]. Nearly all radioisotope technologies detect and quantify radioisotopes based on the detection of a nuclear decay event. However, for many radioisotopes, this is an inefficient process, resulting in the need for high levels of radioactivity that are costly and require extensive safety precautions. For this reason, many investigators have avoided using radioisotopes in their research [5].

Radiocarbon-labeled drugs are often used to study absorption, distribution, metabolism, and excretion (ADME) in animals and humans. The resulting data aids in defining the metabolic fate of the drug and informs the usefulness of comparing pharmacokinetics (PK) and metabolism in animal models to humans. Radiocarbon (^{14}C) is often the label of choice because it is stable enough to be incorporated into virtually any carbon position on a given molecule using standard organic chemical synthesis methodology, yet it is active enough of a β-emitter to be detected by standard techniques, such as liquid scintillation counting (LSC) [6]. The use of ^{4}C has limitations due to the long half-life of this isotope (5740 years), which renders LCS counting of less than 5 to 10 picomoles ^{4}C/mL of plasma or urine impractical [2], which renders the assessment of drugs at microgram doses impossible (for highly potent drugs or microdose studies).

Accelerator mass spectrometry (AMS) is a highly sensitive technology for quantifying radioisotopes that has more recently been applied to the biomedical sciences. Quantification of radioisotopes with AMS does not rely on the nuclear decay, but rather on the direct quantification of the isotopic nuclei through mass spectrometry (reviewed in [7]). This provides much greater sensitivity for isotope detection (10^3 to 10^9 times greater than decay counting), leading to the use of lower chemical and radioisotope doses and the analysis of smaller samples, which enables studies to be performed safely in humans, using exposures that are environmentally or therapeutically relevant while generating little radioactive waste. Most biomedical AMS studies have employed radiocarbon (^{14}C) as the radiolabel, although the capability exists for detecting other isotopes, including ^{3}H, ^{26}Al, ^{41}Ca, ^{10}Be ^{36}Cl, ^{59}Ni, ^{63}Ni, and ^{129}I. Both ^{14}C and ^{3}H are commonly used in tracing studies because they can be readily incorporated into organic molecules, either synthetically or biosynthetically. This review will focus on the direct detection of ^{14}C from biological samples, either using AMS or laser-based systems that do not rely on decay counting. Technologies that are reviewed herein include the conversion of samples to graphite or carbon dioxide, followed by direct AMS analysis, combined AMS/ion trap mass spectrometry, and laser-based quantitation, along with examples of applications of these technologies to molecular toxicology, cell turnover, metastasis, and chemotherapy drug resistance.

2. Technology

2.1. Graphite

The most important considerations in preparing samples for radiocarbon analysis are the amount of radioisotope in each sample, preventing contamination, and knowing the sources and amounts of any carbon introduced during processing. Numerous precautions need to be in place throughout the procedure to ensure the amount of isotope present is within the dynamic range of the spectrometer and to minimize the potential for contamination to ensure that the isotope detected in the sample is associated with the labeled compound under investigation. The sample of interest, for example, blood, DNA, or high-performance liquid chromatography (HPLC) fractions, must be converted to a form that is compatible with the ion source of the instrument. The standard procedure for the preparation of ^{14}C-labled biological samples for AMS analysis is the conversion of the biological sample to graphite. This homogeneous state ensures uniformity and comparability between samples and standards that are compatible with the AMS ion source [8]. Graphite samples for AMS quantification are most commonly prepared by the reduction of carbon dioxide by hydrogen onto a catalytic iron or cobalt surface at temperatures around 500 °C [9–13]. Reduction of CO_2 to a filamentous graphite using septa sealed vials proceeds rapidly and yields of >95% are routinely obtained. Samples containing as little as 20 μg of carbon can be converted to graphite [14]. The graphite produces intense, long-lasting negative ion beams upon introduction to the cesium sputter ion source with extremely small isotopic and mass fractionation. The graphite-preparation stage is the rate-limiting step in AMS studies and precludes real-time analyses, such as those possible with typical LC/MS methods. However, the recent development of a gas-accepting ion source has eliminated the need for graphite in some applications.

2.2. Gas Ionization

The elimination of matrix effects in the ^{14}C-AMS analysis of biochemicals requires the physical and chemical equivalence for all measured carbon atoms. Additionally, almost all AMS systems use a cesium sputter ion source to generate negative ions and, consequently, methods for preparing biochemical samples for analysis were adapted from well-developed techniques used for geochronology studies. Subsequently, biochemical samples for AMS analysis are first combusted to CO_2, followed by reduction to graphite. While these methods have proven to be very successful, the technique suffers from low throughput and requires significant and adroit human handling. The routine preparation of graphitic biochemical samples requires at least 0.5 mg of carbon and limits sensitivity to ~2 amol ^{14}C/mg C (~0.3 mdpm/mg C). This requires that the analysis of biochemical mixtures separated using U/HPLC must be collected as discrete fractions with each treated as an individual sample. Samples from a single 30-minute LC trace can require several days to prepare and take over eight hours of AMS analysis time, costing several thousands of dollars, all of which increases if a higher resolution and/or duplicate analysis is required. In some instances, the number of samples from an LC trace can be reduced by collecting only fractions containing the peak(s) of interest and pooling fractions of "uninteresting regions". However, this is not always a viable option, especially in instances where an entire metabolite profile is required.

Analysis systems that are compatible with the direct input of biochemical separation instrumentation, such as liquid chromatography, would allow real-time analysis, leading to increased resolution, minimal handling, and the ability to do molecule-specific tracing of small samples. This approach involves the direct introduction of carbon as CO_2 into the ion source. This sample form is more efficient for the small samples common to biochemical research and allows for the direct interfacing of separation instrumentation to AMS. A moving wire interface, developed at Lawrence Livermore National Laboratory, provides one solution [15–17]. Briefly, the output of the HPLC is jetted onto a moving wire, which is pulled through a drying oven to remove the volatile solvent before introduction into a high temperature oven where the remaining analyte is combusted. The resultant CO_2 gas is then carried in a helium stream to the ion source of the AMS spectrometer for ^{14}C-quantification of the separated analyte.

The moving wire interface was used to measure human plasma and urine metabolism profiles following environmentally relevant exposures to the polycyclic aromatic hydrocarbons, dibenzo[def,p]chrysene (DBC) (Figure 1) [18] and benzo[a]pyrene [19]. Due to their toxicity, doses to healthy human volunteers were required to be kept as low as possible to minimize risk and the metabolite levels recorded following HPLC separation were so low that they could only have been measured as a CO_2 gas.

Aside from the direct coupling of HPLC-AMS, CO_2 gas-capable ion sources have also been coupled with commercially available combustion furnaces for higher throughput analysis of discrete tracer biochemical samples. In the system described by van Duijn et al., discrete samples containing at least 70 μg carbon and at least 0.52 amol of ^{14}C are combusted using an elemental analyzer, with the resultant CO_2 captured and transferred to a gas-tight syringe for subsequent metering into a gas-accepting hybrid ion source on a 1 MV HVEE AMS spectrometer [20]. Up to 200 samples may be measured automatically in sequence, limited by the capacity of the sample target wheel of the ion source.

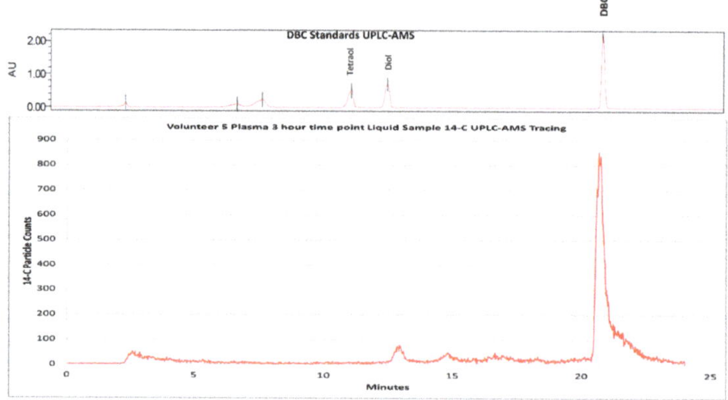

Figure 1. UHPLC-AMS of [^{14}C]-DBC and putative metabolites from human plasma 3 h after and oral dose of 29 ng ^{14}C-DBC. The use of unlabeled parent DBC and standards (DBC-(±)-11,12-diol and DBC-(±)-11,12,13,14-tetraol) (top graph) allow for the identification of ^{14}C peaks in plasma (bottom graph).

2.3. Parallel Accelerator and Molecular Mass Spectrometry (PAMMS)

Historically, AMS measurements have required the use of off-line orthogonal analytical techniques to speciate analytes measured by AMS. This limitation was based on the need to convert analytical samples to graphite prior to AMS measurement, destroying all chemical information in the process [13]. The recent development of the AMS liquid sample interface has enabled measurement of liquid samples without the need for graphitization [15,16,21]. The ability to measure liquid samples by AMS has also made it possible to integrate quantitative analysis with AMS sample measurements. The historical requirement for off-line speciation has been overcome by the recent development of a novel analytical technology that couples AMS directly with accurate mass spectrometry to enable real-time analysis of samples separated by high-performance liquid chromatography (HPLC). Based on the naming convention proposed by Sacks et al., this combined analytical method is referred to as parallel accelerator and molecular mass spectrometry (PAMMS) [22]. The LLNL PAMMS instrumentation is composed of a Waters Acquity H Class HPLC system, a Waters Xevo G2-XS QTOF instrument, an adjustable post-column flow splitter, a custom-built readout device, and the LS-AMS interface as depicted in the block diagram in Figure 2.

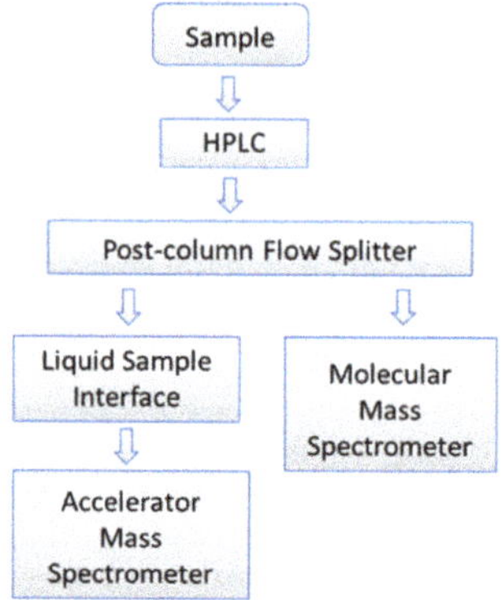

Figure 2. Block diagram showing the PAMMS instrument configuration.

PAMMS provides accurate mass measurement and tandem mass spectrometry for structural elucidation of individual analytes separated by HPLC, but also measures stable carbon and carbon-14 in each separated analyte. PAMMS can therefore enable definitive identification, as well as quantitation, of each separated analyte. For example, Figure 3 shows separation of glutamic acid in a mixture of ^{14}C-labeled amino acids, and identification based on the exact mass and MS/MS fragmentation pattern.

PAMMS represents a significant innovation that takes advantage of the ability of AMS to measure extremely low levels of [14]C. The ability to use low concentrations of radiolabeled substrates in cells and organisms at concentrations accessible by AMS allows quantification of metabolites without perturbing normal metabolism and leads to more relevant quantification of metabolic rates and pathways. Coupling the sensitive isotope detection abilities of AMS with accurate mass spectrometry to identify analytes makes PAMMS a very powerful technique capable of providing both qualitative and quantitative metabolic measurements. Such measurements can improve risk assessment for toxicants and new therapeutic entities, deepen our understanding of xenobiotic and intermediary metabolism, help understand interactions between critical molecular pathways, and improve efforts to model and predict various metabolic and biological states when coupled with biocomputational methods.

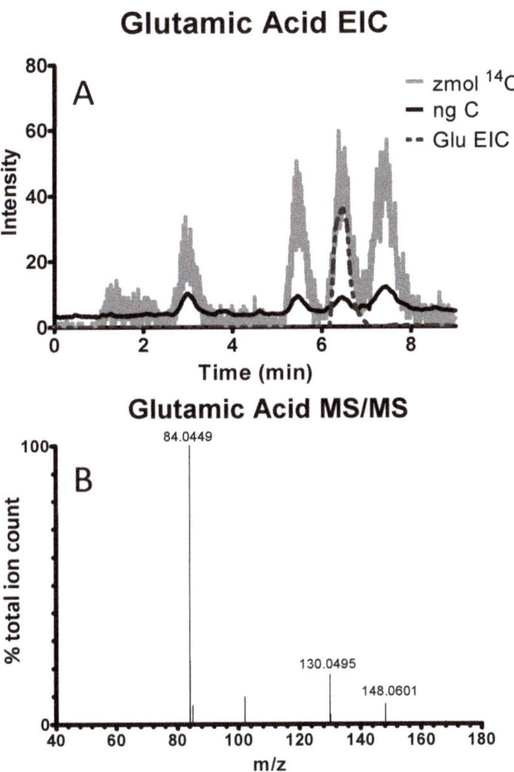

Figure 3. PAMMS analysis of carbon-14 labeled amino acid standards, showing (**A**) extracted ion chromatogram (EIC) and (**B**) mass spectrum for glutamic acid.

2.4. CRDS

AMS's complexity, large size, time consuming sample processing, and relatively high cost have been an obstacle to the scientific community's adoption of the method and its applications [23]. This has led to several scientific groups exploring new ways to measure [14]C [24–28]. One of these methods, cavity ringdown spectroscopy (CRDS), has demonstrated ^{14}C sensitivities below contemporary levels. Saturated-absorption cavity ring-down spectroscopy has achieved the greatest sensitivity of the CRDS techniques, with a minimum-detection limit 60-times smaller than the requirement for basic-biological studies [29].

Compared to AMS, CRDS is a simpler laser absorption technique for ^{14}C detection, which leverages a high-finesse optical cavity constructed with high-reflectivity mirrors (<1 ppm losses). This setup permits gas–laser interaction path lengths equivalent to tens of kilometers and therefore, increased sensitivity. A measurement starts by coupling resonant laser light into the optical cavity. This light is then interrupted, and an exponential decay, or "ring-down", is recorded on an optical detector. Differences between the characteristic decay time of empty and sample filled cavities are used to quantify the target species.

For biological studies utilizing ^{14}C, carbonaceous analytes are combusted into CO_2 and introduced into the cavity. While CRDS does not have the sensitivity of AMS, several groups have demonstrated the sensitivity to resolve natural background ^{14}C levels [29–31]. Furthermore, validation studies have been conducted, demonstrating CRDS produces results congruent with AMS when applied to duplicate samples [30,32].

3. Applications

3.1. Radiocarbon Dating

^{14}C is produced naturally in the upper atmosphere by nuclear reactions between cosmic radiation and atmospheric gases, notably ^{14}N. This natural ^{14}C production rate varies slightly over time as the Earth's magnetic field changes and the cosmic ray fluxes fluctuate, but it has remained relatively constant over most of recorded history, producing a natural source of $^{14}CO_2$ that subsequently labels every living thing on Earth as carbon moves through the food chain and carbon cycle. As long as a plant or animal is alive, it is replenishing or increasing its carbon either directly from the atmosphere (plants) or indirectly through consumption of plants and other animals. When an organism dies, the carbon replacement ceases. Since ^{14}C is radioactive (half-life $T_{1/2} = 5730$ y), the decrease in the $^{14}C/C$ concentration in tissue or biological structures compared to the atmospheric record can be used to determine how long an organism has been dead. Willard Libby was awarded the 1960 Nobel Prize in Chemistry for the development of radiocarbon dating [33].

3.2. Bomb Pulse Dating

Above-ground testing of nuclear weapons produced an anthropogenic spike in atmospheric $^{14}CO_2$ and consequently produced a small, but measurable excess ^{14}C label in every living thing on the planet. This spike in ^{14}C is generally called the radiocarbon bomb pulse. The pulse nearly doubled the natural atmospheric ^{14}C between 1955 and 1963, when the Limited Test Ban Treaty ended atmospheric, under water, and outer space detonations by the United States, Soviet Union, and the United Kingdom. Since the peak in 1963, atmospheric $^{14}CO_2$ has been decreasing as carbon moves into the biosphere and marine reservoirs and the burning of ^{14}C-free fossil fuels drives the atmosphere to pre-bomb ^{14}C in 2018 to 2019. The date of biological molecule synthesis is highly correlated to atmospheric $^{14}CO_2$, so the atmospheric record is a chronometer of the molecular age or carbon source. Figure 4 depicts the ratio of atmospheric radiocarbon to total carbon ($^{14}C/C$) from 1900 to 2015 for the northern and southern hemispheres based on compiled data [34]. The $^{14}C/C$ differs between the hemispheres since the weapons tests were conducted at relatively few locations, mostly in the northern hemisphere. Before 1955 and after 1970, there is little difference in the annual averages of the hemispheres.

3.2.1. Carbon Source Determination

Since fossil-derived carbon is devoid of ^{14}C, chemicals produced from petroleum do not contain ^{14}C. Chemicals, vitamins, or food additives from a "natural" biological source possess the atmospheric

$^{14}CO_2$ signature. The $^{14}C/C$ of an "all-natural" or "real" product should, therefore, be consistent with 100% biologically sourced material. If fossil-derived carbon is added to the product, the $^{14}C/C$ is depressed and easily measured by AMS. For example, in a 2011 study, real vanilla extracted from vanilla beans possessed a contemporary radiocarbon signature of $F^{14}C = 1.059 \pm 0.004$ modern while imitation vanilla had a $F^{14}C = 0.038 \pm 0.001$ modern [35,36]. The analysis of ^{14}C can determine if a food or personal care product has been adulterated with synthetic compounds. The technique can also be used to determine if packaging contaminates a food product with an unintended or undesirable compound. When a compound, such as phthalate, is found in a food product, ^{14}C AMS analysis of the purified compound can determine if it is naturally occurring in the food product, a result of leaching from packaging, or a combination of these. Analyses of stilton cheese and butter for bis(2-ethylhexyl) phthalate (DEHP) found that about 24% and 16% of the DEHP in the dairy products were of biological origin and not from packaging materials [36,37].

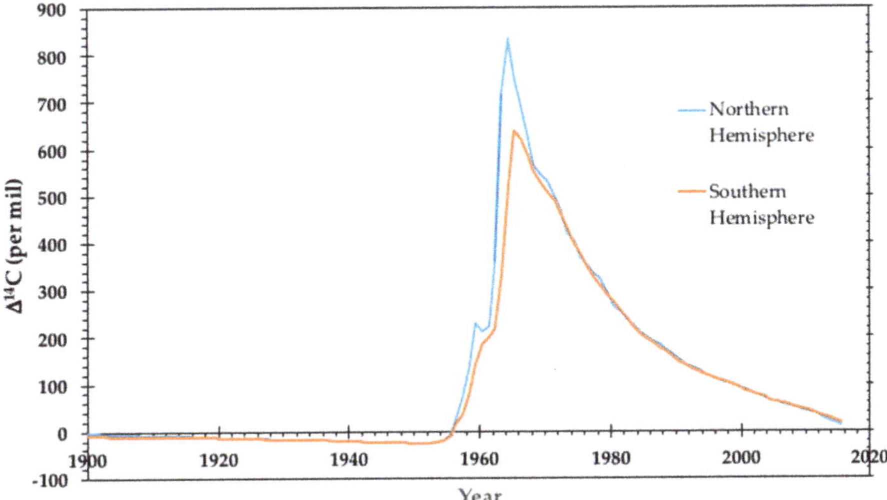

Figure 4. Annual averages of atmospheric $^{14}C/C$ for the northern and southern hemispheres. Data before 1959 is derived from plant material while data from 1959 to present is derived from atmospheric CO_2 collections and plant material. Data is reported in the $\Delta^{14}C$ convention described by Stuiver and Polach [38].

3.2.2. Structural and Pathological Protein Dating

The standard method to date bone found at archeological sites is to demineralize the bone, extract the collagen, and use an ultrafiltration procedure to exclude smaller fragments, typically under 30 kD. Collagen extraction works well because it is resistant to diagenesis (mineral exchange) of the mineral component of bone while being protected by the mineral structure. Collagen is not static in bone, it turns over as bone is remodeled throughout life, with the rates of turnover varying with age, type of bone, position within a bone, physical activity, etc. Hence, the ^{14}C of bone collagen is an integration over a lifetime of variable inputs and outputs. The changing $^{14}C/C$ in the bomb pulse enables studies to determine the approximate turnover in bone [39] and opens forensic applications for approximating the ages of skeletal remains [40].

Many structural proteins are found outside of bone. Many of these structural proteins are found in the extra-cellular matrix (ECM) of organs, muscle, cartilage, ligaments, and vasculature. Human lung parenchymal elastic fibers were shown to be the age of the person using ^{14}C analyses and aspartate racemization [41]. Radiocarbon dating of collagen from Achilles tendon and human articular cartilage shows growth and turnover through adolescence, but virtually no turnover during adulthood [42,43]. The eye lens continues to grow throughout life adding new cells to the outside in series of layers. The crystallin proteins that provide much of the structure of the lens have been shown to have very little turnover throughout life [44,45].

The chronological deposition of pathological structures is approachable by bomb pulse dating. In the progression of Alzheimer's Disease, the pathological structures of neurofibrillary tangles (NFT) and senile plaques (SP) accumulate over time. Post mortem analyses of the separated structures provided an average age, weighted by the rate of accumulation [46]. The average ages of the SP and NFT predated clinical symptoms of Alzheimer's disease in half the cases, indicating significant accumulation before cognitive deterioration. In another example of long-term accumulation, arterial plaques that restrict blood flow have also been shown to develop over decades [47,48]. The collagen extracted from excised cerebral aneurysms tends to be about 3 years old, even for aneurysms followed by imaging for years [49,50]. Additionally, subjects with risk factors of smoking, cocaine use, and hypertension had collagen within a year old, indicating very rapid carbon turnover [49].

3.2.3. Cell Lifetime and Turnover

Genomic DNA only acquires significant new carbon at cell division, so the ^{14}C/C of nuclear DNA is a metric of the cell birth date [51]. Cell or nuclei surface markers can be utilized using fluorescence activated cell sorting to isolate specific cell types for analyses. DNA is isolated from specific cell populations, rinsed thoroughly to remove residual solvents, checked for purity using UV/Vis absorbance, processed for AMS analyses using high precision natural radiocarbon preparation techniques, and measured by AMS. Genomic DNA dating has been used to investigate neurogenesis throughout many regions of the human brain [51–55]. The lack of bomb pulse carbon in neuronal DNA of subjects born before 1955 indicated that DNA repair provides an insignificant amount of new carbon after cell division [51]. It has also been used to determine that adipocytes turnover approximately every 10 years [56] while the lipids they hold cycle every 1.5 years [57] and cardiomyocytes turnover at a low rate [58]. Using bromodeoxyuridine (BrdU) and ^{14}C/C analyses of DNA from pancreatic β-cells, it was determined that insulin producing β-cells turnover at a 1% to 2% annual rate through early adulthood and then cease to turnover after the age of 30 [59]. Antibody-secreting plasma cells have been found to persist for decades in the intestines although antibodies last only several weeks in circulation [60]. An alternative to dating DNA, dating histones established histone turnover as a critical regulator of cell type-specific transcription and plasticity in the mammalian brain [61].

3.3. Tracking the Fate of Cells Labeled with [^{14}C]Thymidine

Using a similar principle, we can also quantify the number of cancer cells that colonize distant sites to form metastatic tumors [62]. In a recent publication, we have taken cancer cell lines with varying metastatic potential, and labeled them in vitro with ^{14}C-thymidine, such that we could identify by AMS a single labeled cell among 1 million unlabeled cells. These labeled cells were introduced in mice via various routes known to produce metastatic tumors (tail vein, TV; intracardiac, IC) and after 2-weeks or 12-weeks post injection, all organs were examined for the presence of metastatic tumors. Whether visible tumors were present or not, total DNA was isolated from each organ, and the total carbon was examined for the presence of ^{14}C. The amounts of ^{14}C detected per organ were referred back to the amount of ^{14}C present per cell, prior to injection into the mice, to determine how many cells traveled to distant sites and initiated a metastatic tumor (Figure 5A).

Using this approach, we determined that less than 5% of human cancer cells injected into immunodeficient mice form subcutaneous tumors, and even fewer cells initiate metastatic tumors. Comparisons of metastatic site colonization between a highly metastatic (PC3) and a non-metastatic (LnCap) prostate cancer cell line showed that PC3 cells colonize target tissues in greater quantities at 2 weeks post-delivery, and by 12 weeks post-delivery, no ^{14}C was detected in LnCap xenografts, suggesting that all metastatic cells were cleared (Figure 5B). The ^{14}C-signal correlated with the presence and the severity of metastatic tumors. AMS measurements of ^{14}C-labeled cells provides a highly-sensitive, quantitative assay to experimentally evaluate metastasis and colonization of target tissues in xenograft mouse models. In the future, this approach could potentially be adapted to evaluate tumor aggressiveness and assist in making informed decisions regarding treatment, towards a more informed personalized therapy regimen.

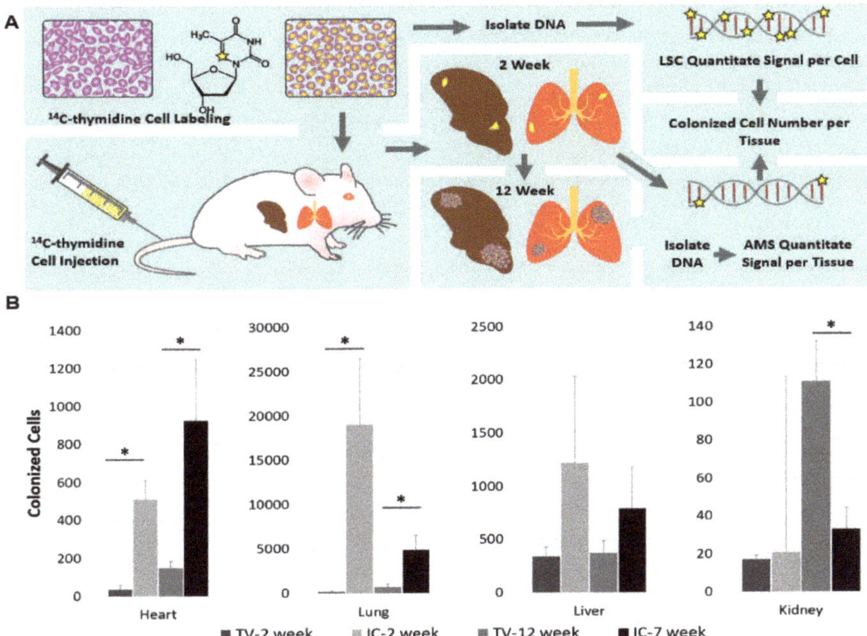

Figure 5. Workflow and validation of ^{14}C-labeling cancer colonization assay. (**A**) Schematic of colonization assay workflow. Cells were first cultured with ^{14}C-thymidine media to achieve single cell resolution and injected into NSG mice via the tail vein (TV), heart (IC), or subcutaneous (SQ) routes of delivery. Injected cells were allowed to metastasize for up to 12 weeks. Tissues were harvested at early (2 weeks post injection) and late (12 weeks post injection) time points and DNA was isolated and quantified using AMS. In parallel, the activity of ^{14}C-thymidine label in cultured cells was quantified using liquid scintillation counting (LSC). AMS measurements and LSC readings were combined to calculate the number of colonized cells per each organ examined. (**B**) Tail vein and intracardiac injected cancer cell colonization. Profile of colonized cells in target tissues calculated from ^{14}C signal in DNA from target tissues isolated at 2-weeks post injection, 7-weeks for intracardiac (IC), or 12-weeks for tail vein (TV) (*n* = 5).

3.4. Low Dose Toxicity

One of the biggest advantages of AMS is the ability to perform low dose toxicity studies, which allow for the assessment of chemicals at environmentally relevant dose levels. Numerous studies have used AMS to investigate the biodisposition of chemicals at low-dose human exposure levels. Below are examples of some of these studies.

3.4.1. Naphthalene

Naphthalene (NA) is ubiquitous in both the indoor and outdoor environment. Common sources of NA exposure include combustion products from vehicle emissions, biomass, cigarettes and wildfires, mothballs, and house-hold block deodorizers. Naphthalene metabolites have been detected in the urine of nearly all children and adults tested, regardless of locale or occupation [63]. Further, studies in children have shown increased chromosomal aberrations that correlate with urinary markers for NA exposure, but these studies cannot establish a cause and effect relationship [63]. Furthermore, NA exposure caused bronchiolar alveolar carcinomas in female mice and neuroblastomas in the nasal epithelium of rats in the National Toxicology Program carcinogenesis bioassays [64,65]. The mechanism of cancer initiation is unclear, however, so investigations of protein adducts, DNA adducts, and repair tolerance [66,67] have been conducted using NA and its metabolite, 1,2 naphthoquinone (NQ). Using well established techniques to obtain metabolically active, live tissue samples [68], freshly micro-dissected respiratory tissues were incubated with NA or NQ at 250 µM, calculated as equivalent to the tissue concentration obtained from exposure to the 10 ppm OSHA exposure limit for NA [69]. DNA adducts of NA and NQ are present in low levels, but protein adducts are much more common [66,67]. The technique of ex vivo exposure of metabolically active tissue avoided making an aerosol of a ^{14}C-lableled toxic chemical and is applicable to other inhalation hazards.

3.4.2. Triclocarban

In a recently published study by Enright et al., the potential of an environmentally relevant concentration of the antimicrobial, triclocarban (TCC), to transfer from the mother to the offspring during development was evaluated using AMS [70]. Triclocarban is an antimicrobial found in many personal care products (i.e., deodorants, soaps) and is among the top 10 most commonly detected wastewater contaminants [71,72]. Given its prevalence in the environment, bioaccumulation of TCC has been observed and reproductive effects have been noted as a result from exposure [73]. Exposure to compounds, such as TCC, during development may have deleterious consequences to the developing embryo and fetus, given their heightened sensitivity to perturbations in hormone levels and immature protective mechanisms (i.e., liver metabolism, DNA repair mechanisms).

In this study, ^{14}C-labeled TCC (100 nM) was administered to CD-1 mouse dams through their drinking water up to gestation day 18, or from birth through to postnatal day (PND) 10.

Using AMS, the concentration of TCC was determined in both offspring and dams after exposure; TCC transferred from mother to offspring both trans-placentally (0.005% ± 0.001% ingested dose/gram (%ID/g) and through lactation (0.015% ID/g ± 0.002%) (Figure 6). The three-fold higher concentration in offspring after exposure through lactation (p = 0.003) demonstrated that TCC readily transfers through breast milk. After exposure through lactation, TCC exposed offspring were heavier in weight than unexposed controls (p = 0.016 for PND21–56), with females more affected (11% increase) than males (8.5% increase) (data not shown). Tissue accumulation was also quantified using AMS at 6 weeks post exposure. TCC-related compounds were detected in tissues with higher concentrations observed in the brain, heart, and fat. Quantitative real-time polymerase chain reaction (qPCR) of liver and fat tissue suggested alterations of lipid metabolism in exposed female offspring; this was further supported by an increase in fat pad weights and hepatic triglycerides. This was the first report quantifying the translocation of an environmentally relevant concentration of TCC from mother to offspring; this study was enabled by the high sensitivity of AMS. Taken together, our findings suggest that TCC readily transfers from the mother to the offspring and that early-life exposure may interfere with lipid metabolism, which can ultimately have implications for human health.

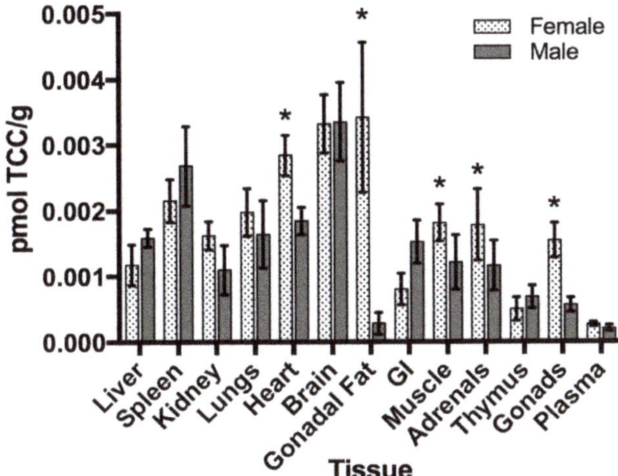

Figure 6. Tissue distribution of 14C-TCC in exposed offspring at postnatal day 42. Data is expressed as pmol of TCC/gram of tissue ± SEM (n = 5/sex). * p < 0.05, when comparing female to male offspring.

3.4.3. Benzo[a]pyrene

Benzo[a]pyrene is a widely studied polycyclic aromatic hydrocarbon (PAH) that has been shown to induce cardiovascular, developmental, immunological, and reproductive disorders in model systems [74–76]. It has also been implicated as a human carcinogen and is an environmental chemical of concern for human exposure according to the Agency for Toxic Substances and Disease Registry [77]. Using the recently developed on-line UPLC-AMS interface with the gas accepting ion source, human pharmacokinetic and metabolite profiles were determined from microdose exposures of BaP. Five human volunteers were exposed to an oral dose of 46 ng (5 nCi) of ^{14}C-BaP. Blood was collected at given time intervals and pharmacokinetic and metabolite parameters were quantified by AMS. At the dose used, BaP was fairly rapidly eliminated from the plasma and very little parent compound was present in the plasma even at the earliest time point examined, indicating extensive metabolism of BaP in these human subjects. The use of the UPLC-AMS together with the on-line gas accepting ion source provided exquisite sensitivity (zepto-mole[14]C in biological samples), allowing for the quantification of BaP plasma metabolites of BaP from exposure levels that were 5 to 15 times lower than the estimated daily exposure to BaP [19].

3.5. Diagnostic Microdosing: Using Drug-DNA Adducts as Biomarkers of Chemotherapy Response

Chemotherapy drugs that modify DNA are a cornerstone of modern cancer treatment and are used in nearly half of all cancer patients [78,79]. However, their efficacy is limited by severe side effects and intrinsic or acquired drug resistance, eventually causing treatment failure [78,80–82]. AMS has been used over the past 15 years for the measurement of drug–DNA interactions, and the resulting data have been correlated with cell sensitivity and/or tumor response in mice and humans [83–96]. The overarching hypotheses of this work are that a threshold level of drug–DNA adducts are required for cell killing and clinical response, and that microdose induced drug–DNA adduct levels are predictive of the cellular capacity to achieve such a threshold upon therapeutic dosing. This approach, known as "diagnostic microdosing", has been initially demonstrated for platinum-based chemotherapy for the treatment of solid tumors and induction chemotherapy for leukemia. Based on this and other work, it is clear that for some chemotherapeutics, microdose-induced drug-DNA adducts are predictive of drug sensitivity and response in cell culture and mouse tumor xenograft studies, along with an extension of this effort to two pilot clinical studies focused on platinum-based chemotherapy (clinicaltrials.gov identifier NCT01261299 and NCT02569723) and a retrospective study on viably cryopreserved human acute myeloid leukemia (AML) cells.

The diagnostic test protocol consists of four steps: (1) Creation of the individualized biomarkers in patient cells by exposure to ^{14}C-radiolabeled drugs, (2) isolation of DNA containing the biomarkers, (3) determination of the ^{14}C associated with the DNA via AMS analysis, and (4) comparison of the patient's drug-DNA adduct levels to a database of clinical responses in order to assign a predictive score that indicates the probability of response.

Three key observations have included: (a) The level of drug-DNA adducts is generally low in resistant cancers and high in responsive cancers, (b) microdosing predicts the level of therapeutic-induced drug-DNA adducts, and (c) microdose-induced DNA adduct frequencies correlate with cellular drug sensitivity and patient response. Selected published examples of these results are summarized below.

3.5.1. Predicting Response to Platinum-Based Therapy with Microdosing

Figure 7 shows cell culture data for a set of six cancer cell lines dosed with [^{14}C]carboplatin. The drug interacts with DNA to form [^{14}C]carboplatin-DNA "monoadducts" that are measurable by AMS (Figure 7A). The monoadduct levels formed by microdoses were linearly proportional to those formed by therapeutically relevant concentrations of the drug in the media over 24 hours (Figure 7B). This is an important observation, since it implies that monoadduct levels formed from microdoses are likely to be predictive of those induced by therapeutic doses in patients. Half of the cell lines tested had carboplatin IC$_{50}$ values below 100 μM (approximately the in vivo C$_{max}$ in humans) and were assigned as "sensitive". The remaining cell lines were designated as "resistant". The sensitive and resistant cell lines could be significantly differentiated based on microdose-induced carboplatin monoadduct levels (Figure 7C), establishing proof of concept and justifying in vivo studies in mice and humans.

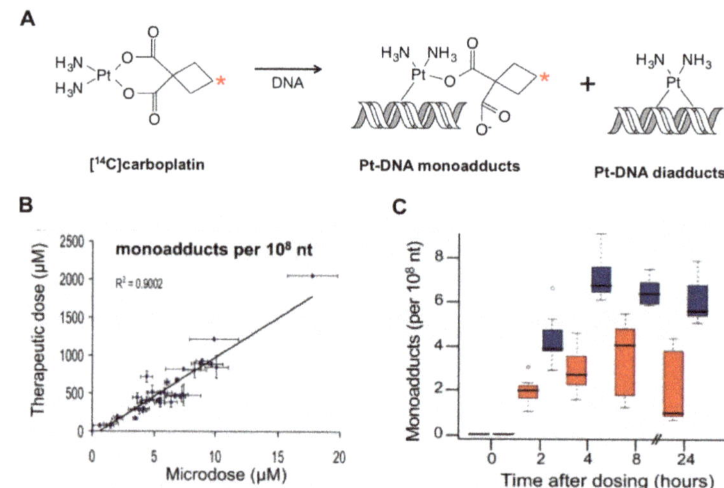

Figure 7. Correlation of microdose-induced [^{14}C]carboplatin-DNA adduct levels to therapeutic dose-induced adduct levels in cancer cell lines. (**A**) Diagram of carboplatin-DNA adduct formation. (**B**) Linear regression of microdose-induced versus therapeutic dose-induced carboplatin-DNA adducts. (**C**) Sensitive cell lines (blue) have significantly higher carboplatin-DNA adduct levels than resistant cancer cell lines (red).

Figure 8 shows data supporting the correlation between drug-DNA adduct levels and in vivo treatment response in mice and humans. Mice bearing patient derived bladder tumor xenografts (PDX) were established as depicted in Figure 8A. There was a significant correlation between microdose-induced carboplatin-DNA monoadduct levels and tumor growth inhibition of platinum-based chemotherapy (Figure 8B). The pilot clinical trial accrued 10 bladder cancer patients (stage II and higher), for whom platinum-based chemotherapy was administered. Patients were administered approximately 1% of the therapeutic dose of [^{14}C]carboplatin (a microdose), followed by blood sampling within 24 h. DNA isolated from peripheral blood mononuclear cells (PBMC) was assessed for carboplatin-DNA adduct levels by AMS. Within three months of the microdosing assay, patients began standard of care chemotherapy regimens consisting of either cisplatin or carboplatin in combination with other drugs (typically gemcitabine (GC) or methotrexate, vinblastine, and doxorubicin (MVAC)). Patient response was determined at the time of cystectomy—typically after three cycles of chemotherapy. A patient whose tumor burden in the bladder was reduced to pT1 or less was considered a responder and a patient with a pT2 or greater tumor was considered a non-responder, based on standard RECIST criteria [97]. The main endpoint of the study was to demonstrate a significant difference in the mean drug-DNA adduct levels in PBMC (a surrogate for tumor tissue) between responders and non-responders. Seven patients responded to chemotherapy (green circles), whereas three patients showed disease progression (red squares) (Figure 8C). Responders exhibited approximately 2-fold higher mean monoadduct levels (green line) than non-responders (red line) (0.741 ± 0.346 vs. 0.283 ± 0.202 monoadducts/10^8 nt, respectively, $p = 0.069$). The drug-DNA adducts were distributed in two distinct groups; high (0.941 ± 0.030 adducts per 10^8 nt) and low (0.266 ± 0.158 adducts per 10^8 nt) drug-DNA adduct levels with a statistically significant difference ($p < 0.001$). All five patients in the high adduct level group responded to chemotherapy, which is a 100% positive predictive value (PPV) for this group. The low adduct level group included three non-responders and two responders. The results of this clinical trial show a clear trend for responders to have higher Pt-DNA adduct levels 24 h after microdose administration, which supports the feasibility of patient stratification by the diagnostic microdosing approach.

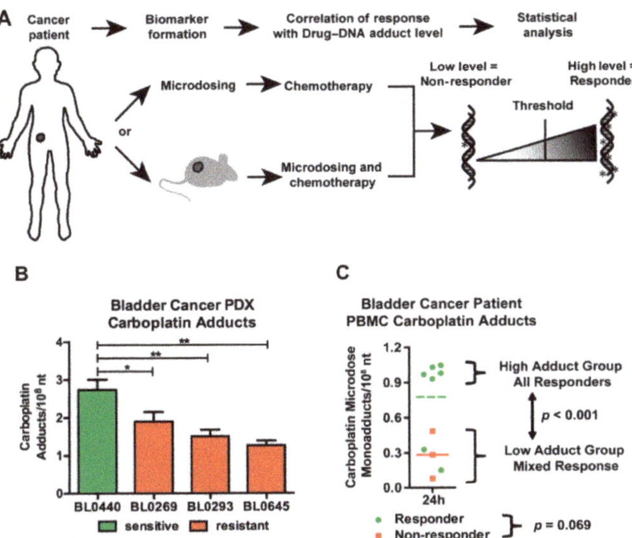

Figure 8. Correlation of microdose-induced [^{14}C]carboplatin-DNA adduct levels to therapy response in bladder cancer PDX and patients. (**A**) Clinical study design. (**B**) Carboplatin-DNA adduct levels are significantly higher in the sensitive bladder cancer PDX model (green) than in the resistant models (red). (**C**) Correlation of PBMC [^{14}C]carboplatin-DNA adduct levels to the response in 10 bladder cancer patients (green = responder, red = non-responders, line = mean adduct level).

3.5.2. Ex Vivo Diagnostic Microdosing for Predicting Response to 7 + 3 in AML Patients

In contrast to our previous efforts that focused on administering microdoses to patients, our more recent work has focused on establishing proof-of-concept for a lab-based test in which biobanked or fresh patient leukemia patient samples are dosed ex vivo. This change will allow us to overcome our past difficulty in accruing patients that were unwilling to undergo IV administration of a radiolabeled drug that would not provide a direct benefit (a non-interventional study). Furthermore, ex vivo microdosing allows us to analyze multiple drug regimens on each patient sample. AML is ideal for this concept, since it is a "liquid tumor" that can easily be accessed via a blood draw or bone marrow biopsy, and is predominantly treated with two drugs that both interact with DNA.

The most effective therapy for AML is treatment with "induction chemotherapy" with two DNA damaging drugs, including the antimetabolite, cytarabine (ARA-C), and an anthracycline, such as daunorubicin (DNR) or idarubicin (IDA)—structures shown in Figure 9A. This regimen is known as 7 + 3 (7 days of continuous infusion ARA-C and 3 days of bolus DNR or IDA), and is the standard of care for up to two thirds of AML patients. Treatment is started as soon as possible, typically within 5 to 7 days of diagnosis [98–101]. In addition, a subset of patients, including eligible younger patients and relapsed or refractory (R/R) patients, can be treated with a combination of high-dose bolus ARA-C and DNR or IDA, known as 3 + 4 [102–105]. Patients who are not eligible for 7 + 3 are typically placed on a less toxic, ARA-C-containing regimen.

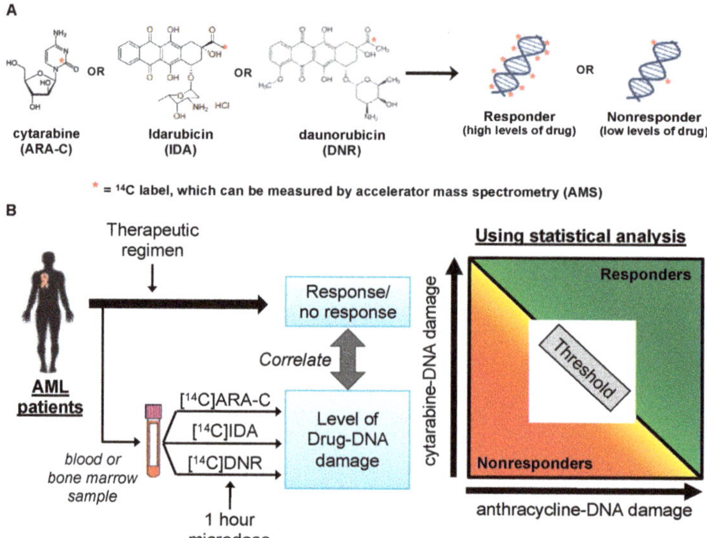

Figure 9. Overview of the ex vivo "diagnostic microdosing" strategy. (**A**) Radiocarbon-labeled cytarabine (ARA-C), idarubicin (IDA), or daunorubicin (DNR) bind to or are incorporated into AML DNA in proportion to the cellular sensitivity to each drug. The resulting drug-DNA "adducts" can be quantified by accelerator mass spectrometry (AMS). (**B**) Strategy for using in vitro microdosing to predict AML patient response to 7 + 3 chemotherapy. Cells isolated from a blood draw or bone marrow (fresh or viably cryopreserved) are briefly exposed to microdoses of each drug (triplicate wells per drug) and assessed by mass spectrometry for quantitation of drug-DNA adduct levels as biomarkers of clinical response to 7 + 3 induction chemotherapy.

Basic research into the significance of drug–DNA adducts in patients treated with anthracycline derivatives and antimetabolites similar to IDA [106] and ARA-C [107] chemotherapies has been reported, but none of these findings have been translated into clinical use [108–113]. Several reports have documented associations between drug–DNA adduct levels and clinical response and overall survival [93,114]. These reports support the concept that a predictive stratification strategy can be used to personalize chemotherapy if the capacity for cells to form high levels of drug-DNA adducts can be predicted prior to the initiation of therapy. This body of work is being adapted to predicting AML patient responses to 7 + 3 induction chemotherapy by implementing a diagnostic microdosing test (Figure 9B).

After optimization of the protocols using cell culture experiments, we performed the diagnostic microdosing protocol on 19 clinically annotated viably cryopreserved primary human AML samples, including 10 responders and nine nonresponders to 7 + 3 induction chemotherapy, and DOX protocol on 10 primary AML samples. When the primary samples were grouped based on patient response, the responsive patients had higher mean drug–DNA adduct levels compared to the nonresponders for all dosing regimens (Figure 10A–C). Statistical differences between the responders and nonresponders were determined by unpaired *t*-tests with $p < 0.05$ as the statistically significant cutoff. The ARA-C- and DOX-DNA adduct levels from each primary AML sample when plotted together showed a complete separation in responders and nonresponders (Figure 10D—for 10 patients, since DOX- and ARA-C combined adduct data are currently only available for 10 patients).

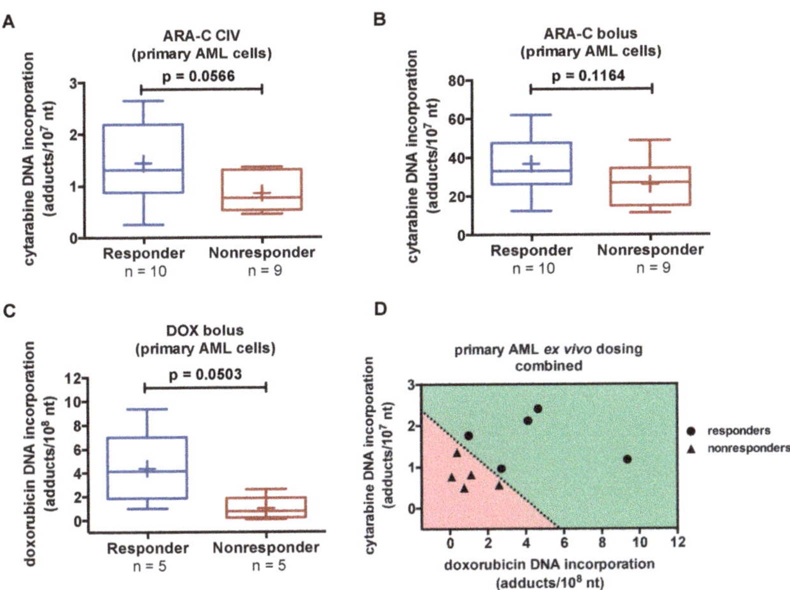

Figure 10. Correlation of ARA-C- and DOX-DNA levels to 7 + 3 response after in vitro dosing of 20 primary AML samples. PBMC were exposed to either exposed to a microdose of [^{14}C]ARA-C or [^{14}C]DOX at ~1% of the approximate plasma C_{max} obtained with ARA-C CIV (**A**), ARA-C bolus (**B**), or DOX bolus (**C**) observed in patients. Cells were dosed for 1 h followed by DNA isolation and AMS analysis. These data show proof of principle that diagnostic microdosing is useful for predicting patient response to 7 + 3 chemotherapy, but a larger confirmatory study is necessary. Furthermore, the ARA-C and DOX-DNA adducts can be plotted together to differentiate responders and nonresponders—average adduct levels for each patient are shown for simplicity (**D**).

In summary, the ability to quantitate the DNA incorporation of ARA-C and DOX in paired sensitive and resistant cell lines was demonstrated, and higher drug incorporation rates in the more sensitive cell lines was observed. Furthermore, similar correlations were observed in AML cell lines and with patient response primary AML samples. These preliminary data show that the adduct levels indeed correlate with resistance to doxorubicin and ARA-C.

4. Conclusions

Although much of the scientific demonstrations of the technological advances reviewed herein for radiocarbon tracing are outside the scope of drug development and toxicology, there are clear implications for such studies. Drug discovery, early clinical development, and studies to support regulatory approval can all benefit from higher sensitivity and low-cost, high throughput radiotracer technologies [115–118]. For example, in human studies, with much lower radiocarbon doses that are currently used for HPLC-LSC, studies are not only technically desirable, but are also more ethical since healthy human volunteers would have reduced radiation exposures. Absolute bioavailability for oral compounds is another obvious application that would benefit from improved radiotracing technologies. Although "standard" LC-MS is sufficient for PK and ADME studies of most drug candidates, there will always be a subset of efficacious lead compounds that possess very high potency at very low doses or with unusual PK or biodistribution that require radiolabel studies. Therefore, drug development is likely to be more successful as radiotracer technologies become more accessible and accepted. Radiocarbon tracer technology continues to evolve in order to increase throughput, reduce cost and the amount of radioactivity needed, and broaden the types of applications that can be accommodated. Basic science, environmental and molecular toxicology, and clinical translational applications continue to be uniquely enabled by radiotracer studies, which are critical for advancing our understanding of fundamental biology, protecting the environment, and improving human health.

Funding: The UC Davis Comprehensive Cancer Center is supported by Cancer Center Support Grant P30CA093373 from the NCI and SBIR contracts to Accelerated Medical Diagnostics Phase I HHSN261201000133C (P.T. Henderson), Phase II HHSN261201200048C (P.T. Henderson), NIH/NCI K12 award K12 CA 138464 (B.A. Jonas), CTSC Pilot Program/NIH award ULT1TR001860 (B.A. Jonas), UC Davis Academic Senate grant FJ161B1 (B.A. Jonas), UC Davis School of Medicine Dean and American Cancer Society award IRG-95-125-16 (B.A. Jonas), LLNL grants LDRD 08-LW-100 (P.T. Henderson and M. Malfatti), and the Knapp Family Fund (P.T. Henderson). 1R01CA176803-01 (Pan), VA Merit 1I01 BX003840 (Pan) and the Knapp Family Fund (P.T. Henderson).

Acknowledgments: This work was performed in part under the auspices of the U.S. DOE by LLNL under Contract DE-AC52-07NA27344 nd supported by the National Institute of Health General Medical Sciences Sciences (2P41GM103483-16). The contents do not represent the views of the U.S. Department of Veterans Affairs or the United States Government.

Conflicts of Interest: Henderson, Zimmermann, and Pan are shareholders of Accelerated Medical Diagnostics, Inc.

References

1. Abramson, F.P. Crims—Chemical-Reaction Interface Mass-Spectrometry. *Mass Spectrom. Rev.* **1994**, *13*, 341–356. [CrossRef]
2. Rosler, H. The Impact of W. K. Rontgen's Discovery on the Use of Internalizable Sources of Ionizing Energy in Diagnostic and Therapeutic Nuclear-Medicine. *Experientia* **1995**, *51*, 686–702. [CrossRef]
3. Devries, R.A.; Debruin, M.; Marx, J.J.M.; Vandewiel, A. Radioisotopic Labels for Blood-Cell Survival Studies—A Review. *Nucl. Med. Biol.* **1993**, *20*, 809–817. [CrossRef]
4. Young, V.R.; Ajami, A. Isotopes in nutrition research. *Proc. Nutr. Soc.* **1999**, *58*, 15–32. [CrossRef] [PubMed]
5. Mayer, A.; Neuenhofer, S. Luminescent Labels—More Than Just an Alternative to Radioisotopes. *Angew. Chem. Int. Ed.* **1994**, *33*, 1044–1072. [CrossRef]
6. Garner, R.C.; Barker, J.; Flavell, C.; Garner, J.V.; Whattam, M.; Young, G.C.; Cussans, N.; Jezequel, S.; Leong, D. A validation study comparing accelerator MS and liquid scintillation counting for analysis of 14C-labelled drugs in plasma, urine and faecal extracts. *J. Pharm. Biomed. Anal.* **2000**, *24*, 197–209. [CrossRef]
7. Turteltaub, K.W.; Vogel, J.S. Bioanalytical applications of accelerator mass spectrometry for pharmaceutical research. *Curr. Pharm. Des.* **2000**, *6*, 991–1007. [CrossRef] [PubMed]
8. Vogel, J.S.; Turteltaub, K.W.; Finkel, R.; Nelson, D.E. Accelerator mass spectrometry. *Anal. Chem.* **1995**, *67*, 353A–359A. [CrossRef] [PubMed]

9. Vogel, J.S.; Southon, J.R.; Nelson, D.E.; Brown, T.A. Performance of Catalytically Condensed Carbon for Use in Accelerator Mass-Spectrometry. *Nucl. Instrum. Methods B* **1984**, *5*, 289–293. [CrossRef]

10. Vogel, J.S.; Nelson, D.E.; Southon, J.R. C-14 Background Levels in an Accelerator Mass-Spectrometry System. *Radiocarbon* **1987**, *29*, 323–333. [CrossRef]

11. Vogel, J.S.; Southon, J.R.; Nelson, D.E. Catalyst and Binder Effects in the Use of Filamentous Graphite for Ams. *Nucl. Instrum. Methods B* **1987**, *29*, 50–56. [CrossRef]

12. Vogel, J.S. Rapid Production of Graphite without Contamination for Biomedical Ams. *Radiocarbon* **1992**, *34*, 344–350. [CrossRef]

13. Wilson, A.T. A Simple Technique for Converting CO_2 to AMS Target Graphite. *Radiocarbon* **1992**, *34*, 318–320. [CrossRef]

14. Ognibene, T.J.; Bench, G.; Vogel, J.S.; Peaslee, G.F.; Murov, S. A high-throughput method for the conversion of CO^2 obtained from biochemical samples to graphite in septa-sealed vials for quantification of ^{14}C via accelerator mass spectrometry. *Anal. Chem.* **2003**, *75*, 2192–2196. [CrossRef] [PubMed]

15. Thomas, A.T.; Ognibene, T.; Daley, P.; Turteltaub, K.; Radousky, H.; Bench, G. Ultrahigh efficiency moving wire combustion interface for online coupling of high-performance liquid chromatography (HPLC). *Anal. Chem.* **2011**, *83*, 9413–9417. [CrossRef] [PubMed]

16. Thomas, A.T.; Stewart, B.J.; Ognibene, T.J.; Turteltaub, K.W.; Bench, G. Directly coupled high-performance liquid chromatography-accelerator mass spectrometry measurement of chemically modified protein and peptides. *Anal. Chem.* **2013**, *85*, 3644–3650. [CrossRef]

17. Ognibene, T.J.; Thomas, A.T.; Daley, P.F.; Bench, G.; Turteltaub, K.W. An Interface for the Direct Coupling of Small Liquid Samples to AMS. *Nucl. Instrum. Methods Phys. Res. B* **2015**, *361*, 173–177. [CrossRef]

18. Madeen, E.P.; Ognibene, T.J.; Corley, R.A.; McQuistan, T.J.; Henderson, M.C.; Baird, W.M.; Bench, G.; Turteltaub, K.W.; Williams, D.E. Human Microdosing with Carcinogenic Polycyclic Aromatic Hydrocarbons: In Vivo Pharmacokinetics of Dibenzo[def,p]chrysene and Metabolites by UPLC Accelerator Mass Spectrometry. *Chem. Res. Toxicol.* **2016**, *29*, 1641–1650. [CrossRef]

19. Madeen, E.; Siddens, L.K.; Uesugi, S.; McQuistan, T.; Corley, R.A.; Smith, J.; Waters, K.M.; Tilton, S.C.; Anderson, K.A.; Ognibene, T.; et al. Toxicokinetics of benzo [a] pyrene in humans: Extensive metabolism as determined by UPLC-accelerator mass spectrometry following oral micro-dosing. *Toxicol. Appl. Pharm.* **2019**, *364*, 97–105. [CrossRef]

20. van Duijn, E.; Sandman, H.; Grossouw, D.; Mocking, J.A.J.; Coulier, L.; Vaes, W.H.J. Automated Combustion Accelerator Mass Spectrometry for the Analysis of Biomedical Samples in the Low Attomole Range. *Anal. Chem.* **2014**, *86*, 7635–7641. [CrossRef]

21. Ognibene, T.J.; Salazar, G.A. Installation of hybrid ion source on the 1-MV LLNL BioAMS spectrometer. *Nucl. Instrum. Methods B* **2013**, *294*, 311–314. [CrossRef]

22. Sacks, G.L.; Derry, L.A.; Brenna, J.T. Elemental speciation by parallel elemental and molecular mass spectrometry and peak profile matching. *Anal. Chem.* **2006**, *78*, 8445–8455. [CrossRef] [PubMed]

23. Vogel, J.S.; Palmblad, N.M.; Ognibene, T.; Kabir, M.M.; Buchholz, B.A.; Bench, G. Biochemical paths in humans and cells: Frontiers of AMS bioanalysis. *Nucl. Instrum. Methods B* **2007**, *259*, 745–751. [CrossRef]

24. Labrie, D.; Reid, J. Radiocarbon Dating by Infrared-Laser Spectroscopy—A Feasibility Study. *Appl. Phys.* **1981**, *24*, 381–386. [CrossRef]

25. Murnick, D.E.; Dogru, O.; Ilkmen, E. Intracavity optogalvanic spectroscopy. An analytical technique for ^{14}C analysis with subattomole sensitivity. *Anal. Chem.* **2008**, *80*, 4820–4824. [CrossRef]

26. Galli, I.; Pastor, P.C.; Di Lonardo, G.; Fusina, L.; Giusfredi, G.; Mazzotti, D.; Tamassia, F.; De Natale, P. The ν_3 band of $^{14}C^{16}O_2$ molecule measured by optical-frequency-comb-assisted cavity ring-down spectroscopy. *Mol. Phys.* **2011**, *109*, 2267–2272. [CrossRef]

27. Genoud, G.; Vainio, M.; Phillips, H.; Dean, J.; Merimaa, M. Radiocarbon dioxide detection based on cavity ring-down spectroscopy and a quantum cascade laser. *Opt. Lett.* **2015**, *40*, 1342–1345. [CrossRef] [PubMed]

28. McCartt, A.D.; Ognibene, T.; Bench, G.; Turteltaub, K. Measurements of Carbon-14 with Cavity Ring-Down Spectroscopy. *Nucl. Instrum. Methods Phys. Res. B* **2015**, *361*, 277–280. [CrossRef]

29. Galli, I.; Bartalini, S.; Ballerini, R.; Barucci, M.; Cancio, P.; De Pas, M.; Giusfredi, G.; Mazzotti, D.; Akikusa, N.; De Natale, P. Spectroscopic detection of radiocarbon dioxide at parts-per-quadrillion sensitivity. *Optica* **2016**, *3*, 385–388. [CrossRef]

30. McCartt, A.D.; Ognibene, T.J.; Bench, G.; Turteltaub, K.W. Quantifying Carbon-14 for Biology Using Cavity Ring-Down Spectroscopy. *Anal. Chem.* **2016**, *88*, 8714–8719. [CrossRef]

31. Fleisher, A.J.; Long, D.A.; Liu, Q.N.; Gameson, L.; Hodges, J.T. Optical Measurement of Radiocarbon below Unity Fraction Modern by Linear Absorption Spectroscopy. *J. Phys. Chem. Lett.* **2017**, *8*, 4550–4556. [CrossRef]

32. Kratochwil, N.A.; Dueker, S.R.; Muri, D.; Senn, C.; Yoon, H.; Yu, B.Y.; Lee, G.H.; Dong, F.; Otteneder, M.B. Nanotracing and cavity-ring down spectroscopy: A new ultrasensitive approach in large molecule drug disposition studies. *PLoS ONE* **2018**, *13*, e0205435. [CrossRef]

33. Taylor, R.E. Dating Techniques in Archaeology and Paleoanthropology. *Anal. Chem.* **1987**, *59*, A317. [CrossRef]

34. Graven, H.; Allison, C.E.; Etheridge, D.M.; Hammer, S.; Keeling, R.F.; Levin, I.; Meijer, H.A.J.; Rubino, M.; Tans, P.P.; Trudinger, C.M.; et al. Compiled records of carbon isotopes in atmospheric CO_2 for historical simulations in CMIP6. *Geosci. Model Dev.* **2017**, *10*, 4405–4417. [CrossRef]

35. Buchholz, B.A.; Sarachine, M.J.; Zermeno, P. Establishing Natural Product Content with Natural Radiocarbon Signature. In *Progress in Authentication of Food and Wine*; Ebeler, S.E., Takeoka, G.R., Winterhalter, P., Eds.; ACS Books: Washington, DC, USA, 2011; p. 27.

36. Nelson, M.A.; Ondov, J.M.; VanDerveer, M.C.; Buchholz, B.A. Contemporary Fraction of Bis(2-Ethylhexyl) Phthalate in Stilton Cheese by Accelerator Mass Spectrometry. *Radiocarbon* **2013**, *55*, 686–697. [CrossRef]

37. Tong, T.; Ondov, J.M.; Buchholz, B.A.; VanDerveer, M.C. Contemporary carbon content of bis (2-ethylhexyl) phthalate in butter. *Food Chem.* **2016**, *190*, 1064–1068. [CrossRef]

38. Stuiver, M.; Polach, H.A. Reporting of C-14 Data—Discussion. *Radiocarbon* **1977**, *19*, 355–363. [CrossRef]

39. Hedges, R.E.M.; Clement, J.G.; Thomas, C.D.L.; O'Connell, T.C. Collagen turnover in the adult femoral mid-shaft: Modeled from anthropogenic radiocarbon tracer measurements. *Am. J. Phys. Anthropol.* **2007**, *133*, 808–816. [CrossRef]

40. Buchholz, B.A.; Alkass, K.; Druid, H.; Spalding, K.L. Bomb Pulse Radiocarbon Dating of Skeletal Tissues. *New Perspect. Forensic Hum. Skelet. Identif.* **2018**, 185–196. [CrossRef]

41. Shapiro, S.D.; Endicott, S.K.; Province, M.A.; Pierce, J.A.; Campbell, E.J. Marked Longevity of Human Lung Parenchymal Elastic Fibers Deduced from Prevalence of D-Aspartate and Nuclear-Weapons Related Radiocarbon. *J. Clin. Invest.* **1991**, *87*, 1828–1834. [CrossRef] [PubMed]

42. Heinemeier, K.M.; Schjerling, P.; Heinemeier, J.; Magnusson, S.P.; Kjaer, M. Lack of tissue renewal in human adult Achilles tendon is revealed by nuclear bomb ^{14}C. *FASEB J.* **2013**, *27*, 2074–2079. [CrossRef]

43. Heinemeier, K.M.; Schjerling, P.; Heinemeier, J.; Moller, M.B.; Krogsgaard, M.R.; Grum-Schwensen, T.; Petersen, M.M.; Kjaer, M. Radiocarbon dating reveals minimal collagen turnover in both healthy and osteoarthritic human cartilage. *Sci. Transl. Med.* **2016**, *8*, 346ra90. [CrossRef]

44. Lynnerup, N.; Kjeldsen, H.; Heegaard, S.; Jacobsen, C.; Heinemeier, J. Radiocarbon Dating of the Human Eye Lens Crystallines Reveal Proteins without Carbon Turnover throughout Life. *PLoS ONE* **2008**, *3*, e1529. [CrossRef]

45. Stewart, D.N.; Lango, J.; Nambiar, K.P.; Falso, M.J.S.; FitzGerald, P.G.; Rocke, D.M.; Hammock, B.D.; Buchholz, B.A. Carbon turnover in the water-soluble protein of the adult human lens. *Mol. Vis.* **2013**, *19*, 463–475.

46. Lovell, M.A.; Robertson, J.D.; Buchholz, B.A.; Xie, C.S.; Markesbery, W.R. Use of bomb pulse carbon-14 to age senile plaques and neurofibrillary tangles in Alzheimer's disease. *Neurobiol. Aging* **2002**, *23*, 179–186. [CrossRef]

47. Goncalves, I.; Stenstrom, K.; Skog, G.; Mattsson, S.; Nitulescu, M.; Nilsson, J. Dating components of human atherosclerotic plaques. *Eur. Heart J.* **2010**, *31*, 794.

48. Hagg, S.; Salehpour, M.; Noori, P.; Lundstrom, J.; Possnert, G.; Takolander, R.; Konrad, P.; Rosfors, S.; Ruusalepp, A.; Skogsberg, J.; et al. Carotid Plaque Age Is a Feature of Plaque Stability Inversely Related to Levels of Plasma Insulin. *PLoS ONE* **2011**, *6*, e18248. [CrossRef]

49. Etminan, N.; Dreier, R.; Buchholz, B.A.; Beseoglu, K.; Bruckner, P.; Matzenauer, C.; Torner, J.C.; Brown, R.D.; Steiger, H.J.; Haggi, D.; et al. Age of Collagen in Intracranial Saccular Aneurysms. *Stroke* **2014**, *45*, 1757–1763. [CrossRef]

50. Etminan, N.; Dreier, R.; Buchholz, B.A.; Bruckner, P.; Steiger, H.J.; Hanggi, D.; Macdonald, R.L. Exploring the Age of Intracranial Aneurysms Using Carbon Birth Dating Preliminary Results. *Stroke* **2013**, *44*, 799–802. [CrossRef]

51. Spalding, K.L.; Bhardwaj, R.D.; Buchholz, B.A.; Druid, H.; Frisen, J. Retrospective birth dating of cells in humans. *Cell* **2005**, *122*, 133–143. [CrossRef]

52. Bhardwaj, R.D.; Curtis, M.A.; Spalding, K.L.; Buchholz, B.A.; Fink, D.; Bjork-Eriksson, T.; Nordborg, C.; Gage, F.H.; Druid, H.; Eriksson, P.S.; et al. Neocortical neurogenesis in humans is restricted to development. *Proc. Natl. Acad. Sci. USA* **2006**, *103*, 12564–12568. [CrossRef] [PubMed]

53. Spalding, K.L.; Bergmann, O.; Alkass, K.; Bernard, S.; Salehpour, M.; Huttner, H.B.; Bostrom, E.; Westerlund, I.; Vial, C.; Buchholz, B.A.; et al. Dynamics of Hippocampal Neurogenesis in Adult Humans. *Cell* **2013**, *153*, 1219–1227. [CrossRef] [PubMed]

54. Bergmann, O.; Liebl, J.; Bernard, S.; Alkass, K.; Yeung, M.S.; Steier, P.; Kutschera, W.; Johnson, L.; Landen, M.; Druid, H.; et al. The age of olfactory bulb neurons in humans. *Neuron* **2012**, *74*, 634–639. [CrossRef]

55. Yeung, M.S.; Zdunek, S.; Bergmann, O.; Bernard, S.; Salehpour, M.; Alkass, K.; Perl, S.; Tisdale, J.; Possnert, G.; Brundin, L.; et al. Dynamics of oligodendrocyte generation and myelination in the human brain. *Cell* **2014**, *159*, 766–774. [CrossRef]

56. Spalding, K.L.; Arner, E.; Westermark, P.O.; Bernard, S.; Buchholz, B.A.; Bergmann, O.; Blomqvist, L.; Hoffstedt, J.; Naslund, E.; Britton, T.; et al. Dynamics of fat cell turnover in humans. *Nature* **2008**, *453*, 783–787. [CrossRef]

57. Arner, P.; Bernard, S.; Salehpour, M.; Possnert, G.; Liebl, J.; Steier, P.; Buchholz, B.A.; Eriksson, M.; Arner, E.; Hauner, H.; et al. Dynamics of human adipose lipid turnover in health and metabolic disease. *Nature* **2011**, *478*, 110–113. [CrossRef] [PubMed]

58. Bergmann, O.; Bhardwaj, R.D.; Bernard, S.; Zdunek, S.; Barnabe-Heider, F.; Walsh, S.; Zupicich, J.; Alkass, K.; Buchholz, B.A.; Druid, H.; et al. Evidence for cardiomyocyte renewal in humans. *Science* **2009**, *324*, 98–102. [CrossRef] [PubMed]

59. Perl, S.; Kushner, J.A.; Buchholz, B.A.; Meeker, A.K.; Stein, G.M.; Hsieh, M.; Kirby, M.; Pechhold, S.; Liu, E.H.; Harlan, D.M.; et al. Significant human beta-cell turnover is limited to the first three decades of life as determined by in vivo thymidine analog incorporation and radiocarbon dating. *J. Clin. Endocrinol. Metab.* **2010**, *95*, E234–E239. [CrossRef]

60. Landsverk, O.J.; Snir, O.; Casado, R.B.; Richter, L.; Mold, J.E.; Reu, P.; Horneland, R.; Paulsen, V.; Yaqub, S.; Aandahl, E.M.; et al. Antibody-secreting plasma cells persist for decades in human intestine. *J. Exp. Med.* **2017**, *214*, 309–317. [CrossRef] [PubMed]

61. Maze, I.; Wenderski, W.; Noh, K.M.; Bagot, R.C.; Tzavaras, N.; Purushothaman, I.; Elsasser, S.J.; Guo, Y.; Ionete, C.; Hurd, Y.L.; et al. Critical Role of Histone Turnover in Neuronal Transcription and Plasticity. *Neuron* **2015**, *87*, 77–94. [CrossRef] [PubMed]

62. Hum, N.R.; Martin, K.A.; Malfatti, M.A.; Haack, K.; Buchholz, B.A.; Loots, G.G. Tracking Tumor Colonization in Xenograft Mouse Models Using Accelerator Mass Spectrometry. *Sci. Rep.* **2018**, *8*, 15013. [CrossRef]

63. Orjuela, M.A.; Liu, X.; Miller, R.L.; Warburton, D.; Tang, D.; Jobanputra, V.; Hoepner, L.; Suen, I.H.; Diaz-Carreno, S.; Li, Z.; et al. Urinary naphthol metabolites and chromosomal aberrations in 5-year-old children. *Cancer Epidemiol. Biomarkers Prev.* **2012**, *21*, 1191–1202. [CrossRef]

64. Abdo, K.; Eustic, S.; McDonald, M.; Jokinen, M.; Adkins, B.; Haseman, J. Naphthalene: A respiratory tract toxicant and carcinogen for mice. *Inhal. Toxicol.* **1992**, *4*, 393–409. [CrossRef]

65. North, D.W.; Abdo, K.M.; Benson, J.M.; Dahl, A.R.; Morris, J.B.; Renne, R.; Witschi, H. A review of whole animal bioassays of the carcinogenic potential of naphthalene. *Regul. Toxicol. Pharm.* **2008**, *51*, S6–S14. [CrossRef]

66. Buchholz, B.A.; Haack, K.W.; Sporty, J.L.; Buckpitt, A.R.; Morin, D. Free flow electrophoresis separation and AMS quantitation of ^{14}C-naphthalene-protein adducts. *Nucl. Instrum. Methods B* **2010**, *268*, 1324–1327. [CrossRef] [PubMed]

67. Buchholz, B.A.; Carratt, S.A.; Kuhn, E.A.; Collette, N.M.; Ding, X.X.; Van Winkle, L.S. Naphthalene DNA adduct formation and tolerance in the lung. *Nucl. Instrum. Methods B* **2019**, *438*, 119–123. [CrossRef]

68. Van Winkle, L.S.; Kelty, J.S.; Plopper, C.G. Preparation of Specific Compartments of the Lungs for Pathologic and Biochemical Analysis of Toxicologic Responses. *Curr. Protoc. Toxicol.* **2017**, *71*, 24.5.1–24.5.26. [CrossRef]

69. Morris, J.B. Nasal dosimetry of inspired naphthalene vapor in the male and female B6C3F1 mouse. *Toxicology* **2013**, *309*, 66–72. [CrossRef]

70. Enright, H.A.; Falso, M.J.S.; Malfatti, M.A.; Lao, V.; Kuhn, E.A.; Hum, N.; Shi, Y.; Sales, A.P.; Haack, K.W.; Kulp, K.S.; et al. Maternal exposure to an environmentally relevant dose of triclocarban results in perinatal exposure and potential alterations in offspring development in the mouse model. *PLoS ONE* **2017**, *12*, e0181996. [CrossRef]

71. Halden, R.U. On the need and speed of regulating triclosan and triclocarban in the United States. *Environ. Sci. Technol.* **2014**, *48*, 3603–3611. [CrossRef]

72. Coogan, M.A.; La Point, T.W. Snail bioaccumulation of triclocarban, triclosan, and methyltriclosan in a North Texas, USA, stream affected by wastewater treatment plant runoff. *Environ. Toxicol. Chem.* **2008**, *27*, 1788–1793. [CrossRef] [PubMed]

73. Geiss, C.; Ruppert, K.; Heidelbach, T.; Oehlmann, J. The antimicrobial agents triclocarban and triclosan as potent modulators of reproduction in Potamopyrgus antipodarum (Mollusca: Hydrobiidae). *J. Environ. Sci. Health A Tox. Hazard Subst. Environ. Eng.* **2016**, *51*, 1173–1179. [CrossRef] [PubMed]

74. IARC. *Some Non-heterocyclic Polycyclic Aromatic Hydrocarbons and some Related Exposures. Monographs on the Evaluation of Carcinogenic Risks to Humans*; IARC: Lyon, France, 2010.

75. World Health Organization (WHO); International Programme on Chemical Safety. *Selected Non-Heterocyclic Polycyclic Aromatic Hydrocarbons*; Environmental Health Criteria 202; WHO: Geneva, Switzerland, 1998.

76. EPA. *Toxicological Review of Benzo[a]pyrene, Integrated Risk Information System, National Center for Environmental Assessment*; Integrated Risk Information System, National Center for Environmental Assessment, EPA: Washington, DC, USA, 2017.

77. Abadin, H.G. The toxicological profile program at ATSDR. *J. Environ. Health* **2013**, *75*, 42–43. [PubMed]

78. Wheate, N.J.; Walker, S.; Craig, G.E.; Oun, R. The status of platinum anticancer drugs in the clinic and in clinical trials. *Dalton Trans.* **2010**, *39*, 8113–8127. [CrossRef] [PubMed]

79. Dilruba, S.; Kalayda, G.V. Platinum-based drugs: Past, present and future. *Cancer Chemother. Pharmacol.* **2016**, *77*, 1103–1124. [CrossRef] [PubMed]

80. Kelland, L. The resurgence of platinum-based cancer chemotherapy. *Nat. Rev. Cancer* **2007**, *7*, 573–584. [CrossRef] [PubMed]

81. Wang, D.; Lippard, S.J. Cellular processing of platinum anticancer drugs. *Nat. Rev. Drug Discov.* **2005**, *4*, 307–320. [CrossRef]

82. Galluzzi, L.; Vitale, I.; Michels, J.; Brenner, C.; Szabadkai, G.; Harel-Bellan, A.; Castedo, M.; Kroemer, G. Systems biology of cisplatin resistance: Past, present and future. *Cell Death Dis.* **2014**, *5*, e1257. [CrossRef] [PubMed]

83. Boocock, D.J.; Brown, K.; Gibbs, A.H.; Sanchez, E.; Turteltaub, K.W.; White, I.N. Identification of human CYP forms involved in the activation of tamoxifen and irreversible binding to DNA. *Carcinogenesis* **2002**, *23*, 1897–1901. [CrossRef]

84. Martin, E.A.; Brown, K.; Gaskell, M.; Al-Azzawi, F.; Garner, R.C.; Boocock, D.J.; Mattock, E.; Pring, D.W.; Dingley, K.; Turteltaub, K.W.; et al. Tamoxifen DNA damage detected in human endometrium using accelerator mass spectrometry. *Cancer Res.* **2003**, *63*, 8461–8465.

85. Hah, S.S.; Sumbad, R.A.; de Vere White, R.W.; Turteltaub, K.W.; Henderson, P.T. Characterization of oxaliplatin-DNA adduct formation in DNA and differentiation of cancer cell drug sensitivity at microdose concentrations. *Chem. Res. Toxicol.* **2007**, *20*, 1745–1751. [CrossRef]

86. Brown, K.; Tompkins, E.M.; Boocock, D.J.; Martin, E.A.; Farmer, P.B.; Turteltaub, K.W.; Ubick, E.; Hemingway, D.; Horner-Glister, E.; White, I.N. Tamoxifen forms DNA adducts in human colon after administration of a single [^{14}C]-labeled therapeutic dose. *Cancer Res.* **2007**, *67*, 6995–7002. [CrossRef] [PubMed]

87. Hah, S.S.; Henderson, P.T.; Turteltaub, K.W. Towards biomarker-dependent individualized chemotherapy: Exploring cell-specific differences in oxaliplatin-DNA adduct distribution using accelerator mass spectrometry. *Bioorganic Med. Chem. Lett.* **2010**, *20*, 2448–2451. [CrossRef]

88. Wang, S.; Zhang, H.; Malfatti, M.; de Vere White, R.; Lara, P.N., Jr.; Turteltaub, K.; Henderson, P.; Pan, C.X. Gemcitabine causes minimal modulation of carboplatin-DNA monoadduct formation and repair in bladder cancer cells. *Chem. Res. Toxicol.* **2010**, *23*, 1653–1655. [CrossRef] [PubMed]

89. Henderson, P.T.; Li, T.; He, M.; Zhang, H.; Malfatti, M.; Gandara, D.; Grimminger, P.P.; Danenberg, K.D.; Beckett, L.; de Vere White, R.W.; et al. A microdosing approach for characterizing formation and repair of carboplatin-DNA monoadducts and chemoresistance. *Int. J. Cancer* **2011**, *129*, 1425–1434. [CrossRef] [PubMed]

90. Jiang, S.; Pan, A.W.; Lin, T.Y.; Zhang, H.; Malfatti, M.; Turteltaub, K.; Henderson, P.T.; Pan, C.X. Paclitaxel Enhances Carboplatin-DNA Adduct Formation and Cytotoxicity. *Chem. Res. Toxicol.* **2015**, *28*, 2250–2252. [CrossRef] [PubMed]

91. Scharadin, T.M.; Zhang, H.; Zimmermann, M.; Wang, S.; Malfatti, M.A.; Cimino, G.D.; Turteltaub, K.; de Vere White, R.; Pan, C.X.; Henderson, P.T. Diagnostic Microdosing Approach to Study Gemcitabine Resistance. *Chem. Res. Toxicol.* **2016**, *29*, 1843–1848. [CrossRef] [PubMed]

92. Wang, S.; Zhang, H.; Scharadin, T.M.; Zimmermann, M.; Hu, B.; Pan, A.W.; Vinall, R.; Lin, T.Y.; Cimino, G.; Chain, P.; et al. Molecular Dissection of Induced Platinum Resistance through Functional and Gene Expression Analysis in a Cell Culture Model of Bladder Cancer. *PLoS ONE* **2016**, *11*, e0146256. [CrossRef] [PubMed]

93. Zimmermann, M.; Wang, S.S.; Zhang, H.; Lin, T.Y.; Malfatti, M.; Haack, K.; Ognibene, T.; Yang, H.; Airhart, S.; Turteltaub, K.W.; et al. Microdose-Induced Drug-DNA Adducts as Biomarkers of Chemotherapy Resistance in Humans and Mice. *Mol. Cancer Ther.* **2017**, *16*, 376–387. [CrossRef] [PubMed]

94. Wang, S.S.; Zimmermann, M.; Zhang, H.; Lin, T.Y.; Malfatti, M.; Haack, K.; Turteltaub, K.W.; Cimino, G.D.; de Vere White, R.; Pan, C.X.; et al. A diagnostic microdosing approach to investigate platinum sensitivity in non-small cell lung cancer. *Int. J. Cancer* **2017**, *141*, 604–613. [CrossRef] [PubMed]

95. Wang, F.; Zhang, H.; Ma, A.H.; Yu, W.; Zimmermann, M.; Yang, J.; Hwang, S.H.; Zhu, D.; Lin, T.Y.; Malfatti, M.; et al. COX-2/sEH Dual Inhibitor PTUPB Potentiates the Anti-tumor Efficacy of Cisplatin. *Mol. Cancer Ther.* **2017**. [CrossRef]

96. Scharadin, T.M.; Malfatti, M.A.; Haack, K.; Turteltaub, K.W.; Pan, C.X.; Henderson, P.T.; Jonas, B.A. Towards predicting AML patient response to 7+3 induction chemotherapy via diagnostic microdosing. *Chem. Res. Toxicol.* **2018**. [CrossRef]

97. Eisenhauer, E.A.; Therasse, P.; Bogaerts, J.; Schwartz, L.H.; Sargent, D.; Ford, R.; Dancey, J.; Arbuck, S.; Gwyther, S.; Mooney, M.; et al. New response evaluation criteria in solid tumours: Revised RECIST guideline (version 1.1). *Eur. J. Cancer* **2009**, *45*, 228–247. [CrossRef]

98. Rai, K.R.; Holland, J.F.; Glidewell, O.J.; Weinberg, V.; Brunner, K.; Obrecht, J.P.; Preisler, H.D.; Nawabi, I.W.; Prager, D.; Carey, R.W.; et al. Treatment of acute myelocytic leukemia: A study by cancer and leukemia group B. *Blood* **1981**, *58*, 1203–1212.

99. Lowenberg, B.; Downing, J.R.; Burnett, A. Acute myeloid leukemia. *N. Engl. J. Med.* **1999**, *341*, 1051–1062. [CrossRef]

100. Sekeres, M.A.; Elson, P.; Kalaycio, M.E.; Advani, A.S.; Copelan, E.A.; Faderl, S.; Kantarjian, H.M.; Estey, E. Time from diagnosis to treatment initiation predicts survival in younger, but not older, acute myeloid leukemia patients. *Blood* **2009**, *113*, 28–36. [CrossRef]

101. Juliusson, G.; Antunovic, P.; Derolf, A.; Lehmann, S.; Mollgard, L.; Stockelberg, D.; Tidefelt, U.; Wahlin, A.; Hoglund, M. Age and acute myeloid leukemia: Real world data on decision to treat and outcomes from the Swedish Acute Leukemia Registry. *Blood* **2009**, *113*, 4179–4187. [CrossRef]

102. Kern, W.; Estey, E.H. High-dose cytosine arabinoside in the treatment of acute myeloid leukemia: Review of three randomized trials. *Cancer* **2006**, *107*, 116–124. [CrossRef]

103. Willemze, R.; Suciu, S.; Meloni, G.; Labar, B.; Marie, J.P.; Halkes, C.J.; Muus, P.; Mistrik, M.; Amadori, S.; Specchia, G.; et al. High-dose cytarabine in induction treatment improves the outcome of adult patients younger than age 46 years with acute myeloid leukemia: Results of the EORTC-GIMEMA AML-12 trial. *J. Clin. Oncol.* **2014**, *32*, 219–228. [CrossRef]

104. Weick, J.K.; Kopecky, K.J.; Appelbaum, F.R.; Head, D.R.; Kingsbury, L.L.; Balcerzak, S.P.; Bickers, J.N.; Hynes, H.E.; Welborn, J.L.; Simon, S.R.; et al. A randomized investigation of high-dose versus standard-dose cytosine arabinoside with daunorubicin in patients with previously untreated acute myeloid leukemia: A Southwest Oncology Group study. *Blood* **1996**, *88*, 2841–2851.

105. Bishop, J.F.; Matthews, J.P.; Young, G.A.; Szer, J.; Gillett, A.; Joshua, D.; Bradstock, K.; Enno, A.; Wolf, M.M.; Fox, R.; et al. A randomized study of high-dose cytarabine in induction in acute myeloid leukemia. *Blood* **1996**, *87*, 1710–1717.

106. Cutts, S.M.; Swift, L.P.; Pillay, V.; Forrest, R.A.; Nudelman, A.; Rephaeli, A.; Phillips, D.R. Activation of clinically used anthracyclines by the formaldehyde-releasing prodrug pivaloyloxymethyl butyrate. *Mol. Cancer Ther.* **2007**, *6*, 1450–1459. [CrossRef] [PubMed]
107. Major, P.P.; Egan, E.M.; Beardsley, G.P.; Minden, M.D.; Kufe, D.W. Lethality of human myeloblasts correlates with the incorporation of arabinofuranosylcytosine into DNA. *Proc. Natl. Acad. Sci. USA* **1981**, *78*, 3235–3239. [CrossRef] [PubMed]
108. Kufe, D.W.; Munroe, D.; Herrick, D.; Egan, E.; Spriggs, D. Effects of 1-beta-D-arabinofuranosylcytosine incorporation on eukaryotic DNA template function. *Mol. Pharmacol.* **1984**, *26*, 128–134.
109. Raza, A.; Gezer, S.; Anderson, J.; Lykins, J.; Bennett, J.; Browman, G.; Goldberg, J.; Larson, R.; Vogler, R.; Preisler, H.D. Relationship of [3H]Ara-C incorporation and response to therapy with high-dose Ara-C in AML patients: A Leukemia Intergroup study. *Exp. Hematol.* **1992**, *20*, 1194–1200.
110. Gervasoni, J.E., Jr.; Fields, S.Z.; Krishna, S.; Baker, M.A.; Rosado, M.; Thuraisamy, K.; Hindenburg, A.A.; Taub, R.N. Subcellular distribution of daunorubicin in P-glycoprotein-positive and -negative drug-resistant cell lines using laser-assisted confocal microscopy. *Cancer Res.* **1991**, *51*, 4955–4963.
111. Coley, H.M.; Amos, W.B.; Twentyman, P.R.; Workman, P. Examination by laser scanning confocal fluorescence imaging microscopy of the subcellular localisation of anthracyclines in parent and multidrug resistant cell lines. *Br. J. Cancer* **1993**, *67*, 1316–1323. [CrossRef]
112. Swift, L.P.; Rephaeli, A.; Nudelman, A.; Phillips, D.R.; Cutts, S.M. Doxorubicin-DNA adducts induce a non-topoisomerase II-mediated form of cell death. *Cancer Res.* **2006**, *66*, 4863–4871. [CrossRef]
113. Coldwell, K.E.; Cutts, S.M.; Ognibene, T.J.; Henderson, P.T.; Phillips, D.R. Detection of Adriamycin-DNA adducts by accelerator mass spectrometry at clinically relevant Adriamycin concentrations. *Nucleic Acids Res.* **2008**, *36*, e100. [CrossRef] [PubMed]
114. Stornetta, A.; Zimmermann, M.; Cimino, G.D.; Henderson, P.T.; Sturla, S.J. DNA Adducts from Anticancer Drugs as Candidate Predictive Markers for Precision Medicine. *Chem. Res. Toxicol.* **2016**. [CrossRef] [PubMed]
115. Swart, P.; Lozac'h, F.; Simon, M.; van Duijn, E.; Vaes, W.H. The impact of early human data on clinical development: There is time to win. *Drug Discov. Today* **2016**, *21*, 873–879. [CrossRef]
116. Morcos, P.N.; Yu, L.; Bogman, K.; Sato, M.; Katsuki, H.; Kawashima, K.; Moore, D.J.; Whayman, M.; Nieforth, K.; Heinig, K.; et al. Absorption, distribution, metabolism and excretion (ADME) of the ALK inhibitor alectinib: Results from an absolute bioavailability and mass balance study in healthy subjects. *Xenobiotica* **2017**, *47*, 217–229. [CrossRef] [PubMed]
117. Husser, C.; Pahler, A.; Seymour, M.; Kuhlmann, O.; Schadt, S.; Zell, M. Profiling of dalcetrapib metabolites in human plasma by accelerator mass spectrometry and investigation of the free phenothiol by derivatisation with methylacrylate. *J. Pharm. Biomed. Anal.* **2018**, *152*, 143–154. [CrossRef] [PubMed]
118. Schadt, S.; Bister, B.; Chowdhury, S.K.; Funk, C.; Hop, C.; Humphreys, W.G.; Igarashi, F.; James, A.D.; Kagan, M.; Khojasteh, S.C.; et al. A Decade in the MIST: Learnings from Investigations of Drug Metabolites in Drug Development under the "Metabolites in Safety Testing" Regulatory Guidance. *Drug Metab. Dispos.* **2018**, *46*, 865–878. [CrossRef] [PubMed]

MDPI
St. Alban-Anlage 66
4052 Basel
Switzerland
Tel. +41 61 683 77 34
Fax +41 61 302 89 18
www.mdpi.com

Toxics Editorial Office
E-mail: toxics@mdpi.com
www.mdpi.com/journal/toxics